Governing Global Electronic Networks

The Information Revolution and Global Politics
William J. Drake and Ernest J. Wilson III, series editors
http://mitpress.mit.edu/IRGP-series

The Information Revolution and Developing Countries
Ernest J. Wilson III, 2004

Human Rights in the Global Information Society
edited by Rikke Frank Jørgensen, 2006

Mobile Communication and Society: A Global Perspective
Manuel Castells, Mireia Fernández-Ardèvol, Jack Linchuan Qiu, and Araba Sey, 2007

Access Denied: The Practice and Policy of Global Internet Filtering
edited by Ronald J. Deibert, John G. Palfrey, Rafal Rohozinski, and Jonathan Zittrain, 2008

Governing Electronic Global Networks: International Perspectives on Policy and Power
edited by William J. Drake and Ernest J. Wilson III, 2008

Governing Global Electronic Networks

International Perspectives on Policy and Power

Edited by William J. Drake and Ernest J. Wilson III

The MIT Press
Cambridge, Massachusetts
London, England

For information about special quantity discounts, please e-mail special_sales@mitpress.mit.edu

This book was set in Stone Serif and Stone Sans on 3B2 by Asco Typesetters, Hong Kong.
Printed and bound in the United States of America.

Library of Congress Cataloging-in-Publication Data

Governing global electronic networks : international perspectives on policy and power / edited by William J. Drake and Ernest J. Wilson, III.
p. cm. — (Information revolution & global politics)
Includes bibliographical references and index.
ISBN 978-0-262-04251-2 (hbk. : alk. paper) — ISBN 978-0-262-54197-8 (pbk. : alk. paper)
1. Telecommunication policy. 2. Telecommunication—International cooperation. I. Drake, William J. II. Wilson, Ernest J., III.
HE7645.G68 2008
384—dc22 2007037983

10 9 8 7 6 5 4 3 2 1

Contents

Preface

The purpose of this book is to offer the reader nontraditional perspectives on the global governance of global information and communication networks. Most work on this subject concentrates on the largest, most powerful players in the world system. Through their lens, hegemonic states and large multinational corporations are the center of attention. In this volume we broaden the focus and consider the concerns of those with less power and less influence—the *nondominant* actors, most notably the developing countries and civil society. In other words, this book views the global governance of networks more from the bottom up, and the outside in. Not surprisingly, the view from the bottom and the outside is not the same as the view from the top down and the inside out. Substantive priorities vary, as do interpretations of the value and fairness of the institutionalized global processes that lead to substantive outcomes. From the perspective of Washington, DC, and London, policy priorities adhere around efficiency and market access, and institutions like the World Trade Organization (WTO) and the World Intellectual Property Organization (WIPO), while the view from Pretoria and Sao Paulo is more likely to emphasize digital divides and to seek leverage points in the International Telecommunication Union (ITU) and international conferences like the World Summit on Information and Society (WSIS).

Scholars and practitioners who seek to analyze this second perspective, like those in this volume, are different in other ways too. They are more interested in explaining the economic and political origins of the evolving rules of the game that structure the production and distribution of communications in the global system, and how those rules reinforce global power disparities while leaving some room for maneuver for the nondominant actors. This last point is critical. Our colleagues recognize that the global governance of electronic networks greatly constrains nondominant actors, but it also leaves them some freedom of action. Theirs is not an either/or dichotomy; nondominant actors are neither fully free nor hopelessly controlled.

This book grew from the editors' frustration from attending conference after conference billed as addressing "global" information and communications technology (ICT) issues at which speaker after speaker devoted comparatively little attention to the conditions of the five billion people who live in developing and postcommunist societies. The expert descriptions of the complex rules of the international communications regimes, and their global impacts, rarely included the perspectives of nondominant actors. Furthermore, their accounts of the ways that global networks were governed concentrated mainly on matters of efficiency and Pareto optimality more than on matters of equity and distribution. By contrast, the contributors to this volume concentrate as much on equity as on efficiency, and on the implications of governance arrangements for nondominant actors and the global public interest.

As a complement to describing their structural positions within an inherently unequal system, all the authors also point out the existing spaces for maneuver and leverage that nondominant actors possess to improve their situation individually and through collective action. We believe this offers a much more action-oriented and ultimately optimistic view of power relations in the transition toward a knowledge society, than simply another depressing catalogue of structural inequalities that submerge any possibility for human agency.

A major goal of the book is to uncover the politics that lie beneath global rules and regulations that may seem at first glance to be mainly technical. The authors search for the political and institutional origins of the rules that govern global electronic networks, and the patterns of winners and losers those arrangements create. In this sense, the volume is central to the MIT Press series of which it is a part, The Information Revolution and Global Politics. It is certainly true that good analysts of the information revolution must master the basics of the technology, and the ways they limit what is possible. At the same time, good analysts must appreciate how some stakeholders have more access than others to technology making and to rule making. In general, privileged stakeholders design and enforce governance mechanisms that tend to favor their material and ideological interests, and governance in the global (and national) ICT sector is no exception. How information and communications resources are deployed, how they should operate, and who pays for what are critical negotiation issues in which actors bring to bear all the assets they can in order to gain the distributive outcomes they seek. In the process, some clearly benefit more than others.

At its heart, global governance is about big issues like property rights, the definitions of equity and efficiency, and who gets to write the rules of the game. We concentrate on governance because at this moment in the transition toward knowledge societies, it is a critically important but contested concept and process. In this period of extreme turbulence about the ways basic norms, rules, and regulations guide human and institutional behaviors it is not surprising that all stakeholders are deeply concerned about the character of governance.

Concerns about meanings and definitions and their links to power and agenda setting came to the fore around the WSIS, held in Geneva in December 2003 and in Tunis in November 2005. There and in the lengthy preparatory meetings there were debates over how properly to frame the discussions of information and society. The excitement and energy generated by the WSIS process helped to bump ICT global governance onto a wider and more visible world stage. The editors of this volume attended both meetings, and many of our authors played notable roles in the process. The WSIS debates reinforced our perspective that one needs to devote much more attention to the actions, words, and interests of nondominant actors.

As indicated in the acknowledgments section of this volume, the diversity of participants in the dialogues surrounding this project ensured we would have an antitechnocratic take on the global governance of electronic networks. The initial workshops involved participants with a variety of real-world practical experiences, from grassroots organizers and corporate managers, to public officials and staff from multilateral organizations. The thread of practicality they brought to our deliberations provided a solid grounding in practice that stands in sharp contrast to other projects that are either entirely academic or mainly practical.

Beyond our particular perspective—bottom up, outside in—the volume provides careful explications of the main *concepts* of governance, and the levels at which governance is typically exercised, such as multilateral, minilateral, and private sector governance, each of which has its own inherent strengths and weaknesses for nondominant actors. In addition, the volume presents the reader with rich *empirical descriptions* of what is happening in the governance of a range of substantive topics, from third-generation mobile networks to Internet domain names.

The volume is organized as follows. In the introduction (chapter 1), William Drake sets the stage by providing a historical overview of ICT global governance and mapping its contemporary architecture. He demonstrates that since 1850, we have progressed through three distinct NetWorld Orders (NWOs), each of which has been characterized by a particular blend of dominant technologies, ideas, interest configurations, and institutional arrangements. He concludes that despite the diversity of issues and institutions involved, scholars and policy practitioners alike could usefully pursue holistic analytical approaches to the field of ICT global governance.

We have divided the subsequent essays into three thematic parts. Part I deals with the global governance of infrastructures, or the networks, services, applications, and resources that make communication and information sharing possible. Don MacLean (chapter 2) explicitly addresses matters of rule making, power, and the allocation of resources that lie at the heart of governance. He takes up many of the thorny issues of reform in an organization that was long central to governance but is of less importance today—the ITU. MacLean shows that the ITU is beset by new challenges of radically changing technologies and business models, and suggests new avenues of reform that would, among other things, better accommodate the perspectives of nondominant actors.

Rob Frieden (chapter 3) provides a guide to the ITU's complex governance of international radio frequency spectrum and geostationary satellite orbital slots. While quite technical, when the technical layers are peeled away, Frieden shows that the governance framework tends to favor first movers over latecomers. He describes current pressures on the framework and suggests ways that the interests of nondominant players can be enhanced.

Looking at how self-interested actors jockey for political influence and market position in a very complicated global arena is also the topic of Peter Cowhey, Jonathan Aronson, and John Richards (chapter 4). The third generation of wireless networks represents a huge and hugely valuable territory on which the giants of international commerce battle for market shares, and powerful governments and regional bodies jostle to influence the rules of the game and their interpretations. The authors describe how this market works, the players that now dominate the field, and steps that nondominant actors can take to advance their interests.

Boutheina Guermazi (chapter 5) turns our attention to another key multilateral body—the WTO. She assesses the WTO's Agreement on Basic Telecommunications, and the General Agreement on Trade in Services (GATS) of which it is part, from the standpoint of developing countries' interests. She concludes unequivocally that developing countries possess more leeway that they have exercised under these agreements, but that they have suffered not only because of constraints imposed internationally, but also because they have failed to organize themselves effectively and to pursue sound bargaining strategies.

Part II of the book deals with the global governance of the information, communication, and commerce flowing over the networks. Byung-il Choi (chapter 6) bridges the first two parts of the book by assessing the WTO's GATS from another angle, namely its treatment of international trade in audiovisual services. The author shows that the negotiations in the WTO have pitted proponents of a *trade perspective* (which holds that audiovisual services should be subject to progressive trade liberalization like any other sector) against proponents of a *cultural perspective* (which holds that they should not be so governed due to their special cultural significance). Choi details the politics of audiovisual trade negotiations in the WTO and other settings (bilateral, regional, multilateral) and argues that support for cultural industries should be pursued in a manner that does not unduly distort international trade.

Cees Hamelink (chapter 7) surveys the global governance battles over traditional mass media that have played out in multiple multilateral forums over the past century. His treatment of the timeless tensions between governments' interest in protecting their populations from content they consider harmful, on the one hand, and the (evolving) rights to communicate and seek information freely, on the other hand, demonstrates how old issues reappear again and again even as the technology evolves. On this contested terrain, Hamelink argues, none of the combatants are entirely pure:

sovereign states have their own raisons d'etat for defining which content is harmful, and international media corporations have their own profit imperatives for championing free speech.

Peng Hwa Ang (chapter 8) provides a parallel assessment of the tensions between freedom of speech and state regulation in the context of the contemporary Internet environment. Insisting on the critical role of the state in the evolution of the Internet, his chapter reflects the view of other intellectuals from Asia that collective responsibilities in communications must be taken seriously, even as one respects individual rights. Surveying various initiatives to establish rules on the circulation of content, Ang argues that the inadequacies of industry self-regulation will lead governments to assert themselves more, perhaps including through international cooperation.

Ian Hosein (chapter 9) explores the increasingly central realm of security and cybercrime as the Group of Eight and the Council of Europe have addressed it. The author finds fault with the approaches pursued in these bodies, which raise significant problems with respect to privacy and other key values. Moreover, their efforts have constituted instances of policy laundering because they circumvent national democratic discourses in favor of comparatively closed international decision-making processes.

Henry Farrell (chapter 10) considers governance mechanisms that are designed to protect citizens' privacy rights. He concentrates his attention on the relationship between the United States and the European Union (EU), which have pursued very different approaches to the question, with the former preferring weaker international rules than the latter. The author views power relations as central to the transatlantic accommodation that has been reached, and to the EU's efforts to push third-party countries toward higher levels of privacy protection.

Christopher May (chapter 11) examines the international politics of intellectual property protection within the WTO and WIPO. Emphasizing power dynamics and the distributive issues of winners and losers, he argues that the current trajectory toward strict and expansive intellectual property rules has been driven by the industrialized countries and their industries and is contrary to the needs of nondominant actors, especially the developing countries. May covers a range of issues before outlining ways in which nondominant actors can exploit the flexibilities provided by the relevant arrangements to promote developmental and public interest objectives.

The third and final part of the book deals more intensively with the problems nondominant actors confront in seeking to participate in ICT global governance processes. David Souter (chapter 12) writes from the perspective of one of the authors of the pathbreaking international report *Louder Voices* (2002). Souter assesses the continuing relevance of the report's main findings regarding the domestic and international institutional issues that can limit the effectiveness of developing countries' participation. In general, he argues that the most pressing problems concern domestic constraints and capacity building, and that sophisticated leadership will be needed if developing countries are to participate and defend their interests effectively in international negotiations.

Tracy Cohen and Alison Gillwald (chapter 13) demonstrate how the sort of issues highlighted by Souter play out in a specific case—the highly politically charged context of post-apartheid South Africa. The authors present a fine-grained analysis of the country's participation in the ITU and the WTO, demonstrating that international power dynamics and institutional factors can interact with domestic conditions to limit developing countries' influence in governance processes. They argue that these institutions do provide developing countries with some flexibility, but that global governance reform is needed nevertheless.

The same themes of power, institutions, and participation emerge with considerable clarity in the contribution by Milton Mueller and Jisuk Woo (chapter 14). The authors examine South Korea's involvement in the Internet Corporation for Assigned Names and Numbers (ICANN). They show that Western governments (most notably the United States) and transnational firms (such as intellectual property interests) have dominated ICANN at the expense of the developing and transitional economies, which the authors dub the "rest of the world." Mueller and Woo propose a series of changes to ICANN in order to enhance the rest of the world's effectiveness in that crucial international body.

Wolfgang Kleinwächter (chapter 15) turns our attention from developing countries to another set of nondominant actors—global civil society. He traces the evolution of civil society participation in the WSIS process and the development of "multistakeholderism" as a new principle that may come to inform more of ICT global governance. However, he cautions that to make this scenario viable, civil society actors will have to become better organized and more adept at securing governmental recognition of their legitimacy and importance as partners.

In the volume's conclusion (chapter 16), I synthesize some of this project's main lessons by posing and answering four guiding questions: Is there a Washington consensus separate from the preferences of nondominant actors? Are the current ICT governance mechanisms working well or are they broken? What is the impact of the current GGEN arrangements on nondominant actors? What can scholars and researchers do to help practitioners in the field of ICTs?

We believe the perspective of those who stand on the outside looking in, and at the bottom looking up, is an important corrective to some of the work on ICT global governance. Not every chapter in this book analyzes this overarching theme to the same extent, or even from the same angle. But the authors' analyses help to situate the "outside-in" discussion within a broader range of substantive and institutional issues not typically incorporated into discussions of governance. The reader is invited to reflect on how these contributions help enrich our understanding not only of those on the outside and at the bottom, but also how these insights may change our understanding of the behaviors of the privileged nations who sit inside, and at the top, of the governance system. Much of the scholarly work on international regimes, for example, starts with the assumption that global governance structures are positive sum arrangements for rich and poor alike. Our authors call this assumption into question.

Not every volume can cover every topic. Some issues we did not include in this volume still deserve more critical attention. For example, scholars should pay more attention to the weight of private sector actors in global governance, both in their efforts to influence government bodies, and also in their own firm-level commercial and long-term strategic choices. Still, as this book's authors demonstrate, there are more than enough important issues to engage scholars for years to come in the changing dynamics of the governance of global electronic networks.

Ernest J. Wilson III

Acknowledgments

This volume is the product of an extended process of dialogue and collaboration. In November 2002, we organized a pair of workshops at the Central European University in Budapest, Hungary, under the aegis of the Information Technology and International Cooperation (ITIC) Program of the Social Science Research Council (SSRC). These initial brainstorming workshops explored public policy and research issues related to the global governance of the ICTs involved in electronic networking. Participating in the workshops were thirty-three experts from nineteen countries drawn from international organizations, governments, businesses, civil society organizations, academia, and research institutions. They included Carlos Afonso, Casey Anderson, Byung-il Choi, Tracy Cohen, Darius Cuplinskas, William Drake, Anriette Esterhuysen, Henry Farrell, Laura Forlano, Francois Fortier, Vera Franz, Robert Frieden, Victor Gao, Alison Gillwald, Boutheina Guermazi, Ian Hosein, Tim Kelly, Meelis Kitsing, Andrey Korotkov, Nino Kuntseva-Gabashvili, Robert Latham, Don MacLean, Kaz Maekawa, Tattu Mambetalieva, Marta Mateo, Christopher May, Milton Mueller, Saskia Sassen, Sidharth Sinha, Motohiro Tsuchiya, Ernest Wilson, Jisuk Woo, and Robert Valantin. Twelve of the scholars in attendance provided initial concept memos on the dynamics of global governance in some key ICT issue areas, most of which subsequently developed into book chapters. Additional scholars were recruited at different points in time in order to expand our coverage of newly emerging issues, and dialogue among project participants continued both online and in various meetings held under ITIC and other auspices. The evolutionary assembling of authors and topics, coupled with rapid change in the subject matter being addressed, resulted in some discrepancies in the chronological endpoints of the draft chapters; this was particularly true with respect to the United Nations' 2002–2005 World Summit on the Information Society (WSIS) negotiations, in which some of the authors were quite involved. As such, all but one of the chapters were revised and updated in 2006–2007.

In addition to the authors and other project participants, we would like to thank the following people. Craig Calhoun, the president of the SSRC, and Saskia Sassen, who served as chair of the ITIC Steering Committee, supported the project from the outset, and created a facilitative intellectual and institutional environment. Robert Latham served as director of the ITIC Program and played a key role in developing the project concept in collaboration with the editors. Laura Forlano served as project manager of the ITIC Program and provided important intellectual input and logistical support. Becky Lentz, a program officer in the Knowledge, Creativity and Freedom Program at the Ford Foundation, provided financial support for the ITIC Program and, by extension, the initial workshops in Budapest. Finally, Darius Cuplinskas, director of the Information Program at the Open Society Foundation, provided additional financial and logistical support for the meetings in Budapest. Our heartfelt thanks to them all.

1 Introduction: The Distributed Architecture of Network Global Governance

William J. Drake

The burgeoning use of global electronic networks and related information and communication technologies (ICT) is widely recognized to be one of the defining features of contemporary world affairs. Electronic networks underlie and enable the relational networks linking individuals and organizations that are catalyzing economic, political, and sociocultural change on a worldwide basis. So deep and widespread is the change in key domains of social organization that it is difficult to disagree with Manuel Castells's observation that, "Networks constitute the new social morphology of our societies, and diffusion of networking logic substantially modifies the operation and outcomes in processes of production, experience, power, and culture."[1] Among other things, as Robert Latham and Saskia Sassen point out, networks are giving rise to significant new digital formations, "communication and information structures largely constituted in electronic space. Examples are electronic markets, Internet-based large-scale conversations, knowledge spaces arising out of networks of nongovernmental organizations (NGOs), and early conflict warning systems, among others. Such structures result from various mixes of computer-centered technologies and the broad range of social contexts that provide the utility logics, substantive rationalities, and cultural meanings for much of what happens in these electronic spaces."[2] As the Internet and digital convergence continue to evolve and computation becomes increasingly ubiquitous in the years ahead, the centrality of electronic networks at the national and global levels and the creation of such new social forms will only increase.

Of course, electronic networks do not simply appear holus-bolus or have some intrinsic, transcendent properties. Instead, their capabilities, utilization, and impact result from social shaping processes that reflect human agents' objectives and interactions. Governance, or social steering, is a configurative force in this context, and it is exercised at multiple levels, from the intraorganizational up to local, national, regional, and global spheres. This book is concerned with the last of these levels—the global governance, particularly by intergovernmental institutions, of networks and related ICT.[3] The electronic networks of interest here generally are planetary in scope, but they may be less geographically extensive and still have significant and configurative implications for both global networking and global policy. When electronic networks and the transactions they facilitate have crossed national borders, governments and the private sector usually have sought to establish shared rule systems, procedures, and programs that would guide the behavior of the actors involved.

The global governance of electronic networks has a very long pedigree. Indeed, international telecommunications was the first field in which nation-states established a multilateral intergovernmental organization—the International Telegraph Union, now the International Telecommunication Union (ITU), formed in 1865. Over the nearly century and a half to follow, global governance has evolved through three distinct phases (described later in this chapter), each of which has been characterized by a particular blend of dominant technologies, ideas, interest configurations, and institutional arrangements. In the third, current phase, we have seen a proliferation in the number and forms of governance mechanisms, as well as a deepening shift in social purpose away from restrictive state control and regulation and toward promoting globalized markets, private sector control, and security. Global governance mechanisms today include not only arrangements negotiated by governments, the private sector, and multistakeholder collaborations, but also arrangements imposed by powerful governments and companies possessing monopoly or oligopoly power in particular global markets. In addition, the coordinated convergence of national policies and corporate practices is becoming a significant source of global ordering, even if it is not codified in collective agreements. This book could not explore these latter forms of global governance, but they are important parts of the contemporary mix nonetheless.

The transformations in governance underway today have been driven by a number of factors. The most important of these has been the material and ideational power of the major industrialized countries, most notably the United States, and of transnational corporations (TNCs). In contrast, nondominant actors—such as the developing countries, small- and medium-sized enterprises, and civil society organizations (CSOs) —generally have found themselves to be in the positions of governance takers, rather than governance makers. To be sure, in some cases these actors have influenced governance decision making to varying degrees or have benefited from its results. But in many others, the current trajectory has been adverse to their interests, at least as they define them.

As the last point may suggest, the stakes in this arena are high. How and to whose benefit global governance is configured raises significant questions from a global public interest perspective. For example, many observers have heralded the Internet age as an era of almost unlimited possibilities for human empowerment and the weakening or dissolution of restrictive power centers. Yochai Benkler captures this spirit when he suggests,

> What characterizes the networked information economy is that decentralized individual action—specifically, new and important cooperative and coordinate action carried out through radically distributed, nonmarket mechanisms that do not depend on proprietary strategies—plays a much greater role than it did, or could have, in the industrial information economy.... The declining price of computation, communication, and storage have, as a practical matter, placed the material means of information and cultural production in the hands of a significant fraction of the world's population—on the order of a billion people around the globe.[4]

But in practice, how easily and to what ends that billion people can utilize ICT, whether their empowerment will be offset by other dynamics, and where this leaves the billions more who lack the same opportunities, will all be directly affected by the character of global governance and related patterns of social ordering.

The high stakes involved were made particularly clear in the course of the United Nations' 2002–2005 World Summit on the Information Society (WSIS) process. That process involved thousands of representatives from stakeholder groupings, including a number of the authors in this volume. It comprised major summits held at Geneva in December 2003 and at Tunis in November 2005, and a series of very lengthy preparatory negotiations and regional conferences. The WSIS tackled a broad range of issues pertaining to the global information society and adopted four instruments containing general principles and norms that, while nonbinding, impacted the global policy discourse on ICT and the programmatic work of many organizations going forward.[5] In the course of the process, global Internet governance—what it is, and who should control it—became a key point of contention. Had the summit resulted in the sort of significant changes to Internet governance that were being proposed by many developing and transitional countries, the consequences for the future evolution of the Internet and the global information society more generally would have been profound indeed.

With these considerations in mind, this volume places at center stage the questions of power and social purpose in network or ICT global governance. Of course, explaining outcomes requires due attention to the interests and negotiating behavior of the dominant players that shape them—what might be called governance "from above." But where possible, the contributors also try to emphasize how global governance looks "from below," most notably from the perspectives of the developing countries and civil society advocates of public interest objectives. Moreover, it should be noted that our intended audiences in this project included not only scholars, but also policy practitioners working in international organizations, governments, the private sector, and civil society. The cases included herein were selected because they are practically important, rather than in accordance with a particular scholarly methodology. Given this orientation, after describing the governance mechanisms in question and assessing the power dynamics and issues in play, each of the chapters concludes with a set of recommendations for action by either particular stakeholder groupings or the international community as a whole. Some are broadly framed principles, while others address specific problems currently under discussion in international forums. Taken together, they may be seen as constituting a starting point for the development of a holistic global public interest agenda.

To set the stage for the discussions to follow, this chapter takes a somewhat unorthodox path for an introduction to an edited volume. As the authors do not employ a singular theoretical approach that needs to be introduced and the thrust of their chapters is previewed in Ernest Wilson's preface, I will instead provide an overview of governance mechanisms that is intended to complement and contextualize the authors' more detailed treatments. Accordingly, the first section addresses the definitional question of what we mean by global governance, particularly with respect to networks and related ICT. The second section highlights the main stages in the evolution of network global governance from 1850 to the present. The third section provides a survey of some of the major global governance mechanisms pertaining to network infrastructure, both physical and logical, as well as related transport services. These include the frameworks for telecommunications regulation and standardization, radio frequency spectrum management, satellite systems and services, trade in telecommunications services, trade in ICT goods, and Internet identifiers. The fourth section provides an overview of governance mechanisms pertaining to the information, communication, and commerce conveyed over electronic networks. These include the frameworks for information flow and content, trade in content services, intellectual property, electronic commerce, cybersecurity, and privacy protection. Finally, the concluding section offers some brief thoughts about the potential value of viewing this range of mechanisms from an integrative, holistic analytical perspective.

The Nature of ICT Global Governance

The term *global governance* gained widespread currency in the discourses of international relations scholars and practitioners during the 1990s. The term fit well within a zeitgeist shaped by the end of the Cold War, globalization, the Internet stage of the information revolution, the growth of private corporate authority, the mobilization of CSOs, and the alleged erosion of territorial sovereignty as the primary organizing principle of world politics. These and related trends seemed to increase the demand for global-ordering mechanisms created through not only traditional forms of intergovernmental cooperation, but also industry self-governance, multistakeholder partnerships, and transgovernmental relations. The global governance rubric seemed to encompass the new issues and collective responses that were emerging, and it had the added benefit of being nicely alliterative and hence catchy and marketable as well.

In consequence, global governance became the raison d'être for a cottage industry of scholarly researchers and policy analysts. For example, in 1995, a high-level Commission on Global Governance released a wide-ranging and influential report calling for new initiatives to manage global problems and reform the United Nations. In the same year, the Academic Council on the United Nations System (ACUNS) and the United Nations University launched a new scholarly journal, *Global Governance: A Review of Multilateralism and International Organizations*. In parallel, academic institutions and think tanks established a number of research programs on the subject, and many international organizations and CSOs adopted the term as an organizing construct for their work. Interestingly, and exemplified in particular by the ACUNS journal, global governance has served as a vehicle for policy-oriented dialogue between scholars and policy practitioners from all sectors. Indeed, one could argue that the concept has become so widely institutionalized across professional environments and incorporated into so many actors' thought and work that it has become for them a social episteme, or "the background intersubjective knowledge—collective understandings and discourse—that adopt the form of human dispositions and practices that human beings use to make sense of the world."[6]

But what do we really mean by global governance? Given the passage of time since its rise in the lexicon and the extent to which it has been internalized by analysts and practitioners, one might expect there to be a widely agreed on understanding of the term. Nevertheless, no standard meaning is evident in the relevant academic and policy literatures and discourses. A particularly unhelpful source of dissensus is the tendency of some people to use the term in a highly normative manner, and to conflate it with particular instanciations of governance they do or do not favor. For example, there are liberal internationalists who equate the term with "good governance" and efforts by the international community to advance worthy goals like peace, development, and environmental stewardship; conservatives who equate it with unduly politicized and bureaucratic intergovernmentalism or even ominous schemes to establish world government; and progressives who equate it with neoliberal economics, corporate control, and the dominance of the rich over the poor. In reality, of course, governance can be done well or poorly and can serve any number of social purposes.

Even among the scholars and practitioners who construe the term in an appropriately value-neutral manner, there are differences, sometimes substantial, in its interpretation. To note some examples, governance and global governance have been defined as follows:

> This study operationalizes the amorphous term *governance* by defining it as the ability of a government to exercise public policy.[7]

> Governance is characterized by decisions issued by one actor that a second is expected to obey.[8]

> There is good reason to use the global governance concept in reference to the salience of globally oriented epistemic elites and authorities.[9]

> Global governance is governing, without sovereign authority, relationships that transcend national frontiers. Global governance is doing internationally what governments do at home.[10]

One of the best ways to explore *global governance*, what world government we actually have had, is to consider the history of *world organizations*, those intergovernmental and quasi-governmental global agencies that have (nominally) been open to any independent state (even though all states may not have joined.[11]

Governance refers to all the ways in which groups of people collectively make choices.[12]

[The authors] define "global governance" as collective efforts to identify, understand, or address worldwide problems that go beyond the capacity of individual states to solve.[13]

Governance, at whatever level of social organization it may take place, refers to conducting the public's business: to the constellation of authoritative rules, institutions and practices by means of which any collectivity manages its affairs.[14]

At the most general level, governance involves the establishment and operation of social institutions (in the sense of rules of the game that serve to define social practices, assign roles, and guide interactions among the occupants of these roles) capable of resolving conflicts, facilitating cooperation, or, more generally, alleviating the collective-action problems in a world of interdependent actors.[15]

The Centre understands global governance not as government but as a minimum framework of principles, rules and laws necessary to tackle global problems, which are upheld by a diverse set of institutions, including both international organisations and national governments.[16]

Governance is the sum of the many ways individuals and institutions, public and private, manage their common affairs. It is a continuing process through which conflicting or diverse interests may be accommodated and cooperative action may be taken. It includes formal institutions and regimes empowered to enforce compliance, as well as informal arrangements that people and institutions either have agreed to or perceive to be in their interest.[17]

By global governance is meant not only the formal institutions and organizations through which the rules and norms governing world order are (or are not) made and sustained—the institutions of state, intergovernmental cooperation and so on—but also those organizations and pressure groups—from [multinational corporations], transnational social movements to the plethora of non-governmental organizations—which pursue goals and objectives which have a bearing on transnational rule and authority systems.[18]

As a point of departure, governance is here conceived at a very abstract level as spheres of authority (SOAs) at all levels of human activity—from the household to the demanding public to the international organization—that amount to systems of rule in which goals are pursued through the exercise of control... Governance, in other words, encompasses the activities of governments, but it also includes any actors who resort to command mechanisms to make demands, frame goals, issue directives, and pursue policies.[19]

To meet the requirements of a broad conception, governance is here regarded as sustained by rule systems that serve as steering mechanisms through which leaders and collectivities frame and move toward their goals. In the state-centric world some of the rule systems are resided over by states and their governments, while international institutions and regimes maintain others. In the multi-centric world numerous steering mechanisms are to be found in NGOs, and still others consist of informal SOAs that may never develop formal structures.[20]

By governance, we mean the processes and institutions, both formal and informal, that guide and restrain the collective activities of a group. Governance need not necessarily be conducted exclusively by governments and the international organizations to which they delegate authority. Private firms, associations of firms, nongovernmental organizations (NGOs), and associations of NGOs all engage in it, often in association with governmental bodies, to create governance; sometimes without governmental authority.[21]

Analytically, we define global governance by three criteria... First, we see global governance as characterized by the increasing participation of actors other than states, ranging from private actors such as multinational corporations and (networks of) scientists and environmentalists to intergovernmental organizations ("multiactor governance")... Second, we see global governance as marked by new mechanisms of organization such as public-private and private-private partnerships, alongside the traditional system of legal treaties negotiated by states... Third, we see global governance as characterized by different layers and clusters of rule-making and rule-implementation, both vertically between supranational, international, national and subnational layers of authority ("multilevel governance") and horizontally between different parallel rule-making systems.[22]

International regimes... are institutional arrangements whose members are states and whose operations center on issues arising in international society... Transnational regimes, by contrast, are institutional arrangements whose members are nonstate actors and whose operations are pertinent to issues that arise in global civil society... I use the phrase *global governance*... to refer to the combined efforts of international and transnational regimes.[23]

In the broadest possible definition, "governance" relates to any form of creating or maintaining political order and providing common goods for a given political community on whatever level ... "new modes of global governance" would refer to those institutional arrangements beyond the nation-state that are characterized by two features: the inclusion of non-state actors, such as firms, private interest groups, or nongovernmental organizations (NGOs) in governance arrangements (actor dimension); [and] an emphasis on non-hierarchical modes of steering (steering modes).[24]

These eighteen formulations by leading scholars and practitioners obviously raise numerous questions. For present purposes I will highlight just three issues that bear directly on the treatment of global governance in this chapter. First, a few of these definitions emphasize social actors and their characteristics and interrelationships, while most of the others emphasize social institutions and rule systems. As is often noted in the literature, the etymology of the word *govern* derives from the Greek words "kybenan" and "kybernetes," meaning "to steer" and "pilot of helmsman," respectively. Actor-centric understandings derived from the latter term were long the norm in ordinary language; governance was equated with the exercise of control by an authoritative actor, particularly a government. But from the 1990s onward there has been a general evolution away from this approach, and rightfully so.

Governance is best defined in terms of the act of governing—establishing prescriptions and proscriptions that steer or guide—rather than the particular governors that engage in the act. Given that governance exists in many social settings and can be exercised by a variety of actors, this avoids having to load into the definition variable listings of the actors and their characteristics and relationships in accordance with different analysts' preferences. Moreover, while actor-centric definitions tend to explicitly or implicitly invoke hierarchical authority relationships, it is entirely possible to have rule systems that instead derive from symmetric, horizontal relationships. Consequential rules must be authoritative in the sense of being credible and of widely recognized applicability, but they need not be vertically imposed or backed by any particular actor's power or authority. In short, who governs is a separate matter from what it means to govern, and maintaining this separation also comports with normal language usage; if we do not define verb/noun pairings like dominate/dominance, resist/resistance, or tolerate/tolerance in terms of who is involved or their attributes, why should we do so with govern/governance?

Second, and conversely, most of the definitions emphasize process and the role of institutions, such as decision-making procedures and substantive rule systems, in steering social action. This approach is preferable, but some of these definitions add elements that seem extraneous or based on particular instanciations that reduce their generalizability to other cases. For example, global governance need not by definition provide common goods; be nonhierarchical, necessary to tackle global problems, or concerned only with truly worldwide problems that are beyond the capacity of individual states to solve; include particular organizational forms like public-private partnerships; or be capable of accommodating diverse interests, resolving conflicts and collective-action problems, and enforcing compliance. In many cases, these conditions may be absent. Nor does it seem right to apply the term to any and all the ways in which groups of people make choices; to mere efforts to identify and understand problems; or to the activities of pressure groups seeking to influence governance decision making.

Third, many of the definitions emphasizing process and the role of institutions seem to equate the term to collective decision making. However, it is also possible for a dominant actor to unilaterally establish global governance mechanisms. Global ordering may result from such actors using power to impose rules, or it may simply arise post hoc as other actors opt to align their behavior. Either way, what matters in such cases is that governance mechanisms are collectively recognized to be applicable to rather than negotiated by a globally significant range of actors.

These comments suggest that a definition of global governance should be action-oriented rather than actor-oriented, concerned with steering mechanisms or institutions, generalizable across cases, and reasonably concise. As such, for the purposes of this chapter, I suggest the following working definition: *global governance is the develop-*

ment and application of shared principles, norms, rules, decision-making procedures, and programs intended to shape actors' expectations and practices and to enhance their collective management capacities in world affairs. A few words of elaboration follow.

The word *development* is used here in the sense of an intentional activity, irrespective of who engages in it. Some scholars consider general customs or common patterns of behavior emerging in a spontaneous, decentralized manner to be instances of governance. This view is particularly common among libertarian scholars of the Internet.[25] But ordering arising from uncoordinated action lacks both the steering character and the injunctive weight of governance, and it is in any event often rather ephemeral. The word *application* is meant to suggest that generally only prescriptions and proscriptions that have been articulated and are recognized to be in force count. As previously noted, merely identifying and debating problems that require steering is not the same as actually engaging in governance.

Relatedly, *shared* does not necessarily mean collectively agreed. But it does mean recognized as applicable by the relevant actors in a given arena. Given the variability across cases, exactly how many actors this must be for a mechanism to be considered an example of global governance cannot easily be specified within the terms of a definition, and indeed none of the formulations listed previously specify the domain of the "global". If global is construed as meaning only universal or planetary, we would be talking about a rather limited range of governance arrangements, which leaves out many that have a substantial configurative impact on world affairs. I would argue that unilateral, bilateral, regional, and plurilateral mechanisms can be instruments of global governance if the actors recognizing them as authoritative account for a majority of the behavior in the relevant global political space. An example would be if the member countries of the Organization for Economic Cooperation and Development (OECD) account for 70 percent of a given type of economic transaction and adopt a common policy framework governing it; that countries that are involved or marginally involved in such transactions are not parties to the agreement does not detract from its globally ordering character. Moreover, such less-than-universal arrangements may set standards and create conditions to which nonparties must adapt or comply either in the near term or later when they do begin to engage in the transactions in question. In these circumstances, governance *in* the global may effectively serve as governance *that is* global.[26]

Since *principles, norms, rules, and decision-making procedures* are the main elements of the now standard definition of an international regime, it might seem preferable to simply say that. However, it is possible to have principles, norms, rules, or decision-making procedures that facilitate steering but are not the elements of a fully developed rule system or regime. This is certainly the case in the global network environment, and particularly for the rather fluid realm of Internet governance. Parties engaged in problem solving often devise a general principle or rule that helps them to sort out the immediate problem at hand but which is not embedded in some larger institutional framework.

None of the preceding definitions from the literature specifically mentions *programs*. Doing so captures institutionalized activities that shape social actors' expectations, practices, and interactions but do not consist of developing and applying rule systems that regulate their daily conduct. Obviously, not all programs help bring order to some global realm or significantly impact actors' capacities to participate in collective decision making, but a definition should encompass those that do. For example, while international organizations have been somewhat marginalized as actors of interest in the contemporary scholarly literature on international cooperation, their secretariats may operate programs for the management of shared resources and facilities, production and dissemination of information, monitoring of events, building of members' capacities, and so on that have a demonstrable impact on the management of global problems.[27] Other actors do so as well, particularly in fields like global electronic networks where roles and responsibilities are diffused across social sectors and levels of organization. Finally, the phrase *intended to shape actors' expectations and practices and to enhance their collective management capacities in world affairs* is meant to underscore again the role of intentionality and what the principles, norms, rules, procedures, and programs actually do.

This definition seems generalizable to a wide variety of institutional arrangements. It can apply equally to governance mechanisms irrespective of their institutional context (whether they are negotiated under the aegis of formal organizations or are freestanding); form, that is, intergovernmental (treaties, recommendations, guidelines, declarations, memorandums of understanding, custom) or private sector (contracts, memorandums of understanding, codes, custom) agreements; strength (formal or informal, binding or voluntary); decision-making procedures (voting or consensus, dispute resolution and the like); the range and interrelatedness of issues covered; participants (public sector, private sector, civil society, or multistakeholder participation or application, and universal or smaller-n groupings); compliance mechanisms (centralized or decentralized monitoring and enforcement); distributional biases (equitable or inequitable allocations of rights, responsibilities, and benefits); and so on.

Turning then to the focus of this volume, governance has generally been applied to network infrastructures and to the information, communication, and commerce conveyed over such infrastructures. The simple binary distinction between "carriage" and "content" has long been employed in national and global communications policies, with different kinds of rules being applied to each level. In recent years it has become common to observe that the binary is breaking down due to technological convergence and regulatory change, and that issues pertaining to infrastructures are often integrally interwoven with those pertaining to transactions and content. Moreover, with the growing centrality of data communication architectures, it has become common to discuss networks using a more differentiated topology, for example, the four-layered model of Internet Protocol (IP)-based networks (applications, transport, Internet, and network access layers), and so on.

For present purposes, it is not necessary to differentiate between the various technical functions performed at different layers of the infrastructure. To characterize the focus of global governance mechanisms, it is sufficient to employ the traditional binary categories recognizing that "carriage" or infrastructure (networks, services, and applications) and "content" or networked information, communication, and commerce are just handy simplifying categories, and that governance mechanisms that are intended to focus on issues at one level may also involve by extension issues normally addressed at the other level.

Accordingly, we can define network/ICT global governance as *the development and application of shared principles, norms, rules, decision-making procedures, and programs intended to shape actors' expectations and practices and to enhance their collective management capacities concerning global electronic networks and the information, communication, and commerce they convey*. Such governance mechanisms have been devised to perform a variety of functions. These include generic functions common to rule systems, such as constraining actors from undertaking certain courses of action they might otherwise choose; empowering actors to undertake other courses of action with community assent; reducing the transaction cost of forming agreements; reducing the cost of creating, acquiring, or distributing information related to the rules of the game; establishing liability rules and property rights; and generally, facilitating collective learning and management.

In addition, governance mechanisms of course perform more substantively specific functions concerning the management of global communication and information issue-areas. Examples here include ensuring that ICT based in different countries can be physically interconnected and logically interoperable; managing the allocation of common pool resources, such as radio frequency spectrum or Internet domain names; specifying the terms and conditions according to which traffic will be passed from one interconnected network to the next; facilitating the commercial exchange of goods and services conveyed in such traffic; proscribing certain types of information flow; protecting network security and guarding against vulnerabilities to disruption or corruption; protecting information security and fighting cybercrime; protecting intellectual property and personal privacy; and so on.

Historical Evolution

The nexus of global governance mechanisms assessed in this volume are the product of a long process of international institutionalization. While tracing in detail the evolution of that process would be well beyond our objectives here, it is useful to highlight its main contours to establish the background for the survey of mechanisms that follows. In broad terms, the global governance of electronic networks can be said to have developed in three main stages since the mid-nineteenth century. Each of these stages has constituted what I would call a NetWorld Order (NWO), a distinctive global governance architecture characterized by a particular blend of dominant technologies, ideas, interest configurations, and institutional arrangements. In the latter years of the first two stages, the emergence of disruptive technologies set off social pressures that drove the transition to the next stage. Whether the same will happen with the current, third NWO remains an open question.

The first NWO lasted from the mid-nineteenth century until the early 1980s. It commenced with the development of multilateral institutions for telegraphy in continental Europe that were predicated on each participating state having similar domestic institutional arrangements and a shared social purpose that would guide their international cooperation. The core institution at the national level was a government ministry of posts and telegraphs (and later, also of telephones—a PTT), around which were arrayed a set of supportive societal constituencies like manufacturers, political parties, labor unions, and customers—a nexus Eli Noam has dubbed the "postal industrial complex."[28] Beginning from this base, the state-led organization of networking spread through colonialism and emulation to much of the world over the century to follow, albeit with a few notable exceptions where state-regulated private carriers, usually monopolies, were maintained instead. As new technologies like the telephone, radio, and telex came online, they generally were folded into the frameworks of state service provisioning or regulation of private providers. Hence, the project of state building was in most places a defining feature of the period, alongside the rationalization and integration of industrial economies and circuits of production and distribution that were typically driven, in noninfrastructural sectors, more by the private sector.

By extension, the core feature of global governance in the period was the social construction of national sovereignty as the baseline requirement with which global communications had to comport. With respect to international institutions, governments generally interpreted sovereignty as meaning state control, or in the exceptional cases, strong state authority over private carriers. But toward the end of the period, the merging of computers and of communications became a disruptive technology around which a new interest configuration arrayed to press for change. In consequence, the first NWO ended with the erosion of the global consensus equating sovereignty with state control, and the onset of privatization and market liberalization.

To flesh out these broad generalizations a little, let us begin with the PTT. On the European continent, many states had established postal monopolies that took control of the telegraph from the outset, arguing that electronic messaging was just a different way of doing the same thing and required comparable treatment in order to achieve integrated, nationwide systems. The same reasoning would later be applied to the telephone. While in some countries the state building process was conflictual because entrepreneurs resisted nationalization, most followed that path between the late nineteenth century and the early post–World War II period, depending on their domestic developmental and political trajectories. For its part, Britain nationalized its domestic telegraph and telephone systems in 1868 and 1911, respectively, but retained a robust set of private cable companies that dominated intercontinental connections between national systems into the early twentieth century. In colonized countries, the national carriers effectively were branches or affiliates of the European administrations that became national PTTs after independence.

The PTTs generally played the triple roles of national policymaker, monopoly provider of networks and services, and monopsony purchaser of privately produced equipment. In the richer countries that had equipment manufacturing industries, this status gave the state influence over the direction and pace of technological development; for example, the rollouts of international radio telegraphy and telephony frequently were slowed and shaped by the PTTs' desires to recover their investments in predecessor technologies. When radio emerged in the first decade of the twentieth century, it was controlled by the PTTs and used as an alternative or complement to wireline telecommunications. When radio broadcasting came online in the 1920s, most countries established separate government agencies and service suppliers, sometimes allowing private stations to acquire less desirable spectrum slots.

The exceptions were as interesting and important as the rule. As with soccer and socialism, the most exceptional case was the United States. Western Union was an early power in telegraphy, and the United States retained private carriers and went through cycles of competition and industry concentration until after World War I, when the American Telegraph and Telephone Co. (AT&T) consolidated its power as a near-monopolist in major telephone markets alongside an array of small independents providing local service that often interconnected with AT&T's network. The federal government encouraged this consolidation for reasons that were in some cases similar to the rationales for PTT control abroad. The partnership between AT&T and the government deepened after the Communications Act of 1934 created the Federal Communications Commission (FCC) and the first semicoherent national policy framework for telecommunications. AT&T achieved dominance on international voice routes alongside some international record carriers that had entered the market earlier to provide telegraph and later telex services.

Partial exceptions to the rule involved mixed public/private sector models. For example, Canada combined government monopolies in certain provinces with regulated private firms, while Finland and Denmark had private local operators alongside PTTs that controlled everything else.[29] In addition, carriers from the United States and Britain sought to serve some domestic markets abroad but were frequently forced out via nationalization or other means. However, in other places—parts of Latin America, Hong Kong, the Caribbean and Oceana Islands, and Spain until 1945—companies like International Telegraph and Telephone or Cable & Wireless stayed on as national private monopolists.

In addition, despite the global spread of the PTT model, private carriers were key players on many intercontinental routes. Some of the more interesting stories from the early history of global networking revolve around the struggle between the British cable cartel that dominated in the nineteenth century and the U.S.-based upstarts that supplanted it in many places during the twentieth century. As David Headrick has documented, the governments of the two countries got actively involved in these squabbles due to a mix of national security and commercial considerations, which included a U.S. campaign to break the British grip on communications to the countries and territories within its empire, which bore fruit in the early post–World War II era.[30] Also influential in some contexts were the major telecommunications equipment manufacturers and large corporate users of networks, including the press and financial and petroleum firms.

Prior to World War II, the ideational realm was heavily influenced by and supported the dominant players. Many arguments were advanced at different points in time to justify national monopolies in general and governmental monopolies in particular. For example, from an economic standpoint, it was said to be more efficient to have single providers of infrastructure than to have duplication of investments on routes. Monopolies could reap economies of scale and scope, and manage the research and development costs, capital requirements, and risk involved in building networks. In the early twentieth century, the theory of natural monopoly was embraced as an ex post rational that made such arguments "scientific." Further, it was suggested monopolies were necessary technologically, since network elements must all work together as systems were upgraded asymmetrically over time; socially, in order to facilitate the cross-subsidization of rates for low demand or income regions and users and achieve universal service; and politically, so as to build integrated national systems and identities, as well as to preserve national sovereignty, avoid vulnerability dependence on presumably predatory foreign suppliers, and ensure easy international coordination. The notion that radio frequency spectrum was scarce and required centralized management and allocation also supported state authority. Looking back through the analytical lenses of late twentieth and early twenty-first-century academia, some analysts seem to dismiss such notions as mere ideological covers for avaricious, rent seeking behavior. However, a close reading of historical records suggests that these and related ideas often were believed and foundational to the self-conceptions of systems building, public service-oriented, reformist state technocrats and their counterparts in private monopolies like AT&T, or "Ma Bell."

These social forces had a strong configurative influence on the development of global governance mechanisms for electronic networks. The first such mechanism to be erected was the multilateral regime for telecommunications networks and services, which provided a framework for the noncompetitive organization of international markets, interconnection of national networks and equipment, and cross-border exchange of traffic. Its core principles were first set out in the 1850 Treaty of Dresden, which created the four-member Austro-German Telegraph Union. In 1855, Belgium, France, Sardinia, Spain, and Switzerland formed the West European Telegraph Union along essentially the same lines. After several years of coordination between the two groups, the governments involved decided to devise a unified and broader multilateral regime. Hence, in 1865, Napoleon III invited twenty European governments to Paris to create what would become the world's first formal international organization—the International Telegraph Union. The Treaty of Paris comprised a convention that laid out the political principles governing telegraph relations between members, and an annexed set of regulations that established more detailed guidelines for the technical and economic organization of international networks and services. Technical committees for telegraph and telephone standardization were launched in the 1920s. As the membership expanded and telegraph issues became increasingly intertwined with telephone and radio matters, governments in 1932 decided to replace the telegraph union and the related International Radiotelegraph Union (discussed next) with a unified International Telecommunication Union (ITU) as of January 1934. The ITU was restructured, expanded, and brought into the United Nations by a diplomatic conference held in 1947.[31]

The second governance mechanism to be erected was the international regime for radio frequency spectrum. The regime established a framework for the distribution and utilization of radio frequency spectrum for terrestrial telecommunications and broadcasting services and, since 1963, of spectrum used by satellites and of related satellite parking slots in the geostationary orbit (GSO). In 1906, twenty-nine governments met in Berlin to sign a convention launching a collaboration that was referred to as the International Radiotelegraph Union and organizationally housed in and managed by the International Telegraph Union's Berne Bureau.[32] In 1927 they revised the treaty and with private sector participants created a committee to develop voluntary technical standards for radio. In 1932, the radio operations were folded into the new ITU. The aforementioned 1947 diplomatic conference revised the ITU's core treaties and inter alia strengthened and clarified members' rights and obligations regarding spectrum management. In particular, it created the International Frequency Registration Board (IFRB), which was charged with compiling and publishing member governments' frequency assignment notifications and judging their conformity with regime rules. These innovations constituted an instance of rule-governed regime change; the regime's guiding principles and objectives remained unaltered, but were now buttressed by a stronger and more elaborate nexus of norms, rules, and programs. Adding to this nexus was the establishment of complementary principles in a non-ITU instrument, the 1967 Outer Space Treaty negotiated through the United Nations' Committee on the Peaceful Uses of Outer Space.

A third governance mechanism was the international satellite regime. Like the telecommunications regime, the satellite regime played a central role in the development of global communications in accordance with a state-centric and anticompetitive orientation. The regime comprised both multilateral rules on the organization of the global satellite services market, and a major programmatic activity—the creation and operation of what was long the dominant service supplier in the market. A U.S.-led initiative, the International Telecommunications Satellite Organization (INTELSAT) was launched in 1964 as a consortium of twenty governments. The governments were designated as parties to the founding treaty and selected representatives (for most countries, the national PTTs) that served as shareholders and signatories to a special operational arrangement. Subject to their oversight, satellite systems management was provided by the dominant shareholder, a private firm called the Communications Satellite Organization (Comsat) that was created by the U.S. Congress in 1962. In 1971, eighty-five governments approved a treaty to make INTELSAT a permanent intergovernmental organization based in Washington, DC, and managed by a general secretariat rather than Comsat; these changes became effective in 1973. Weighted voting ensured that the industrialized countries, especially the United States, maintained overall control of the organization even as the developing country membership blossomed.

While the mechanisms just mentioned all focused on the collective management of infrastructure, they sometimes contained elements related to the content and cross-border flow of information over networks as well. For example, in the ITU treaties and their predecessors, priority was (and still is) given to government messages and emergency communications, and in the latter case, there were additional obligations like mandatory distress signals and ship-to-shore intercommunication irrespective of the telegraphic apparatus used (following the Titanic disaster of 1912). In addition, as a general matter, ITU members undertook to preserve the secrecy of private international correspondence, at least with respect to third parties. But from the 1850 Treaty of Dresden up to today's ITU Constitution, members also have reserved a broad right to stop any private communications that "might appear" dangerous to their security or otherwise contrary to national laws, public order, and decency. Obviously, the right as stated is rather sweepingly permissive, and it implies a corollary right (and ability), in accordance with national laws, to monitor and inspect the substance of messages in order to make such determinations. Many if not all states have given themselves that authority in some form, and some have found abundant occasions to exercise it.

For example, in the era of the British intercontinental cable cartel, messages were frequently monitored and censored, much to the chagrin of other countries. But in general, the right in principle to project territorial authority onto cross-border communications was not contested, as all states considered it to be integral to the preservation of national sovereignty.

With the development of radio broadcasting in the 1920s, governments began to consider the need for additional rules concerning the acceptable content and cross-border flow of mass media. But due to political differences over the desirability and scope of regulation and the weakness of the functional incentives relative to those in telecommunications, governments did not attempt to establish a broad multilateral organization or regime. A group of European governments launched an International Broadcast Union (IBU) in 1925, but it remained a rather limited affair and in 1950 was replaced by the more aptly named European Broadcasting Union (EBU), the broadcaster members of which concentrated more on technical matters and program exchanges. Other regions of the world developed similar unions of national broadcasting organizations. Hence, debates on global governance mechanisms devolved to multiple organizations like the League of Nations, the UN General Assembly, the ITU, and the United Nations Educational, Scientific and Cultural Organization (UNESCO). In general, the prevalence of political divisions and the frequent employment of non-binding instruments resulted in ambiguous, weak, and fragmentary global ordering, which in effect left matters in the hands of individual states and large media companies.

Given the national monopolies' tight control over networking, the first real challenge to the global order would originate outside their domain with the development of data communications in the United States. After having been nurtured by massive U.S. military contracts, a nexus of manufacturers dominated by International Business Machines Corp. (IBM) built a commercial industry by selling mainframe computers and related systems to large organizational customers. Of particular importance in this regard were IBM's System 360 and System 370 families of compatible computers, launched in 1964 and 1970, respectively. For geographically distributed organizations, interconnecting their in-house systems via telecommunications lines to perform distributed data processing and file management became an attractive prospect. The possibilities had been demonstrated by IBM and other contractors to the Department of Defense in the 1950s with the Semi-Automatic Ground Environment (SAGE) air defense system project, which "was essentially the first wide-area computer network, the first extensive digital data communications system, the first real-time transaction processing system."[33] SAGE concepts were then transferred to the business world with IBM's 1964 Semiautomatic Business Research Environment (SABRE) airline reservations system. In the years to follow, the data communications industry took off with computer manufacturers adopting competing proprietary platforms like IBM's Systems Network

Architecture (SNA) and laid the foundation for the development of related markets for specialized value-added networks (VANs), information services, online databases, and the like.

Collectively, these trends blurred the boundary line between "in-house" computing and telecommunications. New technological possibilities set off a learning process in which telecommunications was increasingly reconceptualized as an extension of corporate users' internal management information systems to be customized for competitive advantage, rather than as a plain vanilla public utility to be procured from monopoly providers on whatever terms they cared to offer. Large users in the airline, financial, petroleum, automobile, and other industries began to see that data communications could vastly enhance the company-wide management of their operations, both domestically and internationally.

AT&T's monopolistic practices therefore became a problem for a widening array of U.S. companies. Like the PTTs abroad, AT&T restricted customers' ability to attain, configure, and use leased circuits; to transition from discreet point-to-point leased circuits to interconnected private networks; and to attach specialized customer premise equipment (CPE) to private lines in order to enhance their flexibility and control. As such, corporate users began to complain to the FCC and demand greater freedom, a cause that IBM and other computer firms joined. Similarly, large corporate users demanded the right to procure systems and services from other suppliers, while potential competitive suppliers began to step up and demand entry into the emerging market niches. From the 1960s onward, this new corporate interest configuration grew into an informal political movement that generated a steady stream of calls for the FCC to curtail AT&T control first over the emerging specialized markets, and later over the public switched telephone network (PSTN) and provision of basic services.

Moreover, technological change created the incentives for an intellectual sea change regarding the optimal governance of telecommunications amid what was being called the information revolution. By the mid-1960s, a growing number of economists and industry analysts were questioning the continuing applicability of the old rationales for protecting AT&T from competition, such as the theory of natural monopoly. Increasingly, telecommunications was seen by many observers as being both key to the vitality of the economy as a whole and too important to be left in the hands of a monopolist. While AT&T and its supporters could dismiss the demands of potential competitors as being driven by narrow self-interest, this was more difficult with respect to the claims of independent analysts. In short, over time the combination of corporate demands and new ideas in the new technological environment led the FCC to reevaluate the wisdom of constraining innovation and entrepreneurship to benefit one company; to allow the market entry of competitors in telecommunications, particularly in the fields of specialized business services and long-distance telephony; and to pursue an incremental process of deregulation that fed into AT&T's divestiture in 1984 and helped create a highly diverse and competitive market for communications systems and services.

As these dynamics deepened within the United States and spread beyond its borders to other industrialized countries in the early 1980s, global governance lurched into a second NWO that would last until 1995. Unlike its predecessor, in which there was usually broad concordance among governments (with, in some cases, the notable exception of the United States) on the fundamental purposes and principles shaping collective governance, the second NWO was a period of growing conflict over these matters. Nevertheless, a new overarching orientation eventually took hold that effectively decoupled the dominant intersubjective understanding of sovereignty from state-controlled monopolies and favored market liberalization and privatization. This orientation would eventually spread from telecommunications to mass media as well.

The new approach favored by the U.S. private sector and growing array of analysts gathered significant momentum under the Reagan administration. The increasing freedom to build or lease and configure telecommunications infrastructure helped catalyze the explosive growth of computer networking and internetworking not only in the business arena, but in academia and civil society as well. Alongside TCP/IP, which was largely developed in the 1970s under U.S. military sponsorship and became the foundation of "the" Internet, grew a number of protocol suites like the International Organization for Standardization's (ISO)[34] Open Systems Interconnection (OSI) and offerings from individual firms like Xerox and IBM; public data networks like Telenet, TYMNET, UUNet, USENET, and BITNET; commercial online and conferencing services like Compuserve and the WELL; and on and on. With time lags and constraints due inter alia to PTT policies, parallel systems would arise elsewhere around the world, most notably within the OECD region, resulting in a worldwide matrix of computer networks at times linked together via gateways.[35]

Early in its tenure, the Reagan administration decided to make promoting international competition in communications a key economic objective.[36] The U.S. government began to undertake bilateral negotiations with foreign governments with the objective of winning not only better treatment of corporate users, but also market access for U.S. suppliers of telecommunications services and equipment. Moreover, the Office of the U.S. Trade Representative (USTR) joined forces with an emerging community of service industry specialists to argue that jointly provided telecommunications services in fact constituted trade, and as such should be covered by the General Agreement on Tariffs and Trade (GATT) organization. The notion that international services exchange had trade-like properties first emerged in the early 1970s, and by the early 1980s the United States was pressing other governments to set trade in services rules as part of what became the 1986–1994 Uruguay Round negotiations. The very act of revisualizing telecommunications as part of a larger category of services transactions that are "traded" created a strong conceptual bias toward openness, and set a new yardstick for evaluating telecommunications regulations as nontariff barriers (NTBs) to be removed.[37] Hence the GATT organization was an attractive venue in which to push for a new multilateral framework that would deal with the economic dimensions of international correspondence, as well as a means of pressuring administrations in the ITU to reform the telecommunications regime.

The United Kingdom, Japan, and Canada decided early in the decade to undertake varying degrees of privatization and liberalization in the hope of energizing their markets. But other industrialized countries' PTTs and their domestic supporters opposed the American approach, claiming that the Americans' new discourse about "restrictive trade barriers," "abuse of dominant position," and "excessive regulation" simply reflected the interests of large American firms poised to swoop down on their presumably vulnerable markets. They stepped up efforts launched in the 1970s to build monopoly-controlled public packet switched data networks that purportedly obviated the need for the competition and private networking demanded by TNCs. The PTTs and their manufacturer partners worked in the ITU to devise the requisite standards for such offerings as X.25 packet networks, X.400 message handling systems, and Integrated Services Digital Networks (ISDNs), as well as recommendations that justified the restrictive treatment of private lines and networks. These efforts were coupled with the adoption of other regulatory, trade, and industrial policies that the new interest configuration condemned as protectionism. Hence, the 1980s were marked by growing discord and drift in bilateral and multilateral policy discussions. PTT managers were aghast at being described as cartel managers conspiring against the free market, since commercial considerations had never been acknowledged criteria for evaluating standards and regulations.

Nevertheless, by the late 1980s the PTTs' positions were becoming untenable. Local corporate users, especially those in financial and other service sectors, found themselves competing with American-based counterparts that were benefiting from the efficiencies and enhanced range of choice in systems and applications associated with liberalization. For these users, market incentives pointed to the desirability of achieving similar gains with their home PTTs, and of extending these gains to cross-border services. Further, a conceptual realignment accompanied their shift to more globally oriented profiles. They were coming to see themselves as having similar interests as American users in relation to states, insofar as they were more concerned with accessing the best resources than with buying nationally. Hence, the regulatory preferences, negotiating agendas, and intellectual orientations of large users across the industrialized world began to converge, which substantially strengthened the campaigns of the International Chamber of Commerce (ICC), the International Telecommunications Users Group (INTUG), and other U.S.-inspired efforts by industry alliances to promote liberalization. Individually and collectively, companies began to press PTTs in the industrialized countries for greater flexibility with respect to leased circuits, private networks, CPE attachment, computerized information services, and a host of related issues that mattered to businesses but did not yet threaten directly the monopolists' control over PSTNs and basic services.

A parallel shift was occurring on the supply side of the market. The increasing globalization and differentiation of demand generated new opportunities that could best be realized in a competitive environment. Traditional telecommunications manufacturers and new entrants, whether medium-sized startups or large computer and electronics firms crossing market niches, increasingly sought out foreign sales, inter alia, to help recover the rising research and development costs of advanced CPE and network equipment. Similarly, emergent suppliers of private networks and services often needed to offer international reach in order to lure corporate customers and hence favored global liberalization. Where states were slow to change, some companies devised novel solutions to get around market access barriers, such as international corporate alliances and gray markets.

In addition, the emerging reconceptualization of telecommunications' role in economic activity called into question whether PTTs should retain their exclusive jurisdictions and, indeed, the nature of the PSTN itself.[38] The spread of American-style intellectual frameworks and the growing debate about trade in services in the GATT helped to redefine how industry analysts and government officials across the industrialized world regarded their national monopolies. High-level politicians and trade ministers alike began to believe that liberalization would energize their economies and be in the national interest. This trend gained additional momentum in 1987 when the Commission of the European Communities (EC) launched an ongoing initiative to liberalize the European market in support of its regional integration agenda.

In short, in the second half of the 1980s, PTTs across the industrialized world came under domestic as well as international pressure to open their markets and allow corporate customers significantly greater freedom in configuring and using network resources. In the years to follow, they undertook—with varying degrees of enthusiasm—liberalization programs that focused in particular on specialized business-oriented markets and stimulated the development of a wide range of new entrants like capacity resellers, systems integrators, and providers of specialized VANs. Whereas the integrated, analog, plain vanilla PSTN was the paradigmatic technology of the first NWO, their distributed, digitized, and specialized systems and services took on the same status in the second NWO.

Going further, many PTTs were broken up in the 1990s, with the national telecommunications administrations separated from the postal services and ministerial policy organs. The telecommunications administrations were privatized, often with the governments retaining 51 or 49 percent of shares of the stock at the outset and then releasing more to the market over time. The newly privatized entities came to be referred to as public telecommunications operators (PTOs). With time lags, a substantial number of developing countries subsequently moved down the same path. Major PTOs from the industrialized world aggressively globalized, becoming transnational "super carriers" that invaded national markets and formed interfirm alliances to provide corporate users with integrated, end-to-end solutions on the most lucrative and high-volume routes. Inevitably, these changes began to alter many governments' preferences regarding the international order.

The shape of global governance was redefined in consequence. On the one hand, existing arrangements like the telecommunications and satellite regimes became the sites of heatedly contested policy discourses and difficult negotiations that eventually resulted in promarket reforms sought by governments and industries from the industrialized world, above all the United States. On the other hand, new arrangements were created to carry the momentum forward and lock in binding commitments to liberalization. The key innovation in this regard was the creation of the General Agreement on Trade in Services (GATS) as a result of the Uruguay Round negotiations launched in September 1986. The agreement signed in April 1994 covered a wide range of issues, and among other things replaced the GATT organization with a stronger World Trade Organization (WTO), in effect January 1, 1995. It also established a third international regime under the WTO's institutional aegis—the Trade-Related Aspects of Intellectual Property Rights (TRIPS) agreement. The TRIPS included the guiding principle of binding minimum standards, the implementation of which could be subject to WTO dispute resolution and sanctions for noncompliance. Going forward, any deals concluded among WTO members in other contexts, from bilateral or regional agreements up to the multilateral level in the World Intellectual Property Organization (WIPO), could only establish higher levels of protection for property owners or reduce existing limitations on their rights.

Another important governance initiative, which is frequently overlooked in accounts of global liberalization's spread and consolidation, pertained to point-to-point information flows. With the blossoming of data communication and processing activities in the 1970s, many governments and analysts began to raise concerns about what was happening within these transnational corporate cyberspaces. Accordingly, at a conference organized by the OECD in 1974, an expert group dubbed the phenomena "transborder data flows" (TDF), which in contrast to "international" data flows invoked a mental image of corporate activities unmediated by territorial authority, and raised the question of whether it "constituted a problem sufficiently important in its implications for national sovereignty for governments to propose regulatory action."[39] For the next ten years, a wide-ranging international debate raged that, in the intergovernmental context, played out not in the ITU, which had no mandate to discuss content matters, but rather in the OECD and the Intergovernmental Bureau of Informatics (IBI), an organization of over forty members, mostly developing countries. In the end, the initial calls from a few industrialized countries and many developing countries for new multilateral regulations gave way to the adoption of regional and plurilateral instruments that effectively locked in a rough consensus among the key players that corporate TDF should generally occur without governmental impediments.

As had happened with the first NWO, pressures that built up during the latter years of the second NWO catalyzed the shift to a new governance architecture. But in this case, the dynamic was different; the third NWO that emerged from 1995 represented a very substantial deepening of rather than a striking departure from the fundamental orientation of its predecessor. The key driving force in this period, which continues today, has been the Internet.

Major developments on two fronts made 1995 a turning point. First, the GATS and the rest of the new WTO framework came into effect at the year's outset and created a legal and political context for the progressive liberalization of global telecommunications and ICT markets in the years ahead. Second, the U.S. government, which was the key source of governance via contract during the early noncommercial development of TCP/IP-based internetworking, withdrew its support for the National Science Foundation's NSFNet backbone and transitioned the Internet to a new architecture in which commercial network service providers interconnect to exchange traffic and also sell bandwidth to Internet Service Providers (ISPs) that cater to end users. In addition, the government authorized the sale of domain names, Netscape released improved versions of its pioneering Navigator browser, and Microsoft launched its Internet Explorer browser. Together, these developments greatly catalyzed the emerging commercialization and mass popularization of the Internet and laid the foundation for the boom to come. In time, the Internet's takeoff as a global mass medium was complemented and further stimulated by two further megashifts related to liberalization and digitization: the worldwide boom in mobile communications, and the advent of media convergence.[40]

The overarching features of global governance in this period have been twofold: substantively, the continuing reform and strengthening of prior arrangements geared toward the facilitation of markets and innovation, and the creation of new mechanisms designed to facilitate markets and promote security; and institutionally, a significant increase in the range and variety of global governance mechanisms, in which context private sector "self-governance" figure prominently. In addition, in the early twenty-first century, a new set of overarching political considerations have become an important part of the mix: demands by nondominant stakeholders, especially developing countries and civil society, for greater transparency, accountability, and inclusive participation in decision making; and the increasing importance of new stakeholders, including the multistakeholder technical and administrative communities that develop and operate much of the Internet, and individual users.

The configurative social forces and trends of the contemporary period are sufficiently familiar that they can be treated briefly here. Not long after being ensconced in Washington, DC, the Clinton administration launched a series of initiatives to promote the Internet and global electronic commerce as industry-led catalysts of an information-based "new economy." The new economy zeitgeist was powerful and widely internalized to the point that traditional public policy arrangements widely came to be seen as antithetical to wealth creation and progress. In this intellectual context, and working in close cooperation with industry associations, analysts, and even civil society actors, the administration launched an active and multidimensional campaign to promote an open and competitive "global information infrastructure" with the Internet at its core. Parallel networks of like-minded policy entrepreneurs, industry players, academics, and civil society actors—Ernest Wilson refers to those active in the developing countries as "information champions"—sprang up around the world to push forward Internet development and related activities within their respective countries, sometimes in the face of significant governmental and PTT opposition.[41]

As with the initial responses to pressures for liberalization and privatization in the second NWO, many governments and national PTOs or PTTs greeted the Internet's emergence with a mix of fear and disdain. The Internet constituted a cultural challenge to the established models of governments and PTOs, as the business and technical communities involved favored more bottom-up, rapid, and flexible problem solving and innovation. Moreover, the personalization of computing meant that users were growing accustomed to configuring information resources to suit their diverse interests, and when PTO practices or government policies got in their way, these became the targets of ridicule and anger by increasingly vocal and libertarian users who depicted them in global electronic forums as the actions of hopelessly out-of-touch dinosaurs. These shifts in the ideational realm translated into a significant broadening of informed public opinion against the self-preserving policies of entrenched telecommunications incumbents.

The Internet also represented a technological challenge to the dominant players, in that its technical standardization, model of internetwork intercommunication, and modes of innovation undermined time-honored practices and organizational models subject to their control. And the Internet's data communication and later, voice telephony applications presented an economic challenge to PTT/PTO revenues from high-priced services like international fax and telephony. Seen against the backdrop of now-established challenges from a burgeoning range of specialized private providers, particularly for business-oriented services, the Internet appeared to many governments and incumbent operators as just another American-born threat to their known universe that should be contained.

Nevertheless, its burgeoning mass popularity and increasing importance to global communications and commerce made accommodation of this subversive force the only option. In the industrialized world, the PTOs eagerly if belatedly plunged into the emerging markets by providing access services in competition with the newly emergent independent ISPs, building broadband capacity, cultivating mobile Internet markets, aggressively adapting their network architectures and service portfolios, shifting from traditional PSTNs to IP-based networks, and even supplying Internet telephony services to corporate customers in competition with their traditional voice services. In large parts of the developing world the reorientation has been slower and more contested. While many developing country PTOs, especially in the middle-income countries, have moved up the learning curve, others, especially in the least-developed countries, have been slower to adjust.

These tectonics shifts have had a profound effect on the shape of global governance. The established multilateral regimes for telecommunications, satellites, and trade in network-related goods and services were reformed or strengthened to more actively facilitate the development of globalized private markets. In parallel, new regimes were erected for the same purpose. For example, WIPO, which had been launched by a 1967 convention and became a member of the UN system in 1974, greatly expanded its work program to include the promotion of intellectual property rights in the digital environment. From the mid-1990s, WIPO embarked on the negotiation of new treaties designed to firm up and extend online protections. International organizations also took up market-enabling rulemaking on matters like global electronic commerce, and related programmatic work in the development policy field. And in parallel with this deepening market orientation, governments also began to devote significant attention to the burgeoning problem of cybercrime threats to network and information security. The most important multilateral instrument erected in this arena is the Council of Europe's (CoE) Convention on Cybercrime, which was established in 2001 and came into force in 2004. The convention is a binding treaty designed to harmonize national laws regarding substantive criminal and procedural laws, and to promote international cooperation and mutual legal assistance among states.

An especially important development in this era has been the proliferation of industry "self-regulatory" or "self-governance" arrangements, the effects of which generally reached far beyond the selves involved. These pertain in particular to the Internet—both its logical infrastructure and related services, and the communications and global electronic commerce it facilitates. They include industry agreements with respect to such items as global electronic commerce, network and information security, content regulation, personal privacy protection, and above all, the management of the Internet domain name and IP address allocation systems.

Regarding the latter, in many ways the paradigmatic event of the period has been the creation in 1998 of the Internet Corporation for Assigned Names and Numbers (ICANN). A California-based nonprofit corporation, it has operated under contracts, memoranda of understanding, and now a joint project agreement with the U.S. Department of Commerce. ICANN represents a complete departure from the tradition of intergovernmental, sovereignty-based governance that preceded it, and its contested performance of key functions such as the governance of the domain name system (DNS) and other core resources in the service of an amorphously defined transnational "Internet community" and other stakeholders has been a constant source of friction with many governments, most notably in the developing world.[42]

Against the backdrop of this schematic historical overview, the next two sections of this chapter provide an integrated survey of some of the key global mechanisms that collectively comprise the distributed architecture of contemporary ICT global governance. In the case of familiar mechanisms of recent pedigree, the treatment is somewhat brief. In contrast, those arrangements with longer histories that extend across multiple NWOs receive somewhat more extended coverage.

The Global Governance of Infrastructure

International Telecommunications

The ITU-based international telecommunications regime created in 1865 promoted three overarching objectives. The one that was most definitive of the first NWO was to consolidate and buttress national sovereignty. Under the regime, states could configure and govern their domestic networks and industries however they please, but were to conduct their international correspondence in accordance with the mutual consent of the other sovereign states involved, for example by not creating competition between their national administrations. During the telegraph union era, the internal dimension of sovereignty was more a matter of the letter than the spirit of the law, as strong community pressures were exerted on countries to conform to the continental European norm. As George Codding has observed, "Nationalization or complete control over telegraph was always an unwritten prerequisite for membership," and even the United Kingdom, whose firms were the driving force in intercontinental cable development, was not admitted until its domestic network was taken over by the state.[43] The United States (along with Canada and some Latin American countries) never joined the telegraph union because it did not want to nationalize or impose treaty obligations on its private carriers. Subsequently, a more permissive intersubjective understanding of acceptable domestic practices became necessary as the ITU's membership steadily expanded beyond its European roots to include countries that were slow to or did not nationalize their systems.

Despite this shift to accommodate countries that had private carriers rather than government administrations, until the 1980s the overwhelming majority of members continued to interpret the sovereignty principle in a manner that buttressed monopoly control domestically and precluded international competition. While the regime included provisions for special arrangements between correspondents that deviated from the norm but did not affect third parties, in general states followed restrictive rules of the game and imposed these on private firms from the United Kingdom,

United States, or elsewhere that provided service up to their electronic borders. Similarly, they adopted rules controlling the ways in which large corporate users of telecommunications like the airlines, banks, and news agencies could employ private leased circuits and networks. Among other things, these rules set prices and established technical constraints that limited the ability of companies to evolve their private circuits into competitive threats to the PTTs' principal revenue bases.

A second objective of the regime was to promote interconnection between national networks via technical standardization. While the early telegraph treaties specified the precise kinds of terminal apparatus and lines to be used, it was later recognized that to facilitate technical progress, standards should be set in more flexible instruments. Among other things, this meant that from the 1920s onward, standards were developed in specialized committees and set out in voluntary recommendations.

Finally, the third overarching objective was to facilitate international correspondence without undercutting sovereignty. Services generally were treated as "jointly provided" by outbound and inbound countries, with the revenues divided between them and any transit countries in between. Over the years, ITU members utilized a number of methods for managing the division of revenues, the most important of which—the international accounting and settlements system—attained dominance in the 1960s. The joint service provision principle, and the complex norms and rules operationalizing it, established an international order in which public or private sector monopolists could exchange traffic without competing in each other's markets. This was the fundamental political glue that held the system together and ensured broad intergovernmental support for and compliance with the regime.

For over a century, the regime evolved in a largely stable manner with new technologies and services being slowly incorporated into the dual system of national monopoly control and international joint service provisioning. The principal political challenges involved accommodating and constraining the United States and the private sector. With regard to the former, the United States has had a difficult and sometimes hostile relationship with the ITU and its predecessor, but since both sides had a strong interest in having the United States participate in the system, compromises were necessary. For example, after the United States joined the new telecommunication union in 1932, differential obligations were employed to facilitate its involvement, such as the division between a strict "European" system for tariffs and accounting and a more permissive "Extra-European" one, and members' acceptance of the United States' refusal to sign or sign without significant reservations the telegraph and telephone treaty regulations.

With regard to the private sector, although in the nineteenth and early twentieth centuries the vast majority of intercontinental submarine cables were controlled by a cartel of firms based in particular in the United Kingdom, the PTTs were able to use their control of landing rights and regulatory authority to bring these companies and subsequent private carriers into the telecommunications regime largely on their terms. The decision-making procedures evolved to enhance the role of the private sector in technical and operational standardization, but overall its independent influence remained constrained by procedural rules and state power.

This situation changed substantially during the second NWO. The information revolution, U.S. domestic deregulation, corporate demands for worldwide market liberalization, and the spread of new ideas about telecommunications governance progressively undermined the political foundations of the old regime. During the 1980s, the variable international spread of these forces generated substantial conflict between proliberalization and promonopoly governments, especially between the United States and Europe. These disagreements came to a head at the 1988 World Administrative Telegraph and Telephone Conference (WATTC), which had been called to devise a new set of International Telecommunication Regulations (ITRs) for the emerging technologies and markets. A PTT effort to erect new multilateral restrictions was abandoned under intense pressure from the United States and U.S. business interests in particular, and the WATTC instead signed a treaty that effectively allowed competitive supply and enhanced user control. As countries liberalized and the private sector increasingly found ways to enter markets in the years ahead, the new regulations opened the door for operators and users of private networks to extend their operations internationally, thereby facilitating the transition to a multiprovider world. Hence, the 1990s witnessed a worldwide explosion in the number and diversity of international networks and service providers that operated "special arrangements" outside the scope of the ITRs.

Building on the WATTC's outcomes, key regulatory recommendations on the configuration and use of private leased circuits and networks were relaxed in the early 1990s. At the same time, technical standardization for interconnection was reformed to be more responsive to an increasingly heterogeneous industry and moved to a significant degree out of the ITU nexus. The cumulative result of these dynamics was a fundamental transformation of the regime's social purpose and of some of its key instruments. The three overarching principles that shaped the regime during the long century of monopoly control remained in place, but they were now interpreted and implemented in a manner that put far fewer restrictions on market forces and private control.[44]

In the third NWO, the regime's salience to the organization and operation of global markets has decayed significantly. Externally, the GATS and the Basic Telecommunications Agreement (BTA) reached under it (described later in this section) have redefined the social purpose and consequential rules of the game of multilateral governance and have shifted the locus of market-related decision making down Geneva's Avenue de la Paix to the WTO. The Internet, which had initially been derided as inferior to ITU-approved networks, has rapidly reshaped the global environment with relatively little regard to the ITU arrangements and their traditional proponents. Globalizing PTOs and specialized operators have been employing "new modes of operation" like callback, refile, Internet telephony, and international simple resale to move traffic around the world in ways that circumvent the spirit or practice of the regime's system for joint service provisioning. And internally, the market-enabling revision of some of the regime's key provisions, coupled with deepening national and regional liberalization (particularly within the EU), has meant that it guides a declining share of the universe of activity. Moreover, in 1997 the FCC decided to unilaterally lower international accounting and settlement rates between the United States and corresponding countries, most notably in the developing world, which had benefited from high charges on inbound calls. Other countries and the EU enacted their own reforms, and soon member compliance and the dominance of the ITU-based system waned. The net effect of these developments has been to push the world's oldest international regime into a state of drift and decline, with its instruments remaining legally in force but actually governing less and less of the global industry's actual behavior.[45]

As Don MacLean, David Souter, and Tracy Cohen and Alison Gillwald all discuss in their contributions to this volume, many governments in the developing world have found the erosion of the regime to be a difficult pill to swallow. Even for those that have pursued liberalization and privatization, the ministries and dominant carriers responsible for shaping national strategies frequently favor the stability and control promoted by multilateral regulation. In consequence, they at times have pushed for agreements intended to put the genie back in the bottle, but which arguably would be contrary to a broadened and more vibrant approach to development. For example, some have sought to use ITU forums to build consensus on restricting Internet telephony, which can eat into incumbent revenues already hard-hit by declining accounting and settlements rates, but which is highly beneficial to users and small and medium-sized enterprises and can contribute to growth and development.

Another example concerns international Internet interconnection. Carriers in developing countries are required by the big global backbone operators to pay the full cost of the connections, irrespective of the bidirectional nature of the traffic flow, and the hefty fees imposed can have negative effects on Internet deployment and use in the global South. Hence, some countries have advocated mutual compensation schemes —based on traffic flows, the number and geographical coverage of routes, transmission costs, and the like—akin to the eroding accounting and settlements system for telephony. Indeed, in 2000, ITU members took the unprecedented action of adopting a recommendation to this effect, over U.S. and corporate objections, that called for mutual compensation. While the desire for a shared governance solution is understandable, this outmoded prescription has not found favor among the Northern-based carriers that dominate the world market. The developing countries also have insisted that the ITU hold a World Conference on International Telecommunications in 2012 to review the 1988 ITRs in light of the contemporary, Internet-driven age. While the industrialized countries presumably would refuse to accept a treaty revision that clearly extended strict regulatory limitations into the Internet environment, it is possible that some developing countries would try to interpret any accord in a manner that reinforces their traditional national carriers at the expense of other stakeholders' interests.

That said, despite the erosion of its multilateral regulatory framework, the ITU remains the principal global forum for international telecommunications policy matters. With 191 nations as members and 517 corporations as members of its standardization, radio, and development sectors, the ITU is uniquely able to bring together all states and many firms to tackle issues within its mandate.[46] While many Internet-related companies in particular refuse to participate and civil society is largely excluded, work in the ITU can contribute to mutual understanding and the coordinated convergence of national policies and corporate practices. For example, the ITU is becoming an important venue for work on network security issues that has fed back into national and corporate programs, even if this has not yet been codified in a treaty framework. As governments around the world struggle to define approaches to the convergence of digital media around IP-based networks, the ITU may serve as an important vehicle for promoting convergence and coordination on such thorny questions as identity management and law enforcement, public safety, service reliability, universal service obligations, Internet television, integrated telephone/Internet Electronic Numbering (ENUM), identifying physical objects connected to "the Internet of things," and so on.[47]

In this context, it remains possible that states and powerful firms will at times be able to leverage ITU mechanisms to promote restrictive policy frameworks that reinforce their power at the expense of other players and their visions. Some critics argue that the current work on technical standards and policies for IP-based Next Generation Networks—which are conceived of as IP-based platforms for voice, data, and video convergence—could lead to implementations that restrict the Internet's flexibility. This could prove to be particularly true if efforts to recoup investment costs and provide differentiated levels of quality of service result in the sort of discrimination that has been at the heart of the U.S. debate over "net neutrality." In light of these and related trends, the ITU is still an important forum on policy matters, irrespective of the regime's erosion.

Technical Standardization

During the first NWO, the technical standardization of networks was heavily configured by the international telecommunications regime of which it was an integral part. Governments, national carriers, and major manufacturers controlled the ITU standards process and often were able to influence the rate, direction, and diffusion of technical change. The recommendations they adopted on standardization reflected the regime's structuring of Internet work and intermarket relations and hence functioned as regime rules by which the higher principles of sovereignty and interconnection were balanced and operationalized. For example, during the first century of cooperation, the standards concentrated when possible on the gateways, signaling, and transmission between national extensions, leaving countries free to employ internationally incompatible systems at the domestic level as NTBs to equipment trade.[48]

From the late 1950s onward, automation, the growing volume and diversity of traffic, and then the digitization of networks and connection of information systems pushed standardization from the international edges deep into national domains; and from minimalist and voluntary conformity requirements toward detailed specifications that in some cases were referenced in treaties and viewed as essentially binding. These shifts required a more complex interpretation of the regime's principles and political bargains across remapped political space. So did two related changes associated with the merging of computing and communications: first, ITU committees increasingly had to collaborate and work out turf arrangements with nongovernmental standards bodies active in computer and electronic systems, such as the International Electrotechnical Commission (IEC) and the ISO. Second, the ITU and its partners increasingly opted to pursue some standards on an a priori basis, before products had been introduced at the national level, so as to attenuate new industry rivalries and promote common objectives. This approach was taken in particular with two large-scale projects conceived in the 1970s and undertaken in the 1980s: ISDN, and the ISO-ITU collaboration on OSI.

As the telecommunications regime loosened and became less determinant of global market organization during the second NWO, technical standardization outgrew its confines. Liberalization and the deepening integration of computing and telecommunications meant that standards conceived to support dominant incumbents could not as easily be imposed on dynamic and competitive markets. ISDNs originally were intended to serve as single, integrated national high-speed networks that could replace private lines and networks, but antimonopoly forces subsequently derided them as "Innovations Subscribers Don't Need." With the spread of liberalization, ISDNs ended up being just another service offering in some markets, one now being eclipsed by broadband Internet access. Similarly, OSI was supposed to be a vendor-neutral alternative to IBM's SNA and related offerings that would strengthen the hands of the industry players involved, but users opted for the Internet's TCP/IP model and OSI became marginalized.

Liberalization and convergence also brought into the game a heterogeneous array of actors from interrelated ICT industries that viewed standards as integral to their competitive strategies. Frustrated by the slow pace of work and dominance of incumbents in the ITU, the new players frequently opted to pursue standardization in a rapidly proliferating array of more market-driven national, regional, or industry/technology-specific forums. In response, the ITU reformed its procedures to speed the pace and increase the market-friendliness of its work; established a corporate advisory council on standardization; and sought to reposition itself as a central node in an interinstitutional network that would give a multilateral stamp of approval to standards developed elsewhere. But despite these reforms, the ITU's historical dominance continued to slip, and the organization now had to compete for industry's attention. While hundreds of privatizing carriers and other entities participated in its work and generated a wide range of standards, a substantial share of the total network-related standards pie was increasingly unrelated to or at most influenced indirectly by the regime.

The devolution of technical standards activity from the ITU to an expanding universe of organizations and collaborations has accelerated substantially during the third NWO. The Internet environment has been a major driver in this process, and much of the key work is being done in bodies indigenous to the technical and operational community like the Internet Engineering Task Force (IETF) and the World Wide Web Consortium (W3C), the operations of which are widely seen as more flexible than the ITU. In parallel, many other standards groups have converged into the Internet space from myriad ICT-based industries to work on specific technologies and applications employed in global electronic commerce, mobile communications, audiovisual services, security, and so on. Where necessary, some of them collaborate with the ITU (which, since 2001, no longer requires governmental approval of standards), but the varying working styles, architectural orientations, and material interests of their participants can be sources of friction. As is underscored by the fissures between the traditional telecommunications industry's "bellheads" and the Internet's indigenous "netheads" over Next Generation Networks and related matters, at stake are fundamental choices about the design principles that will guide the future evolution of the electronic networks.

While the environment is infinitely more diverse and robust than in the days of the PTT monopolies, ensuring that standardization processes are not subject to capture by particular industry factions and that key standards remain open and flexible remains a problem. Individually or in blocs, large and powerful firms still aggressively push to get their technologies annointed as the basis for global standards in ways that could limit innovation and opportunities for other suppliers or tinkering users. As Peter Cowhey, Jonathan Aronson, and John Richards demonstrate in chapter 4, in cases like third-generation mobile wireless standards, the combination of competing interests and universal, consensus-based decision making in the ITU can slow the development and roll-out of important technologies. Moreover, some of the native Internet standards bodies, which are generally seen as models of effective bottom-up community decision making, have experienced significant pressure from major firms in recent years.

International Radio Frequency Spectrum

As with the telecommunications regime, preserving national sovereignty was a core concern of the radio regime launched in 1906. ITU members could organize their domestic systems as they pleased as long as their operations were consistent with the regime's rules with respect to transborder issues. Nation-states were exceptionally keen to ensure that they had total control over the spectrum's distribution and use within their territorial borders, and that they did not encounter harmful interference or other problems from their counterparts' operations. Pursuing these objectives involved periodic intergovernmental treaty negotiations that produced thousands of pages of highly detailed provisions, as well as equally painstaking work defining voluntary technical and operational standards. Moreover, in addition to social and commercial objectives, national security considerations figured prominently in the policy process and made the concern with sovereignty all the more acute. Particularly for states with internationally dispersed and mobile military and intelligence operations, securing access to choice pieces of the spectrum and GSO drew the close attention of national security planners in and out of the ITU.

Balancing the concern with sovereignty were general principles specifying its acceptable exercise. The ITU treaties and the UN's 1967 Outer Space Treaty established that spectrum and orbital slots are scarce natural resources and the common property of mankind. As such, they are not subject to national appropriation, and efforts to claim otherwise—for example, a 1976 declaration by eight equatorial countries asserting sovereignty over portions of the GSO above their territories—were rejected. Instead, usage rights could be vested in one of two ways. Under the a posteriori or "first-come, first-served" procedure, a state notified the ITU of its planned frequency assignments and satellite positionings and these were assessed to ensure conformity with the regulations, so that, for example, there was no harmful interference with existing, legally registered users. If the judgment was favorable, the assignments were listed in a master register of frequencies and enjoyed legal status; if it was unfavorable, the parties involved were supposed to undertake coordination to resolve the issue, although the ITU had no power to force a member to forgo an unapproved activity. Alternatively, under the a priori procedure, ITU conferences could adopt plans assigning frequencies to stations or allotting frequency channels to countries for use in a given service, which governments then assigned to particular stations within their jurisdictions. At a broader level, ITU conferences also allocated frequency bands to particular services or sets of services on a regional or global basis, and these were then listed in the table of frequency allocations. Finally, as with the telecommunications regime, ITU study groups developed a wide array of technical and operational standards specifying how the treaty principles were to be carried out.[49]

During the first NWO, the international politics of the radio regime were quite different from those of the telecommunications regime. In this case, the United States was a major player and supporter of the multilateral framework. Indeed, all the industrialized countries, including the Soviet Union, were actively engaged in deploying terrestrial and later satellite technologies and benefited from a system that allowed them to stake claims, exploit their technological capabilities, and reduce unwanted signal interference. The main challenge to the regime came from the developing countries, which from the 1970s decried the prevalence of first-come, first-served claims under which the industrialized countries locked up the desirable spectrum real estate. They used their numerical majority in the ITU to impact a series of diplomatic conferences by pushing for greater reliance on a priori planning that would take into account their special needs and geographical situations and be in keeping with treaty provisions calling for "equitable access." The industrialized countries resisted generalizing the planning approach, noting that the ITU treaties also specify that spectrum and the GSO must be used efficiently and economically, that is by countries that currently have the capabilities and requirements. Nevertheless, a partial accommodation was reached in the form of plans that set aside orbital slots and some portions of spectrum for future use by all countries, including those not yet in a position to do so.

The institutional fundamentals of the radio regime remained unaltered in the second NWO. Irrespective of liberalization and the development of new approaches like spectrum auctions at the national level, governments remained firmly wedded to the same state-centric framework for international spectrum they had employed since 1906. What did change, though, were the forces driving the intergovernmental process. The rapid growth and differentiation of markets for new spectrum-hungry technologies gave an expanding range of corporations direct interests in the outcomes of ITU negotiations. Technological advances like digital compression and fiber optic networks freed up some spectrum, but the flood of proposed technologies and uses generally offset the potential flexibilities. In response, the private sector substantially increased its involvement in both the technical standards study groups and the big diplomatic conferences, in the latter instance by joining and (in the U.S. case) even heading national delegations and by launching large-scale lobbying campaigns. The 1992 World Administrative Radio Conference was a watershed in this shift to commercially oriented negotiations, in which governments from the industrialized world essentially became evangelists for their national champions' respective technologies. Hence, if the United States wanted spectrum reallocated for its companies' low earth orbit and medium earth orbit satellites, the Europeans wanted reallocations that favored their companies' new terrestrial mobile systems, and the negotiations became focused on finding compromises to accommodate both. As was illustrated by the case of non-geostationary satellite services, the markets for which ultimately failed to blossom as projected, this industry-driven reorientation of the process could have downsides. But more to the point, the main axis of contention had shifted from North–South debates on equitable access to North–North contests between firms and technologies. The politics of the radio regime had changed dramatically, even if its guiding principles and modalities had not.

Despite the sweeping changes occurring in other global governance environments, these conditions have persisted in the third NWO. The ITU still manages international spectrum through seemingly continuous and painstakingly slow intergovernmental negotiations that produce treaties numbering thousands of pages. If one compares the ITU's discussions with the national-level debates and policy reforms of the Internet era concerning, inter alia, spectrum auctions, unlicensed spectrum and the notion of a spectrum commons, municipal Wi-Fi, and all the rest, it is difficult not to concur with Don MacLean's understated observation in chapter 2 that the ITU model "appears curiously old-fashioned."[50]

Nevertheless, while change of the regime itself has not occurred, change within the regime has been ongoing. First, the above-mentioned trends toward increasing commercialization and private sector involvement have grown significantly; to note just one example, there were three times as many nongovernmental observers at the 2000 World Radiocommunication Conference as there were at the 1995 conference. Second, the range of technologies and uses vying for spectrum, and hence of the issues to be managed, has continued to grow at a seemingly exponential rate due to demands of the mobile telephone and Internet booms, broadband networking, global positioning systems, military and related national security systems, aeronautical services, intelligent transportation systems, emergency and disaster communications, and the like. And third, the regime has been confronted with some significant political challenges arising from the confluence of changing incentives structures and its traditional procedural modalities. As Don MacLean (chapter 2) and Rob Frieden (chapter 3) both note, one of the most difficult examples of this has been the struggle to contain speculative registrations of so-called "paper satellites" that do not reflect real communications requirements. The rapid growth of such registrations generated costly backlogs and forced ITU members to devise new administrative and financial rules but not to reconsider the fundamentals of the sovereignty-based architecture.

International Satellite Systems and Services

The United States conceived INTELSAT as a government-initiated monopoly provider, a sort of meta-PTT in the sky that would operate in a manner similar to regulated domestic public utilities. The organization served as a "carrier's carrier," providing international transmission to the earth station gateways of signatory operators' national networks, rather than a competitive service directly to end customers. While its tariffs were supposed to be nondiscriminatory among users of the same service, it is widely believed that there was considerable cross-subsidization of the prices charged to developing countries, which of course made the latter enthusiastic supporters of the organization.[51] Over time, its fleet of satellites would provide international service to much of the planet and prove particularly useful in integrating into the global network of networks developing countries and territories for which terrestrial links were difficult or costly to construct. In addition, it provided domestic services for some developing countries unable to construct sufficient terrestrial facilities.

During the first NWO, the international politics of INTELSAT mainly revolved around two major sets of issues. First, while in the 1960s the United States sought to ensure an INTELSAT monopoly, by the time the definitive arrangements were negotiated in 1971 it had become clear that other countries wanted to have domestic, regional, and specialized satellite systems under their local control. The compromise reached allowed separate systems if they successfully underwent a coordination procedure to ensure technical compatibility with and avoid "significant economic harm" to INTELSAT. While this key term was left undefined to provide leeway, it is widely believed to have been interpreted to mean that a new system's traffic would not have been carried on INTELSAT or would not reduce its share of the market by more than 3 percent. This compromise facilitated the creation of regional systems like EUTELSAT, ARABSAT, and PALAPA and specialized systems like the International Maritime Satellite Organization (INMARSAT). The Soviet Union, which had opposed INTELSAT's creation, established the INTERSPUTNIK system in 1971 to provide services to countries under its influence.

Second, the overwhelming dominance of the United States and the managerial role of Comsat caused frictions during the interim agreements period of the 1960s. A particular source of tension between the United States and Europe was the procurement of equipment and launch services, which the Americans largely monopolized. Accordingly, European governments demanded a "fair share" or quota-based approach to purchasing. While such requirements were not formalized in the 1971 definitive agreements, the Europeans ultimately were able to get enough of the action to acquire expertise and build up an aerospace industry that could compete effectively for contracts as time went by. The regime faced a much more significant challenge during the second NWO. In 1984, the Reagan administration abandoned the United States' longstanding defense of INTELSAT's dominance in favor of fostering competition by separate private systems. A number of U.S.-based companies had by now developed the technology and resources needed to crack the global market, so the government moved unilaterally to license their competitive entry on the ground that no one competitor would immediately cause significant economic harm. In response, in 1985 INTELSAT adopted a resolution on a 108-to-1 vote urging members not to participate in linking with any separate international systems serving the lucrative Atlantic market, and the organization's leadership soon embarked on a crusade against the decision that included lobbying the U.S. Congress—despite the fact that INTELSAT's budget included a substantial U.S. contribution. These efforts were not well received; the United States moved ahead in authorizing new entrants, and in 1988 the Pan

American Satellite Corporation (PanAmSat) launched a satellite with Peru as its first correspondent. Moving beyond denunciations of American unilateralism, other industrialized countries abandoned ship and authorized their own competitors, and by 1999, "more than 200 commercial geosynchronous satellites were in orbit above the earth, of which approximately 73 served the United States. Of these, only 17 satellites belonged to INTELSAT, of which just 13 served the United States."[52] At the same time, the rapid spread of fiber-optic cables also was eating into INTELSAT's market share. The regime's formal constitution had not yet changed, but its collective interpretation and implementation clearly had.

The shift to market-facilitating governance was also evidenced by a parallel development. In the 1970s and 1980s there was much debate about the use of specialized remote-sensing satellites that gather photographic and other information about the earth and its environment by detecting and measuring radiation. While governmental programs had generated concerns among countries that were being sensed by others and lacked the technology to do it themselves, the commercialization of sensing raised additional issues when viewed in the context of concurrent debates on global communications. As usual with transterritorial information flows, the overarching framing of the issue was in terms of national sovereignty. The United States and other industrialized countries that had the technology and were developing commercial industries advocated the unrestricted use of sensing satellites and distribution of collected imagery, whereas the developing countries, Soviet bloc, and even some industrialized countries insisted on the primacy of sovereignty and prior consent. After over a decade of debate on the matter, in 1986 the UN General Assembly adopted a resolution, Principles Relating to Remote Sensing of the Earth from Outer Space, in which prior consent was abandoned in favor of a recognized right of the sensed countries to have nondiscriminatory access at reasonable cost to information gathered concerning their territories.

Finally, truly sweeping international regime change occurred during the third NWO. As private operators moved into the market and ate away at INTELSAT's position, they increasingly argued that the organization's legal immunities and coordination requirements gave it an unfair competitive advantage, and that a "level playing field" was therefore required. INTELSAT rescinded its "no significant economic harm" test, but U.S.-based companies remained voracious critics and pressed for legislation privatizing the organization. Given the new environment, INTELSAT's management and signatories became converts to the American view that privatization not only would benefit its competitors, but also would free the organization to compete more effectively as well. Hence, in 1999, they resolved to pursue privatization by spinning off a commercial entity. At the same time, they agreed to retain a residual intergovernmental body that would not provide services, but would ensure through a public services agreement that the new company would preserve global connectivity and affordable access for developing countries. Following the same logic, INMARSAT members decided to privatize, beating INTELSAT to the punch by completing the process in 1999.

In 2000, the U.S. Congress passed the Open-market Reorganization for the Betterment of International Telecommunications (ORBIT) Act to encourage INTELSAT's privatization. The new entity was to be a publicly traded company (although governments retained shares); operate without the legal immunities of its predecessor; forgo the requirement that separate systems coordinate with it; and secure orbital slot registrations and spectrum from governments that subscribed to the WTO's BTA (a description of which follows), rather than through the ITU system. The following year, INTELSAT's assets were transferred to the new company, Intelsat Ltd., and the residual treaty-based International Telecommunications Satellite Organization (ITSO) was established. Also in 2001, the intergovernmental European Telecommunications Satellite Organization, EUTELSAT, was privatized. Analysts are divided on the public interest implications of these developments. While some analysts charge that privatization undermines the traditional treaty commitment to affordable and universal access and leaves countries vulnerable to the vagaries of U.S. foreign policy, others maintain that ITSO's policies and U.S. law provide sufficient protections. Either way, that intergovernmental treaty organizations have been turned into private companies competing for markets is a rather remarkable example of fundamental international regime change.

International Trade in Telecommunications Services

The GATS regime created by the Uruguay Round negotiations embodies a fundamental sea change from the way international telecommunications had been visualized and governed for almost a century and a half. Telecommunications plays a dual role under the regime: as a lucrative service sector in its own right, with global revenues of $1.4 trillion in 2006; and as a means of delivering other types of services that are embodied in information, such as financial, management consulting, audiovisual, and advertising services. In the latter role, telecommunications is the leading form of cross-border delivery, one of four designated "modes of supply" for services. The others include the movement of the consumer to the producer's country; the movement of a producer that is a natural person to the consumer's country; and the "commercial presence" of producer firms in consumer countries, including—if a country wishes to grant this concession—by foreign direct investment (FDI).

The GATS is assessed in detail in this volume by Boutheina Guermazi (chapter 5), Byung-il Choi (chapter 6), and Tracy Cohen and Alison Gillwald (chapter 13). Hence, for present purposes it is sufficient to note that it has three main elements. The first is the Framework Agreement, which includes fifteen principles or General Obligations and Disciplines, such as most-favored-nation (MFN) status, which usually apply to national commitments, and Specific Commitments, which are negotiated undertakings by governments to liberalize the provision of service sectors or subsectors. A given category of service transaction is opened to competitive supply only insofar as a government agrees to do so; it can pick and choose what to liberalize or not (e.g., by allowing the cross-border supply of a particular service but not its supply via other means). There are three such commitments: to provide market access by removing quantitative restrictions, to ensure national treatment, and to undertake any additional commitments that governments choose. The second main element is a set of eight annexes that clarify or modify how the general obligations apply to issues unique to certain service sectors and modes of supply and open the way to further negotiations on some of them. And the third main element comprises the National Schedules in which governments list their commitments. Comprising several thousand pages, the schedules include a wide range of market-opening measures in different service sectors, including telecommunications.

Of particular interest in the present context are the parts directly related to telecommunications. To address the service delivery role, WTO members added a Telecommunications Annex to the agreement that runs directly counter to the traditional thrust of ITU instruments. The annex obliges governments that have made market access commitments in various sectors—value-added telecommunications services, financial services, professional services, and so on—to ensure that foreign providers of such services have access to and use of public telecommunications transport networks and services controlled by the dominant national carriers on a reasonable and nondiscriminatory basis. Moreover, governments must ensure that foreign service suppliers also have access to and use of private leased circuits, and that they can purchase or lease and attach terminal or other equipment interfacing with public networks; interconnect private leased or owned circuits with public networks, or with circuits leased or owned by another service supplier; and use computer protocols of their choice, provided that this does not disrupt the provision of telecommunications transport networks and services to the public generally. In addition, governments must ensure that foreign service suppliers can use these networks and circuits to transfer information without undue impediments within and across national borders, and that they can access information contained in databases held in any member country. In addition, the annex requires that governments apply no conditions on access and use other than is necessary to safeguard the dominant carriers' public service responsibilities, protect the technical integrity of public networks, or ensure that foreign service suppliers only provide services that have been designated as open to competition in their market access commitments. However, if they meet these criteria, governments may adopt certain specified market limitations. Given the liberalization of both the relevant ITU rules and the wider telecommunications environment, and with the possibility that disagreements can lead to conflicts with powerful players and trigger the WTO dispute settlement process, these provisions rarely have given rise to significant problems for foreign service suppliers.

The establishment of the GATS regime arguably was the biggest change in the governance of global telecommunications since the founding of the ITU. Never before has there been a broad-based, multilateral regime that actively promotes international competition as a way to organize the world market. The GATS establishes procompetitive principles to which, if countries make relevant commitments, national policies must be adapted; institutionalizes mechanisms of mutual surveillance and binding dispute resolution; and sets a disciplined baseline for progressive liberalization in the future. This is almost the exact opposite of the world market envisaged and organized by the traditional telecommunications regime.

But despite its long-term significance, the immediate impact of the agreement was limited by the fact that only a few small-market countries made partial market access commitments on basic telecommunications services, such as telephony, which constitute the biggest chunk of the global market. Accordingly, urged on by the United States in particular, countries willing to contemplate more substantial basic telecommunications commitments launched a new negotiation. In February 1997, sixty-nine governments that collectively accounted for over 90 percent of the global market concluded the aforementioned BTA, thereby bringing basic telecommunications into their national schedules of commitments (at the time of writing, just over one hundred countries have accepted the BTA or made similar commitments). In addition, fifty-seven of these countries endorsed the Reference Paper (today, about eighty countries have done so), which established six overarching principles for the redesign of national regulatory rules and institutions:

- Competitive safeguards Governments must ensure that major PTOs do not engage in anticompetitive cross-subsidization; use information gathered from competitors with trade-restricting results; or fail to make available, on a timely basis, technical information about their facilities and operations that competitors need to enter the market.
- Interconnection PTOs must provide market entrants with interconnection at any technically feasible point in the network. Interconnection is to be offered under nondiscriminatory terms and conditions no less favorable than the provider gives its own services. Interconnection rates are to be cost-oriented, transparent, and (where economically feasible) unbundled. A dispute mechanism administered by an independent national body is to handle disputes over interconnection terms and other issues.
- Universal service Such obligations are to be administered in a transparent, nondiscriminatory, and competitively neutral manner that is no more burdensome than is required to meet the policy objectives.
- Public availability of licensing criteria Where licenses are needed, information and decision-making procedures are to be transparent.

- Independent regulators Regulatory bodies are to be separated from service providers and not accountable to them.
- Allocation and use of scarce resources Procedures for allocating and using frequencies, numbers, and rights-of-way are to be carried out in an objective, timely, transparent, and nondiscriminatory manner.

Collectively, these principles represent a substantial departure from the old multilateral order. Under the Reference Paper, a carrier seeking market access in a country that has made the necessary commitments to allow commercial presence can extend its network directly into the country, interconnect at a point of presence, and terminate traffic there at cost-oriented rates. If direct investment is allowed, it could opt to build or buy its own facilities within a foreign country. These types of "beyond the borders" deep integration represent a significant shift from the ITU's traditional joint service model, which involved connecting and revenue sharing between carriers from mutually exclusive national domains.

In November 2001, WTO members launched the Doha Round of multilateral negotiations. Due to battles over unrelated issues, most notably agriculture, the round has stalled and teetered on the brink of collapse. Should the round ultimately come to a successful conclusion, further liberalization of telecommunications networks and services can be expected. The United States is pressing for all WTO members to undertake full basic and value-added telecommunications commitments; ensure full adherence to the Reference Paper; and consider the full privatization of their national telecommunications operators and networks. The European Union is making broadly parallel demands. With the two leading WTO powers pushing in the same general direction, the pressure on other countries to make significant changes to their domestic institutional arrangements is substantial.

International Trade in ICT Goods

During the first NWO, governments did not attempt to subject global ICT equipment markets to strong multilateral trade disciplines. In telecommunications, the market was heavily concentrated; indeed, as late as 1983, OECD members still accounted for 90 percent of world production, with twelve companies from the United States, Britain, France, Germany, Italy, Sweden, the Netherlands, and Japan accounting for 80 percent of all sales.[53] The market also was highly fragmented: in the global North, many of the major manufacturers were focused on serving monopsonistic carriers at home rather than abroad, and their home markets were protected from foreign entry by nationalistic procurement policies, proprietary intranetwork standards, and other trade barriers. In parallel, colonial and postcolonial ties often were leveraged to maintain separate spheres of influence for these firms in parts of the global South. The remaining national markets that lacked indigenous producers typically were supplied via trade or FDI by the more internationally oriented firms like ITT, GTE, Philips, and Ericsson.

In short, the configuration of interests and institutions in the regime-making states and the lack of powerful countervailing political or ideational forces precluded market-oriented global governance, such as the application of GATT rules. In fact, telecommunications was specifically excluded from the plurilateral Agreement on Government Procurement signed by twenty-eight GATT members in 1979 (although the United States did pressure Japan to apply the rules as part of a bilateral deal). Hence, what intergovernmental steering there was derived principally from the international regulatory regime for telecommunications networks and services. While the early telegraph treaties specified the terminal apparatuses that could be used and hence favored the rather few manufacturers of those systems, the development over time of technical standardization processes cleared the way for an expanded set of large manufacturers able to design to the specifications for transmission, switching, and terminal equipment. Even so, the focus on systems requiring standardization for international interconnection and traffic exchange often left national monopolists leeway to employ proprietary standards or implementations within their national networks, which inter alia served as NTBs to foreign entry. Finally, one might add that the situation in telecommunications was broadly parallel to that in the mainframe computer industry, where concentrated markets (after a period of shakeouts and consolidation), preferential procurement policies, and national champion strategies in the major producer countries combined with a nexus of tariffs and NTBs to distort trade patterns and limit open competition.

Conditions began to change in the second NWO. The asymmetric liberalization being pursued by key countries and regional blocs not only increased the market access of foreign suppliers, most notably in the United States, but also stimulated technological change and the growth of a robust range of suppliers of diverse products like CPE (such as terminals and private branch exchanges) and data networking and customized business systems. Both established large manufacturers with new incentives to innovate and smaller entrants targeting niche markets were now looking to export, and corporate users were seeking to acquire the best systems available with less regard to their national origins. Ideational change and political shifts like the Reagan administration's trade campaign also supported business demands for greater competition in international markets. Given these forces and the gathering momentum of telecommunications liberalization and privatization, varying levels of market opening were achieved in bilateral agreements among industrialized countries. Moreover, international standardization deepened across a wider array of network elements, thereby facilitating participation in more markets by a broader range of suppliers. And the coverage of the plurilateral Agreement on Government Procurement was significantly expanded, inter alia, to include telecommunications equipment, even if most OECD countries continued to exempt purchases by certain governmental entities or public utilities.

The ground has shifted further during the third NWO. Many governments have been reconsidering the case for trade barriers due to an array of factors, such as the liberalization and privatization of telecommunications, the globalization of ICT production and export activity, the increasing importance to national economic development of broadly diffused and applied ICT, the expanded base of consumer demand in an age of personalized technology and the Internet, ideational change, and so on. The cumulative result has been to stimulate movement at multiple levels of governance toward market-opening initiatives. The key development at the multilateral level has been the WTO's Information Technology Agreement (ITA). In 1996, twenty-nine member countries, mostly from the industrialized world, agreed to eliminate tariffs and other duties and charges on a wide range of ICTs. In addition, the ITA provides for the review of NTBs and consultations with an eye toward encouraging their reduction or elimination. And importantly, given the long history of discrimination in this arena, ITA commitments are made on an MFN basis and hence can benefit all WTO members. The subsequent ITA II talks have encountered some difficulties and continue in the Doha Round, but as is the ITA covers a fairly broad range of network-related technologies. Moreover, the agreement now has seventy members, including key markets like China and India, that the WTO calculates account for 97 percent of world trade in ICT. Governmental trade barriers and anticompetitive corporate practices undoubtedly remain in various market segments, and some governments, most notably in the least developed countries, still seem to favor the short-term gains from high tariffs to the longer-term gains from technology diffusion. Nevertheless, viewed in historical terms, there clearly has been a significant transformation here in the social purpose and institutional mechanics of global governance, and the ICT goods sector has been normalized and brought into the mainstream of trade diplomacy.

While the rules of the game for trade in physical ICT goods are moving toward greater openness, the situation with respect to immaterial goods remains cloudier. In the Doha Round negotiations, governments are thus far at an impasse with regard to the treatment of digital products like software, music, videos, and e-books. The United States argues that products that are normally sold in the format of physical goods should be subject to the strong nondiscrimination disciplines of the GATT. In contrast, the EU insists that anything that moves over a network should be viewed as a service and subject to the weaker requirements of the GATS. As one close observer has concluded, "The gap between proponents of GATT-like treatment and those that propose a GATS classification seems unbridgeable."[54] Moreover, there has been no progress in the Doha Round on closely related e-commerce issues, such as the valuation of digital products.

A related issue concerns the application of duties to digitized products. The WTO's 1998 ministerial conference adopted a declaration on global electronic commerce that included a pledge that members would continue the general practice of not applying customs duties to products embodied in electronic transmissions across borders. The United States pushed this symbolic initiative to help create a consensus for liberalized treatment of e-commerce. However, developing countries have expressed strong concern about the loss of customs revenues on products that were previously embodied in material form and subject to duties and hence have balked at making the moratorium permanent. In sum, while trade in immaterial products is expanding, uncertainties about the applicable rules of the game may be reducing the pace.

Internet Core Resources/Names and Numbers

ICANN is often depicted in the popular press as the "government of the Internet" because it performs a set of functions key to the management of Internet identifiers, that is, domain names and IP numbers. The organization is responsible for generic (gTLD) top-level domain name management and coordinates with country code (ccTLD) registries; performance of the Internet Assigned Numbers Authority (IANA) functions, such as delegating local registrations of IP addresses to Regional Internet Registries (RIRs), administering the root nameservers at the top of the DNS tree, and administering protocol parameters; and making policy on a host of related issues, many of which go beyond its nominally narrow and technical mission. Previously, the technical functions had been performed under U.S. government contract by one man, Jon Postel. ICANN regulates much of the namespace marketplace through a web of contractual relationships with domain name providers, but remains ultimately answerable to and dependent upon the U.S. government. For example, ICANN can give preliminary approval to modifications, additions, or deletions to the root zone file, but ultimate approval is up to the U.S. government. In addition, ICANN also established the Uniform Domain-Name Dispute-Resolution Policy (UDRP), a WIPO-administered alternative dispute system that has widely been used by large corporate trademark holders to defend their brands and to suppress cybersquatting.

Despite ICANN's centrality to the management of the Internet's core resources, responsibility for other key governance activities is distributed among a host of players. Principal ones include the U.S. government, which exercises nominally "light touch" control over ICANN through its contracts and other arrangements, and also maintains legal authority over the master databases or root zone file that contains the authoritative listing of gTLDs and ccTLDs; Verisign Inc., which maintains the server on which resides the zone file, as well as control over the .com gTLD; other commercial registries like Afilias, which is responsible for the .info gTLD, as well as the Public Interest Registry (PIR) that maintains the .org gTLD for noncommercial entities; the RIRs—AfriNIC (for Africa), APNIC (for the Asia and Pacific region), ARIN (for Canada, many Caribbean and North Atlantic islands, and the United States), LACNIC (for Latin America and parts of the Caribbean), and RIPE NCC (for Europe, parts of Asia, and the Middle East)—not-for-profit, membership-based organizations that oversee the allocation and registration of Internet number resources; the commercial registrar industry; managers of ccTLDs per country or territory; the operators of the thirteen root servers (ten in the United States, two in Europe, and one in Japan) that receive updates of the zone file, which are distributed twice daily, and are now replicated via Anycast techniques by servers on all continents in order to bring them closer to users and allow a better distribution of traffic; and so on. So while ICANN is the linchpin for the global system, it is hardly alone in providing governance of Internet naming and numbering.

By any measure, ICANN embodies a truly revolutionary model for the governance of a central element of the global networking environment. Aside from being a private nonprofit entity subject to California law and contractual oversight and approval by the U.S. government, it has many distinctive features, for example, a federal, multistakeholder structure comprising subgroups of entities from all organized segments of the Internet addressing industry, such as trademark holders, ISPs, and domain name managers; representatives of individual users and civil society; a Governance Advisory Committee (GAC), through which governments participate on what is nominally a purely advisory basis; an appointed board of directors that has final authority on policy issues; a powerful staff that enjoys considerable latitude in interpreting policy and undertaking programmatic work; generally open meetings that are webcast (save, notably, board and GAC gatherings) that are held around the world to facilitate participation of regional interests; various means of remote, online participation options like public listservs, blogs, and requests for comment; and so on. This sort of comparatively freewheeling and nominally "bottom-up" process represents a stark contrast with most intergovernmental and industry-only forums for ICT governance decision making.

Despite (or perhaps in part because of) these attributes, ICANN has from its inception given rise to an unprecedented degree of controversy. An early experiment with direct elections to the board of directors by citizens around the world was abandoned to much public outcry by those favoring truly participatory and democratic governance. Successive boards have demonstrated a penchant for opaque and seemingly ad hoc decisions affecting many organizations and individuals around the world. The decision-making processes have often appeared to be captured by major corporate constituencies like famous-brand trademark holders, so it has effectively worked to limit the introduction of new global domains that would purportedly dilute their brands. This has had the adverse effects of limiting freedom of speech, slowing business development, and creating artificial scarcities that engender conflict among stakeholders. Moreover, critics charge that ICANN has lacked a clear and consistent process for selecting among gTLD proposals, relying instead on painfully slow and opaque "beauty contests" that typically reward well-connected and well-established applicants. Indeed, as U.S. Representative Edward Markey once quipped, Vatican decisions on the selection of a new pope are more transparent than ICANN's process for selecting new gTLDs. And as Milton Mueller and Jisuk Woo detail in chapter 14, ICANN has posed significant problems for participants and would-be participants from the developing world. In short, the list of complaints from members of the ICANN community that have arisen since its birth goes on and on.

More recently, ICANN has been the focus of vociferous criticism from national governments. The U.S. government's decision to give responsibility for the coordination of naming and numbering to a private sector entity was driven to a significant extent by its desire to keep control of the Internet out of the hands of the ITU and its traditionalist members. This has never rested well with many developing country governments, who argue that the Internet is to a significant extent built on top of PTO/PTT facilities; that governments should have authority over public policy issues, which they define broadly, pertaining to a communications medium that is increasingly central to their economies and societies; and that the ITU is the logical multilateral organization through which decisions should be made. Accordingly, they have for years pressed through various diplomatic conference resolutions and communications with the ITU secretariat for the ITU to take on a leading role in this arena, up to and including replacing ICANN entirely. At the same time, many European governments as well as the European Commission have had their own lengthy list of complaints about ICANN and have sought to assert greater multilateral public authority over its operations, albeit not necessarily in the ITU.

For the United States and the global business and technical communities in particular, the notion of intergovernmental control over ICANN or the naming and numbering systems more generally has always been a nonstarter. But after dying down a bit, the issue was revisited and catapulted to the top of the global ICT agenda by the United Nations' 2002–2005 WSIS process. In lengthy and extremely heated debates that stretched over three years of preparatory negotiations of the texts, both the EU and a broad coalition of developing countries criticized ICANN's nature and performance. The EU did not seek to replace the organization with something else, but rather to have governments agree on a set of public policy principles that would shape ICANN's decisions on some key matters. For the developing countries, the complaints were more fundamental, and pertained to the legitimacy of having a private sector entity subject to U.S. laws making key decisions about the Internet; the relegation of governments to an advisory role in relation to a private entity controlled by other stakeholders; and hence their desire to replace U.S. authority over both the zone file and ICANN with some sort of intergovernmental framework, if not under the ITU then under a new, custom-built entity. In addition, they feared (incorrectly) that by virtue of its control over the zone file, the U.S. government could somehow "flip a switch" and cut a country off from the Internet; expressed grave concern about the slow pace of introduction of multilingual domain names; worried that with the expected exhausting in a few years of the supply of IPV4 numbers, and the difficult transition to IPV6 numbering they would be marginalized; and so on.

In the end, the United States made it clear that it was unwilling to contemplate any of the radical changes being proposed by either bloc of countries. Instead, it nominally agreed to engage in a process of "enhanced cooperation" with governments and relevant international organizations and entities, most notably ICANN, with an eye to addressing the concerns about private authority over policy decisions many regard to be the rightful preserve of governments. In reality, little has happened along these lines at the intergovernmental level since the WSIS concluded, leaving many governments deeply frustrated and angry. For its part, ICANN has responded to the WSIS debate by undertaking a series of reforms intended to increase its transparency and accountability, including increasing the influence of the GAC. Early signs indicate that governments may use this new momentum to advance policy "requirements" that could politicize the process of introducing new gTLDs, assert sovereign authority over the independent operators of ccTLDs, and so forth.

The Global Governance of Networked Information, Communication, and Commerce

The Flow and Content of Information

In the first NWO, the social construction of national sovereignty was the overarching consideration in global governance debates on the information conveyed over electronic networks. This was especially true in the case of mass media wherein, as Cees Hamelink explains in chapter 7, sovereignty was framed as being at odds with the free flow of information. Sovereign states' rights under ITU instruments to stop cross-border transmissions applied to broadcasting as well as private communications because the ITU defined telecommunications broadly as any transmission or reception of signals and messages via wired or wireless media. However, this right alone was insufficient for states that equated sovereignty with inviolable borders because of the promiscuous nature of some radio signals and the technical and financial demands of stoppage. Many authoritarian governments therefore desired collective commitments to curtail outbound transmissions from other countries that could spill across their borders.

The competing principle of the "free flow" of information reflected Western liberal and especially American values, and was supported by post–World War II developments in international human rights law. The term was often construed as equivalent to or an extension of the protections enunciated in Article 19 of the 1948 Universal Declaration of Human Rights (UDHR): "Everyone has the right to freedom of opinion and expression; this right includes freedom to hold opinions without interference and to seek, receive and impart information and ideas through any media and regardless of frontiers." A number of international instruments on human rights and related matters have since invoked and reinforced these freedoms in varying formulations. Even so, they are somewhat circumscribed by sovereignty-based provisions within the UDHR and other instruments. Moreover, the free-flow principle has been contested not only by authoritarian states opposed to the free exchange of ideas, but also by governmental and nongovernmental stakeholders that saw the term providing cover for U.S. commercial interests and cultural imperialism.

Given the sharply divided interests involved, no singular and comprehensive solution to the "sovereignty vs. free flow" question was possible. When international radio broadcasting began in the 1920s, some governments expressed concern about the purported need to "protect" their publics from dangerous foreign ideas. But these concerns were not universally shared; by the mid-1920s, in Europe alone there were over two dozen stations providing international service, and the originating countries saw this as *their* sovereign prerogative. Hence, an international regime for radio broadcasting involving limitations on cross-border transmissions was not pursued, and sustained cooperation in the IBU and related bodies tended to focus instead on technical matters, including as inputs into ITU activities. In consequence, some authoritarian states took matters into their own hands and built expensive facilities to jam inbound signals. Agreement on the legality of jamming proved elusive save in the ITU, which proscribed spectrum usage that caused harmful interference with the signals of third parties.

The growth of television broadcasting after World War II did not spur discussion of a multilateral regime on information flow and content because of the limited range of terrestrial signals. As such, neighboring countries dealt with signal spillovers and related matters in bilateral or regional contexts. But the subsequent development of direct broadcasting satellite (DBS) did set off divisive and protracted debates in the ITU, UNESCO, and the UN General Assembly from the 1960s into the 1980s, when widespread commercial service finally commenced. Each of the three bodies adopted a principle of "prior consent," which ultimately was formulated to mean that a state that intends to initiate or authorize a DBS service is to notify the proposed receiving states and enter in consultations to receive permission if requested. Only the ITU adopted a treaty instrument, as the United States and other countries with concerns about multilateral restrictions "were prepared to accept principles in the ITU context—in which technical problems can cover up for political issues—principles that they were not ready to accept in the UN context."[55] In the end, the narrow technical specification of the ITU's approach and the soft-law or unratified status of other instruments created ambiguity, and the few countries that proceeded without obtaining prior consent did not suffer notable consequences.

While only for DBS did governments manage to set generally applicable sovereignty-based principles, they enjoyed somewhat greater success in agreeing to limit the dissemination of specific substantive types of messages that were deemed to be especially problematic. Efforts to do so did not begin with the rise of electronic media. Probably the first initiative was the Carlsbad Decrees of 1819, in which a group of German states agreed to help each other suppress the expression, inter alia in print publications, of opinions that were hostile to each other's institutions. In addition, the early twentieth century saw the establishment of treaties suppressing the production, possession, and circulation of obscene print materials. And as international broadcasting took off, heightened concerns about the rapid and widespread distribution of potentially incendiary information gave rise to a number of initiatives. For example, in 1925, the IBU passed resolutions asking governments and broadcasters to suppress transmissions intended to prejudice good international relations. Similarly, in 1936, the League of Nations adopted the International Convention Concerning the Use of Broadcasting in the Cause of Peace, which enjoined members to prohibit or stop transmissions that could incite populations to commit acts incompatible with each others' internal order or lead to war. However, as Germany, Italy, Japan, and the United States were not among the twenty-eight signatory nations, the practical effects of the agreement were negligible. Arguably more noteworthy were the United Nations' post–World War II resolutions and conventions on war propaganda, racial discrimination, the protection of minors, crimes against humanity, cultural exchange, and so on that included provisions applicable to international communications irrespective of the medium used. While these agreements were carefully crafted not to impinge on sovereignty too much and were certainly violated, most notably in times of war, they did open the door to normative pressuring and have a somewhat civilizing effect on the use of global networks.

The balance between sovereignty and free flow was also at the heart of the most controversial challenge to mass media during the first and early second NWOs—the proposed New World Information and Communications Order (NWICO). Throughout the 1970s, developing countries and the Soviet bloc pressed principally in UNESCO for a NWICO in parallel with the debate on a New International Economic Order that was taking place in particular in the UN Conference on Trade and Development (UNCTAD). In brief, the campaign "was an outgrowth of third-world resentment of the imbalances in international news flows, as summarized in the phrase, 'one-way-flow'; the lack of respect for third-world peoples' cultural identity that such imbalances reflected; the monopoly positions of TNCs, which were perceived as a threat to the countries' national independence; and the inequitable distribution of communications resources in the world."[56] In response, the challengers pressed themes that came to be known as the "four Ds" (democratization, decolonization, demonopolization, and development) and advocated measures that would strengthen the hands of states, regulate media organizations, promote a collective "right to communicate" and a balanced flow of information, expand development assistance and technology transfers, and so on. These proposals engendered virulent opposition from Western governments, media companies, and journalists that resulted in the substantial watering down of the negotiated texts. The hotly contested Mass Media Declaration adopted in 1978 called for the "free flow and wider and more balanced dissemination of information," a greater diversity of information sources, responsible journalism, respect of the rights and dignity of all nations, expanded development assistance, and so on, and an international commission convened to study the situation and make recommendations adopted a similar approach. By 1980 the challengers' regulatory agenda had been effectively replaced by the consensual creation of a new development assistance program, but the United States and the United Kingdom remained disgruntled and subsequently withdrew from UNESCO, throwing the organization into an extended period of financial and political difficulties.

The transition from the first to the second NWO also witnessed a parallel development with respect to point-to-point computer communications—the previously mentioned TDF debate. From the mid-1970s to the mid-1980s, governments argued about a number of potential challenges to national sovereignty and related values stemming from the rapid growth of intercorporate and especially intracorporate TDF. Many of these same issues are currently being debated again with respect to the Internet—albeit not in the context of purely private cyberspaces.[57] On one side of the debate were those who feared that TDF could have negative consequences and hence merit collective regulation, since purely national action would be ineffective. Many developing countries were firmly of this view, and on some issues the Canadians, Scandinavians, French, and other continental Europeans shared their concerns to varying degrees. On the other side of the debate were those who feared that the concerns being raised were simply veiled rationales for digital trade protection and burdensome regulations directed in the first instance at U.S.-based TNCs, which were the biggest and most advanced users of TDF. As one might expect, the principal proponents of this view were the U.S. government and business community. Also an important factor in the debate was a lively expert community comprising scholars, consultants, privacy advocates, and the like, the members of which were not united in their principled and causal beliefs. The experts' influence was magnified by the newness and complexity of the issues and by many states' uncertainties about whether the potential problems were real and what would be in their national interests.

Under these circumstances, the policy process was heavily ideational in character, and the strategic control of information and discourse was central to the outcomes. U.S.-based TNCs mobilized their counterparts in a number of multinational business lobbying groups around the notion that international regulation would endanger all businesses that were users of transnational networks and would stifle innovation and competitiveness. Moreover, because most of the information about what TNCs were doing with TDF and how this might matter was proprietary and not publicly accessible, it proved easy to argue that the concerns being raised were mere shibboleths with no factual basis, and that imposing regulations would be too technically difficult and draconian. The private sector's perception-management campaign, control of the technology and markets, and structural power over national economic destinies turned the discursive tide away from the possible need for regulations and toward the need to maintain business confidence in an increasingly network-dependent world economy. Wavering governments in the OECD fell into line, and settled the matter by agreeing to two plurilateral instruments.

Hence, in 1980, the OECD adopted Guidelines on the Protection of Privacy and Transborder Flows of Personal Data, a fairly mild and nonbinding instrument that was acceptable to the United States and global business. The other OECD instrument adopted pertained to the broader range of economic, legal, and social issues. Here too, far from imposing regulations as some had originally envisioned, the 1985 Declaration on Transborder Data Flows acknowledged that TDF was largely circulating without restriction, held that this was beneficial, and urged governments to avoid the creation of unjustified barriers to the international exchange of data and information. For its part, the developing country-dominated IBI relented on the issue, but was later disbanded. In short, the thrust of a debate begun in the sovereignty-oriented first NWO had been decisively reversed in the second. The OECD declaration, while nonbinding, reflected an understanding among governments that the issue had run its course, and that there would be no collective legitimation of new general regulations on corporate data transfers based on such considerations as the location of data processing.

During the third NWO, the perceived urgency of governance challenges has vastly increased due to the Internet. With over a billion individual users able in principle to globally disseminate information on a mass scale, sometimes anonymously, proregulatory governments' traditional concerns about penetration by foreign media organizations (and their own inability to disseminate the information needed for a "balanced flow") have been overtaken. Not surprisingly, the first responses have been at the national level. Across the world, governments have adopted laws and regulations and courts have made rulings curtailing speech they deem undesirable, requiring the use of filtering technologies, compelling ISPs and websites to monitor and take down noncompliant content, and so on. Major technology companies have helped to facilitate such restrictions in both democratic and authoritarian countries. In addition, there have been a number of celebrated cases in which governments and courts have sought to force information providers outside their borders, such as Yahoo!! and YouTube, to remove or block local access to undesired information. In the aggregate, the growth of uncoordinated extraterritorial governmental censorship and industry's own anticipatory or other restrictions are producing balkanization or, perhaps more accurately, a neomedieval condition of multiple, overlapping, and competing layers of authority.

In response to or anticipation of such political problems, the private sector has undertaken a number of efforts to develop self-governance mechanisms. In terms of collectively defined rules, in some countries and regions fairly effective industry codes of conduct have been agreed to by Internet service providers, advertisers, electronic game providers, and the like.[58] With the globalization of markets, such rules are being extended transnationally through contractual relations, trade association coordination, and other means to provide a measure of transaction-specific governance. But as Peng Hwa Ang argues in chapter 8, efforts to develop effective opt-in international private rule systems for content labeling like the W3C's Platform for Internet Content Selection and the Internet Content Rating Association initiative have encountered significant difficulties.

In light of the shortcomings of self-regulation and politicians' need to appear responsive, it is reasonable to expect fresh calls from some quarters for intergovernmental frameworks that address speech questions. Nevertheless, the complexity of the issues and variations in national legal systems and traditions make it difficult to reach broad international agreement on either restrictive or permissive measures. For example, in the first phases of the 2002–2005 WSIS process, it was a major struggle to even get the governments assembled to include a full reference to article 19 of the UDHR in the Geneva declaration, much less to seriously consider the relevance of other internationally recognized human rights to access the Internet and other modes of communication.[59] With universal governance mechanisms for speech per se impossible, the main action probably will remain at the regional and plurilateral levels, and even there difficulties exist. Among the most active and important bodies is the CoE. Its cybercrime convention includes a protocol requiring signatories to criminalize racist and xenophobic speech, but it is optional and leaves governments with considerable freedom. Similarly, at the time of writing, the CoE is drafting a convention to protect children against sexual exploitation and abuse that will include provisions on pornography, but it is expected to allow signatories latitude in setting the age below which its provisions become relevant.

Despite the international political differences on speech per se, there are some arenas in which governments have succeeded in establishing regimes affecting the distribution of content. These pertain to such issues as trade, intellectual property, privacy and consumer protection, and so on, and are taken up in the sections to follow.

International Trade in Content Services

Of potentially increasing significance has been the inclusion of culturally oriented content services under the WTO's GATS. Byung-il Choi examines this development in detail in chapter 6. During the Uruguay Round negotiations of the second NWO, some countries—most notably, France, and by extension, the EU—were hesitant to bring international trade in audiovisual services under the aegis of the GATS as the United States was urging. Ultimately, a compromise was reached to include these services and hence allow future bargaining over progressive liberalization, but opponents were freed to withhold MFN treatment and avoid substantial market access and national treatment commitments.

The issue has been resurrected in the ongoing Doha Round, with the United States pushing for governments to make commitments in at least some subsectors. This time, the effort is receiving varying degrees of support from countries, including in the developing world, that either believe they would benefit from more open markets, or are willing to exchange support on this issue for American help on other negotiation items. Nevertheless, many governments remain reluctant to undertake trade commitments regarding cultural content like movies and television programming, the Internet-based diversification of content and distribution channels notwithstanding.

In response to the commodification and globalization of corporate cultural products, the demands for expanded GATS commitments, and related trends, countries led by France and Canada in particular have worked in UNESCO to shore up their defenses. The results were the 2001 Universal Declaration on Cultural Diversity, and the 2005 Convention on the Protection and Promotion of the Diversity of Cultural Expressions. A binding legal regime, the latter was adopted by a vote of 148 in favor, 2 opposed (the United States and Israel), and 4 abstentions (Australia, Nicaragua, Honduras, and Liberia). As Choi also explains, the convention addresses the creation, production, distribution/dissemination, and access to and enjoyment of cultural activities, goods, and services; and reaffirms states' sovereign right to establish cultural policies and the need for international cooperation to support the cultural expressions of all countries. While it does not directly address the behavior of transnational media conglomerates, it does contain language that could be construed as allowing flexibility in the interpretation of GATS commitments, and as such could lead to tension with the WTO framework.

As with audiovisual services, the GATS covers all other commercially provided services of a "content" nature that are embodied in information and may be delivered via cross-border telecommunications networks or other modes of supply. This includes, inter alia, accountancy, financial, advertising, architectural, engineering, computer, legal, educational, environmental management, health, and tourism services. The only question is whether in practice, governments choose (or are pressured) to undertake market access commitments in a given sector. During the Uruguay Round, networked service delivery was generally thought to involve private corporate networks based on proprietary technologies. Governments were selective in making substantial commitments on cross-border services supply, preferring instead that providers like banks and consulting firms enter their markets via commercial presence (which generates local taxation and employment and boosts local businesses that sell goods and services to foreign suppliers), or that customers visit their territories (as with tourism) for related reasons.

With the Internet's subsequent emergence as a medium for commercial exchange, the level of global electronic commerce has radically increased and its composition has changed. Now a potentially unlimited range of individuals and organizations, large and small, can directly engage in unmediated international trade, and policing the bit stream or erecting direct barriers to transmission is difficult. However, in many cases indirect barriers can be maintained, for example, by preserving licensing and accreditation requirements that do not recognize foreign-sourced services or by limiting related services. Accordingly, the United States is pressing for all WTO members to make full commitments in complementary services that are increasingly integrated into network-based transactions including distribution, computer services, express delivery, advertising, and certain financial services. The EU proposal pertaining to the networked delivery of services is also quite sweeping.

Beyond the depth of commitments under the existing arrangement, the Internet's emergence as a vehicle for networked trade also has raised questions about the adequacy of the framework itself. As the previously mentioned debate over the classification and treatment of digital goods indicates, there are reasons to believe that some provisions of the GATS, and perhaps also of the GATT and TRIPS agreement, are not properly configured to accommodate the specificities of global electronic commerce. Accordingly, in 1998 the WTO adopted a work program on e-commerce and mandated the organs responsible for each agreement to examine areas in which issues may arise and require adjustments to the texts. Several years of debate ensued that satisfied members that the existing treaties were sufficient with respect to some of the issues, but as noted previously, for other important matters like digital goods no agreement has been reached. Unless these matters are taken up and resolved more effectively, it is possible that the Doha Round, should it conclude successfully, will end up leaving some key and conceptually challenging matters to be worked out in any dispute resolution cases that may arise, which is hardly the normatively optimal way to make governance decisions.[60]

A related topic in the current negotiations, in which the prospects for agreement seem brighter, concerns domestic regulation. Since trade in services takes place "behind the borders" and many service industries—education, financial, accounting, medical, and the like—are subject to extensive regulation, domestic regulatory institutions and policies intended to promote social objectives or merely to protect incumbent providers can have a strongly inhibiting impact on trade. Indeed, even when governments undertake market access commitments on paper, restrictive regulations often negate their value in practice. To address this dilemma, WTO members have devoted a substantial amount of energy in recent years to developing additional GATS disciplines on the conduct of domestic regulation. While all WTO members, including the United States, are adamant about preserving their sovereign regulatory rights, the negotiations are intended to ensure that these are exercised in a manner that is nondiscriminatory, transparent, and no more trade-restrictive than is necessary to preserve legitimate social objectives. Current trends indicate that a deal could be reached geared toward safeguarding regulatory authority and recognizing the special challenges faced by developing countries, but prohibiting the use of regulation as a disguised trade barrier.

Intellectual Property

The establishment of new national laws and intergovernmental regimes for intellectual property protection on the Internet has been one of the defining features of contemporary global governance. It has also been one of the most controversial from the standpoints of developing countries and public interest advocates in civil society, as it threatens to significantly undermine opportunities for access to knowledge and creative uses of digital content by citizens around the world in order to benefit politically connected transnational firms in the United States and other industrialized countries.[61]

Intellectual property has long been an important issue with respect to ICT equipment and software. Conflicts between national patent, copyright, and trademark systems were addressed through legal proceedings and bilateral trade negotiations, such as between the United States and Japan, against the facilitative backdrop of long-standing multilateral regimes like the 1883 Paris Convention for the Protection of Industrial Property and the 1886 Berne Convention for the Protection of Literary and Artistic Works. But the emergence of private and later public cyberspaces raised a host of new issues because of the ease of copying and circulating digital information and the jurisdictional uncertainties raised by transterritorial economic activity. Some of the emerging issues were considered in the OECD during the TDF debates, but multilateral solutions rapidly emerged as the preference of business interests.

The Uruguay Round negotiations concluded just as the Internet was beginning to take off as a global mass medium. Nevertheless, the TRIPS agreement, which is critiqued by Christopher May in chapter 11, does contain provisions of direct relevance to the online world. For example, article 10 confirms that computer programs are "literary works" and extends protection to compilations of data; article 11 requires member countries to provide authors or their successors with exclusive rights to authorize or prohibit the commercial rental of computer programs and cinematographic works; and articles 12 and 14 set standards for the term of protections afforded to copyrighted and sound-recording works, respectively. Moreover, under its rules, "any intellectual property agreement negotiated subsequent to TRIPS among and/or involving WTO members can only create higher standards. Higher standards, which could result from bilateral, plurilateral or multilateral treaties, have come to be commonly referred to as 'TRIPS-plus'... the concept covers both those activities aimed at increasing the level of protection for right holders beyond that which is given in the TRIPS Agreement and those measures aimed at reducing the scope or effectiveness of limitations on rights and exceptions under the TRIPS Agreement."[62]

In the third NWO, multilateral action has proceeded down two interrelated tracks. Work has continued in the WTO on strengthening and implementing the TRIPS agreement; as with the rest of the Doha Round, the ultimate resolution of these efforts is uncertain at the time of writing. And WIPO has launched a number of initiatives focusing specifically on the networked environment. In 1996, WIPO members agreed on a Copyright Treaty and a Performances and Phonographs Treaty. Under these instruments, governments are required to tighten their national copyright laws and extend or clarify the rights of owners to control communication and the "making available" of their creations, as well as the distribution of copies. In addition, governments must curtail the circumvention of technological protections for rights management and any tampering with the tags and codes associated with copies of protected works and phonograms. Together with the United States' 1998 Digital Millennium Copyright Act, the EU's 2001 Copyright Directive, and related national and regional actions, WIPO treaties ushered in the beginning of a new and globalized enclosure movement for the digital domain. The slow pace of accession has frustrated the treaties' proponents; indeed, while the instruments finally came into force in 2002, just over sixty member countries have joined them, a large share of which are the industrialized countries. As such, the United States and other proponents continue to pressure developing countries to accede.

In parallel, the United States and the EC pushed in 1996 for the adoption of a proposed WIPO treaty on intellectual property and databases. Their proposals would have inter alia extended strong protections to collections of data that are not sufficiently original to qualify for copyright, but may involve corporate investments in their creation and maintenance. Such a shift from protecting creativity to protecting investments and establishing rights over such data could greatly reduce the public's access to information that is currently in the public domain. Concerted opposition to the proposal by developing countries and segments of the business community (including electronics and computer companies) forced its abandonment in 1996, but the issues could be revisited in WIPO and have been addressed by other instruments like the EU's Database Directive.

Going further, in 1999 WIPO adopted an ambitious Digital Agenda. In line with the stream of activity it laid out, in 2001 WIPO adopted a Joint Recommendation Concerning Provisions on the Protection of Marks, and Other Industrial Property Rights in Signs, on the Internet. But probably the most significant and controversial initiative to be undertaken has been the effort to negotiate a new broadcasting treaty. Some of the competing draft texts to emerge could permit broadcasters to control information already in the public domain, extend property rights from twenty to fifty years, create new ownership rights to signals, further curtail the use of anticircumvention technologies, require signatories to join the prior Internet treaties, give broadcasters much greater rights than artists and performers, and (in the U.S. version) apply well beyond the broadcast arena to encompass webcasting as well. The loose and expansive framing of the U.S. proposal on the latter area would have significant detrimental effects on users and the technological and economic development of the Internet.

Catalyzed by this and related developments, a strong coalition of CSOs has come together and worked with key developing countries, led by Brazil and Argentina, to oppose the broadcast treaty. Moreover, other business factions that could be negatively affected by its overly expansive language on webcasting, including elements of the computing and Internet service industries, have been mobilized in support of the cause. As such, the treaty's fate now appears to be highly uncertain. Moreover, the developing countries and the CSO coalition have built on this momentum to promote a sweeping development agenda that would reorient WIPO, a UN agency, away from a narrow focus on placating business interests and toward the promotion of access to knowledge and the preservation of the public domain of ideas and information. Probably in no other domain of network global governance have developing countries and civil society activists been so successful in pushing a change of direction toward public interest concerns.

Given these developments and various organizational problems within WIPO, the United States and other ardent champions of strict intellectual property protection are stepping up their efforts on other tracks. For example, the United States has been building into a variety of bilateral and regional trade agreements intellectual property requirements that go beyond what has been achieved in WIPO. Here the possibility of binding dispute settlements under the WTO's TRIPS agreement and of resulting trade sanctions becomes a factor in pressuring governments toward higher levels of protection. Hence, in the absence of what they consider to be sufficient movement in WIPO on digital rights, the proponents of increasing the scope, strength, and duration of global intellectual property protections are piecing together a pattern of global governance in which the accumulation of small-n agreements and the virtual projection of national laws and policies figure prominently.

Electronic Commerce

As with network security, the global governance of electronic commerce reflects the accumulation of activities distributed across multiple institutions and levels of social organization. Most of the heavy lifting has been done during the third NWO. As mentioned previously, work conducted in the WTO since the end of the Uruguay Round has determined that the GATT, GATS, and TRIPS are fully applicable to networked trade. But beyond market access and related WTO matters, there are a number of e-commerce issues for which shared solutions have been sought.

Private sector contracts are a key source of order in the global ICT environment. In the case of e-contracts, while these are naturally very heterogeneous in form and substance, there are some common problems for which market participants have desired internationally harmonized solutions. For example, contracting over networks among globally dispersed parties via Electronic Data Interchange and other services raises questions about when a contract can be said to have been concluded, the real locations of the parties, the incorporation of terms by reference, the transfer of rights, differences between national laws, legal barriers posed by existing international agreements, and so on. To enhance predictability and reduce transactions costs, industry associations like the ICC have developed collective rules and tools in the form of guidelines, codes, model contracts, and standardized terminologies for such issues as contracting, identity management, and authentication. In the latter case, individual firms like Microsoft, with its Passport system, have been an important force. In parallel, a broad coalition of firms participating in the Liberty Alliance are pushing a competing, open standards approach intended to promote federated, multilateral trust relationships and greater user control and choice among diverse identity management tools.

At the same time, governments have recognized that self-governance approaches are not entirely sufficient for all purposes and have sought to provide a facilitative legal environment. A principal forum for these efforts has been the UN Commission on International Trade Law (UNCITRAL), in particular its Working Group on Electronic Commerce. In 1996, UNCITRAL adopted a Model Law on Electronic Commerce to be used as a national legislative basis for the elimination of legal barriers to e-commerce. After some changes to mollify industry criticisms that it was overly broad and inflexible, a United Nations Convention on the Use of Electronic Communications in International Contracts was adopted by the General Assembly in 2005. The convention seeks to enhance legal and commercial certainty by addressing such issues as the determination of a party's location in an electronic environment, the time and place of electronic communications, and the criteria used in establishing equivalence between electronic communications and paper documents and between electronic authentication and handwritten signatures. Other related initiatives have been undertaken by the ITU, the UN Centre for Trade Facilitation and Electronic Business (UN/CEFACT), the International Institute for the Unification of Private Law (UNIDROIT), and the OECD, which in 1998 adopted a Declaration on Authentication for Electronic Commerce.

Also important in the e-commerce context is the question of legal jurisdiction. Probably the most important initiative to achieve some harmonization has been taking place in the Hague Conference on Private International Law. The organization is working to set rules on the jurisdictional aspects of international lawsuits and facilitate the recognition and enforcement of foreign judgments with respect to business. In 2005 a convention on choice of court agreements was negotiated, but no country has signed it at the time of writing. It could apply not only to the sixty-seven member states, but potentially many others; over sixty additional countries have previously opted to become parties to other Hague Conference conventions. The convention would apply to business-to-business (B2B) commerce cases in which the parties have chosen a court in their contract. It would make these provisions enforceable if the parties entered into a legal dispute, and would also require that other national courts enforce any judgments rendered by the courts selected in the contract.

Taxation is of course a preeminently national-level issue, but governments also maintain an elaborate mesh of bilateral tax treaties that establish some baseline harmonization on a host of technical issues such as location of taxation, transfer pricing and sourcing, valuation and computation of profits, and sharing of revenues. The borderless nature of e-commerce transactions poses challenges to the treatment of many of these issues, so the most densely interlinked countries have been working to develop shared solutions. At the plurilateral level, the OECD has been at the forefront of efforts to think through the e-commerce issues and promote cooperation among tax authorities. In 1998, the organization set out some basic principles and since then has pursued an active work program to elaborate on the details. Among these principles are that the approaches used for conventional commerce are applicable but should be clarified; that taxation should occur in the jurisdiction where consumption takes place; that reporting requirements and collection procedures should be neutral and fair to level the playing field; and that digital products should be treated as services for tax purposes. The EU has drawn on the OECD work in setting its regional policy, which includes a uniform approach to the application of value-added taxes. This has resulted in tensions between the EU and United States over the former's insistence that externally located companies should collect tax on B2B sales into Europe on behalf of its member countries.

Consumer protection is a key concern in global e-commerce, but there is no broad multilateral organization or regime focused specifically on these issues. Instead, they have been tackled through a combination of bilateral, regional, and plurilateral cooperation, particularly among the industrialized countries, and industry standards. The bilateral context has been a source of tensions between the United States and Europe, with the former arguing that disagreements over transnational transactions should be addressed under the law of the supplier's home country, while the EU has opted for a more differentiated model under which if an external seller specifically targets consumers in member countries, the laws of those countries apply. At the plurilateral level soft-law agreements have been the tools of choice: the OECD has adopted Guidelines for Consumer Protection in the Context of Electronic Commerce, and Guidelines for Protecting Consumers from Fraudulent and Deceptive Commercial Practices Across Borders; the APEC has adopted Voluntary Consumer Protection Guidelines for the Online Environment; and the International Consumer Protection and Enforcement Network (ICPEN), which comprises the trade practices law enforcement authorities of more than two dozen mostly industrialized countries, works to share information about e-commerce activities affecting consumer interests. In parallel, industry groups like BBBOnLine, the ICC, and the GBDe have adopted their own codes of conduct and guidelines.

A related issue is Internet spam. While many of those involved in the spam industry would argue that it is a legitimate business practice and simply a form of e-commerce, spam is also routinely employed for phishing and fraud, and raises a host of other problems as well. Estimates seem to vary widely, but it appears that spam may account for up to 80 percent of global Internet traffic. The cost to the world economy in terms of lost productivity, wasted bandwidth, and so on is enormous. The burden falls especially hard on developing and transitional countries, where slow and expensive connections, limited access to anti-spam technology and expertise, and related factors create disincentives to Internet adoption and use. Despite the calls by various developing countries for multilateral action in the ITU, the industrialized countries and the global business community have maintained that technical fixes, national laws, mutual assistance and information sharing and the like, and industry self-governance provide the best solutions to the problem. Hence, Australia and South Korea have signed a memorandum of understanding on information sharing and mutual support in the enforcement of national laws; Australia, the United Kingdom, and the United States have done likewise; the European Union has adopted rules applicable to its members; and in 2006, the OECD's Task Force on Spam developed an anti-spam toolkit comprising recommended policies and measures that can be taken. Complementing these governmental efforts are the programs of technical standards bodies; organizations like the Spamhaus Project; and self-governance collaborations or industry associations like the Anti-Spam Technical Alliance, the Anti-Phishing Working Group, the ICC, and others.

Cybersecurity

The generic terms *cybersecurity* and *cybercrime* encompass a range of issues that are sometimes distinct and other times closely interrelated. Accordingly, some global governance mechanisms are designed to address one or a few of them, while others are of wider scope. The CoE cybercrime convention disaggregates the issues within its broad scope into four categories: offenses against the confidentiality, integrity, and availability of data and systems, such as spreading viruses and other kinds of malware, or illegally accessing or interfering with information infrastructures; computer-related offenses, such as using ICT to engage in traditional types of criminality like forgery and fraud; content-related offenses, such as disseminating child pornography or other prohibited types of substantive information; and offenses related to infringements of copyright and related rights (a topic addressed previously). Another approach taken in many policy discussions is to cluster the issues into two broad categories: network security/crimes and information security/crimes. For narrative purposes it is easier to treat these two sets of issues together in this section rather than distribute them here and in the preceding section; furthermore, the reasons for protecting or attacking systems generally relate to the specific information they handle.

Network security and information security were not a major focus of global governance mechanisms for most of the first NWO. In the era of analog and monopoly-controlled networks, the ITU treaties simply required that members avoid harmful interference in the radio sphere and ensure the maintenance within their territories of the channels and installations needed for uninterrupted international telecommunications. More politically charged security issues like the construction and use of networks to strengthen military capabilities and imperialist campaigns, military attacks on cables and other enemy facilities and the treatment of neutral lines during wartime, the interception of communications (and in the case of telegraphy, the breaking of codes and ciphers), and so on were thought to be beyond the ITU's technical mandate.

By the second NWO, liberalization, digitization, and networked information systems began to pose new kinds of threats that elicited international debate and action. Hence, the ITU adopted numerous standards meant to strengthen the resilience and reliability of networks, including by constraining customers from attaching equipment that the carriers said could cause technical harm to their networks. In addition, following the celebrated incident of the Morris worm that disrupted the early Internet, negotiators at the 1988 WATTC included in the new ITRs a provision stating that any special arrangements on matters not of concern to member governments generally (such as private networks) should avoid technical harm to the facilities of third countries. Ten years later, members' responsibility to protect other countries from disruptions emanating from within their jurisdictions would be brought into the ITU's organizational convention as well.

In a similar vein, during the TDF debate a number of security issues were considered, with particular attention being devoted to the vulnerability of networked information systems to disruptions from abroad. Further work in the OECD led to the adoption of the Guidelines for the Security of Information Systems in 1992. These groundbreaking guidelines advanced principles applicable to both public and private sector entities and called for worldwide security technical standards, rules for the allocation of risks and liabilities, national laws that are backed by sanctions and harmonized across member countries, and collaboration on matters of national jurisdiction, mutual legal assistance with criminal cases, and so on.

Concerns about technology-based threats to networks and the information they contain or convey have increased dramatically in the third NWO. The Internet's design is optimized for openness and ease of communication rather than security, which makes it a ready vehicle for an endless array of individuals and organizations that are dedicated to probing and exploiting its weaknesses. The worldwide reliance on insecure Microsoft products poses related problems. The consequences for individual users, organizations, critical infrastructures, national security, and the world economy have included waves of viruses, worms, trojans, phishing, zombie botnets, spoofing attacks, denial of service attacks, identity thefts, information warfare gambits, and so on, as well as the rapid blossoming of a lucrative security industry in response. As the 2005 report by UNCTAD summarizes, "The global information security market is estimated at around $40 billon, half of which is represented by the United States. The corresponding estimates of economic damage caused by security breaches in 2003 vary from $12.5 billion for viruses only to over $200 billion for all forms of digital attack."[63]

Governance responses to cybersecurity challenges are highly distributed across multiple institutions and levels of social organization. Involved in various pieces of this terrain are national-level bodies like the National Infrastructure Protection Center of the U.S. Department of Homeland Security, as well as the various intelligence and law enforcement agencies; specialized industry organizations like the International Systems Security Certification Consortium, the Information Systems Security Association, the Center for Internet Security, and the Internet Security Alliance, as well as more general organizations like the ICC and Global Business Dialogue on Electronic Commerce (GBDe); regional organizations like the EC and the European Network and Information Security Agency; technical standards bodies like the IETF and European Telecommunications Standardization Institute; ICANN, through its Committee on Security and Stability; plurilateral organizations like the Group of 8, the OECD, and the Asia-Pacific Economic Cooperation (APEC); universal multilateral organizations like the ITU; and many more. A particularly interesting and understudied aspect of this complex nexus is the growing role of the private sector in tracking communications over its infrastructure and supporting lawful intercept and digital forensics activities.

Much of the important cybersecurity activity is programmatic in nature and involves trend monitoring, information sharing, and responding to outbreaks and attacks. Organizations involved include the many national and subnational Computer Emergency Response Teams (CERTs) like the CERT Coordination Center, a federally funded program at Carnegie Mellon University that studies Internet vulnerabilities, serves as an international information clearinghouse, assists websites that have been attacked, and publishes security alerts; the Forum of Incident Response Teams (FIRST), which comprises more than 170 member organizations from governments, the private sector, and academia in the Americas, Asia, Europe and Oceania; and parallel bodies like the Asia Pacific Security Incident Response Coordination (APSIRC), and the Information Sharing and Analysis Centers (ISACs) in the United States. Moreover, there are the various information sharing and mutual assistance agreements between law enforcement agencies, including via Interpol's High Tech Crime Unit; and the secretive ECHELON program for electronic surveillance that is said to be run by the intelligence agencies of the United States and four other English-speaking nations.

International rulemaking and policy coordination are a growing part of the governance mix. Some notable intergovernmental examples include the OECD's 2002 Guidelines on the Security of Information Systems and Networks, which strengthened and expanded the 1992 edition; the EC's 2001 Communication on Creating a Safer Information Society by Improving the Security of Information Infrastructures and Combating Computer-related Crime; APEC's 2002 Statement on the Security of Information and Communications Infrastructures and its Program of Action; the UN General Assembly's 2003 resolution urging the creation of a global culture of cybersecurity; the G8's work program on shared principles and action items; the OECD's 1997 guidelines on cryptography; the Wassenaar Arrangement regarding export controls on cryptographic and other technologies; and the activities of the International Law Enforcement Telecommunications Seminar, a collaboration of law enforcement agencies from the industrialized countries that has set "requirements" for surveillance capabilities that have been incorporated into national laws, manufacturing designs, and technical standards.

Certainly the most important regime instrument adopted to date is the CoE cybercrime convention, which is closely and critically assessed by Ian Hosein in chapter 9. As we have noted previously, the agreement requires the elaboration of substantive and procedural laws at the national level, provides for mutual legal assistance between countries, and addresses a range of network security and content-related issues. The convention is open to accession by non-CoE states, and has been signed by such countries as Canada, Costa Rica, Japan, Mexico, South Africa, and the United States. As such, it is the broadest intergovernmental regime in place to deal with security matters, and is for better or for worse informing much of the work being done in other bodies. In the years ahead, it is possible that the ITU will seek to advance a universal agreement that builds on these foundations. In sum, the architecture of collective governance with respect to the network security arena is highly distributed, rapidly evolving, and deserving of far more attention by scholars than it has received to date.

Privacy Protection

In the context of the previously mentioned TDF debate, in 1980 the OECD adopted Guidelines on the Protection of Privacy and Transborder Flows of Personal Data. This nonbinding instrument set forth a set of principles for fair information practices, for example, that individuals should be notified when their data is being collected; data should be used only for specified purposes and not be disclosed without its subjects' consent; data should be kept secure from abuse and made accessible to its subjects for correction; and data collectors should be accountable to subjects for following the principles. The guidelines' weakness was due in particular to strong pressure from the United States and global business, which claimed that privacy protection was somehow intended to serve as a covert trade barrier against U.S.-based transnational firms. Following their establishment, U.S.-based firms pretended to follow them and European governments pretended to believe that they were, post hoc studies indicating otherwise notwithstanding.

The following year, the Council of Europe adopted a binding Convention for the Protection of Individuals with regard to Automatic Processing of Personal Data. The convention specifies that signatory governments shall not, for the sole purpose of protecting privacy, prohibit or subject to special authorization TDF of personal data going to the territory of another party. However, they can derogate from this obligation if their legislation includes specific regulations for certain categories of personal data, except if the other party provides an equivalent protection; or when the transfer is made to the territory of a nonsignatory through the intermediary territory of another party. In principle, these provisions could have allowed the blocking of TDF to countries lacking sufficient privacy protections, and as such many in the U.S. business community roundly denounced them. In practice, there is little evidence that the agreement was implemented in a manner that caused notable problems for industry. The difficulty of monitoring transmissions, problems of administrative capacity, and so on may have been factors in this regard.

In 1995, the EU adopted a stronger Directive on the Protection of Personal Data that, inter alia, gives data subjects the right to access and correct data about them and limits flows to countries with inadequate protections. Concerns of the United States and U.S. businesses that these rights would interfere with corporate operations led to the negotiation in 2000 of a face-saving "safe harbor" agreement with the EU, according to which companies from nonmember countries can claim to provide equivalent levels of protections via other means, such as self-regulation. Henry Farrell argues in chapter 10 that the EU has pushed weaker trading partners to adopt stronger data protection laws, and hence may be promoting a convergence of national policies in the absence of a broad international regime. APEC and the United Nations have established other intergovernmental frameworks, although these are soft-law agreements that have been influenced by the United States.

The private sector also has devised a number of self-regulatory instruments pertaining to general data collection and use, for example, in the ICC and GBDe. There are also programs designed specifically for the Internet environment, such as TRUSTe and BBBOnLine, and the W3C's Platform for Privacy Preferences Project. In general, the flowering of self-regulatory arrangements has done much more to head off governmental efforts to develop strong rules and verifiable implementation than it has to protect the privacy of individuals. While an investigation of governance based on market power is beyond the scope of this book, one could add that the internal policies of individual companies like Google and Microsoft certainly have rule-like effects on large swaths of cyberspace as well.

Conclusion: Toward the Holistic Analysis of Distributed Governance

As the preceding survey indicates, the realm of global networks and related ICTs is replete with global governance mechanisms. Unlike some international policy spaces, there is no one or even principal international regime or set of programs that covers the whole range of issues involved. Nor is there an overarching meta-regime, defined by a singular logic or social purpose, into which all the issue-specific mechanisms we have mentioned and the many that we have not are somehow nested. Instead, what has developed over the past century and a half is a highly distributed and heterogeneous architecture comprising an array of arrangements that generally were created on a piecemeal, stand-alone basis to deal with individual functional and political problems. Within some issue-areas there is interinstitutional coordination, but in the aggregate there is nothing like a tightly interwoven system.

The diversity of this governance architecture presents some analytical challenges. Academic and policy research efforts in areas like international security, trade, or environmental policy constitute fairly coherent "fields" that are well institutionalized and supported within traditional academic disciplines. In contrast, ICT global governance is a horizontally cross-cutting "field of fields" that is generally not recognized to be such, including by many who are actually in it. For a number of reasons, there has been little effort to view international institutions and cooperation in the various domains of ICT global governance as comprising a unified terrain meriting integrative or comparative inquiry.

For example, the scholars involved hail from the disciplines of political science, sociology, economics, law, business and management, communication studies, information studies, and even technical fields like engineering and computer science. They often gear their work toward their home disciplines rather than toward the multidisciplinary array of people working on similar questions with respect to other ICT governance arenas. In consequence, most of the literature does not employ a shared analytical vocabulary in the same manner as political science work on international regimes and cooperation, and it is often inadequately cumulative. Moreover, a good deal of the interesting research is focused on specific policy issues rather than the larger regimes or cross-cutting field in which they are situated, and is generated by people working in international organizations, think tanks, consulting firms, and CSOs rather than in academia. And the experts on each topic tend to embrace varying priorities, publish in different journals, go to different meetings, and eschew efforts to generalize or draw linkages across cases. In short, as a terrain of study, the global governance of global electronic networks is highly fragmented along multiple axes.

Developing a more holistic vision of network global governance could yield a number of benefits. For example, assessing the historical development of the contemporary landscape of governance arrangements in this manner could enrich our understanding of how and why networks have been constructed and employed as they have and, by extension, how they have helped to shape the trajectories of world politics, economics, and society. In some of the major, zeitgeist-shaping scholarly and popular works on global networks and globalization that have been published in recent years, it at times seems that the technology is simply there, a given with fixed properties from which the analysis departs without looking back, when in fact their precise configuration, embedding in social structures, and utilization are the result of complex historical dynamics and power relationships that mattered. A holistic approach also would lend itself to the assessment of generalizable causal dynamics with respect to international power, counter-power strategies, the roles of the private sector and civil society, the influence of ideas and discourse, leadership, bargaining, and so on. The sorts of explanatory theories favored by international relations scholars (such as those in the rationalistic and constructivist traditions), communication studies scholars (such as those in the critical and administrative traditions), and so on, across the academic topography, could be systematically worked through via progressive research agendas if ICT global governance were recognized to be a legitimate and coherent field of inquiry.

A holistic approach would be useful as well in comparing governance mechanisms and considering generalizable questions of interest to scholars and practitioners of institutional design.[64] That is, in addition to assessing the governance architecture as a whole, one could consider the parts in relation to both the whole and each other. One could then ask such questions as: Which issues have or have not given rise to what kinds of collective rules and programs, and why? When have intergovernmental or private sector self-regulatory arrangements been used, and to what effect? Where intergovernmentalism is needed, what have been the costs and benefits of bilateral, plurilateral, regional, or multilateral configurations? What have been the relative merits of treaties and hard-law vs. soft-law mechanisms like guidelines, declarations, opinions, or light coordination and information sharing? When is it better to anchor cooperative rules in a formal organizational setting, to distribute functions among networked institutions, or to have freestanding agreements? What would constitute best practices in the agenda setting, negotiation, and compliance enforcement stages of cooperation? How has compliance been monitored and noncompliance sanctioned, or dispute resolution pursued? Which mechanisms have performed more or less well than the others in promoting technological innovation and new markets, or in terms of institutional efficiency? Which adapt most effectively to changes in their operational environments? Which have done better or worse with respect to transparency, accountability, inclusion, and social justice? Which most effectively empower nondominant stakeholders in terms of knowledge acquisition, negotiating capabilities, and national capacity building and implementation? Are there lessons that intergovernmental, private sector, and multistakeholder approaches could learn from one another? What are the costs and benefits of governance mechanisms that cover a range of interrelated issues vs. those that are narrowly targeted? How can procedural transparency and accountability and a good substantive balance between competing objectives be achieved? In short, examining governance options in relation to what has or has not worked in related issue-areas might help answer these and related questions more effectively.

Similarly, a holistic analysis might help to identify weaknesses and gaps in the coverage of important issues. A few examples of problems that have not given rise to strong collective responses or have fallen between the cracks of existing mechanisms may include access to knowledge, privacy protection, certain aspects of network security, Internet spam, and problems associated with market concentration like international Internet interconnection pricing, competition policy, and restrictive business practices. Of course, whether some form of collective response to any of these challenges is really needed is a matter of debate, but if one takes as given that the nexus of existing institutions and the issues agreed to be under their mandates exhausts the range of possible cooperation, it may be more difficult to address some important outstanding problems.

Finally, a holistic approach would direct our attention to the possibility of procedural and substantive tensions between governance mechanisms, particularly in the case of multidimensional issues. As the existing arrangements have been created in a piecemeal manner in response to individual problems, it is possible that some may at times clash a bit with others. For example, there have been tensions between ITU and WTO instruments, between instruments concerning the free flow of information and national sovereignty, between different technical standardization processes (e.g., between the ITU and the Internet Architecture Board), between organizations and rules involved in managing names and numbers, and between arrangements pertaining to personal privacy and security or cybercrime, intellectual property, and freedom of speech, and so on. Examining how well the various institutions mesh into a whole might facilitate the identification and resolution of problems and unleash latent value through interinstitutional synergies.

This sort of enterprise is most definitely not just of academic interest. Policy practitioners in international organization secretariats, governments, business, and civil society already have confronted at least some of these issues, although perhaps not as starkly framed as they are here. The WSIS process, which is addressed by David Souter (chapter 12) and Wolfgang Kleinwächter (chapter 15), provided examples of this at two levels. First, it was the first extended worldwide effort to assess the policy aspects of global information society generally. Never before had persons from all the major stakeholder groupings—international organization secretariats, governments, industry, civil society, and the technical and administrative community—had an opportunity to debate at length the broad range of issues and governance mechanisms involved in global ICT that cut across the turfs of vertically segmented international organizations and national government ministries. It would be politically difficult for any one international organization to hold discussions of issues pertaining to the conduct and mandate of multiple other international organizations, much less the activities of industry self-governance mechanisms and multistakeholder collaborations. It took a UN summit with a nominally unrestricted mandate to have this sort of interaction.

For over three years, participants gathered in lengthy preparatory meetings and regional consultations to debate and agree on language pertaining to a plethora of topics. This included such issues as access to ICT and knowledge, Internet interconnection charging, Internet identifiers, telecommunications regulation, spectrum management, Internet names and numbers, privacy protection, intellectual property, security, culture and content spam, voiceover IP, intellectual property, network security and information security, development policy, trust issues, human rights, mass media, the public domain, cultural and linguistic diversity, free and open-source software, community-level ICT, e-health, e-education, e-government, e-commerce, e-ethics, e-everything, as well as the special challenges relating to gender, age, minorities, indigenous peoples, and marginalized social groups.

In the process, collective, transformational substantive learning took place. This was demonstrated by the widespread recognition, reflected both in the debates and outcome documents, of the integral interrelationships between nominally discrete issues, policies, and institutions. The intersubjective construction of "the global information society" as an overarching global policy space within which individual issues are remapped as interrelated parts of a whole meriting coordinated action shared some features with other UN summit outcomes, such as the construction and institutionalization on the global agenda of climate change or sustainable development.[65]

A second example concerns the heated debate on Internet governance during the WSIS. The nature, conduct, and potential reform of Internet governance was explored at length, and the discussions included the sort of institutional design questions posed previously within a fairly holistic framing. This was especially true within the Working Group on Internet Governance (WGIG), a forty-person multistakeholder body that was appointed by UN Secretary General Kofi Annan to study Internet governance, develop a working definition of it, identify the public policy issues that are relevant, and develop a common understanding of the respective roles and responsibilities of governments, existing international organizations, and other forums as well as the private sector and civil society from both developing and developed countries. The WGIG, on which some of the authors in this book (Ang, Drake, MacLean, and Kleinwächter) served as members, adopted a holistic approach to assessing Internet governance mechanisms in relation to each other and collectively, and put forward recommendations on that basis.[66] The WGIG advanced a broad and holistic definition of Internet governance akin to the definition of ICT global governance offered earlier in this chapter, namely that "Internet governance is the development and application by Governments, the private sector and civil society, in their respective roles, of shared principles, norms, rules, decision-making procedures, and programmes that shape the evolution and use of the Internet."[67] The definition was ultimately adopted by the international community, included in the final summit's Tunis Agenda for the Information Society, and helped to defuse the battle over what Internet governance was and who should do it.

A holistic approach was also evidenced by the WGIG's principal operational recommendation, to create a new UN Internet Governance Forum (IGF). The WGIG proposed a mandate for the IGF that maps with the points made above, was amplified by the preparatory committee meeting held in Tunis on the eve of the November 2005 summit, and was adopted by the governments and other parties assembled. The mandate was as follows:

We ask the UN Secretary-General, in an open and inclusive process, to convene, by the second quarter of 2006, a meeting of the new forum for multi-stakeholder policy dialogue—called the *Internet Governance Forum* (IGF). The mandate of the Forum is to:

a) Discuss public policy issues related to key elements of Internet Governance in order to foster the sustainability, robustness, security, stability and development of the Internet;
b) Facilitate discourse between bodies dealing with different cross-cutting international public policies regarding the Internet and discuss issues that do not fall within the scope of any existing body;
c) Interface with appropriate inter-governmental organisations and other institutions on matters under their purview;
d) Facilitate the exchange of information and best practices, and in this regard make full use of the expertise of the academic, scientific and technical communities;
e) Advise all stakeholders in proposing ways and means to accelerate the availability and affordability of the Internet in the developing world;
f) Strengthen and enhance the engagement of stakeholders in existing and/or future Internet Governance mechanisms, particularly those from developing countries;
g) Identify emerging issues, bring them to the attention of the relevant bodies and the general public, and, where appropriate, make recommendations;
h) Contribute to capacity-building for Internet Governance in developing countries, drawing fully on local sources of knowledge and expertise;
i) Promote and assess, on an ongoing basis, the embodiment of WSIS principles in Internet Governance processes;
j) Discuss, inter alia, issues relating to critical Internet resources;
k) Help to find solutions to the issues arising from the use and misuse of the Internet, of particular concern to everyday users;
l) Publish its proceedings.[68]

It is too early to tell at the time of writing whether the IGF will be able to fulfill this mandate, particularly in the face of reluctance in practice of some powerful stakeholders to create a process that is as robust as the mandate implies. Nevertheless, it is perhaps a hopeful sign that the global community was able to sign off on such a framework, recognizing that viewing Internet governance mechanisms holistically with an eye to promoting improvements is explicitly contemplated. In the meanwhile, it would be useful if academics and other analysts were to carry such work forward. Hopefully, this volume will help lay the foundation for a progressive research program along these lines.

Notes

1. Castells 1996, 469.

2. Latham and Sassen 2005, 1.

3. Many ICT issues are not now and do not need to be subject to global governance as it is conceptualized in this volume, but those that are generally pertain to electronic networking. Given this, and the common practice of most people in the field of using the term *ICT* when referring to network-related technologies, the terms *ICT global governance* and *network global governance* are used interchangeably in this volume.

4. Benkler 2006, 3.

5. See in particular, World Summit on the Information Society 2003a, 2003b, and 2005, as well as civil society's alternative declaration to the Geneva summit, *Shaping Information Societies for Human Needs* (2003).

6. Adler and Bernstein 2005, 295.

7. Reinicke 1998, 4, emphasis in the original. However, the author later adds that governance is "a social function . . . [that] does not have to be equated with government" (87) and spends most of the book demonstrating this point.

8. Kahler and Lake 2003, 7–8.

9. Hewson and Sinclair 1999, 10.

10. Finkelstein 1995, 369.

11. Murphy 1994, 1.

12. Florini 2003, 5, emphasis in the original.

13. Thakur and Weiss. See the United Nations Intellectual History Project web page for their forthcoming book, "The UN and Global Governance: An Idea and its Prospects." Available at www.unhistory.org/publications/globalgov.html.

14. Ruggie 2004, 9.

15. Young 1994, 15.

16. The Centre for the Study of Global Governance at the London School of Economics and Political Science, "Research Themes," available at www.lse.ac.uk/Depts/global/2research.htm.

17. Commission on Global Governance 1995, 2.

18. Held, McGrew, Goldblatt, and Perraton 1999, 50.

19. Rosenau 1997, 145.

20. Rosenau 2003, 393.

21. Keohane and Nye 2000, 12.

22. The Global Governance Project (a joint research program of eight European research institutions), "About the Project," available at www.glogov.org/front_content.php?idcat=3.

23. Young 1999, 11.

24. Risse 2005, 165, 167.

25. For example, David Johnson and David Post have argued that, "de facto rules may emerge as a result of the complex interplay of individual decisions by domain name and IP address registries (regarding what conditions to impose on possession of an online address), by sysops (regarding what local rules to adopt, what filters to install, what users to allow to sign on, and with which other systems to connect) and by users (regarding which personal filters to install and which systems to patronize)." See Johnson and Post 1997, 67–68. Similarly, Johnson, Susan Crawford, and John Paltry suggest that "The aggregation of numerous individual decisions about who to trust and who to avoid will create a diverse set of rules that most accurately and fairly serves the interests of those who use the online world. In other words, we can use 'peer production of governance' to address the collective action problems that arise in the online context." See Johnson, Crawford, and Palfry 2004, 9.

26. This distinction has been introduced by Robert Latham, as follows: "Governance that is global refers to the steering at the global level. Governance in the global refers to all the governance that occurs throughout the global order." See Latham 1999, 28. Some but certainly not all of the latter may effectively rise to the level of the former.

27. In so doing, international organizations' secretariats often enjoy enough autonomy and leeway that their efforts cannot be viewed as simply an epiphenomena of members' desires; for a discussion, see Barnett and Finnemore 2004.

28. See Noam 1992, 23.

29. For an analysis of the politics that sustained the decentralized exceptions to the European rule, see Davies 1994.

30. See Headrick 1991.

31. The state technocrats involved in the international telecommunications regime brought to bear the same sort of systems building, public service-oriented perspectives they employed at the national level; for a discussion, see Murphy 1994.

32. The name "International Radiotelegraph Union was not used in the radiotelegraph conventions that preceded the establishment of the ITU by the 1932 conference, perhaps because the name too closely resembled that of its administrative home, the International Telegraph Union. Instead, it referred to the countries that were signatories or adherents to these conventions.

33. Flamm 1988, 89.

34. The counterintuitive acronym, ISO, derives from the Greek *isos*, meaning equal.

35. For an overview of the computer networking matrix, see Quarterman 1990.

36. For a classic statement of the new U.S. policy, see National Telecommunications and Information Administration, Government of the United States 1983.

37. For a discussion, see Drake and Nicolaïdis 1992.

38. See Noam 1987.

39. Gassman and Pipe 1976, 27.

40. On the mobile revolution, see Castells et al. 2006.

41. See Wilson 2004.

42. The scholarly literature on ICANN is abundant and growing. For an early but still instructive analysis, see Mueller 2002.

43. Codding 1952, 42.

44. For varying assessments of the nature and causes of this transformation, see Cowhey 1990, Drake 1994, and Zacher with Sutton 1996.

45. The legal preservation but decaying salience of the regime's instruments speaks to the intersubjective nature of regimes and the centrality of expectations. For a discussion of these trends, see Drake 2000.

46. Nevertheless, the move from multilateral regulation to looser coordination and cross-fertilization among national policies also has increased the role of nonuniversal institutions. National entities in key states, most notably the U.S. FCC, can be very influential in this context. Regional bodies like the EU, the European Conference of Postal and Telecommunications Administrations, the Asia-Pacific Economic Cooperation, the Asia-Pacific Telecommunity, and the Inter-American Telecommunication Commission coordinate national policies to varying degrees, with the EU framework being the strongest and most elaborate. Plurilateral bodies like the OECD also promote international policy consensus and national policy convergence.

47. The term *Internet of things* refers to the use of the Internet to connect inanimate objects via electronic tags and sensors. For a discussion, see International Telecommunication Union 2005.

48. As a leading analyst and ITU participant has observed, "For many years [ITU standards committees] were well occupied with work on the details necessary for compatible systems that would carry the traffic over the international lines connecting the national-international transfer points. The domestic side of these transfer points and the whole domestic network, including the subscriber terminals, were exempted from the purview of [ITU recommendations]. Thus, the interference with state prerogatives was limited to the international service from the gateway points." Wallenstein 1990, 73.

49. Here as elsewhere I am simplifying greatly in describing the workings of the relevant instruments. Unfortunately, the scholarly literature on the radio regime is exceptionally thin; the best treatment, which is now forty years out of date, is still Leive 1970.

50. See MacLean, Chapter 2, 97.

51. On the difficulty of assessing INTELSAT's tariffs, see Snow 1987. The uncertainty about this and other policies is due to intentional opacity; as Heather Hudson notes, "There is no mechanism for external analysis of costs, tariffs, internal cross-subsidies, and the like. Critics have argued that Intelsat's costs are higher than necessary... Intelsat denies these allegations, but is not required to make public any documentation of its costs and rate-setting procedures." See Hudson 1990, 148.

52. Katkin 2005, 9.

53. OECD 1883, 20–21, 130.

54. Wunsch-Vicent 2006, 173.

55. Ploman 1979, 159.

56. Carlsson 2005, 11–12.

57. Among the issues considered in the TDF debate were: the protection of citizens' privacy when data is transferred to countries with weak protections; the weakening of national cultural and linguistic integrity; deepening divisions between the "information rich" and the "information poor;" the stifled development of indigenous online cultural production; difficulties in ensuring access to data held abroad; national security conflicts, as with the American effort to control East-West TDF; the extraterritorial application to data of national laws; the vulnerability to disruption of foreign systems on which countries depend; liability for errors in transmission and processing; the preservation or overextension of intellectual property rights; computer-based crime; the location and concentration of corporate decision making; the location, level, and quality of TDF-related production and employment; outsourcing; the policy treatment of intercorporate commercial transactions, or what later came to be understood as network-based international trade in services; the preservation of sectoral regulatory policies, such as in banking and finance, professional services, and computer services; the commodification of information, including governmental information; the valuation and taxation of information and its transfer; and ultimately, the erosion, as with other kinds of cross-border flows, of what some participants called "information sovereignty." For more on the debate and its outcomes, see Drake 1993.

58. For a good overview of some of the self-regulatory initiatives with respect to Internet content, see Price and Verhulst 2005.

59. The role of human rights in the global information society generally and in the WSIS specifically is addressed in Drake and Jørgensen 2006.

60. The complexities of the issues need not be addressed here; for discussions, see Drake and Nicolaïdis 2000 and Wunsch-Vincent 2006.

61. For an overview of the underlying issues, see Drahos and Braithwaite 2002.

62. Musungu and Dutfield 2003, 3.

63. UNCTAD 2005, 191.

64. For an initial effort along these lines, see Drake 2001.

65. On these and other forms of learning in the WSIS, see Drake 2005.

66. For an overview of the process, see MacLean 2005. For broader perspectives on the nature and politics of Internet governance by participants in this project, see the chapters by Drake, Kleinwächter, and MacLean; and by Mueller, Mathieson, and McKnight in MacLean 2004.

67. See Working Group on Internet Governance 2005a, 4. For a fuller elaboration of the thinking behind this definition, see Working Group on Internet Governance 2005b.

68. World Summit on the Information Society 2005, 11.

References

Adler, Emanuel and Steven Bernstein. 2005. "Knowledge in Power: The Epistemic Construction of Global Governance." In *Power in Global Governance*, ed. Michael Barnett and Raymond Duvall, 294–318. Cambridge: Cambridge University Press.

Barnett, Michael, and Martha Finnemore. 2004. *Rules for the World: International Organizations in Global Politics*. Ithaca, NY: Cornell University Press.

Benkler, Yochai. 2006. *The Wealth of Networks: How Social Production Transforms Markets and Freedom*. New Haven, CT: Yale University Press.

Carlsson, Ulla. 2003. "The Rise and Fall of NWICO—and Then? From a Vision of International Regulation to the Reality of Multilevel Governance." Paper presented at the EURICOM Colloquium, Information Society: Visions and Governance, Venice, May 5–7. Available at www.nordicom.gu.se/common/publ_pdf/32_031-068.pdf.

Castells, Manuel. 1996. *The Information Age: Economy, Society, and Culture, Vol. 1: The Rise of the Network Society*. Oxford: Blackwell Publishers.

Castells, Manuel, and Mireia Fernandez-Ardevol, Jack Linchuan Qiu, and Araba Sey. 2006. *Mobile Communication and Society: A Global Perspective*. Cambridge, MA: The MIT Press.

Codding, George A. Jr. 1952. *The International Telecommunication Union: An Experiment in International Cooperation*. Leiden: E. J. Brill.

Commission on Global Governance. 1995. *Our Global Neighborhood*. Oxford: Oxford University Press.

Cowhey, Peter F. 1990. "The International Telecommunications Regime: The Political Roots of Regimes for High Technology." *International Organization* 44 (Spring): 169–199.

Cutler, A. Claire, Virginia Haufler, and Tony Porter. 1999. "Private Authority and International Affairs." In *Private Authority and International Affairs*, ed. Cutler, Haufler, and Porter, 3–28. Albany: State University of New York Press.

Davies, Andrew. 1994. *Telecommunications and Politics: The Decentralised Alternative*. London: Pinter Publishers.

Drahos, Peter, and John Braithwaite. 2002. *Information Feudalism: Who Owns the Knowledge Economy?* New York: The New Press.

Drake, William J. 1993. "Territoriality and Intangibility: Transborder Data Flows and National Sovereignty." In *Beyond National Sovereignty: International Communications in the 1990s*, ed. Kaarle Nordenstreng and Herbert I. Schiller, 259–313. Norwood, NJ: Ablex Publishing Corp.

Drake, William J. 1994. "Asymmetric Deregulation and the Transformation of the International Telecommunications Regime." In *Asymmetric Deregulation: The Dynamics of Telecommunications Policies in Europe and the United States*, ed. Eli M. Noam and Gerard Pogorel, 137–203. Norwood, NJ: Ablex Publishing Corp.

Drake, William J. 2000. "The Rise and Decline of the International Telecommunications Regime." In *Regulating the Global Information Society*, ed. Christopher T. Marsden, 124–177. London: Routledge.

Drake, William J. 2001. "Communications." In *Managing Global Issues: Lessons Learned*, ed. P. J. Simmons and Chantal de Jonge Oudraat, 25–74. Washington, DC: Carnegie Endowment for International Peace. Available at www.ceip.org/files/pdf/MGIch01.pdf.

Drake, William J. 2005. "Collective Learning in the World Summit on the Information Society." In *The World Summit on the Information Society: Moving from the Past into the Future*, ed. Daniel Stauffacher and Wolfgang Kleinwächter, 135–146. New York: United Nations Information and Communication Technologies Taskforce. Available at www.unicttaskforce.org/perl/documents.pl?id=1544.

Drake, William J., and Kalypso Nicolaïdis. 1992. "Ideas, Interests and Institutionalization: 'Trade in Services' and the Uruguay Round." In *Knowledge, Power and International Policy Coordination*, ed. Peter Haas, a special issue of *International Organization* 45 (Winter): 37–100.

Drake, William J., and Kalypso Nicolaïdis. 2000. "Global Electronic Commerce and the General Agreement on Trade in Services: The 'Millennium Round' and Beyond." In *GATS 2000: New Directions in Services Trade Liberalization*, ed. Pierre Sauve and Robert M. Stern, 399–437. Washington, DC: The Brookings Institution.

Drake, William J., and Rikke Frank Jørgensen. 2006. "Introduction." In *Human Rights in the Global Information Society*, ed. Rikke Frank Jørgensen, 1–49. Cambridge, MA: The MIT Press. Available at www.mitpress.mit.edu/books/0262101157/rev2intro.pdf.

Finkelstein, Lawrence S. 1995. "What is Global Governance?" *Global Governance: A Review of Multilateralism and International Organizations* 1, no. 3 (Sept.–Dec.): 367–372.

Flamm, Kenneth. 1988. *Creating the Computer: Government, Industry, and High Technology*. Washington, DC: The Brookings Institution.

Florini, Ann. 2003. *The Coming Democracy: New Rules for Running the World*. Washington, DC: Island Press.

Gassman, Hans-Peter, and G. Russell Pipe. 1976. "Synthesis Report." In Organization for Economic Cooperation and Development, *Policy Issues in Data Protection and Privacy: Concepts and Perspectives—Proceedings of the OECD Seminar 24th to 26th June 1974*, 12–41. Paris: OECD.

Headrick, Daniel R. 1991. *The Invisible Weapon: Telecommunications and International Politics, 1851–1945*. New York: Oxford University Press.

Held, David, and Anthony McGrew, David Goldblatt, and Jonathan Perraton. 1999. *Global Transformations: Politics, Economics, and Culture*. Cambridge: Polity Press.

Hewson, Martin, and Timothy J. Sinclair. 1999. "The Emergence of Global Governance Theory." In *Approaches to Global Governance Theory*, ed. Hewson and Sinclair, 3–22. Albany: State University of New York Press.

Hudson, Heather E. 1990. *Communication Satellites: Their Development and Impact*. New York: The Free Press.

International Telecommunication Union. 2005. *ITU Internet Reports 2005: The Internet of Things*. Geneva: ITU.

Johnson, David R., and David G. Post. 1997. "And How Shall the Net Be Governed? A Meditation on the Relative Virtues of Decentralized, Emergent Law." In *Coordinating the Internet*, ed. Brian Kahin and James Keller, 62–91. Cambridge, MA: The MIT Press.

Johnson, David R., Susan. P. Crawford, and John G. Paltry Jr. 2004. "The Accountable Net: Peer Production of Internet Governance." *Virginia Journal of Law and Technology* 9, no. 9 (Summer): 1–33. Available at www.vjolt.net/vol9/issue3/v9i3_a09-Palfrey.pdf.

Kahler, Miles, and David A. Lake. 2003. "Globalization and Governance." In *Governance in a Global Economy: Political Authority in Transition*, ed. Kahler and Lake, 1–30. Princeton: Princeton University Press.

Katkin, Kenneth D. 2005. "Communication Breakdown? The Future of Global Connectivity after the Privatization of INTELSAT." *bepress Legal Series, Working Paper 508* (March 12). Available at http://law.bepress.com/expresso/eps/508.

Keohane, Robert O., and Joseph S. Nye Jr. 2000. "Introduction." In *Governance in a Globalizing World*, ed. Joseph S. Nye Jr. and John D. Donahue, 1–41. Washington, DC: The Brookings Institution.

Latham, Robert. 1999. "Politics in a Floating World: Toward a Critique of Global Governance." In *Approaches to Global Governance Theory*, ed. Martin Hewson and Timothy J. Sinclair, 23–53. Albany, NY: State University of New York Press.

Latham, Robert, and Saskia Sassen. 2005. "Digital Formations: Constructing an Object of Inquiry." In *Digital Formations: IT and New Architectures in the Global Realm*, ed. Latham and Sassen, 1–33. Princeton, NJ: Princeton University Press.

Leive, David M. 1970. *International Telecommunications and International Law: The Regulation of the Radio Spectrum*. Dobbs Ferry, NY: Oceana Publications.

Lessig, Lawrence. 1999. *Code and Other Laws of Cyberspace*. New York: Basic Books.

MacLean, Don, ed. 2004. *Internet Governance: A Grand Collaboration*. New York: United Nations Information and Communication Technologies Taskforce. Available at www.unicttf.org/perl/documents.pl?id=1392.

MacLean, Don. 2005. "A Brief History of WGIG." In *Reforming Internet Governance: Perspectives from the UN Working Group on Internet Governance*, ed. William J. Drake, 9–23. New York: United Nations Information and Communication Technologies Taskforce.

Mueller, Milton L. 2002. *Ruling the Root: Internet Governance and the Taming of Cyberspace*. Cambridge, MA: The MIT Press.

Murphy, Craig N. 1994. *International Organization and Industrial Change: Global Governance Since 1850*. New York: Oxford University Press.

Musungu, Sisule F., and Graham Dutfield. 2003. "Multilateral Agreements and a TRIPS-plus World: The World Intellectual Property Organisation (WIPO)." Geneva and Ottawa: Quaker United Nations Office and Quaker International Affairs Programme.

National Telecommunications and Information Administration, Government of the United States. 1983. *Long-Range Goals in International Telecommunications and Information: An Outline for United States Policy*. Washington, DC: U.S. Government Printing Office.

Noam, Eli M. 1987. "The Public Telecommunications Network: A Concept in Transition." *Journal of Communications* 37 (Winter): 30–48.

Noam, Eli M. 1992. *Telecommunications in Europe*. New York: Oxford University Press.

Organization for Economic Cooperation and Development. 1983. *Telecommunications: Pressures and Policies for Change*. Paris: OECD.

Ploman, Edward W. 1979. "Satellite Broadcasting, National Sovereignty, and the Free Flow of Information." In *National Sovereignty and International Communication*, ed. Kaarle Nordenstreng and Herbert I. Schiller, 154–165. Norwood, NJ: Ablex Publishing Corp.

Price, Monroe E., and Stefaan G. Verhulst. 2005. *Self-Regulation and the Internet*. Leiden: Kluwer Law International.

Quarterman, John S. 1990. *The Matrix: Computer Networking and Conference Systems Worldwide*. Bedford, MA: Digital Press.

Reinicke, Wolfgang H. 1998. *Global Public Policy: Governing without Government?* Washington, DC: The Brookings Institution.

Risse, Thomas. 2005. "Global Governance and Communicative Action." In *Global Governance and Public Accountability*, ed. David Held and Mathias Koenig-Archibugi, 164–189. Oxford: Blackwell Publishers.

Rosell, Steven A. 1992. *Governing in an Information Society*. Montreal: Institute for Research on Public Policy.

Rosenau, James N. 1997. *Along the Domestic-Foreign Frontier: Exploring Governance in a Turbulent World*. Cambridge: Cambridge University Press.

Rosenau, James N. 2003. *Distant Proximities: Dynamics Beyond Globalization*. Princeton: Princeton University Press.

Ruggie, John Gerard. 2004. "Reconstituting the Global Public Domain: Issues, Actors, and Practices." Working Paper No. 6 of the Corporate Social Responsibility Initiative, John F. Kennedy School of Government, Harvard University, December. Available at www.hks.harvard.edu/m-rcbg/CSRI/publications/workingpaper_6_ruggie.pdf.

Shaping Information Societies for Human Needs: Civil Society Declaration to the World Summit on the Information Society, December 8, 2003. Available at www.itu.int/wsis/docs/geneva/civil-society-declaration.pdf.

Snow, Marcellus S. 1987. *The International Telecommunications Satellite Organization. INTELSAT.* Baden-Baden: Nomos Verlagsgesellschaft.

United Nations Conference on Trade and Development. 2005. *Information Economy Report 2005*. New York and Geneva: United Nations.

Wallenstein, Gerd D. 1990. *Setting Global Telecommunication Standards: The Stakes, the Players, and the Process*. Norwood, MA: Artech House.

Wilson, Ernest J. III. 2004. *The Information Revolution and Developing Countries*. Cambridge, MA: The MIT Press.

Working Group on Internet Governance. 2005a. *Report of the Working Group on Internet Governance*. Geneva: WGIG. Available at www.wgig.org/docs/WGIGREPORT.doc.

Working Group on Internet Governance. 2005b. *Background Report of the Working Group on Internet Governance*. Geneva: WGIG. Available at www.wgig.org/docs/BackgroundReport.doc.

World Summit on the Information Society. 2003a. *Declaration of Principles—Building the Information Society: A Global Challenge in the New Millennium*. WSIS-03/GENEVA/DOC/4-E. December 12, 2003. Available at www.itu.int/wsis/docs/geneva/official/dop.html.

World Summit on the Information Society, 2003b. *Plan of Action*. WSIS-03/GENEVA/DOC/5-E. December 12, 2003. Available at www.itu.int/wsis/docs/geneva/official/poa.html.

World Summit on the Information Society. 2005. *Tunis Agenda for the Information Society*. WSIS-05/TUNIS/DOC/6(Rev.1)-E, November 15, 2005. Available at www.itu.int/wsis/docs2/tunis/off/6rev1.doc.

Wunsch-Vincent, Sacha. 2006. *The WTO, the Internet and Trade in Digital Products: EC-US Perspectives*. Oxford: Hart Publishing.

Young, Oran R. 1994. *International Governance: Protecting the Environment in a Stateless Society*. Ithaca, NY: Cornell University Press.

Young, Oran R. 1999. *Governance in World Affairs*. Ithaca, NY: Cornell University Press.

Zacher, Mark W. with Brent A. Sutton. 1996. *Governing Global Networks: International Regimes for Transportation and Communications*. Cambridge: Cambridge University Press.

I The Global Governance of Infrastructure

2 Sovereign Right and the Dynamics of Power in the ITU: Lessons in the Quest for Inclusive Global Governance

Don MacLean

Which elements of electronic communication networks require global governance, and which can be left to national regulation—or left unregulated, subject only to the play of market forces? Should the principles and mechanisms of international governance mirror those found at the national level? Or are the two domains so distinct that entirely different principles and mechanisms are required? What role should national governments, the private sector, civil society organizations, and other actors play in the process of governing global electronic networks? How should these roles be organized and coordinated?

Questions of this kind are at the heart of current discussion and debate about the governance of global electronic networks—but they are not new. They were first asked and first answered a century and a half ago, when European states began to recognize that they all would benefit by agreeing on a common approach to governing the exchange of telegraph messages. Among other things, this required states to agree on substantive issues, such as which technology should be the standard for international telegraphy, which languages could be used to compose messages, what tariffs would be charged for different telegraph services, how revenues would be shared between operators in different countries, and even when public telegraph offices would be open. It also required them to agree on organizational and procedural issues, such as the legal form their commitments would take, how they would be enforced, how they could be changed, and which stakeholders could take part in decision-making processes.

Agreements on matters of this kind that were worked out between various European countries during the 1850s led to the establishment of the International Telegraph Union (ITU) on May 17, 1865. Although it was founded by twenty European states, the ITU had worldwide reach since many of its members had colonial possessions. It also had an implicit mission to support what we would now call global development because of the important role telegraph links were beginning to play in facilitating economic activity, promoting trade, and maintaining peace and security. Its global vocation led the ITU to include representatives of colonial administrations and private telegraph companies at an early stage in its work. It was thus not only the first international organization, but also the first to include nonstate actors in its decision-making processes.

The agreements made on international telegraphy in the mid-nineteenth century are now only of historical interest, and without direct relevance to current debates about governance of global electronic networks. However, the general governance questions faced by participants in these long-gone negotiations have endured. Throughout the history of electronic communication networks, these questions have repeatedly arisen at important conjunctures of technological, economic, social, and political change. They are very much with us today and are central to the quest that is underway to develop new forms of governance that include representatives of governments, the private sector, and civil society organizations on a more equal footing, as well as new policy frameworks for governing global electronic networks.

As long as global electronic networks consisted mainly of bilateral arrangements between national monopolies that were owned or controlled by governments, most of these questions were settled in a single forum—the International Telecommunication Union, the direct descendent of the original ITU. However, in the last two decades the nature of global electronic networks has changed utterly as a result of privatization, the introduction of competition in the telecommunications and broadcasting sectors, the negotiation of regional and international agreements liberalizing trade in services, the rise of the Internet, and technological convergence. These changes have brought new issues, players, and forums into the global governance game, have challenged and bypassed many of the ITU's main governance functions, and have moved the organization as a whole from the center of this universe toward its periphery.

In much of the current discussion and debate about the governance of global electronic networks, the ITU is widely viewed—at least in the United States and other developed countries—as a symbol of what must be changed, not as a model for the future. An examination of the ITU's efforts over the past decade to reform itself lends credence to this view. If it continues on its present course, it now seems clear that progressive marginalization is the most the ITU can hope for, and that eventual disappearance "not with a bang but a whimper" its most likely fate. Should this happen, the main losers are likely to be developing countries, particularly the Least Developed Countries (LDCs), many of which still view the ITU as the most appropriate forum for governing the technological and operational aspects of global electronic networks.

This chapter will argue that these results are not inevitable for the ITU, if it can learn the lessons of its recent experience and apply them to find fresh answers to the questions that hang so heavily over the organization as a whole, and with respect to each of its main governance functions.

One of the key challenges facing the ITU is to radically change its governance structures so that representatives of the private sector, civil society organizations, and other nonstate actors can be fully included in decision-making processes in areas where it no longer makes sense to reserve this right for representatives of sovereign states. Another key challenge is to reform the substance of its governance functions so that they are more closely attuned to the needs of a sector no longer dominated by states. These two challenges are closely interrelated. Policy agendas and regulatory regimes do not exist in a vacuum. They reflect the power, interests, and ideologies of the parties that are engaged in governance processes. Those who are not allowed to be present when decisions are made, or who lack the capacity to participate effectively in governance processes, cannot hope to affect substantive outcomes. In the case of the ITU, developing countries, the private sector, and civil society organizations are to varying degrees, depending on the governance issue, absent or not effectively represented in governance processes.

This chapter proposes a new organizational model that would position the ITU to respond to these two challenges. In this model, the ITU's current structure would be broken up. It would be replaced by a new structure, designed according to the principle that governance form should follow governance function. Unlike the current structure, this model would allocate governance rights and responsibilities to representatives of states, the private sector, and civil society on a differential basis in each of the ITU's four main areas of governance, so as to reflect the roles they actually play in the world outside the ITU with respect to policy coordination, regulation of the radio spectrum, technical standardization, and development assistance.

The chapter argues that an organizational change of this kind is the only way to release the governance value that lies buried in the ITU, and that it is a precondition for substantive reform of the ITU's different governance regimes. Unless they are restructured to reflect the configurations of power, interest, and ideology currently at play in what a recent study called "the international ICT decision-making universe," the ITU's experience has repeatedly shown it will be impossible to significantly increase the relevance of these regimes to the real issues of governing global electronic networks.[1]

As well as providing lessons that can lead to its own revitalization, the chapter argues that the ITU's successes and failures, both in recent years and throughout its long history, can provide useful lessons for all parties engaged in the search for inclusive arrangements for governing global electronic networks. These lessons can be applied to specific policy domains that are closely related to the ITU's core governance functions—such as trade in services and Internet governance—as well as to the quest for a general framework for governing global electronic networks that is now underway as a result of the World Summit on the Information Society (WSIS).

In order to draw out these lessons, the first section of this chapter identifies the general patterns that have characterized the development of arrangements to govern global electronic networks since the founding of the ITU in 1865. The second section assesses the fitness of the ITU to continue to play a central role in the governance of these networks in terms of its goals, principles, membership, activities, and institutional structures. The third and fourth sections examine the results of and the lessons that can be learned from the ITU's fifteen-year effort to reform its institutional structures and governance regimes. On this basis, the fifth section proposes a model for radically restructuring the ITU so that it better fits the governance needs of the international community by more effectively including developing countries, the private sector, and civil society organizations in its work. A concluding section looks at the lessons that can be learned from the ITU's experience in the broader quest for inclusive arrangements to govern global electronic networks.

In presenting this analysis and the policy prescriptions that flow from it, the author draws on his experience as head of the ITU strategic planning unit from 1992 to 1999, as well as on his subsequent involvement as an independent consultant in the ongoing effort to reform the ITU, in the work of the G8 Digital Opportunities Task Force (DOT Force), and in WSIS.

The Past as Prologue: The ITU and the Quest for Global Governance

Throughout the history of electronic communication networks, major technological innovations have given rise to new enterprises, transformed economic and social structures, crossed borders, created international rivalries, and led to the development of governance arrangements with almost predictable regularity, accelerating frequency, and an ever-widening circle of economic and social consequences. It is not easy to think of any other field of human endeavor in which the effects of local invention have been so quickly and so frequently felt at the global level, in which the beating of a technological butterfly's wings may indeed shake the foundations of even the most powerful human institutions continents away.

From a technological point of view, the history of global electronic networks can be seen as a series of relatively short cycles—typically of one or two decades' duration—each of which begins with an invention (invariably the subject of dispute as to which individual or country was the true inventor), continues through the stages of application, innovation, and diffusion (usually not for the purpose originally intended by the inventor, and always with disruptive effects), and ends with the construction of governance arrangements designed to ensure that the technology in question is developed, deployed, and operated in the common interest. These arrangements may include all or some of the following features: a policy vision setting out goals and principles; a group of participating parties; a set of activities; legal instruments; institutional structures; procedures; and working methods.[2]

Allowing for the time lag that occurs between the initial demonstration of a new technology and its practical application in a commercial or public service setting—and taking into account the blurring that results from the quickening pace of technological change and the foreshortening of historical vision as we move from past to present—this pattern has repeated in the development of every major new telecommunication network technology, beginning with the telegraph in the 1840s, the telephone in the 1870s, radio telegraphy or "wireless" in the 1890s, radio broadcasting in the 1920s, television broadcasting in the 1950s, geostationary satellite communications in the 1960s, computer communications in the 1970s, optical communications in the 1980s, and the Internet and mobile communications in the 1990s.[3]

From a governance point of view, the history of electronic communication networks also suggests an intriguing, if much more speculative set of hypotheses: that there are patterns roughly akin to "governance long cycles" at the global level, which may last as long as sixty or seventy years; that these cycles alternate between phases of diversification and consolidation in the construction of governance arrangements; that they are triggered by sudden shifts, at the levels of power and policy, in the perceived relationship between electronic communication networks and prevailing economic and social structures; and that the third long cycle in the governance of global electronic networks is now fully under way.[4]

The Long Rise and Sudden Eclipse of the ITU

The first of these long cycles that began with the creation of the International Telegraph Union in 1865 lasted until the 1930s. This was a period of institutional innovation and diversification, which saw international telephony added to the responsibilities of the ITU in 1885, a separate International Radiotelegraph Union established in 1908, and three independent technical bodies set up during the 1920s to standardize telephone, telegraph, and radio communication technologies—the global ICTs of the time.[5]

In the second long cycle, these different governance arrangements were consolidated into a single organization, the International Telecommunication Union. This consolidation process began in 1932 and was completed in 1947 when the ITU took on its modern institutional structure and became a part of the United Nations system. For the next four decades, the ITU was the principal forum for governing electronic networks at the international level, and enjoyed a monopoly of power that reflected the structure of the telecommunications sector within its member states. However, by the late 1980s the ITU's role was beginning to be undermined by the changes that were taking place in the traditional telecommunications industry, as well as by the broader effects of technological change that were captured in concepts such as "the information society" and the "new economy."[6]

At base, a set of fundamental technological advances—in the digitization of all forms of communication, in the development of microelectronics and high-capacity transmission media, and in software design and engineering—had given rise to opportunities to develop new network products and services that competed with the offerings of traditional network operators. In order to capitalize on these opportunities, a worldwide movement began in the most powerful nations and regions of the world to transform the policy and regulatory model that had governed electronic networks at the national level. This movement, which was led by the United States and quickly followed by Japan and Europe, aimed at replacing monopoly with competition, public ownership with private enterprise, detailed regulation with rules for fair and effective competition, and cross-subsidies between profitable and unprofitable services with market-oriented prices and explicit subsidies to achieve social goals.[7]

From a long-term perspective, this transformation in the governance framework of electronic communication networks was undertaken in response to fundamental changes that were taking place in the structure of Organization for Economic Cooperation and Development (OECD) economies, where technological innovation had emerged as a key component of growth, productivity, and international competitiveness; where information-based services had emerged as a leading source of employment; and where new opportunities were arising to use electronic communication networks in the design and delivery of public and social services.

The cumulative effect of these technical, economic, social, regulatory, and political changes quickly undercut the ITU's claim to provide an all-encompassing model for governing global electronic networks, and began to raise questions about its capacity to discharge some of its core technical and regulatory functions. After a decade-long incubation period among OECD countries, a tidal wave of new issues burst onto the global governance agenda—including privatization, competition, deregulation, trade in telecommunication services, convergence, industry self-regulation, intellectual property rights (IPRs) in electronic media, e-commerce, protection of privacy, regulation of undesirable content, network security, cybercrime, the use of information and communication technologies (ICTs) for development, and e-government. Many of these issues fell mainly or entirely outside the ITU's governance mandate and organizational capacity. They brought new players and new forums into the global governance arena from the public, private, and not-for-profit sectors. In addition, many of them were brought into focus for the international community by the Internet, a new kind of electronic communication network that had developed entirely outside of and largely in opposition to the governance model embodied in the ITU.

The Governance Divide

The transition from the consolidated governance model of the second cycle to the diversity of the third had different impacts on developed and developing countries. Although the national administrations that traditionally represented developed countries in the ITU lost power both domestically and internationally in relation to new policy players, developed countries as a whole did not suffer. The agenda of the third cycle was their agenda, not the agenda of the developing countries. Nationally, through the transformation of policy and regulatory frameworks; regionally, through trade agreements; and internationally, through organizations like the OECD and the World Trade Organization (WTO), they had been preparing to play the new governance game for a decade. To a greater or lesser extent, developed countries entered the third cycle with the institutional capacity and the public and private resources needed to engage the new governance agenda, if not in its full scope, at least on matters of highest national interest.

For a number of reasons, most developing countries—particularly the poorest, LDCs—were unprepared for the eclipse of the ITU as the central institution for governing global electronic networks.

During the 1980s, while OECD countries were working together to define a new governance framework based on the presupposition that the building of electronic communication networks should be a private business operating in markets that were regulated to ensure fair competition and protect consumer interests, developing countries were focused on an entirely different agenda. This agenda, which was crystallized in the 1984 report of the Maitland Commission, centered on the twin challenges of modernizing telecommunications infrastructure in the developing world and extending networks so as to provide universal access to basic telephone service in all developing countries.[8] Standing behind this agenda was a policy framework based on the presupposition that telecommunications should continue to be a public service, and that the building of networks in developing countries should be financed largely through public expenditures, supplemented by subsidies and other forms of assistance deriving from solidarity and partnership between developed countries, the private sector, and the developing world.

In addition to these differences of perspective, there were other reasons why developing countries were unprepared for the new agenda that was launched in the 1990s. One was simple lack of awareness and capacity. Given the state of telecommunication networks in most developing countries, their economic structures, and their income levels, most of the new issues simply did not arise; and in cases where they did, there was little real governance capacity to deal with them. Secondly, in the years before the breakup of the Soviet Union in the early 1990s and the subsequent global embrace of capitalism, it was still possible for developing country leaders to believe that there might be alternatives to market-led development. Finally, the hard currency obtained through the ITU system for sharing revenues from international telecommunication traffic, and the incentives that system provided for charging prices that were well above cost, gave developing countries a strong, if short-sighted stake in maintaining the status quo.

Twenty years into the third cycle, the developed countries that initiated the "Big Bang" in global ICT governance—and the many developing countries that have become active participants in the new universe—are justifiably proud of their creation, which has spurred telecommunications innovation, investment, and access on an unprecedented scale throughout the world. From another perspective, however, these benefits have come at a significant governance cost. Where once there was a single forum for governing global electronic networks open to all countries, there now appears to be a global governance void within which a complex and confusing array of local activities take place without any overall coherence or "top-down" coordination.

This is not to say that the new universe is entirely random—far from it. The most powerful government actors are able to exercise a significant degree of policy and regulatory control "from the bottom up" by pursuing national or regional interests or both across a wide range of forums, while the most powerful private actors are able to exercise an equally significant degree of market control by coordinating their activities through private forums or through the exercise of raw market power. But what is often missing are opportunities for the less powerful to be engaged in the discussion of global governance issues, to participate in decision-making processes, to understand the consequences of these decisions, and to adapt their policies, regulations, and practices accordingly. Even with the best of intentions, in the absence of the less powerful their interests are unlikely to be given serious consideration, and the potential benefits of international cooperation not fully realized.

For all these reasons, many developing countries have been slow to accept the ITU's diminished status. Some ITU member states still appear to dream of restoring the union to the center of the governance universe. For many, the shift has left them adrift in the world, without governing institutions in which they feel fully at home.[9]

The Current Quest

Policy, like nature, abhors a vacuum. It was not long before a quest began to put some sort of order into the diverse arrangements that characterize the new governance universe. It is important to be clear about the nature of this quest and how its goals differ from those that guided previous governance cycles. It is not a quest for a new overarching treaty or a new umbrella organization—although that may come in time if there is ever a fourth long cycle to consolidate the present governance diversity. Instead, it is a quest with three less ambitious but nonetheless challenging objectives. The first is to develop a policy vision, along with goals and principles that would apply to all of the diverse governance arrangements that characterize the new cycle, to provide a beacon for guiding and coordinating their activities. The second is to frame these overarching goals and principles in a way that addresses the needs and captures the interests of both developed and developing countries, so that no country is left out of the policy picture. The third is to include partnership among governments, the private sector, and civil society as a fundamental feature of this policy framework and of any coordinating mechanisms that are put in place to give it effect. In other words, the essential goal of this quest is to develop an inclusive policy and action framework that brings together the diverse contributions of all these players—not to establish a new institutional framework based on a new treaty agreement and featuring a new organizational structure.

It is important to note that this quest is not guided by a single vision and is far from a unified movement. For developed countries, it is mainly a quest to universalize the principles and practices required to create and manage private, competitive markets for ICT goods and services. For many developing countries, it is mainly a quest to find a "third way" between the government-owned and controlled monopolies that dominated the sector in the past and a purely market-based approach that would make no special provision for the needs of developing countries.

The first wave in this quest, which was driven by developed countries, was launched at an ITU event, the first World Telecommunication Development Conference (WTDC), which took place in Buenos Aires in 1994. In his keynote address to the conference, U.S. Vice-President Al Gore proposed a set of five market-oriented principles to guide the building of what he called the Global Information Infrastructure (GII).[10] The G7 group of countries built on this proposal and enlarged the scope of the governance quest at a Ministerial Conference on the Information Society hosted by the European Union in Brussels in 1995, which added three additional principles to address social, cultural, and developmental concerns.[11] In 1996, the scope of the quest was further enlarged to include issues of concern to developing countries when, with support from the European Union and at the invitation of South Africa, the representatives from the G7 and forty developing countries met at the Information Society and Development Conference (ISAD) in Midrand, South Africa.[12]

The ITU, which had not been invited to the Brussels meeting and was given only a minor role at Midrand, regained the initiative in 1998 when its Minneapolis Plenipotentiary Conference adopted a resolution proposing that the United Nations should convene a World Summit on the Information Society. The WSIS would involve contributions from UN member states, the private sector, civil society, and international organizations. It would aim at developing a declaration of principles and plan of action that could provide a framework for coordinating the actions of these four stakeholder groups.[13] The United Nations had recently become an active player in the quest for a new governance framework through the activities of its Economic and Social Council (ECOSOC), and readily agreed to the ITU's proposal. In December 2001, the UN General Assembly adopted a resolution authorizing the summit, linking it to the achievement of the development goals set out in the Millennium Declaration and tasking the ITU with the job of organizing the summit.[14]

In parallel with this move, the G8 continued its quest to achieve a substantially similar result. At its 2000 Okinawa summit, the G8 established the DOT Force. This body included representatives from G8 governments, the private sector, civil society, and international organizations, and was given a mandate to recommend objectives and actions that would ensure that ICTs supported the global development agenda. The

DOT Force report was accepted at the 2002 G8 Kananaskis summit, and task force members are implementing its recommendations, in some cases in partnership with the UN ICT Task Force.[15]

As a result of WSIS, these different streams have begun to interact—in some cases to merge, in others to diverge, and in still others to continue on their parallel courses. At this point it is difficult to foresee what will result from all this activity. A consensus appears to be emerging on a set of general principles to guide the governance of global ICTs, as well as on the main lines of action that governments, the private sector, civil society, and international organizations should undertake to give effect to these principles.[16] However, in the current political and economic environment, it is uncertain whether WSIS follow-up activities or any other process will be able to mobilize the political will and financial resources required to implement this agenda. Whatever the outcome of these efforts, it seems likely that the quest to link the governance of global electronic networks with the achievement of the global development agenda and to include the private sector and civil society in global governance arrangements will continue.

Fit for the Future?

How fit is the ITU to continue to discharge its traditional governance activities in a transformed global environment? How useful are the results of the ITU's work to the international community? Is the ITU the right organization to lead the quest for new arrangements to govern global electronic networks, or at least to play a leading role post-WSIS?

This section seeks to answer these questions by analyzing the ITU in terms of the general governance model proposed in the preceding section, in particular by making a distinction between the main elements of the ITU's "policy and action framework" and the principle features of the ITU's "institutional framework," which includes its legal instruments, organizational structures, procedures, and working methods. This analysis suggests that the ITU is reasonably fit—perhaps even surprisingly fit—with respect to the first set of attributes, but woefully unfit with respect to the second set. As we shall see in subsequent sections of the essay, the threats, opportunities, and dilemmas facing the ITU revolve around this fundamental contradiction.

Goals and Principles

The goals and principles that define the scope of the ITU's governance ambitions and guide its activities are set out in article I and in chapter VI of the ITU's Constitution. Allowing for the fact that these provisions are formulated to apply to telecommunications rather than to electronic networks more generally, there is a striking similarity between many of the elements contained in the ITU's policy framework, and corresponding elements in the frameworks for governing global electronic networks that were adopted by WSIS through the 2003 Geneva Declaration of Principles and Plan of Action and the 2005 Tunis Commitment and Agenda for the Information Society.

This is the case, for example, with respect to the ITU's overall objective, which is to promote cooperation among governments, the private sector, and other stakeholders in order to achieve "peaceful relations, international cooperation among peoples and economic and social development." It is also the case with respect to the specific purposes set out in the ITU Constitution that—if translated into contemporary language—would include such goals as closing the digital divide, achieving universal access, applying technology to the economic and social challenges facing developing countries, and achieving an appropriate balance between the protection of privacy and national security.[17]

Membership

One of the central aims of the current governance quest is to devise arrangements that include all the main stakeholder groups—governments, the private sector, civil society, and international organizations. In this regard, the ITU appears to offer a solid base of governance value in spite of the absence of key players. In addition to 189 states, the ITU's membership comprises more than 700 nonstate entities and organizations, including telecommunications, broadcasting, and information service providers; equipment and software manufacturers; scientific, technical, and other nongovernmental organizations; development agencies; and various other bodies with an interest in telecommunications. Over the past few years, the ITU has worked hard at building bridges with the Internet community (the Internet Society is an ITU member, and the ITU takes part in the work of ICANN, the Internet Corporation for Assigned Names and Numbers) and, through the Global Standards Collaboration, with the large number of regional and private sector forums that play an increasingly important role in the development of telecommunication networks and services.[18]

This is not to say that the ITU is completely inclusive, even within the telecommunications domain. Civil society is only represented by a handful of technically oriented organizations, such as the International Amateur Radio Union. An even more striking gap is the formal absence of the independent regulatory authorities that have been spun out of ITU member administrations in more than a hundred countries, and which exercise enormous influence on the development of telecommunications worldwide.

The formal absence of regulatory authorities is largely a consequence of a principle that the ITU itself, among others, has consistently recommended to its member states—that is, that in order to preserve their impartiality, regulatory bodies should be fully independent of both telecommunications operators and government policy makers. However, this principle does not prevent representatives of regulatory bodies from attending ITU meetings as members of national delegations. In fact, this practice is commonplace in ITU radio communication activities where it makes sense because the ITU's Radio Regulations (RRs) provide the foundation for regulating the use of radio frequency spectrum at the national and regional levels. In other areas of ITU activity, however, the inclusion of telecommunications regulators on national delegations tends to be the exception rather than the rule.

Even though independent regulatory authorities are not formally part of the ITU's membership and their representatives are not always included on national delegations, they have become actively involved in the work of the ITU on an informal basis, as a result of initiatives such as the Regulatory Colloquium established by the previous secretary-general, Dr. Pekka Tarjanne, and the New Initiatives Program of the current secretary-general, Yoshio Utsumi. In addition, the ITU's Bureau for the Development of Telecommunications (BDT) has involved telecommunications regulators in its work through a variety of initiatives, including its annual Global Symposium for Regulators, workshops and training courses, T-Reg web site, and series of publications on regulatory trends and issues.

Governance Functions

The ITU's principal source of governance value lies in the activities it undertakes and services it provides, directly to its members and indirectly to the international community as a whole. These functions run on a continuum from "hard" to "soft" governance, in terms of the strength of the obligations they impose, and their impact on the global marketplace for telecommunication equipment, networks, and services. On this scale, the ITU's spectrum regulation activities represent a relatively hard form of governance, since they are based on treaty instruments that have the force of international law, and have very significant impacts on markets for radio communication services. The ITU's development assistance activities, on the other hand, constitute a relatively soft form of governance since they carry no legal weight and represent a very small part of the global development effort. The ITU's technical standardization activities fall somewhere in between these two extremes. They are voluntary and do not have the force of law; but they carry huge commercial weight. In addition, there is a fourth activity—the regulation of international telecommunications—that carried enormous legal and practical weight in the past, since it defined the way in which international telecommunication networks operated. Today, however, it has very little practical effect.

The ITU's governance functions are complex. Some of them have long and convoluted histories. With some notable exceptions that are referenced in this essay, there is relatively little in the public literature to help those who do not have inside knowledge of the ITU to understand and assess them. The aim of the following sections is to provide snapshots of the current state of the ITU's four main governance functions, by focusing on the "flash points" where they have been most critically affected by the changes in the global telecommunications environment mentioned previously.

Regulating International Telecommunications

The ITU's original governance function was to regulate international telecommunications by coordinating national policies and regulations through treaty commitments. This was first done through the International Telegraph Convention, which was adopted at a full-fledged diplomatic conference in 1865, and subsequently continued through various iterations of the International Telecommunication Regulations (ITRs), which were last updated at a World Administrative Telegraph and Telephone Conference (WATTC) in 1988.

The ITRs are now generally considered a dead letter—even by ITU member states, which have unanimously agreed that they are obsolete. They embody the principles and practices of an era that has disappeared in all developed and many developing countries, an era in which international telecommunications was a service jointly provided by corresponding national monopolies. The ITRs provided a treaty-based regulatory framework within which government-owned or government-controlled monopolies prescribed the routes that telecommunications traffic would take to get from one country to another; shared the task of providing circuits; apportioned revenues on the basis of accounting and settlement rates that often bore little relationship to underlying costs; charged customers whatever the market would bear; and precluded any form of competition from public or private network operators.

The regulatory regime established by the ITRs began to unravel in the 1980s. Once competition began to improve service and drive down prices in national telecommunication markets, it proved impossible to keep it out of international markets as well. The 1988 WATTC recognized as much when it included an article on "Special Arrangements" in the ITRs. This article effectively said that two consenting countries could establish whatever competitive arrangements they wanted between them, as long as the results did not cause technical harm to third countries. In the 1990s, technical advances such as callback and Internet telephony further undermined the ITRs. By providing entrepreneurs and ordinary users with low-cost and essentially uncontrollable ways of circumventing monopoly-based arrangements, developments of this kind effectively precluded the possibility of maintaining the ITR regime, even in countries that wished to do so.

The fall of the ITR regime for regulating international telecommunications hit many developing countries hard. The settlement payments they received from developed countries for completing incoming calls at their end of the circuit were important sources of hard currency. However, the fact that this money was not always used to further the growth of telecommunication networks and services in these countries only added to the growing frustration the United States and other developed countries felt with the ITR regime during the 1990s, as competition began to take hold on international routes and the World Trade Organization agreements on trade in telecommunication services opened markets that had previously been closed. A system that had been accepted as a way of subsidizing developing countries during the monopoly era came to be denounced as an unfair tax on U.S. consumers and businesses, and developing countries that failed to meet FCC-defined benchmarks were threatened with unilateral reductions in settlement payments. At the end of the day, developed and developing countries agreed to work together in the ITU to update the accounting rate system, the last remaining practical legacy of the ITRs. In doing so, however, they were clearly backfilling the past, not building the future.

As the ITRs declined toward obsolescence, the ITU launched the World Telecommunication Policy Forum (WTPF) and other informal, nonbinding approaches to coordinating policies and regulations among its member states. However, rather than tacitly agreeing to abandon the ITU's historical role in regulating international telecommunications, some ITU member states—mainly, although not exclusively from the developing world—have taken the view that the ITRs should be revived and updated to reflect the changes that have taken place in telecommunications regulation at the national and international levels (e.g., through the WTO agreements on trade in telecommunication services). Some have even suggested that their scope should be enlarged in order to capture new developments such as the Internet.

In the fourth section, this chapter will examine in more detail which of these approaches to coordinating the policies and regulations of ITU member states is likely to prove the more fruitful.

Regulating Use of the Radio Frequency Spectrum and Satellite Orbital Positions

Today, the ITU's principle regulatory role is to allocate bands in the radio frequency spectrum to telecommunication services, as well as to other services that use radio communications as an integral part of their operations (e.g., broadcasting, aeronautical and marine transportation, public safety, meteorology, astronomy, and other scientific services). In addition, the ITU coordinates the assignment of specific frequencies and satellite orbital positions within these general allocations, so that "harmful interference" between different spectrum services and users is avoided, and registers the results.

Everything connected with the ITU's radio regulation function is massive. Unlike the ITRs, which are contained in a slim volume of a dozen pages, the RRs run to several thousand pages, in four thick volumes. The RRs are continuously updated at the ITU's World Radiocommunication Conferences. These treaty-making events bring together many hundreds of delegates for four weeks, every two to three years, to amend the RRs, by working their way through millions of pages of documentation in the ITU's six official languages. No sooner has one WRC ended than preparations for the next begin, through the work of study groups, conference preparatory meetings, and other intersessional activities. The economic value of many of the decisions that result from the ITU radio regulatory process is great—given the scale of investment required to develop new wireless and satellite communication systems, and the expanding size of the markets for wireless and satellite communications. As a result, there is intense commercial interest in the ITU radio regulation process, even though the private sector is not formally part of it.

Although in many respects it operates at the leading edge of the telecommunications industry, the ITU regime for regulating radio communications appears curiously old-fashioned. The underlying framework of the RRs, like that of the ITRs, is the product of a bygone era when government agencies and public services were the main civilian users of radio frequency spectrum in many countries, when technological limitations required careful separation between different radio communication services and severely restricted the number of spectrum users, and when spectrum rights were assigned by national governments through purely administrative processes, within the framework of an overall spectrum management plan derived from the results of intergovernmental studies and negotiations at the ITU.

In many developed and developing countries, the radio communications picture is now completely different. Government agencies and public services (not to mention the military) remain important users of spectrum—but satellite broadcasting, cellular telephony, and other wireless technologies have created an enormous demand for spectrum to serve business and residential customers. This has increased pressure not only to open new frequency bands, but also to reallocate spectrum from low- to high-demand radio communication services. As in every other area of information and communications technology, the increasing intelligence of radio systems has made it possible for more services and more users to share spectrum resources without harmful interference. In many countries, auctions have replaced administrative processes as the preferred means of allocating spectrum resources to commercial services.

In spite of the dramatic increase in demand for spectrum as a result of the "radio revolution," the introduction of market-based approaches to assigning and managing spectrum resources by many of its member states, the very significant value of some of the spectrum allocation decisions made at WRCs, and the great difficulty it has had responding to requests for its frequency and orbital registration services, the ITU has continued to regulate radio communications in much the same way that it always did. Periodic attempts to reform the ITU radio regulation regime by simplifying the allocation tables and coordination procedures embedded in the RRs have failed. However, the ITU process for regulating radio communications has continued to work well enough that, so far, there has not been a serious attempt to systematically bypass it or to construct an alternative regime. In this sense, the RRs have so far escaped the fate that befell the ITRs as a result of the WTO agreements on trade in telecommunication services. However, the regime has begun to fray around the edges, most notably in connection with the coordination of satellite system filings. Part IV will look in some detail at the different approaches to reforming this aspect of the regime.

Technical Standardization

The bulk of the ITU's governance output, measured in numbers of pages of text, consists of technical standards designed to ensure that equipment can be physically connected to form networks and used without causing harm; that networks operate efficiently and with a high level of quality and reliability; and that different network platforms can interconnect and different services interoperate as seamlessly as possible.

ITU technical and operational Recommendations—which are commonly referred to as standards—are voluntary and do not carry the same formal obligation as ITU regulations. However, their impact on the development of technologies and markets is immense, since they are developed by the companies that manufacture the equipment used to build global electronic networks, and the companies that use these networks to provide global communication services.

As well as developing technical and operational standards, members of the ITU make recommendations in two areas that are closely related to its regulatory responsibilities: the characteristics and frequency requirements of radio systems and services; and international tariff and accounting principles. These recommendations differ from technical standards in two important ways. The ITU considers them as quasiregulatory, since they may be incorporated by reference into regulatory treaty text; and government members participate actively in their development.

The changes in the telecommunications sector outlined previously have affected the ITU's technical standardization activities more profoundly than any of its other governance functions. Representatives of its private sector members do virtually all of the ITU's technical standardization work. Therefore, the ITU can only continue to perform this governance function as long as its private sector members are willing to contribute their time, money, and effort to the organization. To retain the allegiance of these members, the ITU must provide a reasonable return on their investment in comparison to the results offered by other standardization organizations, particularly in terms of the time it takes to develop a standard.

In practice, this means that ITU standardization activities must both compete and collaborate with the numerous standardization bodies active in traditional telecommunications, the broader world of ICTs, and the new world of the Internet. Unlike the ITU, virtually all of these bodies are either private sector organizations or driven by their private sector members. This difference raises very important questions about the ability of the ITU's technical standardization activities to interact effectively with these other bodies. These questions are particularly difficult when they concern cooperation between the ITU and the Internet community, given the different technological visions and governance ideologies that underlie their respective work.

Later, this chapter examines the different approaches that have been taken to enhancing the role of the private sector in ITU standardization activities, so that these activities are better positioned to compete and collaborate in the current governance environment.

Development Assistance

The ITU provides a number of services intended to assist developing countries in modernizing and expanding their telecommunication networks and services, in increasing public access, and in applying information and communication technologies in the service of economic and social development. These activities include technical assistance, training, resource mobilization, support for applications such as e-commerce and distance education, and assistance in developing policy and regulatory frameworks.

In addition to activities specifically targeted at developing countries, the ITU sponsors a number of more general services intended to provide all members of the international community with information on new technologies, networks, and services; the policy, regulatory, and financial issues these developments raise; and opportunities for furthering the growth of global electronic networks through cooperation between the different members of the international community. These activities include the high-profile world and regional TELECOM exhibitions; various policy and regulatory forums and workshops; and publications such as the ITU's World Telecommunication Development Report and statistical indicators.

The ITU's development activities impose no obligations and have relatively little direct impact on the overall global market for telecommunication equipment, networks, and services—even though individual projects may have important consequences for participating countries. Although they account for about one-quarter of the ITU's annual budget, these activities are a "drop in the bucket" compared to the resources available to the World Bank and other international financial institutions, or the financial requirements of developing countries. However, their indirect impact may be considerable. As *Louder Voices*, a study by the Commonwealth Telecommunications Organisation and Panos London demonstrated, in order to strengthen the role of developing countries in the governance of global electronic networks, it is first necessary to build capacity at the national and regional levels. The study found that this requires five things: raising awareness of the links between ICTs and development at all decision-making levels in developing countries; developing technical, regulatory, and policy capacity; providing easy, affordable, and timely access to information on governance issues; strengthening all phases of national policy-making processes, from assessment to implementation and evaluation; and overcoming financial barriers by eliminating distortions in programs that subsidize developing country participation in international ICT decision-making forums.[19]

Modest though they may be, the ITU's development activities address each of these requirements. However, as the *Louder Voices* study found, these efforts have not yet resulted in more effective developing country participation in the organization as a whole. The fourth section of this chapter examines how the ITU's limited resources for development assistance could be more effectively deployed. This is a critical challenge, since the management of information and knowledge in a way that helps all players understand and assess the multiplicity of technical and commercial alternatives, policy and regulatory issues, and decision-making options that confront them in the global ICT universe is a central element in the quest for a policy and action framework for the governance of global electronic networks.

An Interlude on Infrastructure

To sum up the previous sections, the core value of the ITU, to its members and to the international community, lies in the role it plays in facilitating the development of electronic communication networks. In the main, this is highly technical work that is not easily understood or accessible to many of the organizations and groups involved in the quest for a new governance framework, let alone to policy makers and the general public. In the current quest attention is not centered, as it was at the time of the Maitland Commission of the early 1980s, on the development of electronic communications infrastructure. Instead, it is centered on the use of electronic communication networks for economic and social purposes in both developed and developing countries, and on the issues to which this use gives rise in diverse domains that range from trade and commerce to education and health care, privacy and consumer welfare, the delivery of government services, the protection of national security, entertainment, and enlightenment.

As the focus of attention shifts, it is often too easy for actors from the developed world to assume on the basis of their own experience that the marketplace, if wisely regulated, will look after infrastructure and that policy can concentrate on its uses. This version of the "end of history" is shortsighted and is not shared by developing countries, which very clearly understand on the basis of their experience that without modem, efficient, reliable, and pervasive infrastructure nothing else is possible, and that even wisely regulated market mechanisms can only go so far in the developing world—and perhaps in many developed countries as well. In spite of the recent global pause in the construction of telecommunications infrastructure, network technologies will continue to evolve, and the governance functions traditionally performed by the ITU to facilitate their development will continue to be a necessary condition of sustainable global development—whether or not the organization survives.

The Yoke of Sovereignty

If there is still considerable value in the ITU's policy and action framework—in its vision, guiding goals and principles, membership and activities—is there also equivalent value in its institutional framework—its legal instruments, organizational structures, procedures, and working methods?

Whether they entail formal obligations or not, and whether participation is restricted to governments or open to the private sector and civil society as well, all ITU activities are founded on a set of treaty agreements among its member states that set out the principles on which the organization is founded, the goals it aspires to achieve, the rules and procedures that govern its activities, and its organizational structure and working methods. These agreements include the ITU Constitution and Convention and its Administrative Regulations—that is, the Radio Regulations and the International Telecommunication Regulations. In comparison to the treaties founding other, more modern organizations (such as the WTO), the ITU instruments are very detailed—a reflection of the ITU's long history, its habit of frequently amending its treaty texts, and its engineering organizational culture, which is quite effective at fixing things when they break, but rarely pauses to redesign its machinery so that it will not need such frequent servicing.

Appended to these treaties are a series of less formal instruments—decisions, resolutions, recommendations, and opinions—that deal with issues of policy, strategy, and practice that are more specific or time-limited or both than those that are dealt with in treaty text itself.

In the ITU, general governance is provided by the Plenipotentiary Conference, which meets every four years, and by the ITU Council, which is composed of a quarter of the membership and meets annually. Under the ITU's unique federal organizational structure, its Radiocommunication, Standardization, and Development Sectors each has its own governance structure, which mirrors that of the organization as a whole.[20]

The net effect of this treaty edifice is that all ITU activities—including those that impose no obligation on member states and that are largely or exclusively carried out by the private sector and other nongovernmental actors, or by the secretariat—fall ultimately under the formal control of member states. This means that any proposal for change, whether it involves stopping some activity that is no longer useful or launching a new initiative, tends to be assessed not only substantively—in terms of its impact on the goals of the organization and the national interests of member states—but procedurally, in terms of "who does what and how." Essentially, this comes down to assessing which activities, organizational structures, and working methods—and what distribution of rights, obligations, and responsibilities—are consistent with the limits and constraints imposed by "the sovereign right of each state to regulate its telecommunication."

In an ideal world, the representatives of ITU member states might well wish to do everything themselves, much as they used to. In the real world, they accept to varying degrees the need to share power with other actors. However, their failure to satisfactorily reconcile sovereignty concerns with the power dynamics that have reshaped the telecommunications sector by shifting most of the power that once resided in ITU member administrations to the private sector, independent regulators, consumers, and civil society advocates, has seriously eroded the ITU's present and potential role in the governance of global electronic networks.

The ITU Reform Movement: Fifteen Years of Frustration

ITU member states have not been blind to the changes that have reshaped their universe. For the past fifteen years they have been trying to adapt the ITU regime to "the changing telecommunications environment" through a reform program aimed at:

- improving the efficiency and effectiveness of their traditional activities—technical standardization and the regulation of international radio communications;
- putting ITU development activities on the same formal footing as radio communication and standardization;
- enlarging the rights and obligations of private sector members of the ITU;
- developing a role for the ITU as a forum for discussion of global policy and regulatory issues;
- building bridges between traditional telecommunications and the Internet; and
- modernizing the role and management of the ITU secretariat.

Underlying all these issues are two fundamental questions that bear not only on ITU reform, but also on the broader international quest for a new policy and action framework to govern global electronic networks.

The first question concerns the scope of ITU member states' governance ambitions. In view of the erosion of their power and the pressures to do more with less in terms of financial and human resources, should they abandon any hope of continuing to exercise general governance over the telecommunications sector in order to concentrate on their core businesses? Or should they seek instead to reform the ITU with the goal of drawing new actors into the organization, expanding its mandate to address new issues, tapping new resources, and introducing new decision-making processes that would reflect the power shifts that have taken place in the telecommunications sector?

The second question concerns their willingness to share the power they have traditionally enjoyed within the ITU with new actors. This issue arises whatever the scope of the members' different governance ambitions, since it is increasingly difficult either to carry out the ITU's traditional functions or to expand its range of activities without making some accommodation with the new actors that have appeared on the international scene.

Patterns of Power and Preference

It is not easy to succinctly characterize the policy priorities, preferences, capacities, and power of different ITU member states in relation to these two strategic questions, or to the more specific reform issues they underlie. This is all the more the case because there are no easy and simple divisions among the member states of the ITU.

There are significant differences in the preferences of the "ICT superpowers"—the United States, Europe, and Japan. The United States tends to be the most conservative member on questions related to potential enlargement of the ITU's sphere of activity and to the sharing of power with other actors, be they the private sector, nongovernmental organizations, or the staff of the union itself. Japan, on the other hand, has tended to favor an expansion of ITU activity, particularly in the area of coordinating telecommunications policy and regulation, and has been open to enhancing the role of the private sector in some ITU activities. Europe as a whole is somewhere in between—in favor of rationalizing the ITU's regulatory and standards activities and granting a larger role in decision making in the latter area to the private sector, but cautious about seeing the ITU expand into new areas of activity—although individual European states often depart from these positions in one or another direction (e.g., with the United Kingdom often closer to the U.S. position than to some of its European colleagues, while France and Germany are sometimes closer to the Japanese view).

The preferences of developing countries are even more difficult to characterize, given the enormous differences that exist between developing countries and regions. On the whole, though, they tend to support a wider role for the ITU in the new environment, and to be skeptical about giving the private sector or other actors a larger role in ITU decision making unless it is tied to greater financial contributions. Both positions are quite understandable: few developing countries have the resources to pursue their interests in the many intergovernmental and private forums now active in the governance of global electronic networks; few have private sectors capable of supporting their interests in more open decision-making processes; and many regard the nongovernmental organizations that purport to represent their interests with suspicion.

With few exceptions—most notably among the Arab states—developing country members of the ITU tend not to contribute actively to discussion of the big issues of organizational and global governance, but to focus instead on matters of direct concern to developing countries. In the case of Africa and the poorer regions of the Americas and Asia-Pacific, this means focusing mainly on the development assistance activities of the ITU Development Sector (ITU-D). In the case of other developing countries—particularly the "tiger economies" of southeast Asia and the emerging economies of eastern Europe—it means focusing on the technical work of the ITU Standardization and Radiocommunication Sectors.

It is worth calling attention to the policy preferences of a third group of countries—the "governance go-betweens"—which includes both middle-power developed countries (e.g., Australia, Canada, the Nordic countries, the Netherlands, Switzerland) and political leaders from the developing world (e.g., Morocco, South Africa). These countries frequently serve as intermediaries between contending interests.[21] In general, the preferences of this group are moderately progressive on the two key issues of expanding ITU activities and sharing power with other actors, and tend to avoid the extremes of other players. These preferences suit these countries for leadership roles within the various decision-making processes of the organization. However, to date their political skill alone has not been sufficient to resolve the fundamental tensions that exist among other ITU members.

No survey of policy preferences would be complete without mention of a fourth group of countries—the "awakening giants." This group includes countries like China, India, Brazil, and Indonesia, which, while sometimes political leaders in the ITU, do not yet carry the full weight that their market mass and growing technical capacity will surely confer in the coming decades. This group also includes the Russian Federation, which, before the breakup of the Soviet Union, was an ITU superpower on a par with the United States, Europe, and Japan. During the last decade, it has fallen from these heights. However, Russia's underlying technical capabilities, longer-term market potential, and renewed political confidence will likely qualify it as a "reawakening giant."

This group of countries has not been very engaged in the ITU reform process to date, nor in the broader quest for a new global governance framework. Yet without their participation and commitment in the longer term, it will not be possible for either the ITU or the international community to construct anything more than a partial solution to the problem of governing global electronic networks. In an era when markets for telecommunications goods and services are saturated in the developed countries that have traditionally dominated international governance arrangements, the awakening giants of the developing world—countries where market demand remains high and social needs are far from met—are likely to become much more influential players in the global governance game if they can learn to use their power effectively.

The Reform Scorecard: Winners, Losers, and Stalemates

Judging by the results of the series of Plenipotentiary Conferences[22] that have taken place since the ITU reform movement was launched in 1989, member states have not been satisfied with the progress made on the ITU reform agenda. These results have fallen short of the expectations of most developed and developing countries. They have also disappointed the ITU's private sector members, as well as elements of civil society that remain effectively excluded from participating in its activities. Although the ITU still has value in the eyes of many countries and nongovernmental actors (as evidenced by their continued, albeit diminished, willingness to pay their annual membership fees and to contribute to the ITU's work by participating in meetings and conferences), it is clearly caught in a downward spiral that threatens to erode its viability. This is particularly the case because of the financial crisis that has followed the 2002 Marrakech Plenipotentiary Conference, a crisis that was triggered in large part because of the dissatisfaction of the United Kingdom and some other member states with the results of the reform process.[23]

What if the effort to fundamentally reform the ITU finally ends in more or less complete failure at the next Plenipotentiary Conference, which is scheduled to take place in Antalya, Turkey, in 2006—so that the financial and power-sharing constraints imposed through commission or omission by major member states force the ITU to retrench in order to concentrate on its core businesses (principally radio regulation and standardization with a little development on the side), and to abandon the initiatives sponsored by its current and previous secretaries-general to enlarge the "soft governance" activities of the ITU in order to at least partially fill the current void in governance of global electronic networks?[24]

Which countries would be the winners and losers under this scenario? And would this be a good result for global governance?

The overall winner would be the United States, which in recent years has not shown much enthusiasm for fundamental change in the ITU. And why should it? The ITU has generally delivered what the United States has wanted, particularly in terms of access to radio frequency spectrum and satellite orbital resources, and has even made improvements to the accounting rate system for sharing international telecommunication revenues under the threat of bilateral U.S. action.[25] In addition, the United States has been largely successful in preventing the ITU from venturing very far into new areas of activity, particularly in relation to Internet governance and global policy and regulatory coordination.

For Europe and Japan, the results would be mixed. Like the United States they have been winners in terms of what the ITU has delivered through its technical activities, particularly in the area of terrestrial mobile communications. However, they would be losers in terms of the fundamental reforms they sought to make to the ITU—seeking to increase the formal rights of the private sector in the case of Europe, and seeking to develop the ITU as a forum for discussion and harmonization of policies and regulations in the case of Japan.

Assuming that the ITU-D emerged relatively unscathed from this worse-case scenario, it could be argued that developing countries would emerge as survivors—if not outright winners—from the collapse of the ITU reform process. However, from a broader perspective it could equally be argued that the Development Sector is a trap if it continues in its present form and that, in the absence of new and more effective initiatives, developing countries would emerge as the principal losers for several reasons. First, the role of developing countries in what many would see as the real work of the ITU—standardization and radio communication—has not significantly increased as a result of the creation of the Development Sector and the obligations imposed on the other two sectors by the ITU constitution to assist with development. Second, the modest resources of the Development Sector have limited its impact in comparison to the results achieved by many developing countries through participation in alternative development mechanisms, such as the programs of the World Bank and the WTO telecommunications agreements. Third, a decade after the formal creation of the Development Sector, the ITU appears uncertain whether its role is "the development of telecommunications" or "telecommunications for development." Consequently, there is as yet only a limited connection between its development activities, the international development agenda, and the resources available through official development agencies.

The collapse of the ITU reform movement would not likely alter any of these results. Instead, it would probably entrench the divisions that exist between the three sectors, and continue the isolation of the Development Sector in a largely self-contained governance space.

Lessons Learned: What Works and What Doesn't

A result of the kind described in the previous section—although highly likely given the results of the Marrakech conference—is neither inevitable nor desirable, both for the interests of ITU members, and for the broader quest for a policy and action framework for governing global electronic networks. Avoiding this fate and releasing the governance value buried in the ITU will require breaking the sovereignty mold that formed the union in 1947 and still shapes its structures and governance mechanisms today, in spite of the enormous changes that have taken place in telecommunications and in the international environment.

Is there any reason to think that a result of this kind is possible? The experience and observation of the past decade argues that there is, if we consider a set of cases in which ITU reform was systematically frustrated at the formal decision-making level (i.e., in treaty-making processes) by conflicts between the sovereignty-related policy preferences of ITU member states, but in which substantively similar issues were resolved at less formal decision-making levels—where sovereignty concerns could be put "in square brackets"—through cooperation between ITU member states from the developed and developing worlds and a variety of nonstate actors, including the private sector.

Case Study #1: Coordinating National Policies and Regulations

Japan's principal contribution to the ITU reform movement has been the notion that the ITU can provide a useful forum for discussing global telecommunications policy and regulatory issues in order to develop a common vision and shared understanding of their dimensions and implications, and to provide a basis for coordinating action. The United States was very lukewarm to this idea when it was proposed at the 1994 Kyoto Plenipotentiary Conference. However, there is a tradition in the ITU that countries that go to the trouble and expense of hosting these month-long events should get something in return, and the proposal was adopted perhaps as much for this reason as on its merits.

In spite of its initial reluctance, the United States became a committed supporter of each of the three WTPFs (World Telecommunication Policy Forums) that have been held to date—partly for defensive reasons, to avoid an unpalatable outcome, and partly because in each case they offered opportunities to advance the U.S. policy agenda with respect to the issues discussed: licensing and regulation of services provided by low earth orbiting satellites (WTPF-96); implementation of the WTO trade in telecommunication services agreements (WTPF-98); and IP telephony (WTPF-2000).

The substantive results of each of these events was positive, in that they found common ground between the divided interests of developed and developing countries on difficult and controversial policy and regulatory issues; provided a framework for coordinating the activities of the three ITU operational sectors; and on one notoriously difficult issue—reform of the ITU system for sharing revenues arising from international telecommunications traffic between sending and receiving countries—helped overcome a stalemate that threatened to poison relations between the United States and much of the rest of the world as a result of unilateral American action.

The WTPF also introduced new procedures and working methods, which are common in more modern international organizations, into the ITU. The deliberations of each of the policy forums was based on a single working document prepared by the secretary-general with the assistance of a drafting committee drawn from the ITU membership, instead of the scores or hundreds of documents that members contribute to traditional ITU meetings. In addition, because the output of each WTPF was negotiated in advance of the forum, instead of being hammered out during the course of the meeting, it was possible to deal with complex issues in just two or three days, instead of the weeks that it usually takes ITU meetings to make decisions.

In all these respects, although its results are informal and nonbinding, the WTPF has clearly achieved more useful results than the process that was launched at the 1998 Plenipotentiary Conference to review the International Telecommunication Regulations. This process will continue for at least another four years and possibly longer. This review was triggered by the same set of circumstances that gave rise to the WTPF—the emergence of a host of new policy and regulatory issues on the global governance agenda, including the possibility that the obligations some ITU member states have assumed under the WTO agreements contradict their obligations under the ITU treaties. However, because this process has been cast in a treaty-making framework, ITU member states have not been prepared to show the flexibility that is possible in an informal activity, such as the WTPF. No real progress has been made, with some member states wishing to terminate the ITRs, others prepared to absorb the parts that remain useful in other ITU instruments, and still others wishing to update and extend them to address many of the new issues that have appeared on the scene, for example by regulating the Internet and e-commerce activities.

In spite of the contrasting results of these two approaches to coordinating policy and regulation, participants in the 2002 Marrakech Plenipotentiary Conference decided not to hold any WTPFs during the 2003–2006 period and to terminate funding for other informal activities led by the secretary-general to assist members in addressing urgent policy and regulatory issues. The conference did however decide to continue the review of the ITRs during this period and to hold a World Administrative Telecommunication Conference (WATC) to formally revise them, some time in the 2007–2010 period. These decisions, which clearly send the wrong signals to the international community about the direction of ITU reform, were mainly taken for inward-looking reasons stemming from the ITU's current financial crisis, turf concerns related to the respective roles of the general secretariat and the Development Bureau, and the desire of some members to send a negative message to the current secretary-general.

Case Study #2: Industry Self-Governance in Technical Standardization

A second example of how progress can be made through less formal mechanisms in areas where treaty-based approaches have failed is found in the issue that has dominated the ITU reform process for much of the past decade—enhancement of the rights and obligations of nonstate actors, particularly the private sector, in ITU decision-making processes. This has been a major issue in the ITU Telecommunication Standardization Sector (ITU-T) where, as explained earlier, private sector organizations do almost all of the work, and there is increasingly stiff competition from regional standardization organizations, such as the European Telecommunication Standardization Institute (ETSI), and the many standardization forums that have been established by the private sector.

The formal quest to "square the circle" by sharing decision-making power with nonstate actors in an intergovernmental organization has consumed large amounts of energy and significant financial resources over the past decade. Under instructions from the 1994 and 1998 Plenipotentiary Conferences, the ITU Council established three successive task forces to study this issue and make recommendations, which in turn have absorbed large amounts of time at annual council meetings and Plenipotentiary Conference deliberations, most recently in Marrakech. The results of all this effort have been exceedingly meager, with a few amendments to the ITU Constitution and a steadily mounting sense of frustration in the private sector and among the countries that have pressed for reform being the most notable.

The main advocates for this reform have come from Europe, which has consistently pressed for a rationalization of the ITU's regulatory and technical standardization activities, in order to open the way for enhancing the decision-making rights and increasing the financial responsibilities of the private sector in nonregulatory areas. Europe has sought these reforms in order to reflect the changes that took place in the structure of telecommunications in Europe itself and many other countries, to maintain the attractiveness of the ITU as a standardization forum, and to bolster the financial position of the union. For its own reasons, however, the United States has consistently opposed rationalizing the ITU's organizational structure along these lines and power sharing with the private sector. These reasons include opposition in principle to the notion of giving a nonstate actor a decision-making role in an intergovernmental organization; procedural difficulties envisioned in separating technical and regulatory functions, particularly in the Radiocommunication Sector; and concern about the problems that might arise in coordinating and maintaining a national position at ITU meetings, if transnational enterprises were able to vote according to their commercial interests.

While representatives of national governments wrestled unsuccessfully year after year with this conundrum, the ITU Standardization Sector devised a simple and elegant solution to the problem, which nicely avoids the sovereignty issues that have bedeviled the broader effort and gives the private sector an equal role with governments in making decisions in the part of ITU business that is of most interest to them. Until 2001 all standards had to be approved by member states before they became official even though most standards were developed in study groups by and for the private sector, without any government involvement.

This charade, which was a legacy of the past when government PTTs played an active role in the standardization process, and was kept in place to maintain the formal façade of state control, has been effectively abandoned by the adoption of what is known as the Alternative Approval Process. Under this procedure, standards that are deemed to be purely technical at the outset of the standardization process are approved by the members of the study group that developed them—which effectively means that they are approved by the private sector. This device is now used in most Standardization Sector study groups. However, it has not yet been introduced in the Radiocommunication Sector, out of concern that apparently technical recommendations regarding the performance characteristics of radio systems might acquire regulatory status by being incorporated by reference into the Radio Regulations.

Case Study #3: Applying Economic Discipline to the Regulation of Satellite Systems

A third example of how informal approaches to improving governance can overcome impasses created when problems are approached on a formal basis can be found in the story of how the vexing issue of dealing with the huge backlog in satellite system filings that built up in the ITU during the 1990s was finally resolved.

The ITU RRs do more than allocate frequency bands to different services. They also spell out in considerable detail the procedures that member states must follow when notifying frequency assignments and filing plans for satellite systems with the secretariat, as well as the procedures the Radiocommunication Bureau must use in analyzing these filings, in coordinating the proposed assignments with other users in order to avoid harmful interference, and in registering the results once this work has been completed. Finally, in order to supervise the work of the secretariat and assist in resolving coordination problems, the ITU Constitution establishes a part-time International Frequency Registration Board whose members are elected by ITU member states.

The scope and complexity of this regulatory apparatus leaves very little room for managerial discretion on the part of the Radiocommunication Bureau. In addition, the first-come, first-served policy followed by the ITU precludes the application of economic tests and disciplines that might help spectrum managers prioritize or ration access to scarce frequency and orbital resources.

In the 1990s, the explosion in the development of satellite systems that resulted from technological innovation and market liberalization created a huge backlog of satellite system filings. The staff of the Radiocommunication Bureau was unable to process these filings according to the complex procedures spelled out in the Radio Regulations with the resources they had been given by ITU members. This problem was compounded by the fact that a significant number of these filings were for "paper satellites" by operators who simply wanted to stake their claim to a potentially valuable resource that could be traded or sold, and who had no intention of building, launching, and operating a system. However, under the principle of the sovereign right of each state to regulate its telecommunication and the first-come, first-served principle, and in the absence of economic tests or other disciplines, there was no way to prevent filings of this kind.

The satellite backlog added to the image of the ITU as an organization that could not keep up with the times, even in the discharge of one of its most important functions. It also led some satellite system operators to ignore the requirements of the Radio Regulations; in one notorious case, one country simply launched a satellite into an orbital position that had been registered to another; in other cases, operators who did not wish to wait accelerated the process through informal coordination.

Like the issue of enhancing the role of the private sector, the question of what to do about the backlog in satellite filings consumed large amounts of time and energy throughout the 1990s, as successive World Radio Conferences, ITU Council meetings, and even Plenipotentiary Conferences debated ways of coping with the backlog that ranged from the simplification of procedures, to the charging of filing fees, to the adoption of a code of conduct on the part of operators, to cost recovery for the services provided by the Radiocommunication Bureau. The United States played a central role in this drama as, at one and the same time, the main filer of satellite system notifications, the principal complainer about the backlog, and the main opponent of measures that would introduce economic disciplines into the process. The United States drew support in this position from developing and other countries that had no particular stake in the issue since they were not in a position to launch satellites, but which were concerned that the principle of cost recovery might be extended to other services for which they would have to pay.

As in the case of enhancing the role of the private sector in decision making, the satellite backlog problem was finally cracked by a simple solution. On his own initiative the director of the Radiocommunication Bureau prepared and presented an economic analysis of the costs and benefits involved in the satellite notification process, as well as an analysis of the pattern of demand for services. It showed that the satellite systems announced by the United States and a few other administrations were massively subsidized by all other ITU members.

This action was at odds with deep-seated aspects of the ITU's traditional organizational culture, which assigns a passive role to the secretariat and which frowns on public criticism, whether explicit or implied, of member states' behavior. However, with this information on the table, ITU member states had very little difficulty deciding that it was appropriate to introduce cost-recovery mechanisms for this service. This simple exercise of managerial initiative, and the introduction of the mildest possible form of economic discipline to the process of assigning scarce resources, has had the effect of reducing demand, streamlining procedures, and increasing the resources available to the bureau to provide this service.

Case Study #4: Raising Awareness of the Links between ICTs and Development

There has been no clearer illustration of the value potentially present in the ITU, when the constraints of the sovereignty principle are relaxed, than the TELECOM exhibitions and forums that are held on a global basis in Geneva once every four years, and in one or more regions annually. Unlike all other ITU activities, TELECOM events are organized and managed by the secretariat with the assistance of external advisors, experts, and contractors, within the framework of general policy guidelines given by the member states. The original purpose of TELECOM, which was established in 1971, was to inform developing countries about trends and developments in telecommunications, and one of these guidelines is that any profits arising from TELECOM events are to be used to assist developing countries.

After a relatively modest beginning in the waning years of the monopoly era, TELECOM events grew rapidly in response to the changes that took place in developed countries during the 1980s. During the boom years of the 1990s they became truly massive events—among the largest trade shows in the world, accompanied by policy and technical forums that attracted ministers, CEOs, scholars, and practitioners from every part of the world to discuss and debate important issues on the global agenda. They were also often accompanied by the kind of hubris and excess that subsequently plunged the telecommunications industry from market heights into the depths of recession.

TELECOM events have outperformed Development Sector conferences and study groups in raising awareness among developing country decision makers of the role telecommunications can play in national economic and social development, and in stimulating the interest of the press and public in ICTs and the information society. It remains to be seen if this will continue to be the case, or if TELECOMs too will fall victim to the greater financial realism that now characterizes the telecommunication sector. It is unfortunate that the members of the ITU never attempted to find ways to build bridges between TELECOM events and other ITU activities—particularly during the 1990s—in order to benefit from the energy, excitement, and resources they released. Instead of learning from the success of these informal events, member states confined themselves to annual audits of financial performance without ever seriously considering the question of how the ITU could use these events to deliver greater financial and marketing returns to the organization and its developing country members. Nor did they seek to benefit from the fact that TELECOM forums attract sources of influence and expertise from business, academia, policy making, and regulation that are at a higher level than those usually available to all ITU sectors, but to ITU-D in particular.

Recommendations

These four case studies suggest that there is a way forward for the ITU. They show that the central problem facing the ITU does not lie in the policy framework of goals and principles that guide the activities of its members—save in the one principle that has so far been off limit for discussion and debate, the sovereignty principle. They show that the problem does not lie in the ITU's membership: these cases illustrate that developed and developing countries, the private sector, and civil society organizations can work together on even the most difficult issues that have consistently defeated the formal decision-making processes of the ITU, when the constraints imposed by the sovereignty principle are relaxed. These examples also show that the ITU secretariat can play a useful role in supporting more inclusive governance practices if, as a former secretary-general liked to say, "they are allowed to think," instead of being confined to their traditionally passive role.

These examples demonstrate that the fundamental problems facing the ITU, as it seeks to reform itself and contribute to the broader governance challenges of the third cycle, lie in the union's institutional framework—in its legal foundations, organizational structures, formal decision-making procedures, and working methods—which block progress by imposing constraints derived from the sovereignty principle in areas where they no longer make sense, given the changes that have taken place in the global governance environment. The way forward, simply put, is to redesign these elements of the ITU governance model in a manner that allows form to follow function.

It is also clear from some of these examples, as well as from the results of the Marrakech Plenipotentiary Conference, that this is not necessarily the path that ITU member states will choose to follow as they prepare for the 2006 Plenipotentiary Conference. There is every danger that, without some shock mighty enough to shake the introversion and complacency that has characterized much of the ITU reform effort to date, the downward spiral traced in the previous section will continue.

It may be possible that a shock of this kind could be internally generated—that the friction built up as a result of a decade of frustration among the ITU's private sector members and among the countries that have taken leading roles in different aspects of the ITU reform movement, will reach a high enough level to force a redefinition and a reorientation of the reform agenda. At present, however, this does not appear likely to happen. The telecommunications industry has little time, attention, or money to spare for investment in yet another round of ITU reform, and at least some of the countries that have consistently championed the reform movement also appear to be running out of energy and enthusiasm. In the current situation, organizational survival, protection of sectoral interests, and the pursuit of narrow national objectives are likely to be the dominant motivating forces, hardly ideal ingredients for launching an internal movement for radical reform.

Externally, the results of the World Summit on the Information Society, which will unfold in parallel with the next stage of the ITU reform process, may provide opportunities and incentives to reform the ITU that are strong enough to help overcome institutional inertia. The 2003 Geneva Declaration of Principles and Plan of Action, the 2005 Tunis Commitment and Agenda for the Information Society, and the activities that have begun to flow from the WSIS process potentially position the ITU to leverage its still-considerable technical and political assets into "a leading role" in the post-summit world—although not the leading role it often appears to seek. However, this is only likely to happen if real reform takes place—particularly with respect to the more effective inclusion of nonstate actors in ITU activities and governance processes.

The shock needed to change the ITU, so that it continues to effectively discharge the governance functions that are the foundation of its value to the international community at the same time as it begins to leverage its WSIS role into a position of governance leadership, likely can only come through the fusion of internal and external forces for change, coalesced around a new vision for the ITU, a new organizational design, a new governance structure, and new financial arrangements. The time has come for the ITU to learn the lessons of the past fifteen years and to break up its current structure. This is the only way to release the full governance value that is currently buried in the ITU.

The Need for a New Institutional Framework

To escape from its current impasse, to release the governance value that lies buried in its structures, and to prepare the ITU for a possible future role post-WSIS as the institutional home for a new model for governing global electronic networks, it will be necessary to break up the current ITU structure and create a much more loosely affiliated network of four organizations, each of which would assume one of the ITU's current governance roles, and each of which would be governed, operated, and financed on the basis of arrangements tailored to its specific requirements.

In other words, instead of continuing to govern all of the activities currently and potentially performed by the ITU within the confines of an intergovernmental treaty framework, and instead of adopting a "cookie cutter" approach to organizational design—that is, applying the same template to all activities no matter how different their responsibilities—the new organizational and governance model would allow form to follow function.

This would mean that there would continue to be a regulatory agency, responsible for radio matters, which would be founded on a simplified treaty, and a new standardization agency that would be organized and governed under the leadership of the private sector.

Breaking up the current ITU structure and reorganizing its components on the basis of the services they provide to the international community would also mean merging the ITU's Development Sector, TELECOM secretariat, and Strategy and Policy Unit into a new global development agency that would do policy research and analysis, provide training and consulting services, and organize discussion forums and exhibitions in response to client demand. Creating a structure of this kind would make it much easier for civil society organizations to participate in the work of the ITU.

The new model would also mean replacing the Union's general governance and management structures (i.e., the Plenipotentiary Conference, council, and general secretariat) with a much lighter coordinating council that would include members drawn from the three new operating agencies, supported by a central service provider.

A key element of this model would be a plan to put the new network on a solid financial footing by replacing the ITU's current "free choice" financing scheme—in which government effectively subsidizes the private sector and developing countries effectively subsidize the developed world—with a more rational model that would share the cost of ITU operations more equitably and capture some of the economic value inherent in the ITU's governance activities (i.e., the value inherent in ITU standards, radio spectrum and orbital allocations, and information management) in order to help build the technical and policy capacity of developing countries, and support their fuller participation in ITU governance activities.[26]

Without a quid pro quo of this kind that would simultaneously realign governance responsibilities and rebalance financial contributions in a way that would give developed and developing countries, the private sector, and civil society what they really want from the ITU, there is no possibility of real reform.

The appendix to this chapter presents a detailed design, showing how a "new ITU" could be constructed along these lines by restructuring its present elements and drawing on models and practices used in other international organizations.

The Need to Build a Winning Coalition

It would be an enormous challenge to secure the agreement of ITU member states to organizational, financial, and governance changes of the kind recommended above. At this point, it is not clear which countries would have an interest in initiating a reform movement of this kind, although they are more likely to be found among the ranks of the developed and developing country mediators than among the ranks of the countries and regions that have been directly party to the sovereignty-induced stalemates that have impeded ITU reform and undermined its international credibility, financial capacity, and policy creativity. In addition, it would be necessary to have early support from each of the other main groups identified above—particularly from the superpowers and the awakening giants—as well as leaders from the developing world. On the basis of their past performance in the ITU and other international forums, this reform coalition might initially include countries like Australia, Brazil, Canada, Chile, Germany, India, Malaysia, Mali, Mexico, Morocco, the Netherlands, the Nordic countries, Senegal, Singapore, South Africa, Switzerland, Tanzania, and the United Kingdom.

However initiated, it is clear that building a political coalition in support of this vision would require changes in the mindset and behavior of many developing countries. It would mean abandoning forever the idea that the ITU and its member states could be restored to their former position at the center of the global governance universe. It would mean accepting the desirability of plural centers of power operating under different regimes, and adopting a strategic approach to issues of global governance that would use different international forums to pursue national development goals in a consistent and coordinated fashion—just as developed countries have done. Above all, it would mean making changes in policy processes at the national and regional level in the developing world aimed at building capacity through pooling of resources and involvement of all stakeholders.

Changes of these kinds in developing countries will only be possible with active support from developed countries, the private sector, and not-for-profit organizations that share the vision of a new model for governing global electronic networks, are prepared to live with its consequences, and are willing to assist developing countries in taking real advantage of the participatory opportunities it would present.

A key element in building a coalition of this kind would be a strategy to begin to open the ITU up to the light of day. In the view of many, it resembles a closed shop or a highly restrictive, somewhat secretive club. There is little understanding of the importance of ITU decisions at senior levels in the public and private sectors, and little appreciation of its impact on the world. Other organizations, such as the WTO, have been forced to become more transparent, open, and accountable to the international community. They and the process of global governance are arguably the better for it. The great success of TELECOM events and the audiences attracted by ITU policy publications and discussion forums suggest what may be possible if the ITU becomes more open to the world around it.

Whatever the right combination of elements, it is clear that there is very little time for ITU members to create a winning strategy. Over the past five years, the work of the G8 DOT Force, the UN ICT Task Force, and the World Summit on the Information Society have begun to alter the governance universe. Without rapid action, the opportunity to radically reform the ITU's organizational structure—in order to finally resolve the problems that have impeded its performance for the past decade, and to fit the ITU for a leading role in the post-summit environment—will be foreclosed and the ITU will have no option but to continue its descent, in an ever-tightening spiral.

Conclusion

Clearly, the successes and failures of the ITU's attempts over the past decade and a half to develop more inclusive governance arrangements have important implications for the future direction of the ITU reform movement. Are there also lessons that can be learned from this experience and applied to the broader quest for inclusive governance that characterizes the third long cycle? Reflections on the ITU's fifteen-year effort to reform its institutional structures and governance functions, in light of its much longer involvement in the governance of global electronic networks, suggests at least three.

The first lesson concerns the goals of the WSIS follow up process. What should be the lasting legacy of WSIS? At various points in the process, some argued that it should ultimately result in a treaty for regulating cyberspace, or a charter specifying the obligations of developed countries toward the developing world, or a framework convention on Internet governance. The experience of the ITU—over the past decade as well as throughout its long history—suggests that goals of this kind are unrealistic and that the time is not right now for a single, overarching treaty instrument to govern global ICTs. We are clearly in a period of governance diversification and experimentation, similar in some respects to the first long cycle described at the beginning of this essay. This does not mean however that new, focused governance arrangements are not needed, even at the treaty level. A strong case can be made, particularly in light of the changes that have taken place since 9/11, for a convention on cybersecurity, and international arrangements are clearly needed to deal with spam and other abuses of the Internet. A strong case can also be made for the creation of innovative financing mechanisms, based on partnership between government and the private sector, to support the buildout of ICT infrastructure in some developing countries and regions. However, the WSIS process will only succeed if governance issues of this kind are treated separately and on their own merits. Attempts to bundle them into a comprehensive treaty-based package are bound to fail.

The second lesson concerns the need to align the form of governance arrangements with their function, and to avoid "one size fits all" approaches based either on the ideology of national sovereignty or the ideology of industry self-regulation. As well as analyzing the national capacity of developing countries to engage effectively in international ICT decision making, the *Louder Voices* report gave detailed study to the governance processes of the ITU, the WTO, and ICANN, to see how effectively they included developing countries, the private sector, and civil society organizations, both formally and in practice.[27] These three organizations provide very interesting points of comparison, in view of the differences that exist in their structure, functions, working methods, and culture. Neither the WTO nor ICANN is currently as inclusive as the ITU, since the former limits participation in its work to governments, in spite of the enormous impact of its decisions on the private sector and civil society, while the latter limits participation to the private sector, in spite of the strong interest many developing country governments have in its activities. Like the ITU, both of these organizations are seeking ways of becoming more inclusive within the framework of their founding ideologies, principally through the adoption of informal mechanisms that relax the constraints that flow from these ideologies. And like the ITU, both organizations are likely to find that it is not possible to become truly inclusive without more fundamental organizational changes that clearly separate public and private governance responsibilities, at the same time as they create greater synergies between them.

The third lesson confirms two of the principal findings of the *Louder Voices* study: first, that the key to strengthening developing country participation in the governance of global ICTs lies in building technical and policy capacity at the national and regional levels; and second, that without this capacity changes to the governance structures and decision-making processes of international organizations designed to create special spaces for developing countries may mean very little in practice, even if they are potentially valuable.[28] The ITU has a broad range of developing country participants from the public and private sectors, long experience in providing technical assistance, and a separate organizational sector devoted to development activities, and has imposed development obligations on its regulatory and standardization functions. In spite of this, developing countries are far from being fully included in the ITU's principal governance activities, fundamentally because they often lack the capacity to participate effectively at each and every stage of the governance process, which includes technology assessment, issue identification, agenda setting, policy formulation, coalition building, negotiation, policy implementation, and evaluation. Like the ITU, other organizations must direct a larger portion of their energies and resources toward the task of building these capacities, if they are truly serious about achieving the goal of inclusive governance of global ICTs.

Appendix

Transforming the ITU into a Global Telecommunications Organization

This chapter has argued that the time has come to break up the current ITU structure in order to create a much more loosely affiliated network of four organizations, each of which would assume one of the ITU's current governance roles and each of which would be governed, operated, and financed on the basis of arrangements tailored to its specific requirements. The organization of a transformed ITU is depicted in figure 2.1.

The following sections present a schematic design for each of the proposed new structures—in terms of their mission, organization, membership, finance, governance, management, role in providing development assistance, and founding resources.

World Telecommunications Standardization Forum (WTSF) Mission

Mission To develop technical standards for telecommunications equipment, networks, and services.

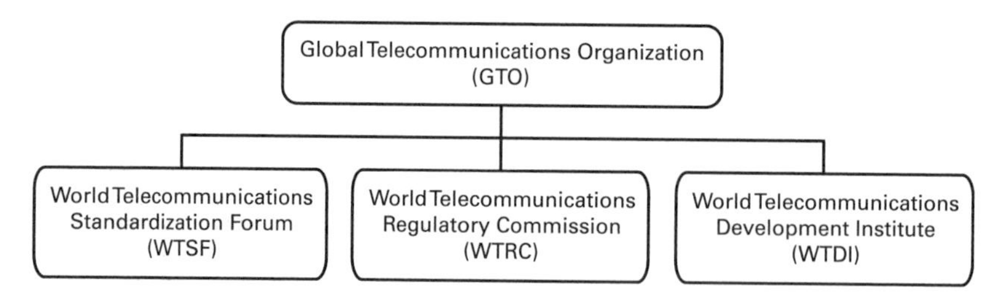

Figure 2.1
The ITU transformed

Organization A "forum of forums" or a "federation of forums" in which a forum is functionally similar to an ITU-T Study Group.
Membership Any entity prepared to pay the costs of joining and participating.
Finance Financed by forum members on a "bottom-up" basis (i.e., there is a fee for being a member of each forum and the total cost of membership = (fee/forum) × (number of forums).
Governance Overall governance provided by a periodic assembly of members, an enhanced version of the current ITU-T Assembly that would have planning and budgetary powers for the WTSF similar to those exercised by the Plenipotentiary Conference for the ITU as a whole under the current structure.

Ongoing governance provided by a council, an enhanced version of the current Telecommunication Standardization Advisory Group (TSAG) that would have planning and budgetary powers for the WTSF similar to those exercised by the ITU Council for the ITU as a whole under the current structure.

Standards approved by individual forums through a process similar to the current ITU-T Alternative Approval Procedure (APP).
Management Each forum headed by a director elected by forum members at the WTSF Assembly, whose responsibilities are similar to those exercised by ITU-T Study Group chairmen under the current ITU structure.

Management and administrative support provided by a secretariat headed by an executive director, elected by the WTSF Assembly and with responsibilities similar to those exercised by the ITU-T director under the current ITU structure.

Terms and conditions of employment for secretariat staff determined by the WTSF Conference and implemented by the WTSE Council.

Administrative support services contracted by the secretariat from the Global Telecommunications Organization (GTO) and other sources on a competitive basis.
Development Assistance WTSF work program and budget incorporates a technical assistance component that provides merit-based meeting fellowships and secretariat internships, as well as outreach activities done in collaboration with the International Telecommunications Development Institute (ITDI).
Founding Resources Startup funding for the WTSF from resources allocated to the ITU Telecommunication Standardization Sector (ITU-T) under the current structure.

World Telecommunications Regulatory Commission (WTRC)
Mission To regulate international telecommunications, including radio communications and whatever regulatory responsibilities result from the World Conference on International Telecommunications (WCIT), which will review the current International Telecommunication Regulations (ITRs) in the 2007–2010 time frame.

To consolidate the registry services provided by different organizational units under the current ITU structure (i.e., spectrum and satellite orbit use, numbering plans, Global Mobile Personal Communications Services (GMPCS).

Organization *Either* one or more conventions and annexes modeled on the practices of the International Civil Aviation Organization (ICAO), which would include a simplified radio regulation convention with annexes for different radio communication services as well as analogous instruments for regulating international telecommunications—if the next WCIT decides to maintain the ITRs in either their current or a modified form.

Or multiple conventions modeled on the practices of the International Maritime Organization (IMO), which would include a general radio regulation convention and individual conventions for specific radio communication services that would be developed by WTRC members on an ongoing basis (i.e., as a "commission of the whole")—as well as analogous instruments for regulating international telecommunications if the next WCIT decides to maintain the ITRs, in either their current or a modified form.

Membership Radio communication and telecommunication regulators from member countries.

Finance UN-style assessments on the basis of GDP or use of spectrum, orbital, and telecommunications resources, or both.

Cost recovery for coordination and registration of spectrum and orbital resource assignments and other registry functions.

Governance Overall governance provided by a periodic conference of members, an enhanced version of the current ITU-R World Radiocommunication Conference that has planning and budgetary powers for the WTRC similar to those exercised by the Plenipotentiary Conference for the ITU as a whole under the current structure.

In the ICAO-based model, ongoing governance provided by a commission whose members are elected by the WTR Conference, and that approves amendments to the WTRC convention annexes in addition to having planning and budgetary responsibilities for the WTRC similar to those exercised by the ITU Council for the ITU as a whole in the current structure.

In the IMO-based model, ongoing governance would be provided by a council that would be elected by the WTR Conference and have planning and budgetary responsibilities similar to those exercised by the ITU Council for the ITU as a whole in the current ITU structure.

Management Management and administrative support provided by a secretariat headed by an executive director elected by the WTR Conference, whose responsibilities are similar to those exercised by the ITU-R director in the current ITU structure.

Terms and conditions of employment for secretariat staff determined by the WTR Conference and implemented by either the WTR Commission (ICAO-based model) or the WTR Council (IMO-based model).

Administrative support services contracted by the secretariat from the Global Telecommunications Organization (GTO) Common Service Provider and other sources on a competitive basis.

Development assistance WTSF work program and budget incorporates a technical assistance component that provides merit-based meeting fellowships and secretariat internships, as well as outreach activities done in collaboration with the International Telecommunications Development Institute (ITDI).
Foundine resources Start-up funding for the WTRC from resources allocated to the ITU Radiocommunication Sector in addition to resource allocated to registry services in other sectors under the current ITU structure.

World Telecommunications Development Institute (WTDI)

Mission To build technical, policy, and regulatory capacity in developing countries.
Organization An institute similar in principle either to the World Bank Institute or to the many public, not-for-profit foundations, institutes, and centers that support development, with a global headquarters and regional offices focused on building capacity in light of local requirements.
Membership Open to government, private sector, civil society, research, academic, international, and intergovernmental entities and organizations.
Finance ITU-style annual voluntary contributions from government and nongovernment members.

Cost recovery for products and services (e.g., data, reports and publications, TELECOM conferences and forums, human resource and organizational development programs, workshops and seminars, consulting, software licensing).

Grants and contributions for specific projects from bilateral and multilateral donor agencies.

Endowments (wealthy individuals, major corporations).

Governance Overall governance provided by an annual meeting of stakeholders.

Ongoing governance provided by a board of directors elected at the annual meeting to represent the interests of stakeholder groups.

Voting and representation weighted in terms of financial and other contributions, with governments retaining a "golden share."

Management Chairman elected by the board of directors.

Executive director of the secretariat appointed by the board.

Terms and conditions of employment for staff determined by the board and implemented by the executive director.

Administrative support services contracted by the secretariat from the Global Telecommunications Organization (GTO) Common Service Provider and other sources on a competitive basis.

Development assistance In addition to overall development of technical, policy, and regulatory capacity, ITDI has a specific responsibility to build the capacity of developing countries to participate effectively in the activities of the World Telecommunications Standardization Forum (WTSF) and the World Telecommunications Regulatory Commission (WTRC).

Founding resources Startup funding for the WTDI from resources allocated to the Telecommunication Development Sector, the TELECOM secretariat, and the secretary-general's New Initiatives Program under the current ITU structure.

The Global Telecommunications Organization (GTO)

Mission Similar to the ITU mission as set out in article 1 of the current ITU Constitution and Convention.

Organization A simplified intergovernmental agreement setting out the purposes and structure of the GTO, including its relationship to the WTSF, the WTRC, and the WTDI, similar in style to the OECD or WTO founding agreements.

Membership Governments, the private sector, and civil society in a tripartite structure similar to the International Labour Organization (ILO).

Finance Cost of GTO governance activities funded through transfer payments ("shareholder dividends") from the WTSF, the WTRC, and the WTDI.

Cost recovery for administrative support services provided to these organizations under contract with the GTO.

Governance Overall governance provided by a periodic general conference of members.

Ongoing governance provided by a council elected by the general conference.

Representation and voting at the general conference and in the council weighted to reflect the GTO's tripartite membership (e.g., in an ILO-style model).

Governance helps steer the activities of the WTSF, the WTRC, and the WTDI by providing a higher-level, independent point of view on their activities—for example, through audit and evaluation activities that measure how efficiently these organizations are performing at the operational level, as well as how effectively they are anticipating and responding to change at the strategic level.

Management GTO activities are supported by a secretariat headed by a secretary-general, who is elected by the general conference and is responsible for overall organizational planning, evaluation, and audit, as well as for providing common administrative services to the WTSF, the WTRC, and the WTDI on a competitive basis.

Development assistance The GTO provides merit-based fellowships and secretariat internships.

Founding resources Startup funding for the GTO from resources allocated to the general secretariat under the current ITU structure.

Notes

1. See MacLean et al., *Louder Voices*, 10–18, for a mapping of the international ICT decision-making universe.

2. When all of these elements are present, this paper will speak of a "governance model." When only some are present, it will refer to "governance arrangements." It will also divide the elements

that make up a full governance model into two groups: a "policy and action framework," which includes goals, principles, participants, and activities; and an "institutional framework," which includes legal instruments, organizational structures, procedures, and working methods.

3. See Michaelis, *From Semaphore to Satellite*, for a useful account of the relationship between technological and governance innovation from the telegraph to the satellite. While there are many sources of information on the governance issues raised by more recent technological developments, the ITU's *World Telecommunication Development Reports* and *Internet Reports*, which are issued on a periodic basis, provide comprehensive, accessible overviews of the links between technological developments, economic and social development, and governance issues at the national and international levels. See the ITU web site, www.itu.int for information on these publications.

4. The term *long cycle* is used in a metaphorical sense in this essay. It is not intended to advance or support any particular economic or social theory.

5. See Codding, *The International Telecommunication Union*, for the early history of the ITU.

6. See Codding and Rutkowski, *The International Telecommunication Union in a Changing World*, and Savage, *Politics of International Telecommunications Regulation*, for accounts of the ITU during this transitional period.

7. See Jussawalla, *Global Telecommunications Policies*; Nordenstreng and Schiller, eds., *Beyond National Sovereignty*; Melody, *Telecom Reform*; and Hudson, *Global Connections*; for contrasting views of these changes.

8. See ITU, *The Missing Link*.

9. For an analysis of the history of the ITU from the perspective of regime theory, see Drake, "The Rise and Decline of the International Telecommunications Regime"; and "Communications."

10. See Gore, "Remarks Prepared for Delivery at the World Telecommunication Development Conference."

11. See G7, *Chair's Conclusions*.

12. See ISAD, *Chair's Conclusions*.

13. See ITU, 2003a. *Collection of the Basic Texts*.

14. See www.itu.int/ITU-D/ for an overview of the activities of the ITU Development Sector.

15. See e-com.ic.gc.ca/epic/internet/inecic-ceac.nsf/en/gv00133e.html for information on the work of the G8 DOT Force and www.unicttaskforce.org for information on the work of the UN ICT Task Force.

16. See www.itu.int/wsis for the results of the World Summit on the Information Society.

17. See ITU, note 13.

18. See www.itu.int/GlobalDirectory/index.html for a list of ITU member states and sector members.

19. See MacLean et al., *Louder Voices*, 19–25.

20. See http://www.itu.int for an overview of the ITU's federal structure.

21. See Doran, *Middle Powers and Technical Multilateralism*.

22. The Plenipotentiary Conference is the ITU's supreme governing body. It meets once every four years to adopt a strategic and financial plan for the next plenipotentiary period; amend the ITU Constitution and Convention (i.e., the basic treaty instrument); and adopt Decisions, Resolutions, Recommendations, and Opinions on specific policy and administrative issues. It also elects the members of the ITU Council, which governs in the period between Plenipotentiary Conferences; the secretary-general, the deputy secretary-general, and the directors of the Radiocommunication, Standardization, and Development Bureaus; and the members of the Radio Regulations Board.

23. See MacLean, "A New Departure for the ITU" and "Open Doors and Open Questions," for an analysis of the decisions of the 1994 and 1998 Plenipotentiary Conferences with respect to the ITU reform agenda.

24. See www.itu.int/osg/spu/ni for information about the ITU secretary-general's New Initiatives Program.

25. See Hudson, *Global Connections*, 423–426, for a succinct account of this very complex and longstanding problem.

26. In the ITU financial system, member states are free to choose the number of units they wish to contribute to the ITU budget from a scale running from forty units at the high end to one-sixteenth of a unit at the low end, instead of being assessed contributions on the basis of national wealth or some other measure of capacity to pay. Sector members contribute to the budget on the same basis, although the monetary value of a sector member unit is only one-fifth the value of a member state unit, a ratio related to the fact that sector members do not have the right to vote and other, equally intangible considerations. The current value of a member state unit is 315,000 Swiss francs, while the value of a sector member unit is 63,000 Swiss francs. Member states currently contribute about 65 percent of the ITU budget and sector members contribute about 13 percent. The remainder comes from the sale of publications, cost recovery for certain activities, and miscellaneous sources of income. With very few exceptions, developed countries contribute less to the ITU under the free choice system than they would under a UN-style system of assessment on the basis of capacity to pay, while developing countries and LDCs contribute more. The contributions of sector members cover only a portion of the cost of supporting their activities in the ITU.

27. MacLean et al., *Louder Voices*.

28. MacLean et al., *Louder Voices*, 26–28.

References

Codding, George A. Jr. 1952. *The International Telecommunication Union: An Experiment in International Cooperation*. Leiden: E. J. Brill.

Codding, George A. Jr., and A. M. Rutkowski. 1982. *The International Telecommunication Union in a Changing World*. Dedham, MA: Artech House.

Doran, Janis. 1989. *Middle Powers and Technical Multilateralism: The International Telecommunication Union*. Ottawa: The North-South Institute.

DOT Force (Digital Opportunity Task Force). 2002. *Digital Opportunity for All*. Available at e-com .ic.gc.ca/epic/internet/inecic-ceac.nsf/en/gv00133e.html.

Drake, William J., ed. 1998. *Telecommunications in the Information Age*. Washington, DC: United States Information Agency.

Drake, William J. 2000. "The Rise and Decline of the International Telecommunications Regime." In *Regulating the Global Information Society*, ed. C. T. Marsden, 124–177. London: Routledge.

Drake, William J. 2001. "Communications." In *Managing Global Issues: Lessons Learned*, ed. P. J. Simmons and Chantal de J. Oudraat, 25–74. Washington, DC: Carnegie Endowment for International Peace.

G7 (G7 Summit "Information Society Conference"). 1995. "Chair's Conclusions." Retrieved February 28, 2006, from europa.eu.int/ISPO/docs/promotion/past_events/isad_conclusioin.doc.

Gore, Albert. 1994. "Remarks Prepared for Delivery at the World Telecommunication Development Conference, Buenos Aires." Available at www.itu.int/itudoc/itu-d/wtdc/wtdc1994/speech/gore.ww2.doc.

Hudson, Heather. E. 1997. *Global Connections: International Telecommunications Infrastructure and Policy*. New York: Van Nostrand Reinhold.

ISAD (Information Society and Development Conference). 1996. "Chair's Conclusions." Available at www.europa.eu.int/ISPO/docs/promotion/past.events/isad.conclusion.doc.

ITU (International Telecommunication Union). 1984. *The Missing Link, Report of the Independent Commission for World Wide Telecommunications Development*. Geneva: ITU.

ITU (International Telecommunication Union). 1989. *The Changing Telecommunication Environment: Policy Considerations for the Members of the ITU, Report of the Advisory Group on Telecommunications Policy*. Geneva: ITU.

ITU (International Telecommunication Union). 1991. *Tomorrow's ITU: The Challenges of Change, Report of the High Level Committee to Review the Structure and Functioning of the International Telecommunication Union*. Geneva: ITU.

ITU (International Telecommunication Union). 2003a. *Collection of the Basic Texts of the International Telecommunication Union Adopted by the Plenipotentiary Conference, Edition 2003*. Geneva: ITU.

Jussawalla, Meheroo, ed. 1993. *Global Telecommunications Policies: The Challenge of Change*. Westport, CT: Greenwood Press.

MacLean, Donald J. 1995. "A New Departure for the ITU: An Inside View of the Kyoto Plenipotentiary Conference." *Telecommunications Policy 19* (April): 177–190.

MacLean, Donald J. 1999. "Open Doors and Open Questions: Interpreting the Results of the 1998 Minneapolis Plenipotentiary Conference." *Telecommunications Policy* 23 (March): 147–158.

MacLean, Donald, David Souter, James Deane, and Sarah Lilley. 2002. *Louder Voices: Strengthening Developing Country Participation in International ICT Decision-Making*. London: Commonwealth Telecommunications Organisation and Panos London. Available at http://www.cto.int/publications/louder_voices_finalreport.pdf.

Mansell, Robin, and Uta Wehn, eds, 1998. *Knowledge Societies: Information Technology for Sustainable Development*. New York: Oxford University Press.

Melody, William H., ed. 1997. *Telecom Reform: Principles, Policies and Regulatory Practices*. Technical University of Denmark, Lyngby: Den Private Ingeniorfond.

Michaelis, Anthony R. 1965. *From Semaphore to Satellite*. Geneva: ITU.

Nordenstreng, Kaarle, and Herbert I. Schiller, eds. 1993. *Beyond National Sovereignty: International Communication in the 1990s*. Norwood, NJ: Ablex Publishing.

Savage, James G. 1989. *The Politics of International Telecommunications Regulation*. Boulder, CO: Westview Press.

3 Balancing Equity and Efficiency Issues in Global Spectrum Management

Rob Frieden

Radio communication spectrum can have great value[1] when serving as the physical means for transmitting desirable content. For example, broadcasters of video programming via terrestrial and satellite networks have accrued hefty stock market valuations in light of their ability to deliver popular content to a large audience. On the other hand spectrum with less geographical reach and bandwidth along with shared accessibility may have little value particularly if a high potential for harmful interference exists.

Radio spectrum has an intangible characteristic much like the air, but also provides a delivery medium for content conferring measurable social and commercial benefits.[2] For spectrum having great value National Regulatory Authorities (NRAs) must determine how best to achieve multiple, possibly conflicting goals[3] that include

- establishing criteria for allocating spectrum uses and deciding who can use spectrum;
- considering whether and how market forces, instead of government decision making, can establish spectrum uses and users;
- finding ways to capture for public benefit at least some of the intrinsic value of spectrum usage;
- dividing spectrum into freely available, shared "commons"[4] and allocations exclusively available to specific licensees; and
- avoiding or resolving conflicts between spectrum users within a country and users in different countries.

Access to radio spectrum resources triggers both equity and efficiency concerns.[5] On the equity side, arguably all citizens in a nation have an ownership interest in and right of access to the spectrum resource. Likewise, two or more nations seeking to use the same portion of the spectrum or the same orbital parking place for a communications satellite may have equally compelling access claims.[6] Most nations consider radio spectrum and satellite orbital slots as resources obligating negotiated sharing arrangements that accommodate as many uses as possible without increasing the potential for interfering uses and higher operational costs.[7]

Most nations have signed treaties foreclosing national ownership or the exercise of sovereignty[8] over space resources,[9] including orbital slots used by communications satellites.[10] Likewise, most nations have signed treaties that bind them to administrative rules and regulations that allocate spectrum for specific uses, establish a process for nations to register uses, and help preempt or resolve disputes.

However, a nation's equal right of ownership and access does not translate into an equivalent level of actual access, because developed nations have acted on their earlier needs for spectrum and satellite orbital slots and have locked up much of the best resources. A global consensus favors international rules that emphasize efficient use of shared radio communication resources by conferring a "first registered, first protected" priority status. This procedure provides a priori specificity about a pending spectrum use, so that even before actual usage private or public enterprises can have assurance that their near-term spectrum use can occur free of harmful interference caused by other existing or future spectrum uses. Developed nations disproportionately benefit in light of their typically earlier registrations, based on existing spectrum requirements and the financial wherewithal to construct and operate the facilities using the registered spectrum.[11] Advocates for an a priori registration system believe that the earliest possible use for spectrum resources will lead to enhanced productivity and welfare.

Many less developed nations consider unfair a first-in-time, first-in-access priority system. These nations do not see optimal global benefits accruing when developed nations can lock up the best spectrum and satellite orbital slots well before less developed nations can generate the demand and financial resources needed to act on a spectrum registration.[12]

Policy makers at both national and multilateral levels need to forge a compromise between equity and efficiency. If they fail, stakeholders may lose patience in the process and resort to unilateral, self-help strategies that could include launching satellites and using spectrum without participating in the multilateral spectrum allocation and registration process. Likewise, individual nations may ignore global policies regarding the preferred uses for a specific frequency band and the procedure for registering and coordinating the launch of satellites into orbit. The absence of a consensus-driven spectrum allocation and orbital slot registration process would raise costs that telecommunication operators and their consumers must bear, because the potential for interference would increase drastically.

This chapter examines the merits of managing spectrum allocations and registrations through a multilateral process organized by the International Telecommunication Union (ITU), a specialized agency of the United Nations. The chapter covers the conflict avoidance and resolution capabilities of the current system and considers the prospect for greater efficiency when prospective spectrum users have to compete in auctions for the right to use spectrum and satellite orbital arcs. The chapter concludes that policy makers need to embrace best practices in spectrum management, including compulsory conservation measures. However, the existing multilateral and intergovernmental coordination process remains a better option than resorting to a fully market-driven system lacking governmental involvement and safeguards.

Longstanding Issues

Efficiency Trumps Equity in the Current Reassessment

Current shortages of spectrum for specific uses, such as wireless access telephony, the Internet, and video program delivery, have prompted governments in both developed and developing nations to rethink how best to manage spectrum. Advocates for marketplace decision making have suggested that NRAs assign spectrum usage rights to the highest bidder, at least for services using frequency bands having limited geographical coverage that would not cross national boundaries. Market advocates point to the overall benefits to society when potential users of public resources have to bid competitively for ownership.[13]

Technological innovations have reduced the potential for harmful interference, making it possible for previously incompatible spectrum uses to occur in the same frequency band and in close geographical proximity.[14] Digital transmissions can be compressed and coded in ways that reduce the amount of spectrum used and facilitate expanded, interference-free communications. Digital signal processing and software-defined radio also provide ways for more simultaneous uses without interference.

For spectrum uses that do not cross borders, marketplace initiatives have great appeal and plausibility, particularly in light of the ability of single nations to implement competitive bidding without adversely impacting spectrum uses and policies in other nations. Absent a cross-border impact, individual nations can choose to foster efficiency, generate revenues for the national treasury, and accrue value for all citizens instead of the lucky few who previously secured spectrum license grants without payment.[15]

Cross-Border Spectrum and Satellite Uses Amplify Equity Concerns

Spectrum uses that cross borders trigger a greater potential for interference and therefore raise more pressing equity and national sovereignty concerns. Stakeholders should agree that one nation cannot foreclose or adversely impact another nation's spectrum access and use opportunities without offering compensation or affirmative efforts to ameliorate the harm. Most nations have signed treaty-level documents that recognize control over spectrum as an element of national sovereignty, but that relinquish some portion of national self-determination based on the view that a multilateral system of coordination will achieve a better outcome.[16] For spectrum uses that routinely operate across borders, such as a communications satellite that can deliver video programming to as much as one-third of the earth's surface, market countervailing equity concerns become more compelling.

ITU rules require nations to share radio communications resources and to coordinate uses. Accordingly, no single nation or private venture can act unilaterally, because most nations have committed to a multilateral approach for shared access to radio communications resources. Any decision by one nation to launch a communications satellite, or to activate a new radio communication facility can have a direct and potentially adverse impact on other nations, particularly ones nearby.

The limitations of physics, politics, jurisdiction, and international treaties complicate the process for sharing. They refute the simple assumption of some analysts that shared global radio communications resources are identical to real estate such that treaties between governments should "extend the property rights system ... into the international realm."[17] Simply analogizing radio communications resources to real estate ignores the fact that privatizing spectrum can foreclose access rights of others, including citizens in nearby countries. A nation can recognize private property ownership rights and can establish binding and effective rules for buying and selling real estate within the country. However, such a marketplace system for radio communications resources would involve the application of rules on an extraterritorial, cross-border basis.

Conflict Avoidance and Resolution through Multilateral Coordination

For radio spectrum and satellite orbital slots, nations collectively and individually have rejected either an absolute market-driven or an equity-driven model for allocating, coordinating, and registering usage. Nations cannot simply occupy spectrum on an as-needed basis, but neither can they expect to have spectrum reserved for their possible future use.

On a multilateral basis, nations look to the ITU[18] to erect and administer mechanisms that balance the conflicting interests of developed and developing nations. While lacking an enforcement mechanism, the ITU provides a forum for both efficiency and equity arguments.[19]

ITU decisions typically occur without a formal vote because the forum seeks to build consensus. However consensus may not easily occur in light of developed countries' disproportionately greater spectrum requirements and their reluctance to conserve bandwidth or to use expensive spectrum-conserving technologies based on speculative, future requirements of developing countries. Furthermore, developed nations may balk at the ITU's system of prioritizing specific uses of spectrum based on a then-current assessment of which services can use which spectrum. With the onset of new sophisticated technologies, an increasing number of previously incompatible spectrum uses and a larger volume of users can operate without interference. Notwithstanding such innovation, the ITU typically allocates only a handful of services for any particular frequency band. Additionally the ITU specifies a primary use, followed by secondary and tertiary uses that must not cause interference even if registered before a primary use.

Enlightened self-interest prompts nations to relinquish a degree of self-determination on spectrum matters. Reaching consensus typically will lower operating costs of transmitters and receivers, speed commercial rollout of new technologies, reduce confusion, and curb the potential for harmful interference. When nations fail to reach a consensus at the ITU, manufacturers may have to produce multiple equipment product lines and consumers may incur higher costs, as has occurred in mobile telephony where no single cellular telephone can operate at all locations. Each nation has to conform its domestic spectrum allocations and license grants to the ITU consensus. Absent an ITU-generated consensus, nations may grant licenses for incompatible uses of the same spectrum and thereby generate greater risk, cost, and congestion for any existing or prospective spectrum use.

The ITU has established shared "rules of the road" with nations generally agreeing on what services shall receive preferential registration rights for particular slivers of allocated spectrum and which nation's satellite has the right to occupy a particular orbital slot. The ITU model provides an effective administrative vehicle to register spectrum and orbital slot usage and to mediate disputes. However it may not prevent "warehousing" of spectrum, that is, registering unneeded uses that can foreclose others, who have near-term requirements, from achieving conflict-free registrations. Also the ITU administrative process cannot foreclose attempts to register "paper satellites," that is, securing orbital slots for satellites with no realistic probability of launch.[20] The ITU has attempted to create impediments to unneeded registrations by imposing financial filing fees and time deadlines for using registered spectrum and orbital slots.[21] However, with the rare exception of prospective satellites offering direct to home video programming in developing nations, the ITU does not deviate from the first-filed, first-registered model by reserving spectrum and orbital slots for future use by operators in developing nations.[22]

Pooling Investment in International Satellites to Promote Access

A possible solution to the inferior, delayed access problem encountered by developing nations may lie in regional coordination by several nations to aggregate funds and telecommunications service demand so that spectrum and orbital slot registrations may proceed earlier than if a single nation sought to construct and operate a costly network.[23] Such pooling of investment helped commercialize satellite technology by making it affordable for civilian use soon after deployment for military, space exploration, and intelligence gathering applications. Multilateralism promoted widespread access to satellite technology, including the opportunities for developing nations to participate in the ownership and management of a global satellite carrier. The pooling of investment among nations helped spread technological and financial risk across a larger group of participants.

Satellite investment pooling by many nations began through the application of a cooperative model similar to one frequently used in agriculture. Where one unit of a product has little if any difference from another unit, as is generally the case with satellite transponder bandwidth, this fungible characteristic supports demand and supply aggregation by numerous producers. For example, all the dairy or wheat farmers in a region can pool their investment and establish a local processing and storage facility for handling all of the farmers' output. This facility helps the farmers fetch the best possible prices for their products and also creates the possibility for some degree of value-adding processing, such as milling the wheat and pasteurizing the milk.

Satellite cooperatives likewise aggregated demand, making it possible to reach efficient scale and scope and to offer a large inventory of transponder capacity. The International Satellite Organization (INTELSAT),[24] European Satellite Organization (Eutelsat), and International Maritime Satellite Organization (Inmarsat) helped expedite the commercialization of space radio communications and made it possible for developing nations to participate with a small initial investment.[25] The satellite cooperatives had charters, negotiating on a multilateral, intergovernmental basis that emphasized the promotion of world peace and understanding through widespread access and use of satellites. These cooperatives operated as businesses, but had missions that emphasized access and service instead of profit maximization.

The global and regional satellite ownership model helped make it possible for developing nations to afford satellite network access. Even the poorest and smallest nations could connect to a global satellite constellation with a minor investment in the cooperative. Expanding membership to include large numbers of developing nations accrued political benefits for the cooperative, but it also made it possible for these nations to acquire satellite capacity at an affordable rate.

However, this model lost most of its financial and policy support as satellite technology evolved and as a competitive marketplace for satellite services developed. Over time the cost of constructing, launching, and operating a satellite network dropped substantially as demand grew for satellite services, particularly delivery of video content to broadcast and cable television networks. Private entrepreneurs saw an opportunity to enter the market. However, incumbent satellite operators, such as INTELSAT, sought to thwart such competition on the grounds that it would cause economic harm and hinder the cooperatives' ability to serve high-cost, rural locations and to facilitate investment and participation by developing nations.

Facing the prospect of facilities-based competition, the satellite cooperatives acted much like cartels intent on preserving their market dominance. Nations with government or private carrier investors in satellite cooperatives executed treatylike documents designed to confer special privileges and immunities so that the cooperatives could achieve their mixed business and political mission. This special status helped reduce the cost of setting up and operating the cooperative, but it also created a semidiplomatic organization insulated from many marketplace forces. For example, the creators of INTELSAT sought to ensure that the cooperative would capture most of the global telecommunications traffic by agreeing not to authorize separate international satellite operators that collectively would cause economic harm to the cooperative. Additionally, the cooperatives and their employees enjoyed special exemptions from tax and other domestic charges imposed by the nation where the cooperatives had their headquarters and where they acquired and launched satellites.

In the mid-1980s, the United States, followed by other nations, permitted commercial satellite alternatives.[26] These systems have achieved marketplace success without adversely affecting the ability of satellite cooperatives to achieve their mission. Management of these cooperatives, having failed in their bid to block competition, turned their attention to finding ways to compete more effectively. In the spirit of "if you can't beat them, join them," the managers of INTELSAT, Inmarsat, and Eutelsat sought to privatize and to become commercial ventures.[27] Efforts were undertaken to ensure "lifeline" access by nations unlikely to benefit from competitive satellite markets.[28]

Current Issues

To understand the reasons behind the push for market resource allocation alternatives, one should consider the strengths and weaknesses in the ITU spectrum management process. At its best, the ITU has balanced equity and efficiency considerations with shared "rules of the road" that have reduced costs by coming up with a global consensus on spectrum allocations and mostly conflict-free spectrum and satellite orbital slot registrations. At its worse, the ITU has forestalled introduction of new technologies and services, helped extend developed nation domination of spectrum and orbital slots, and failed to prevent gaming and manipulation of the registration process.

Paper Satellites

The ITU has not completely closed the loophole that permits the registration of unviable "paper satellites." Another loophole permits a venture in one nation, for example, the United States with its long queue of pending registrations, to secure a favorable and earlier satellite orbital arc registration by using another nation that offers a "flag of convenience" and has no registration backlog. ITU rules do not authorize member nations and their public or private ventures to negotiate financial inducements as a way to resolve more speedily interference and access conflicts. Additionally, the ITU still locks in assumptions about existing technology and the ability of spectrum to accommodate multiple uses by allocating spectrum in service-specific blocks.

Paper satellite filings provide an example of self-help retaliation against the real or perceived inequity in the ITU administrative process. Nations lacking the commercial demand or financial resources to construct, launch, and operate their own satellite network nevertheless can exploit the ITU orbital slot registration system to extract compensation, or at least to vent their frustration by causing processing delays. For example, between 1988 and 1990 the nation of Tonga attempted to register sixteen orbital slots.[29] The principals of Tongasat, the private venture for which the Tonga government filed the satellite registration applications, made several publicized overtures to incumbent satellite operators offering to relinquish some or all of the attempted registrations in exchange for financial compensation. One could easily infer that the consultant advising Tonga's government knew that the ITU registration process lacked sufficient financial and procedural prerequisites, due diligence requirements, and benchmarking that might have stymied paper applications. Only recently has the ITU imposed registration fees to compensate it for the costs incurred in processing a proposed orbital slot registration.[30] The ITU still lacks fully effective due diligence standards and a timetable of deliverables that would remove access rights from registrants who have not demonstrated progress toward a timely satellite launch and spectrum use. Absent these safeguards, even a nation with absolutely no ability to launch satellites could have representatives claim a right to coordinate future interference-free operation of these paper satellites with existing and future satellites operating in the region.

Arguably the instigators of Tonga's satellite registrations had a mind to create a private auction. As a sovereign nation and member of the ITU, Tonga had the right to use the ITU's registration services. While guilty of seeking to register satellites it most likely never would launch, Tonga violated only the spirit of the process. Under current conditions it would take only a few more such registration applications for the ITU administrative process to implode. However, as the first mover in this strategy Tonga staked out orbital slot claims and in effect tried to create a market to be bought out of these claims. In view of Tonga's efforts to "monetize" its orbital stake claims, one can anticipate future scenarios in which more nations might try to extract sizeable nuisance payments or perhaps free or discounted satellite capacity in exchange for abandoning their claims. ITU member nations including Gibraltar and Papua New Guinea have offered to handle the ITU registration process for satellite network proposals based in the United States, but designed to provide services to many nations including the sponsoring registrant.

Spectrum Block Allocations

In addition to encouraging self-help and trafficking, the current ITU spectrum and orbital slot registration process can hamper the efficient, timely, and flexible use of radio communication resources. The ITU spectrum allocation process typically imposes an international template on spectrum uses and registration. The ITU uses a block allocation method for allocating spectrum that subdivides useable spectrum into service-specific slivers of priority use. While the international allocation constitutes a recommendation without the force of a treaty, and nations may opt out by taking a "reservation" to any specific allocation, most domestic regulatory agencies implement the ITU consensus decision.[31]

Allocating spectrum in blocks limits user flexibility, because technological innovations enhance the ability of users to share spectrum even for different services.[32] Heretofore, a consensus decision on spectrum uses reduced costs, promoted single equipment production lines, enhanced connectivity across borders, and supported single or compatible operating standards.[33]

Now the use of service-specific blocks of spectrum can reduce efficiency, flexibility, and the value of spectrum. For example, transceiver miniaturization now makes it possible to use the same satellite radiotelephone when on land, in an aircraft, and on the high seas. However, before the onset of this innovation, the ITU established separate spectrum allocations for land mobile, aeronautical, and maritime satellite mobile services based on the then-appropriate assumption that satellite terminals would remain fixed in only one of the three different locations. The ITU has not yet fully acknowledged the newly achieved ability to use a satellite handset across the three different operating environments, thereby maintaining an unnecessary limitation on the range of frequencies available.

Problems in Competitive Bidding and Spectrum Congestion Remedies

Despite having secured a nearly total national commitment to spectrum sharing, ITU rules do not foreclose individual nations from assigning spectrum based on competitive bidding. Even with glowing endorsements from economists, spectrum auctions and technological innovations do not completely remedy the deficiencies inherent in licensing spectrum for exclusive, private uses. Advocates for "propertizing" spectrum scoff at lofty but vague notions of the public interest and national security, claiming that the "public interest" with respect to the use of spectrum is a vague, ill-defined concept. In their view, under the "public interest" banner the U.S. Congress and the FCC have established far too many protectionist, anticompetitive, anti-innovative, inflexible, output-limiting regulatory regimes.[34]

However defective in implementation, serving the public interest can achieve desirable social outcomes. Put another way, implementation of a marketplace resource allocation can frustrate efforts to achieve social goals. Additionally, in this age of heightened concerns about terrorism and national security it comes across as rather cavalier[35] to suggest that because everything is scarce, government defense, intelligence gathering, homeland security, and public safety agencies should pay for spectrum along with everyone else.[36]

Advocates for competitive bidding also have to recognize the mixed record generated so far. Because governments successfully tailored auctions to extract maximum revenues, in several instances winning bidders could not produce even partial payment, thereby triggering a default and delayed use of heeded spectrum. Conflicts between bankruptcy law and communications law in the United States[37] have resulted in uncertainty whether the defaulted spectrum awards remain in the bankruptcy estate, administered by a court of law, or become available for reauction by the FCC.

Competitive bidding for spectrum earmarked for third-generation mobile telephone and high-speed data services have generated astronomical amounts in the United States and EU nations. However, a change in the overall marketplace attractiveness of telecommunications and information processing ventures has jeopardized operators' ability to recoup amounts bid for spectrum in a timely manner. Spectrum bidding has so raised debt exposure and risk that even blue-chip incumbent telecommunications ventures have incurred significant downgrades in the quality classification of their debt, thereby raising their cost of raising capital. Similarly the substantial near-term increase in debt and financial losses has a direct and substantial impact on the long-term tax liability incurred by these firms.

While competitive bidding advocates may emphasize the potential for scale economies in the ability of firms to aggregate spectrum, opponents note the likelihood for concentration of ownership and control, particularly if regulators waive or eliminate caps on the total amount of spectrum a single operator can control. Spectrum auction advocates note the potential that smaller parcels of spectrum property might become available on the market in the same manner that large parcels of real estate become subdivided.[38] But for spectrum, and in particular highly contested spectrum such as that allocated for third-generation wireless services, the more likely outcome would be zealous consolidation of ownership to achieve a national coverage "footprint." Economists may herald the potential for an up-to-the-minute "spot" market for spectrum, as well as a "secondary" resale market,[39] but such marketplaces have not yet developed to any significant degree even for largely fungible minutes of long-distance calling capacity, or for broadband links between nations. Much of the technological innovations supporting flexible spectrum usage would have to be in place for a spectrum spot market to exist, because access would shift between and among many users in different locations.

The existing spectrum bidding regime in developed nations coupled with corporate mergers and acquisitions and the reluctance of some NRAs to cap the amount of spectrum availability to any single enterprise already have resulted in substantial consolidation and concentration in telecommunications markets. Economists might argue that under a property rights regime, ample spectrum might be converted to mobile radio use to abate allocational scarcity created by the previous ITU and NRA decisions. Again, this ease in conversion presupposes that frequency agile transceivers and other cutting-edge technological innovations will become standard equipment in the near term. This assumption might not prove true, particularly where a developed nation lies physically next to one or more developing countries.

Spectrum auction advocates correctly note that completely domestic spectrum uses would not require coordination with or transfer payments to other nations. However, these advocates underestimate the percentage of spectrum uses that can cross borders. Likewise they do not seem to recognize that nearly all spectrum uses by satellite and all orbital slot occupancies could trigger claims of potential conflicting uses between nations. Arguably each and every nation lying under a satellite footprint might have an ownership claim, just as they now have a right to participate in ITU-administered spectrum and orbital slot coordination with a new usage registrant. Equatorial nations that failed in staking an ownership claim or grew weary of ITU coordination lacking a financial payoff, surely would have renewed vigor if they could extract compensation in exchange for relinquishing their ownership rights.

Extending the competitive bidding process on domestic spectrum to international satellite orbital arc usage has the potential to increase market entry costs substantially. It also could trigger delays in the launch of new satellites until each and every country possibly served by the satellite receives compensation, or otherwise abandons ownership claims to the spectrum used by the satellite as well as its orbital slot.[40]

Property Ownership Violates the Prohibition on National Appropriation of Shared Global Resources

Currently, nations secure priority access to spectrum and satellite orbital arc by successfully maneuvering through the ITU registration process. Priority access results from the voluntary acquiescence of nations without the abdication of possible future access. Furthermore, such access does not constitute an ownership claim, the assertion of jurisdiction over a shared global resource, or a usurpation of another nation's equal right to secure priority access rights through the ITU registration process. The ITU has successfully brokered complex and conflicting spectrum and satellite orbital slot claims, albeit with delays and its increasing need for compensation to shore up its budget. Brokering ownership interests may shorten the time to resolution, even as it raises new complexities.

Empowerment Opportunities for Developing Nations

Notwithstanding increasing stress on multilateral policy making and conflict resolution, new but costly technological innovations offer ways to abate spectrum congestion and interference. Perhaps one way to balance efficiency and equity concerns would be to require developed nations to implement spectrum conservation technologies on an expedited basis, thereby freeing spectrum for use by operators in developed nations. In recognition of their limited access to capital, spectrum users in developing nations might receive a temporary waiver of the requirement to use costly spectrum conservation technologies.

It should come as no surprise that stakeholders, regardless of national residence and wealth, seek access to spectrum at the lowest cost, but also with the greatest degree of certainty of noninterference. The ITU registration process, while not infallible, has provided a degree of certainty that all nations will respect a previously recorded spectrum use. Perhaps developing nations might forego access to spectrum and orbital slots, already in the ITU registration process, and use less desirable orbital slots and spectrum in exchange for compensation, or other types of accommodation.

Because they cannot readily vie for auctioned spectrum, operators in developing nations will seek to maintain the status quo ITU registration process, and possibly opportunities to extract compensation in exchange for not delaying other nations' registrations. One should not underestimate the potential for developing nations to find their voice and extract concessions. They hold a voting majority at the ITU and while most matters do not result in a formal vote, these nations can shape the debate. In light of the fractious nature of previous debates on transborder data flow, and north/south inequality, developed nations should take pains to avoid triggering another round of disputes by refraining from pressing too aggressively for market-based initiatives, or for mandatory use of spectrum conservation technologies.

The degree to which a national government has confidence in marketplace resource allocation constitutes a key factor in the nature of spectrum management for that nation. Nations having confidence in the ability of markets to operate and to maximize private and public benefits typically have a greater propensity to license spectrum through competitive bidding. Nations with less confidence in the utility of market-driven spectrum use, or those that have experienced dissatisfaction with their initial spectrum auction, appear more inclined to use older regulatory models highlighting government oversight.

Recommendations

Developing nations need to learn the best practices in spectrum management. To offset early-mover advantages, developing nations should seek to extract concessions from developed nations in ways that do not come across as extortion.[41] Historically, developing nations have gained little from rhetoric, grand unilateral proclamations, or the private auction strategies of single nations such as Tonga. It does not appear that the ITU will expand the set of frequencies and services reserved for future use by developing nations in the face of compelling current demand by users in developed countries.

The best strategy for disenfranchised nations appears to combine active participation in the ITU process with resumption of cooperative investment pooling and extraction of financial and technological concessions from developed nations. Developing nations should take every opportunity to include equity issues in ITU deliberations, but in a way that does not block progress. Users in developed countries can conserve spectrum and reduce the potential for interference by implementing technological remedies that users in developing nations cannot afford to implement immediately. Accordingly, the ITU might establish a recommendation that couples additional spectrum and satellite orbital slot registrations with an affirmative duty borne by developed nation stakeholders to expedite the use of spectrum conservation technologies.

Another option might involve the partial adoption of market resource allocation techniques. Instead of auctioning off spectrum, nations could engage in the trading of access rights. Pollution abatement strategies provide a helpful case study for spectrum management. Because developed nations cause the most pollution, their producers might pay for the right to exceed a pollution threshold in lieu of having to bear the expense in reducing harmful emissions. In the United States, some less developed states generate comparatively less pollution than other more industrialized states. The United States Environmental Protection Agency and several individual states allow manufacturers and other enterprises the opportunity to secure additional pollution "rights" from other enterprises that generate comparatively less pollution, whether through reduced production or pollution abatement investments.[42] Perhaps a similar sort of transfer payment mechanism might flow from developed nations seeking more telecommunications resources to developing nations unable or lacking the need for spectrum.

Conclusion

When nations apply best practices in spectrum management they can achieve greater operating efficiencies, accommodate more users, and often generate significant new revenues from auctions and other fees. However, best practices also should consider public interest factors that militate against total reliance on marketplace forces to allocate and assign spectrum. New technologies make it possible to conserve spectrum and diversify the number and type of users, but long unresolved issues of fairness and cost persist.

Developed nations with earlier demand for spectrum also have the wealth to pay for spectrum conservation technologies. Compulsory application of spectrum conservation techniques can improve the odds that developing nations will have accessible spectrum at a later date. Similarly, developed nations should respect the sovereignty of nearby nations and make significant efforts to reduce the prospect of future interference caused by transborder spectrum uses.

Technological innovations offer better opportunities than ever for nations to enjoy robust, interference-free spectrum use. To achieve this outcome, developed nations must incur the cost of using spectrum conserving technologies and eschew the simplistic view that all spectrum and satellite orbital slots should flow to the highest bidder.

Notes

1. The fact that parties have bid billions of dollars for the privilege of using spectrum attests to its intrinsic value. The United States Federal Communications Commission has captured billions of dollars for the general treasury from the spectrum auctions it has administered. See United States Federal Communications Commission, undated.

2. Compare Goodman, "Spectrum Rights in the Telecosm to Come" with Hazlett, "Spectrum Tragedies."

3. Benjamin, "Spectrum Abundance and the Choice Between Private and Public Control"; Faulhaber, "The Question of Spectrum"; Frischmann, "An Economic Theory of Infrastructure and Sustainable Infrastructure Commons."

4. In light of technological innovations that make interference-free spectrum sharing feasible, advocates for spectrum commons support unlicensed usage. See Weiser and Hatfield, "Policing the Spectrum Commons"; Werbach, "Supercommons"; Benkler, "Overcoming Agoraphobia."

5. For extensive background on spectrum management issues, see ITU, General Spectrum Management Resources; ITU, Spectrum Reform.

6. Most communications satellites appear to hover above the earth, thereby providing a fixed target to receive signals from earth and send them back down. Satellite services, including the direct-to-home delivery of video programming, can be provided more cheaply if earth-based antennas do not have to track a moving target. Satellites in a "geostationary" condition orbit the earth once every twenty-four hours. For more background on satellite technology and satellite-based businesses, see Parsons and Frieden, *The Cable and Satellite Television Industries*; and Frieden, *Managing Internet-Driven Change in International Telecommunications*.

7. Various treaties and other types of international agreements characterize outer space and radio communication resources located there and on earth as having a "common heritage" character or as being for the "common benefit of mankind." Writes Jefferson H. Weaver: "This characterization of outer space would indicate that the heavens belonged to no one; any resource which could be mined or otherwise appropriated would be available for the taking. This sort of 'first-come, first-served' standard would give the so-called 'spacefaring' powers a powerful incentive to expand their efforts to develop space. At the same time, nations unable to afford attempting even to launch satellites into geosynchronous orbit would find themselves increasingly excluded from these resources. Such a characterization would, however, create a first-come, first-served legal regime. This result would certainly be at odds with the 'common heritage' principle." "Illusion or Reality? State Sovereignty in Outer Space," 221–222. See also Tannenwald, "Law versus Power on the High Frontier."

8. United Nations, Treaty on Principles Governing the Activities of States in the Exploration and Use of Outer Space. The UN's Outer Space Treaty establishes basic principles of space law stating that the exploration and use of outer space shall be for the benefit of all mankind, that outer space is not subject to national appropriation by claim of sovereignty, and that each state party shall authorize, supervise, and be responsible for the space activities of its nationals.

9. See United Nations, note 8. Article II of this treaty establishes that space "is not subject to national appropriation by claim of sovereignty, by means of use or occupation, or by any other means."

10. As part of their commitment to nonappropriation of outer space, nations have agreed to register their space launches with the United Nations. See United Nations, Convention on Registration of Objects Launched into Outer Space.

11. "Though some obligation to accommodate remains when conflicts between early and later registrants arise, early registration affords a measure of legitimacy that supports the first registrant's negotiating position. Because the notification process affords preferential treatment to early registrants, it is often characterized as 'first-come, first-served.'" Roberts, "A Lost Connection," 1112–1113.

12. One camp asserts that governments need to take aggressive steps to compensate for vast differences in access to shared radiocommunication resources: "The digital revolution has transformed the lives of many, but also has left untouched the lives of many others. As a result, a large segment of the world population misses out on the tremendous political, social, economic, educational, and career opportunities created by the digital revolution" (Yu, "Symposium"). Another camp emphasizes the efficiency and productivity gains in relying on marketplace forces to allocate access and use of spectrum: "The rationales for [government] stewardship and for all-encompassing regulation that were offered in 1927 [at the onset of radio broadcasting] were not strong then; they have not grown any stronger with age. There is a better way. I describe it with a new word: 'propertyzing.' By that I mean converting the current system of regulatory permits or licenses to use the spectrum into a full-fledged system of property rights ownership" (White, "'Propertizing' the Electromagnetic Spectrum"). The latest camp has an even more libertarian view and would rely on technological innovations to replace the government role of doling out property rights: "Thus, the auction solution to the problem of FCC regulation may be no better than the previous system of license allocation... I argue that the spectrum might be best governed, at least in part, as a commons," that is, common property available for access by all (Noam, "Spectrum Auctions: Yesterday's Heresy, Today's Orthodoxy, Tomorrow's Anachronism").

13. Hazlett, "Assigning Property Rights to Radio Spectrum User"; Melody, "Spectrum Auctions and Efficient Resource Allocation"; Spiller and Cardilli, "Towards a Property Rights Approach to Communications Spectrum"; Cramton, "The Efficiency of the FCC Spectrum Auction."

14. Hatfield, "The Current Status of Spectrum Management."

15. Kwerel and Williams, *A Proposal for a Rapid Transition to Market Allocation of Spectrum*; Ward, "Secondary Markets in Spectrum."

16. "Sovereignty is the situation of the state which has no political superior over it, but is nevertheless bound by international law." Hoffmann, *International Systems and International Law*, 164.

17. White, "Propertizing," 37.

18. For background on the ITU organization structure and history, see Codding Jr., "The International Telecommunications Union"; White and Lauria, "The Impact of New Communication Technologies."

19. Article 44 of the ITU Constitution states: Members shall endeavour to limit the number of frequencies and the spectrum used to the minimum essential to provide in a satisfactory manner the necessary services. To that end, they shall endeavour to apply the latest technical advances as soon as possible.

"In using frequency bands for radio services, Members shall bear in mind that radio frequencies and the geostationary-satellite orbit are limited natural resources and that they must be used rationally, efficiently and economically, in conformity with the provisions of the Radio Regulations, so that countries or groups of countries may have equitable access to both, taking into account the special needs of the developing countries and the geographical situation of particular countries." See International Telecommunication Union, Use of the Radio-Frequency Spectrum. For background on the ITU spectrum allocation and satellite orbital slot registration process, see Frieden, *Managing Internet Driven Change in International Telecommunications*, chapter 6, "Players in International Telecommunications Policy Making."

20. For background on the ITU satellite orbital slot registration process and the administrative difficulties resulting from paper satellite registration filings, see Tompson, "Space for Rent"; Delzeit and Beal, "The Vulnerability of the Pacific Rim Orbital Spectrum under International Space Law"; and Wong, "The Paper 'Satellite' Chase."

21. International Telecommunication Union, "Processing Charges for Satellite Network Filings and Administrative Procedures"; International Telecommunication Union, "Scrambling for Space in Space."

22. "The most successful application of equity principles to the geostationary orbit arose out of negotiations during the sessions of the Space World Administrative Radiocommunications Conference held in 1985 and 1988. The result was a compromise that produced a hybrid system which combined the 'first-come, first-served' system with an a priori allotment system. Under the plan, each ITU Member was granted an allotment consisting of a nominal orbital position which represented a center point around which to base a maximum ten degree arc on the geostationary orbit, eight hundred megahertz of bandwidth, and a designated service area roughly equivalent to each Member's terrestrial borders. The allotments should not be confused with actual reserved assignments of positions and frequencies for fixed satellite service. They more closely resemble a right of coordination priority. The actual positions and frequencies remain available for use under the traditional allocation process; it is only when a Member begins the process of notification that the allotment plan becomes a factor in the distribution process." Roberts, "A Lost Connection," 1128.

23. Berger, "Proposed Legal Structure for the Silksat Consortium."

24. For background on the formation of INTELSAT and its privatization, see Frieden, "Privatization of Satellite Cooperatives."

25. Additionally U.S. taxpayers and corporations benefited by the formation of INTELSAT as the cooperative established a U.S. headquarters and primarily used U.S.-manufactured and U.S.-launched satellites.

26. United States Federal Communications Commission, Establishment of Satellite Systems Providing International Communications.

27. Lyall, "On the Privatisation of INTELSAT."

28. For example, the U.S. Congress enacted the Open-Market Reorganization for the Betterment of International Telecommunications Act (United States Congress 2000) to ensure that privatized cooperatives do not have the ability to leverage their previous status to secure anticompetitive advantages, but also to ensure that a small residual organization continue to provide core, lifeline services to developing nations, including those lacking access to submarine cable capacity.

29. "From 1988 to 1990, when Tonga made the filings on behalf of Friendly Islands Communications ('Tongasat'), the ITU system permitted a country to register a position for up to nine years before a satellite was launched. Tonga's action 'outraged' the international community because it 'lacked a genuine need' for so many orbital allotments in the Pacific Rim portion of the GSO. Tonga eventually withdrew its request for ten of the sixteen allotments, and, in 1991, it acquired six allotments. But, Tongasat further angered the international community by leasing one allotment to Unicom, a Colorado company, and auctioning off the remaining five allotments." Copiz, "Scarcity in Space," 208.

30. International Telecommunication Union, Plenipotentiary Conference, Resolution 86, Advance Publication, Coordination, Notification and Recording Procedures for Frequency Assignments Pertaining to Satellite Networks (seeking simplification and cost savings in the registration process for satellite networks spectrum use); and International Telecommunication Union, Resolution 88, Processing Charges for Satellite Network Filings and Administrative Procedure (recommending cost-based processing charges for satellite filings).

31. United States Federal Communications Commission, Establishment of Policies and Service Rules (implementing frequency sharing arrangements among different types of satellite operators consistent with policies adopted by the ITU's 2000 World Radiocommunication Conference).

32. Technological innovations also offer ways to reduce the potential for interference and to promote greater spectrum sharing. Digital signal processing, frequency-agile transceivers, and software-managed spectrum use provide unprecedented opportunities to abate spectrum scarcity and congestion. By converting signals into a coded, digital sequence, a format compatible to the language of computers, engineers provide a way to streamline content delivery. A compressed digital signal can fit in a smaller channel, making it possible for more content to be transmitted. Frequency-agile radios hop and skip across various frequencies to avoid interference and to accommodate many users. Software adds intelligence and computation capabilities to transmitters and receivers, expanding total content output and abating the potential for interference.

In an environment where virtually unlimited spectrum access is technologically possible, governments need not micromanage spectrum use, provided they establish technical standards for the equipment and the transmission standards used by radio transmitting equipment. This approach considers spectrum a shared "commons" much like a public park where private ownership and property rights need not exist. For background on the concept of a commons as applied to spectrum use, see Buck, "Replacing Spectrum Auctions with a Spectrum Commons."

33. In economic terms, spectrum allocations by consensus promote positive network externalities in terms of cross-border compatibility of equipment and services. For example, most nations have agreed to the consensus spectrum allocation for satellite frequencies. This means that all nations illuminated by a satellite footprint can access the same satellite having agreed on what frequencies the Satellite will operate and what technical parameters transmitting and receiving earth stations will use. When nations fail to reach such consensus, consumers face equipment and operating frequency incompatibility, as has occurred with cellular radiotelephone service. If the nations of the world had agreed on a single spectrum allocation for cellular radio service, there might have evolved a single transmission standard enabling a single transceiver to operate throughout the world.

34. White, "Propertizing," 35.

35. "Under the property rights system . . . governments would still have the ability to own and use spectrum parcels in ways that taxpayers felt were worthwhile, including defense and public safety, public broadcasting, etc. in the same way (and subject to the same constraints) that public agencies can own and use other forms of property." Ibid.

36. "In principal, we think that government users should acquire spectrum at market prices the same way they acquire other inputs such as oil, real estate and computer equipment. Paying market prices for these other inputs does not diminish the quality of government services." Kwerel and Willams, "A Proposal," 36.

37. "When a licensee goes bankrupt, tension arises if the FCC tries to use its position as a regulator to give it an advantage as a creditor. The question is whether courts should treat the FCC as a creditor or as a regulator in Chapter 11 bankruptcy proceedings involving electromagnetic spectrum licensees. The FCC's dual role has led the Second, Fifth, and D.C. Circuits to reach different and conflicting conclusions regarding the scope of the FCC's regulatory power in such proceedings. The courts' efforts to reconcile the FCC's roles are made difficult because of tension between a primary goal of bankruptcy and the Bankruptcy Code's deference to governmental units acting in their regulatory capacities." Patterson, "The Nature and Scope of the FCC's Regulatory Power," 1375.

38. "[A] system of spectrum property rights would cause spectrum to look much like real estate: Smaller units of spectrum would be available to anyone who could pay the market price." White, "Propertizing," 35.

39. The FCC has endorsed the development of secondary markets. See United States Federal Communications Commission, Promoting Efficient Use of the Spectrum.

40. For a comprehensive argument against satellite spectrum auctions, see Jackson et al., *Public Harms*.

41. Nobuo and Ye, "Spectrum Buyouts."

42. Under a pollution trading system, "a regulatory agency establishes a performance goal for an industry or area and then allocates increments of allowable pollution to each business unit in the industry or area. Because the cost of meeting their performance-based goals will differ among firms, firms with low pollution control costs should invest in a lot of pollution control, and firms with high pollution control costs should invest in less. Marketable permits allow firms to exchange increments of their performance-based goals so that the marginal cost of pollution control is equal across firms, with some firms exceeding their allocated pollution increment and others falling short. An added benefit to the tradable permit system is that it should stimulate investment in cost-effective pollution control technology because firms that can reduce the cost of pollution control can benefit by selling their allocated pollution increments." Blais, "Beyond Cost/Benefit."

References

Benjamin, Stuart Minor. 2003. "Spectrum Abundance and the Choice Between Private and Public Control." *New York University Law Review* 78 (December): 2007–2102.

Benkler, Yochai. 1998. "Overcoming Agoraphobia: Building the Commons of the Digitally Networked Environment." *Harvard Journal of Law and Technology* 11: 287–400.

Berger, Lee. 2001. "Proposed Legal Structure for the Silksat Consortium: A Regional Intergovernmental Organization to Improve Telecommunications Infrastructure in Central Asia and the Trans-Caucasus Region." *Law and Policy in International Business* 33: 99–143.

Blais, Lynn E. 2000. "Beyond Cost/Benefit: The Maturation of Economic Analysis of the Law and Its Consequences for Environmental Policymaking." *Illinois Law Review* Vol. 2000, No. 1, pp. 237–254.

Buck, Stuart. 2002. "Replacing Spectrum Auctions with a Spectrum Commons." *Stanford Technology Law Review* 2. Available at stlr.stanford.edu/STLR/Articles/02_STLR_2/article_pdf.pdf.

Cahill, Susan. 2001. "Give Me My Space: Implications for Permitting National Appropriation of the Geostationary Orbit." *Wisconsin International Law Journal* 9 (Fall): 231–248.

Codding Jr. George A. 1995. "The International Telecommunications Union: 130 Years of Telecommunications Regulation." *Denver Journal of International Law and Policy* 23: 501–512.

Copiz, Adrian. 2002. "Scarcity in Space: The International Regulation of Satellites." *Commlaw Conspectus* 10: 207–226.

Cramton, Peter. 1998. "The Efficiency of the FCC Spectrum Auctions." *Journal of Law and Economics* 41: 727–735.

Delzeit, Albert N., and Robert F. Beal. 1996. "The Vulnerability of the Pacific Rim Orbital Spectrum under International Space Law." *New York International Law Review* 9, no. 1: 69–83.

Faulhaber, Gerald R. 2005. "The Question of Spectrum: Technology, Management, and Regime Change." *Journal on Telecommunications & High Technology Law* 4 (Fall): 123–182.

Frieden, Rob. 1994. "Privatization of Satellite Cooperatives: Smothering a Golden Goose?" *Virginia Journal of International Law* 36: 1001–1019.

Frieden, Rob. 2001. *Managing Internet-Driven Change in International Telecommunications*. Boston: Artech House.

Frischmann, Brett M. 2005. "An Economic Theory of Infrastructure and Sustainable Infrastructure Common." *Minnesota Law Review* 89 (April): 917–1030.

Goodman, Ellen P. 2004. "Spectrum Rights in the Telecosm to Come." *San Diego Law Review* 41: 269–404.

Hatfield, Dale N. 2003. "The Current Status of Spectrum Management," in *Balancing Policy Options in a Turbulent Telecommunications Market: A Report of the Seventeenth Annual Aspen Institute Conference on Telecommunications Policy*, ed. Robert M. Entman. Available at www.aspeninstitute.org/atf/cf/%7BDEB6F227-659B-4EC8-8F84-8DF23CA704F5%7D/BALANCETURBTELECOM.PDF.

Hazlett, Thomas. 1998. "Assigning Property Rights to Radio Spectrum Users: Why Did FCC License Auctions Take 67 Years?" *Journal of Law and Economics* 41: 529–578.

Hazlett, Thomas W. 2005. "Spectrum Tragedies." *Yale Journal on Regulation* 22 (Summer): 242–274.

Hoffmann, Stanley. 1966. "International Systems and International Law," in *The Strategy of World Order*, ed. Richard A. Falk and Saul H. Mendlovitz. New York: World Law Fund, pp. 134–166.

International Telecommunication Union (ITU). Undated. "General Spectrum Management Resources." Available at www.itu.int/osg/spu/ni/spectrum/resources/general/.

International Telecommunication Union (ITU). Undated. "Spectrum Reform." Available at www.itu.int/osg/spu/ni/spectrum/resources/reform/index.html.

International Telecommunication Union (ITU). 2002a. Constitution, chapter VII, Special Provisions for Radio, article 44. Use of the Radio-Frequency Spectrum and of the Geostationary-Satellite Orbit. Available at www.itu.int/publications/cchtm/const/art44.html.

International Telecommunication Union (ITU). 2002b. "Scrambling for Space in Space: ITU Plenipotentiary to Tackle 'Paper Satellite' Problem." Press release, available at www.itu.int/newsarchive/press_releases/2OO2/21.html.

International Telecommunication Union (ITU). 2002c. Plenipotentiary Conference, Resolution 86. Advance Publication, Coordination, Notification and Recording Procedures for Frequency Assignments Pertaining to Satellite Networks. Available at www.itu.int/aboutitu/basic-texts/resolutions/res86.html.

International Telecommunication Union (ITU). 2002d. Plenipotentiary Conference, Resolution 88 (Rev. Marrakesh). Processing Charges for Satellite Network Filings and Administrative Procedures. Available at www.itu.int/aboutitu/basic-texts/resolutions/res88.html.

Jackson, Charles L., John Haring, Harry M. Shooshan III, Jeffrey H. Rohlfs, and Kirsten M. Pehrsson. 1996. *Public Harms Unique to Satellite Spectrum Auctions*, strategic policy research study prepared for the Satellite Industry Association. Available at www.spri.com/pdf/reports/sia/pubharms.pdf.

Kwerel, Evan, and John Williams. 2002. "A Proposal for a Rapid Transition to Market Allocation of Spectrum." Federal Communications Commission, Office of Plans and Policy, Working Paper No. 38. Available at hraunfoss.fcc.gov/edocs_public/attachmatch/DOC-228552Al.pdf.

Lyall, Francis. 2001. "On the Privatisation of INTELSAT." *Singapore Journal of International and Comparative Law* 5: 111–132.

Melody, William H. 2001. "Spectrum Auctions and Efficient Resource Allocation: Learning from the 3G Experience in Europe." *Info* 3, no. 1: 5–10.

Noam, Eli. 1998. "Spectrum Auctions: Yesterday's Heresy, Today's Orthodoxy, Tomorrow's Anachronism: Taking the Next Step to Open Spectrum Access." *Journal of Law and Economics* 41: 765–790.

Nobuo, Ikeda, and Lixin Ye. 2004. "Spectrum Buyouts—A Mechanism to Open Spectrum." Available at www.itu.int/osg/spu/ni/spectrum/presentations/Paper-Ikeda.doc/.

Parsons, Patrick R., and Robert M. Frieden. 1998. *The Cable and Satellite Television Industries*. Needham Heights, MA: Allyn & Bacon.

Patterson, Nicholas J. 2002. "The Nature and Scope of the FCC's Regulatory Power in the Wake of the Nextwave and GWI PCS Cases." *University of Chicago Law Review* 69: 1373–1398.

Lee, Ricky J. 2001. "Reconciling International Space Law with the Commercial Realities of the Twenty-First Century." *Singapore Journal of International and Comparative Law* 4: 194–251.

Roberts, Lawrence D. 2000. "A Lost Connection: Geostationary Satellite Networks and the International Telecommunication Union." *Berkeley Technology Law Journal* 15 (Fall): 1095–1114.

Spiller, Pablo T., and Carlo Cardilli. 1999. "Towards a Property Rights Approach to Communications Spectrum." *Yale Journal on Regulation* 16, no. 1: 53–83.

Tannenwald, Nina. 2004. "Law versus Power on the High Frontier: The Case for a Rule-Based Regime in Outer Space." *Yale Journal of International Law* 29 (Summer): 363–422.

Tompson, Jannat C. 1996. "Space for Rent: The International Telecommunications Union, Space Law, and Orbit/Spectrum Leasing." *Journal of Air Law and Commerce* 62: 279–311.

United Nations. 1967. Treaty on Principles Governing the Activities of States in the Exploration and Use of Outer Space, including the Moon and Other Celestial Bodies. United Nations Treaty Series 610: 205.

United Nations. 1975. Convention on Registration of Objects Launched into Outer Space. United Nations Treaty Series 1023: 15.

United States Congress. 2000. *Open-Market Reorganization for the Betterment of International Telecommunications Act*. Public Law 106–180.

United States Federal Communications Commission (FCC). Undated. Wireless Telecommunications Bureau, Auctions Summary. Available at www.wireless.fcc.gov/auctions/summary.html.

United States Federal Communications Commission (FCC). 1985. Establishment of Satellite Systems Providing International Communications, 101 FCC 2d 1046, modified on reconsideration, 61 Rad. Reg. 2d (P&F) 648 (1986), reconsideration denied, 1 FCC Rcd. 439 (1986).

United States Federal Communications Commission (FCC). 2002. Establishment of Policies and Service Rules for the Non-Geostationary Satellite Orbit, Fixed Satellite Service in the Ku-Band, Report and Order. IB Docket No. 01-96, 17 FCC Rcd. 7841.

United States Federal Communications Commission (FCC). 2004. Promoting Efficient Use of the Spectrum through Elimination of Barriers to the Development of Secondary Markets, 2004. WT Docket No. 00-230, Second Report and Order, Order on Reconsideration, and Second Further Notice of Proposed Rulemaking, FCC 04-167. Available at http://hraunfoss.fcc.gov/edocs_public/attachmatch/FCC-04-167A1.pdf.

Ward, Joseph M. 2001. "Secondary Markets in Spectrum: Making Spectrum Policy as Flexible as the Spectrum Market it Must Foster." *Commlaw Conspectus* 10: 103–132.

Weaver, Jefferson H. 1992. "Illusion or Reality? State Sovereignty in Outer Space." *Boston University International Law Journal* 10 (Fall): 203–222.

Weiser, Philip J., and Dale N. Hatfield. 2005. "Policing the Spectrum Commons." *Fordham Law Review* 74 (November): 663–694.

Werbach, Kevin. 2004. "Supercommons: Toward a Unified Theory of Wireless Communication." *Texas Law Review* 82 (March): 863–973.

White Jr., Harold M., and Rita Lauria. 1995. "The Impact of New Communication Technologies on International Telecommunication Law and Policy: Cyberspace and the Restructuring of the International Telecommunication Union." *California Western Law Review* 32: 1–30.

White, Lawrence J. 2000. "'Propertizing' the Electromagnetic Spectrum: Why It's Important, and How to Begin." *Media Law and Policy* 9 (Fall): 51–75.

Wong, Henry. 1998. "The Paper 'Satellite' Chase: The ITU Prepares for its Final Exam in Resolution." *Journal of Air Law and Commerce* 63: 849–879.

Yu, Peter K. 2002. "Symposium—Bridging the Digital Divide: Equality in the Information Age." *Cardozo Arts and Entertainment Law Journal* 20: 1–52.

4 The Peculiar Evolution of 3G Wireless Networks: Institutional Logic, Politics, and Property Rights

Peter F. Cowhey, Jonathan D. Aronson, and John E. Richards

In 2002 wireless phone connections surpassed the number of wired connections globally and became the primary communications infrastructure for all but the largest firms in many developing countries. New, third generation (3G) wireless networks promised to provide mobile voice and multimedia data to users worldwide. 3G is more advanced than first generation (1G) analog mobile services that provide only voice services and second generation digital services (2G) that handle voice and some text data. The technological advances available using 3G wireless networks could put wireless mobile networks on a par with wired networks for delivering data for households and for small and medium enterprises. To achieve this goal firms invested hundreds of billions of dollars in anticipation of annual revenues in the tens of billions. If 3G succeeded, it would be an important part of tomorrow's global communications infrastructure. However, major problems in the transition to 3G emerged by 1999 and the technology's design and deployment were marked by major controversies even when measured by the hardball politics of rival equipment vendors. What happened? This chapter uses contemporary models of political economy to explain the contested evolution of 3G.

In late 1999 3G seemed ready to take off. The financial community and business press predicted that giant investments in network infrastructure would launch 3G as *the* innovative new consumer service. There was little concern that potential operators paid huge sums in auction fees for the licenses. But when the bubble for technology share prices collapsed, leaving tremendous surplus capacity from the overbuilding of fiber-optic infrastructure, the 3G vision suddenly seemed illusory. Carriers delayed dates for services rollout, equipment vendors admitted to a steady stream of technological glitches, and many content providers abandoned their wireless ambitions. In Europe the projected date for widespread 3G rollout was pushed back to 2004 and then 2006.

The business press advanced numerous explanations for the debacle. Wireless carriers paid too much in auctions for their licenses. Technical glitches caused debt loads to rise even as network launch dates were delayed. There were no really compelling service applications to attract throngs of consumers to the new, higher-speed data services.

In contrast, this chapter argues that the conventional explanations missed the political economic logic of 3G that answers three key questions. First, why did a comprehensive plan for 3G technology deployment become a key goal of global policy even though the level of information technology (IT) coordination and planning required by 3G was unprecedented? Second, why did government policies stumble? Third, what are the lessons for future efforts at global IT coordination?

We argue that the planning goal was ambitious because politicians tried to balance an elaborate set of distributional goals while simultaneously trying to harvest the efficiencies of the new 3G technologies. Reforms required compensation that parsed out the gains from technology innovation between entrenched and new stakeholders.[1] National institutional arrangements tackled this balancing act by creating a policy process dominated by a handful of incumbents that also accommodated some new stakeholders. This truce broke down when ambitious local players had to cooperate in a global coordination process within the context of the International Telecommunication Union (ITU). At the global level, regional compromises could not easily be reached between old and new stakeholders that embraced drastically different business models. The ultimate global compromises delayed the market rollout of 3G while adding more technological diversity and spectrum choices than originally envisioned. This led to market problems that plagued 3G commercialization. The key policy lesson was that the problems of coordination for 3G will probably occur again. So, a different approach to spectrum and standards policy is needed.

The first section of this chapter surveys the dynamics of adjusting stakeholder interests. The second section explains the three sets of policy choices that shaped the design of 2G services. The third section shows how 3G decisions built on these political roots. The fourth section discusses options for reform, while the fifth presents our policy recommendations.

Policy Reform and the Dynamics of Balancing Stakeholder Interests

3G called for a single global plan for technology and spectrum designed to (1) increase the capacity to handle traffic flows for any given amount of spectrum; (2) allow mobile, high-speed data transmission (from 144 Kbps to 2 Mbps) able to handle at least limited motion video capabilities, and thus profitable new services; and (3) facilitate true global roaming of services using a single standard on common radio spectrum. These were ambitious goals, part of a remarkable vision. But planners faced huge coordination challenges, especially given the growing diversity of new stakeholders. Underlying the 3G efforts there also was a daring plan to provide new rewards to key incumbents even as competition was increasing worldwide.

3G is a new technology that raised an old political economy problem. The same factors that induce market innovation also create incentives to distort reform. Economic theory suggests two potential gains from coordinated government intervention in global wireless markets. First, wireless depends on the use of radio spectrum that is subject to crowding and interference problems. Global spectrum coordination could reserve enough spectrum on the same band to allow new global services that benefit from global economies of scale in radio equipment to emerge. Consumers also may benefit from interoperability of equipment (Besen and Farrell 1991; Farrell and Klemperer; Shapiro and Varian 1999). Second, the wireless industry is capital intensive, has large economies of scale, has strong network externalities, and has some path dependency.[2] As a result, incumbent carriers and their equipment vendors seek favorable technology upgrades on a predictable basis. This makes common planning of new technologies, like 3G, attractive (Owen and Rosston 2001).[3] Global network externalities and scale economies in equipment pushed stakeholders to look beyond their borders to arrange global coordination of technology design through standards-setting processes and spectrum allocation for new services.[4] However, if competitive carriers or equipment suppliers can gain from using a superior alternative technology without encountering unacceptable losses on scale economies and network externalities, then incentives for coordination decline. Taken together, the economic realities make it unlikely that there will be large numbers of platforms, but achieving a single platform is difficult. As we shall show, particular market centers (e.g., North America) provide enough scale to permit selections of alternative technology standards.

In short, coordination is attractive, but distributional issues are likely to lead to disagreement over which coordinated solution is best (Krasner 1991).[5] Thus, savvy players often will try to manipulate policy to their advantage in the selection of technology platforms. The double-edged payoff from global coordination became especially challenging because the changing technological foundation of the industry attracts strong political interest. Innovation and the end of monopoly promise huge gains that could be distributed to consumers and new commercial entrants. Speeding up innovation and competition, however, may harm large stakeholders in the industry.

Political choices for 3G revolved around policies that allocated and assigned radio spectrum and technical standards that influenced the choice of technologies, and around institutional processes for regulating markets that shaped how those rights were adjusted over time. These choices influenced the number of competitors in the marketplace for services and equipment, the terms of competition, and the economics of 3G. Politicians usually promoted technological innovation by abolishing a monopoly franchise or otherwise altering property rights in ways that would stimulate wireless competition and create benefits for consumers and new customers for new suppliers.[6] They also tried to ensure significant gains from each new generation of wireless technology to major incumbents. To be successful, the process must include a policy payoff for existing stakeholders, policy solutions that benefit political leaders, and an institutional process that somehow helps to match supply and demand.

The Demand Side of Policy

On the demand side, constituents "bid" for policy favorable to their interests. Some players are more motivated or have more resources to bid for these rights (e.g., more workers who vote). Firms facing large losses from policy changes designed to improve market efficiency are more motivated to act politically than firms that will receive smaller diffuse benefits. This makes optimal reform difficult (Olson 1971).

In telecommunications, an entrenched coalition dominated until the mid-1980s. In each country the traditional monopoly carrier, its well-paid, unionized employees, and the equipment suppliers favored by the carrier worked together (Noam 1993). This coalition finally had to accept greater telecommunications competition because technology created the potential for large efficiency gains that could be redistributed to a new group of prominent stakeholders that advocated market reform (Cowhey 1990a).[7] Nonetheless, the old coalition worked to implement competition in ways that created new sources of market rents for incumbents.

The Supply Side of Policy Reform

On the supply side, politicians in democracies advance their individual careers and their political parties by reforming markets in ways that win credit from voters. In essence, they organize policy initiatives in exchange for votes (Cox and McCubbins 1993). They may seek to improve public welfare, but they also manage a contentious political process with strong stakeholders and imperfect options for matching policy supply and demand. For example, politicians might court business by advocating less government control of wireless markets. But flawed property rights for spectrum that is licensed for a fixed period of time subject to many constraints may move firms to demand extensive government micromanagement of the market.[8]

Political entrepreneurs skew reform by selecting changes that benefit their strongest supporters. At the same time they seek credit for difficult choices from a public that sees the issue as reasonably important, but complicated and obscure. So, political leaders frame the choice in terms of a few clear political "punch lines" to claim credit and limit the potential for critics to mobilize a successful opposing strategy.[9] In particular, politicians emphasize visible benefits from reforms to counter complaints by losers. They may alter reform plans in ways that sacrifice substantial diffuse benefits from competition for "success" on specific visible grounds. For example, European leaders often justify EU initiatives on the basis of creating "good jobs" through the promotion of press-friendly technologies, like 3G. In developing countries attractive measures of success may include highly touted benefits from increased foreign investment and network construction projects. Usually, political leaders also focus on defined consumer benefits—such as the price of a common service like the price of a bundle of minutes on a cell phone—over larger benefits from reductions on less visible prices.

These same political realities explain why regulators frequently create competition that is friendly to large incumbents, rather than push for more vigorous market performance. When carriers run into trouble, their governments often try to ease their pain. Predictably, the carriers most likely to be assisted are the largest firms that employ the most people throughout the country and those that provide the most visible services to voters on a daily basis.

Why Institutions Change Outcomes

Institutional factors further shaped how politicians perform and how they maintain a precarious balance among the interests of their constituents. The reason for this is that institutions that create policies use decision rules and procedures that alter the equilibrium outcome in unexpected ways.

Political leaders grant authority to specialized regulators because these officials possess superior expertise and information and the discretion to act. Such regulators can provide the best combination of improved efficiency within the constraints of implicit political guidelines about the distribution of gains and losses. Today's national policy institutions, including independent regulatory authorities, are designed to throw open the closed doors of the monopoly era.

Regulatory institutions vary in their ability to make decisions when faced with conflicts among key stakeholders. As the ability of any individual player to veto a decision rises and the number of decision points in a policy process increases, the more likely it is that the process will maintain the status quo or produce a decision skewed to serve the needs of players with the strongest veto power (Tsebelis 2002; Austin and Milner 2001 on standards). Most national regulators use some version of majority decision making to limit vetoes by dissenting stakeholders. However, regulatory policy is skewed by due process procedures and legal "safeguards" designed to favor slower, consensus-oriented outcomes.[10] Moreover, their complex procedures may create implicit barriers to smaller entrants participating in the policy process.

At the same time, governments use international institutions to create policies and property rights in global markets, thereby increasing both the efficiency of these markets and the amount of wealth available for domestic redistribution. The tensions between efficiency and redistribution goals, coupled with the special decision properties of global institutions, limit optimization of global reforms (Richards 1999). Many international institutions, such as the ITU, have a large membership and require unanimity in decision making. Although political and economic pressure may induce reluctant parties to compromise, the system is subject to vetoes (Greenstein 1994). Thus, international institutions often deadlock if they do not settle on the lowest common denominator for a decision. These weaknesses shape stakeholder strategies. The ambitious planning for 3G reflected an effort to use a process geared to favor influence by traditional stakeholders to chart a major new technology. New entrants were supposed to compete in 3G on terms defined by a consensus process characterized by a collective veto held by the most powerful players, the traditional corporate leaders. However, the consensus-driven process in the ITU broke down as the range of corporate stakeholders expanded and their interests diverged. The result was stalemate and unexpected compromises.

Defining Global Policies for Wireless Markets

The economics of networks make them somewhat path-dependent. So, to understand the political economy of 3G, it is necessary to examine the political economy of 2G networks. This section begins with a survey of the three key policies for wireless networks and then examines why divergent solutions emerged in 2G.

Standards Setting

The first set of policy choices revolved around the *process for defining and sharing intellectual property (IP) rights and the selection of standards* for global wireless networks. Each new generation of wireless services emerged from a global collaborative planning process between carriers and equipment suppliers coordinated through the ITU and regional and national standards-setting processes. Participation in these processes, the terms of operation, and the conditions imposed on the use of IP in the standards process all shape global technology.

2G technologies emerged in the late 1980s when competition in Europe and Japan was limited and global standards processes reflected this monopolistic legacy. Traditionally, carriers in industrial countries worked with a small set of preferred, nationally or regionally based suppliers in a closed standards process.[11] Significant variations in national standards were common, thereby accommodating various market barriers. For example, developing countries usually were heavily dependent on the counsel of their traditional equipment suppliers. Even efforts to coordinate new 2G services and standards had to plan on these variations because ITU decision making was consensual. Various forms of Time Division Multiplexing Access (TDMA) dominated the market and standards process initially. GSM (Global System for Communications) became especially prominent.

The global decision-making process is complex. The ITU sets wireless network standards in a process that is formally organized around, and fed by, leadership from the major regional standards bodies.[12] The setting of standards and other matters of telecommunications policy are handled in the ITU-T (Telecommunication Standardization Sector), which operates with study groups including many from the private sector, coordinated by the TSAG (Telecommunication Sector Advisory Group) (Besen and Farrell 1991; Schmidt and Werle 1998).

The dynamics of 2G decisions reflected the fact that the shift to competition still was incomplete at the time. Growing economies of scale in the telecommunications equipment industry forced major suppliers to consolidate and become more global by the early 1980s. Moreover, the United States insisted that the opening of its equipment market to imports was contingent on reciprocal opening of other national markets around an open procurement process guided by "open, industry-led, and voluntary" standard-setting processes (Drake and Nicolaidis 1992; Cowhey 1990b). These reforms began to open national standard setting to foreign participation, but during the early 1990s it did not change a key fundamental preference of the largest carriers and suppliers. They still valued a long-term technology planning process for telecommunications that they collectively dominated. This process combined global coordination of standards and industrial policy planning.

IP stakeholders still were mainly incumbents with close ties to service providers and governments. In sharp contrast to the computing industry, their business models reflected their monopoly roots. The traditional equipment firms typically cross-licensed their intellectual property rights for TDMA 2G systems on a cost-free basis while developing major new standards within the ITU system. Everybody needed the IP so, rather than quibble about the precise distribution of payments, the top tier of suppliers gained by using low- or zero-cost licensing to grow the market. They competed on economies of scale, marketing, and systems engineering for large carriers. Recently, to reinforce cross-licensing of an agreed standard among suppliers, large regional bodies only embraced a standard if there was agreement to license the relevant IP to every IP holder under the standard.

Allocating Spectrum

The second set of policy choices revolved around *rules governing the allocation of radio spectrum for specific uses, including the rules of service governing the use of licensed spectrum.* Spectrum allocation refers to the decision about how much spectrum on which frequency ranges to allot to particular services or groups of services.[13] All governments treated the spectrum as a "commons" that required careful licensing to avoid interference problems among rival uses. Even if there were ownership alternatives, political leadership had few incentives to explore them (Hazlett 2001).[14] Revisiting spectrum allocations allowed politicians to earn credit from micromanaging a valuable resource.

Institutional arrangements further skewed market dynamics. Recall that political leadership tried to introduce competitive reform without overly shocking incumbents. Decision processes implicitly served this purpose. Most regulators presumed that new technology should not endanger old users even though there is a strong economic efficiency case for ensuring less than perfect protection (Hazlett 2001). This ensured incumbents strong influence over spectrum planning. For carriers, rules governing the use of the licensed spectrum also created barriers to entry for other forms of wireless networks, such as nonmobile services that might substitute for some mobile service applications: For equipment vendors, the rules made it more difficult for new entrants to deploy novel technologies. Incumbent suppliers therefore played a larger role in shaping new technology markets than, for example, in the computer industry.

This nonmarket environment created an insiders' spectrum game with complex bargaining among government agencies. Officials received input from an advisory process dominated by commercial interests and a few ardent groups such as associations of amateur radio operators. These advocates were highly visible to regulators and to political leaders reviewing regulatory choices, and they had enough staff to work the policy choices in all key global markets. Companies with operational experience also had informational advantages.[15] Smaller and newer companies faced steep entry barriers to participating effectively in the decision-making process.

Global processes predictably reinforced national arrangements. The objective of the ITU's Radio Regulations is "an interference-free operation of the maximum number of radio stations in those parts of the radio frequency spectrum where harmful interference may occur." As supplements to the treaty governing the ITU, the regulations have the "force of an international treaty" (Hudson 1997, 406). Work on designating spectrum for particular uses is undertaken in the ITU-R (ITU Radiocommunication Sector) through a process of study groups that are overseen by the Radiocommunication Sector Advisory Group. Every two years a World Radiocommunication Conference (WRC) makes decisions on new spectrum allocations and other policies to avoid interference among spectrum uses.

The WRC uses a one-country, one-vote system to approve changes in global spectrum allocations and service rules. Although informal polls gauge relative standings of positions, votes are rare. In practice, it is a consensus system that is prone to deadlock. However, government and commercial interests want some measure of certainty about spectrum plans.[16] So, they make compromises at the WRC. The easiest of these involve fewer changes in the existing spectrum plan. Sometimes these outcomes are not to the liking of the United States and other major powers.

It is not surprising that a consensus system reflects the policy roots of key member states. 1G services relied on analog technology in a monopoly era. Despite ITU coordination efforts, the political economy of monopoly resulted in idiosyncratic national spectrum plans in part because of efforts to use spectrum plans to bolster regional suppliers over "out-of-region" suppliers. Usually it was impossible to use a telephone outside of its country of origin because in different countries 1G was deployed on different spectrum bands.[17] Once governments created these disparities in spectrum plans, it required high levels of political commitment and, therefore, political rewards to significantly rewrite spectrum plans. 2G technology revisited the issue of spectrum allocation because everybody agreed that it would require larger allocations in a different band than the previous generation. The European Union and the United States moved in different directions, as described later.

Assigning Spectrum Licenses

The third set of policy choices involves *assigning service licenses.* The number of licenses, the method for selecting licensees, and the sequence of assignment of licenses shape market efficiency. From the early 1980s onward the number of licenses slowly increased, creating more competitive markets. But from the early 1970s, the sequence and methods of licensing decisions provided substantial market rents for the original incumbents and then for their initial challengers.

When wireless, cellular phones appeared around 1983, most governments quickly granted a wireless mobile service license to the incumbent wired network carriers. The incumbents dominated the marketplace and most countries did not even separate the setting of policy from the operation of the national telephone company.

A few governments introduced duopoly in the first generation of services. In the United States, for instance, each of the original seven regional Bell operating companies was awarded one of two wireless licenses in its home territories. Like other early advocates of duopoly, the United States embraced nonmarket-based criteria for awarding the second wireless license. Methods for selecting licensees varied, but "beauty contests" (administrative selection of a sound company promising good performance) and lotteries were popular. Duopoly benefited equipment suppliers that were clamoring for an increase in the number of competitive operators so that they would have more customers to buy their products.[18] The small pool of new entrants rapidly acquired some shared interests with the incumbents because they became prominent players in the regulatory process that determined future spectrum allocation and assignment policies.

Wireless licenses traditionally contained numerous restrictions that weakened them as a form of private property rights for spectrum (Owen and Rosston 2001). This had significant implications for politics and economic performance. In the United States, for example, government spectrum licenses limited the ability of spectrum owners to switch between service types (e.g., from fixed to mobile wireless), the ability of single providers to own more than limited spectrum in a given market (e.g., spectrum caps), and ownership transfer.[19] Licenses also were granted for a set number of years (e.g., fifteen). These conditions could reduce market efficiency by preventing a secondary market in licenses from emerging and reducing flexibility in the services offered by a license holder. Also, stakeholders focused on manipulating government policy, not on creating market alternatives.

Asian and European governments often imposed stricter restrictions, even dictating the type of technology platform that spectrum users could employ to offers services.

Combined with differences in spectrum and challenges of systems integration with the existing national wired network, these conditions effectively limited the range of new suppliers even after the abolition of monopoly supply systems.

In short, the political economy of standards setting, spectrum allocation, and spectrum licensing left a legacy of government micromanagement of wireless markets despite growing levels of competition. The transition to 2G technologies could not escape the consequences of these politics and policies, and they shaped the world market in ways that unexpectedly set the stage for problems involving 3G that are explored later.[20] This section concludes with a review of the regional variations in 2G that influenced the choices about 3G.

Europe The earliest major plan for 2G emerged in Europe where political leaders saw the largest opportunity for taking political credit from market reform by steering it in a specific direction. 2G was seen as a chance to dramatize the benefits of integrating European markets and policy. In 1982 the European Conference of Posts and Telecommunications (CEPT) administrations decided to design a single common standard, the Global System for Mobile Communications (GSM), a variant of TDMA. In 1988 the EU sponsored the creation of the European Telecommunications Standards Institute (ETSI) to create standards for member states in an organization that would be less closely wedded to the traditional national telecommunications monopolists and their suppliers (Hudson 1997, 170–176). However, ETSI used a weighted voting process (requiring a 71 percent majority) based on European market revenues to ensure a prominent role for incumbents.[21] A few non-European firms, like Motorola, also achieved prominence. In contrast to the one-company, one-vote principle of the U.S. Telecommunications Industry Association, the ETSI used weighted voting strongly tied to European market revenues (Gandel, Salant, and Waverman 2003). Predictably, second-tier equipment suppliers complained that the terms for patent pooling for GSM favored the largest European companies (Pelkmans 1998).

The United States successfully urged that ETSI standards be voluntary. However, the EU retained the option of adopting a voluntary ETSI standard as a mandatory European norm and did so by requiring all carriers to use GSM. The EU also bridged differences in national spectrum plans when the EU Council of Ministers issued a directive requiring the use of a single band for GSM.[22] These EU actions built economies of scale around GSM service, allowing it to evolve into the dominant global technology for 2G (Cowhey 1993). The EU considered GSM to be its greatest recent success in industrial policy.

The chance to dramatize telecommunications market reform by the bold GSM scheme explains the enthusiasm of political entrepreneurs. However, given the political influence of incumbents, they needed to see gains also. European operators came to believe that spectrum harmonization would grow the service market, especially for lucrative business users, more quickly on a single band than if the EU adopted a variety of technologies and band plans. This provided a benefit to operators to offset the loss of market protection afforded by idiosyncratic national band plans. Meanwhile, European equipment makers recognized that if they did not create a major new European market for GSM, they would have to lay off large numbers of unionized workers (Sandholz and Zysman 1989; Pelksman 1998; Cowhey 1990b).

Significantly, the EU member states retained general control over spectrum planning and licensing. Although all players saw advantages of unifying the internal market to seize network externalities and scale economies, they still wanted their friendly home governments to control the details of spectrum allocation and licensing. This gap in the powers of the EU ultimately had major consequences for 3G licensing.

On one level the European experiment was a great success. The GSM technology worked. Consumers responded enthusiastically to a true continental service. During the 1980s the market-oriented features of wireless also were appealing when compared to the moribund marketing and expensive prices for traditional telephone service. The European success fueled interest in GSM and, as other countries deployed the technology, strengthened the global standing of the European spectrum band.

Restricted entry limited the amount of competition and bolstered profit margins. In addition, there were no price restrictions on mobile prices, thus allowing premium prices for a popular service that yielded strong margins until the late 1990s. Moreover, the policy of "calling party pays" for those calling to mobile phones meant that lightly regulated mobile operators could charge wireline operators a significant fee for call termination.[23]

The European approach also featured an investment race among the leaders to capture the exploding market for wireless. Most leading carriers were rooted in the wired world because governments gave 2G licenses to the wireline giants. But 2G also spurred traditional carriers, such as Deutsche Telekom, into horizontal cross-entry in 2G services in the traditional territories of other carriers to achieve regional or continental service footprints. These traditional carriers leveraged the large cash flow and business customer base from their original licenses. In addition, entry from major non-European carriers was difficult, thus limiting the pool of competitors. This occurred because most countries had formal or informal restrictions on foreign direct investment until the WTO agreement on basic telecommunications services in 1997.[24]

The success of 2G was a political blessing as regulators tried to introduce competition. Political considerations associated with the high costs and inefficient workforces in their traditional wireline businesses shackled the former telephone monopolists. The introduction of general phone service competition meant that the former monopolists lost markets and their margins declined by more than one half. Most European incumbents saw voice revenues decline from 1998 to 2001 (Jagannathan, Kura, and Wilshire 2003). Competition proved popular with urban consumers and businesses,

Table 4.1
Revenue per employee of major wireline and wireless carriers

Sprint PCS	$1,024,522
Sprint FON	239,368
NTT DoCoMo	2,211,281
NTT	429,045
Telefonica Movile	714,285
Telefonica	200,336
Vodafone AG	185,386
Vodafone Group	691,467
Verizon	285,193
SBC Communications	227,598
Deutsche Telekom	214,819
AT&T Wireless	457,939
AT&T	414,440
France Telecom	206,794
Bell South Corp.	261,292
Industry average	315,629

Source: Multex fundamentals, www.multexinvestor.com/mgi. Accessed on January 6, 2003.

but threatened the many stakeholders in the old incumbents. In this strategic setting, the expansion of former wireline monopolists into 2G eased many political problems because their mobile service subsidiaries earned far more revenue per employee, as table 4.1 shows, with high margins.

In the late 1990s, as carriers looked toward the more competitive future, 3G was appealing because new 3G networks were expected to reenergize market growth as the market for voice-only cell phones matured but those with data connections could grow rapidly (ovum data in *Red Herring*, 2002). Revenue with attractive margins from increased roaming by customers across national borders also was important.

Japan When Japan introduced competition in the mid-1980s, it used the NTT procurement system to produce standards that were just different enough from those of other nations to impede supply by foreign firms. For example, NTY DoCoMo, the dominant Japanese wireless carrier, chose a TDMA variation with idiosyncratic wrinkles. As was usual for NTT's procurement policy at the time, the differences tended to favor a few Japanese suppliers. NTT's procurement policy was opened to international scrutiny when Japan agreed to extend the GATT procurement code to NTT (Noll and Rosenbluth 1995). The Japanese standard made some headway in penetrating the Asian market, but did not generally flourish outside Japan. Still, the large Japanese market provided large-scale economies and high profit margins that financed Japanese suppliers as they adapted their equipment to foreign markets.

In the 1980s, as Japanese equipment exports to America surged and U.S. importers had little success in Japan, noteworthy trade disputes proliferated: The United States negotiated for open procurement by NTT, a process that took years to implement effectively. New competition in telecommunications services also did not help much initially. To manage the competitive market the government organized licensing on the basis of a beauty contest (Noll and Rosenbluth 1995). Each carrier awarded a license had to commit to rapid buildout of the network, thus boosting capital expenditures. Technology plans of carriers were subject to government review. Eventually, one would-be Japanese entrant into the mobile wireless market cut a deal with the U.S. government. It committed to Motorola technology and Washington lobbied for the firm to receive a wireless license that had sufficient spectrum to compete in the vital Tokyo market (Schoppa 1997; Johnson 1989). Despite U.S. success in this negotiation, Japan ventured into 3G with its dominant market share in 2G tied to standards incompatible with Europe and the United States and a continuing tradition of active industrial policy.

The United States The United States began 2G with a more diverse carrier and equipment industry. Due to political incentives created by its federalist system, America's political leaders were traditionally suspicious of granting monopolies. Even the AT&T monopoly rested on a weak, loophole-infested legal foundation (Brock 1981, 89–125, 177–197). By the 1970s a few industry associations, rather than any individual carrier, dominated the standards process. The Telecommunications Industry Association and the Cellular and Telecommunications Industry Association, the key groups, featured open membership and voluntary standards. The Federal Communication Commission (FCC), for its part, adopted a technology-neutral strategy.

Unlike Europe, when 2G came along U.S. suppliers already had a continent-size national market yielding large economies of scale. They had no incentive to compromise on a single standard for creation of a unified market. In addition, they had few expectations that, in the fairly competitive U.S. market for services and equipment, a single standard would primarily benefit only traditional incumbents. As a result, carriers and their suppliers supported technology neutrality in licensing policy and 2G ended up split between two dominant technology camps, CDMA (Code Division Multiplexing Access) and various forms of TDMA for 2G.[25] This initially made it more difficult for users to get seamless coverage in the United States. Over time, the new CDMA technology proved to be much more efficient in the use of spectrum and therefore able to slash costs for carriers by providing more traffic per megahertz of spectrum (see Hjelm 2000; also Lee 2002) (see figure 4.1). An unexpected consequence of the spectrum efficiency of CDMA radios was that the Europeans and Japanese reluctantly concluded that 3G standards should be based on CDMA, even though their manufacturers specialized in TDMA-based technologies. This later created a huge problem for the 3G process, for reasons explained in the next section.

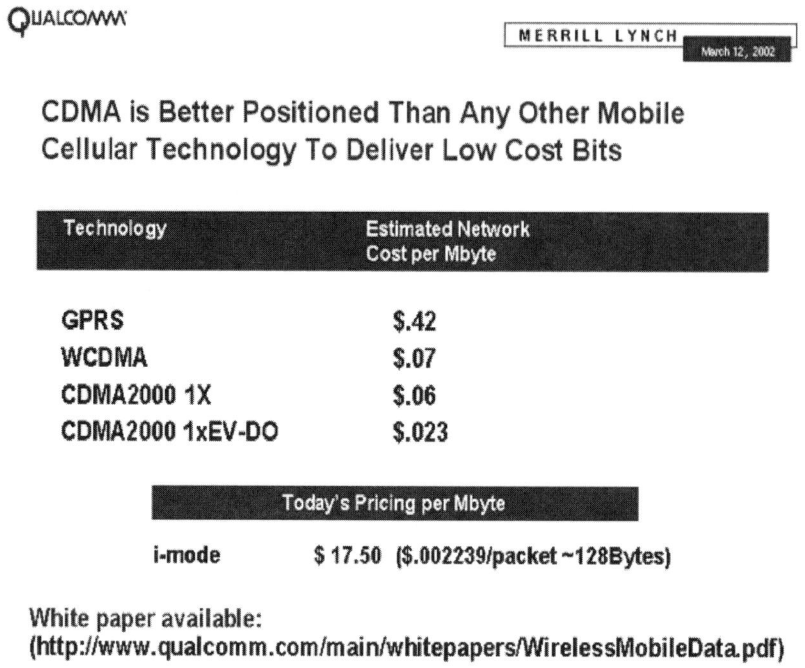

QUALCOMM

CDMA is Better Positioned Than Any Other Mobile Cellular Technology To Deliver Low Cost Bits

Technology	Estimated Network Cost per Mbyte
GPRS	$.42
WCDMA	$.07
CDMA2000 1X	$.06
CDMA2000 1xEV-DO	$.023

Today's Pricing per Mbyte	
i-mode	$ 17.50 ($.002239/packet ~128Bytes)

White paper available:
(http://www.qualcomm.com/main/whitepapers/WirelessMobileData.pdf)

Figure 4.1
3G Lowers the Cost of Data Services

To complement its policy of technology neutrality, the United States also took a different direction with regard to spectrum management. Unlike the EU, as a legacy of a uniform 1G analog network, America already enjoyed unified spectrum band allocations for mobile services. And the U.S. market still sufficed to generate global economies of scale in equipment. Therefore, powerful players, which already occupied spectrum bands used in Europe for 2G, had no compelling reason to abandon them to create transatlantic harmonization. For example, the U.S. satellite industry had ambitious plans for mobile satellite services using low earth orbit systems. These systems needed spectrum that overlapped with possible 2G and 3G systems. These obstacles made political leaders in the first Bush and Clinton administrations reluctant to alter existing spectrum plans (Office of Technology Assessment 1993). So, the United States selected more flexible bands for 2G. Canada followed the U.S. plan because its chief industrial and financial centers were tightly tied to the United States and its flagship equipment firm relied on sales in the United States.

Although spectrum harmonization did not move U.S. politics, 2G was still a hot economic issue. The Clinton administration used it to frame telecommunications reform to show that "New Democrats" were promarket innovators. Thus, the defining political agenda for 2G was a revolution in licensing by the creation of spectrum auctions that both yielded substantial revenues for reducing the government budget deficit and rapidly introduced a more competitive market for 2G services. Combined with the policy of technology neutrality, the Clinton policy set the stage for new national and regional networks, some of which decided to deploy an innovative, "made in America" technology for 2G: CDMA. The takeoff of CDMA thus unexpectedly became part of the success story of auctions for a Clinton White House eager to demonstrate its high-tech-friendly position.

Developing Countries Developing countries benefited enormously from 2G because most had severely underbuilt the wired network compared to demand. Their telephone monopolies suffered from over-staffing, inflated procurement costs, and corruption. They also struggled because their pricing was not related to costs. Governments charged too little for local phone service and too much for long distance. The high profits on long distance services was never enough to build out the local network, but served as a political barrier to realistic pricing of local services (Cowhey and Klimenko 1990). Wireless services provided a political escape from this trap because governments treated 2G initially as a premium service that was entitled to premium rates. It was also faster and cheaper to build out a wireless network, thus allowing operators to meet pent-up demand. To the surprise of most market participants, by 2002 2G made wireless phones more ubiquitous than wired ones. Consequently, developing countries had booming 2G operators that were the stars of the local economy.

The incumbent operator and a few large local firms dominated entry in these markets. As late as 1997 the norm was limited competition and a limited role for foreign investment in carriers. Foreign carriers could only penetrate developing market regions if they spent enormous time and money building a favorable image in the area and cultivating local licensing authorities. Even the partial exceptions were not open markets. According to Pyramid Research in Asia, by mid-1994 Hong Kong, Korea, the Philippines, and Thailand permitted wireless and some wireline competition, but restricted the number of competitors. Two local companies, Hutchinson Whampoa and Wharf, were the challengers in Hong Kong (Bruce and Cunard 1994; Chadran 1994; also on Latin America see Wellenius 1994). In the early 1990s a few of Latin America's larger economies, including Venezuela, Chile, and Argentina, introduced one and occasionally two competitors with caps on levels of foreign investment. These countries opted for modified beauty contests and bargained over the amount charged for the concession and features of the investment and service plans. This process favored a small pool of traditional carriers from Europe and the United States, notably Spain's Telefonica and the U.S. regional Bells, which worked the regional beauty contests and cultivated local partners assiduously.

Developing countries selected spectrum plans influenced by traditional relationships with suppliers. African administrations, long tied to European suppliers, agreed to follow Europe once again on technology standards and band plans. Asia adopted a mixture of band plans and technologies, but the European consumer success in selling GSM led national governments in Asia to tilt toward GSM and the European band plan. The notable exception was Korea's successful decision to advance its technology exports by becoming a major supplier of CDMA. It chose CDMA as a national standard to fuel scale economies for its companies as they prepared to enter the U.S. market for CDMA.

To get along, most countries in the Western hemisphere agreed to follow the U.S. and Canadian allocation decisions, at least in modified fashion. And, by 1997 the introduction of auctions for licensing in Mexico, Brazil, and a few other countries made it easier for new carriers using suppliers other than those traditionally in the market to gain a foothold in the market (MacAfee and McMillan 1996). Equipment suppliers, such as Qualcomm, sometimes even became partial owners of the new entrants to establish their technology in the market.

The Problems of 3G

2G wireless quickly emerged as the shining light of growth for incumbent stakeholders. Market growth soared and service margins reached 20 percent or even more, even in competitive markets. The financial community and traditional carriers became obsessed with mobile wireless. During the 1980s, most countries outside the United States granted a limited number of licenses for wireless and perhaps wired services.[26] As a result, most of the 1990s boasted a perfect climate for profit for wireless carriers—the service was hugely popular and competition was limited enough to concentrate on rapid buildout with high margins.

Still, warning signals surfaced. Even with limited numbers of competitors, margins finally came under pressure at the same time that governments inched toward letting more companies into the market. As markets matured, conventional voice services also grew more slowly. For example, according to data from the Strategis Group and the Cellular Telecommunications and Internet Association, the average price of mobile telephone service in the United States fell from $0.58 per minute in 1993 to $0.21 per minute in 2000. The average U.S. monthly bill fell from $61.49 in 1993 to $45.27 in 2000, as minutes of use per month jumped from 140 to 255. In 2001, 334 million people in Europe owned cell phones, 174 million people in the Asia-Pacific region, and 141 million in the United States. These numbers continued to climb, but at slower rates in mature markets. These figures illustrate the political problem facing government regulators. They planned to increase the number of competitors, but incumbent operators' growth was slowing (Sugrue 2001). Particularly as growth shifted to the developing markets, there was more emphasis on lowering prices by service market innovations (SMS was both popular with the young and cheaper than voice service) and by driving down the price of terminals.

3G planning might have seemed a slightly exotic exercise until the late 1990s when 3G emerged as a promised tonic for reinvigorating growth for incumbents and a few new entrants. This made it possible politically to allow more competitors into the market. Particularly in wealthier markets, technology innovation was supposed to boost the total size of the market while keeping margins high because it would stimulate growth in data traffic and facilitate roaming (a premium service) by high-end users over a few global networks.

Even though, except within Europe and parts of Asia, substantial international roaming was still rare and the investment costs would be gigantic, carriers dreamed of creating global footprints featuring global scale and global branding with seamless international networking. Until 2001, financial markets rewarded these strategies for three reasons. First, global branding was expected to attract large business customers. That would allow global carriers to bargain for better terms from data content providers. Second, it was believed that global scale would increase carriers' bargaining power with equipment suppliers, especially those manufacturing network and handset equipment. Operators normally subsidize handsets sold to their customers and therefore need favorable financing terms on network equipment from equipment providers. In turn, they demand small margins on the handsets they purchase.[27] Third, global operations required deep pockets, and incumbents had substantial financial capacity. In like spirit, although equipment makers knew that some global players would further squeeze margins on handsets, 3G opened up a whole new generation of equipment sales. That was critical for maturing markets in industrial countries.

A timely realization of these goals depended on achieving the original vision of a single global band plan and a single design for technology. This ultimately was a weak point of the 3G strategy. It was blocked by a variety of policy issues.

Managing Intellectual Property to Define a Global Technology Standard

First, high-speed data over mobile networks required substantial spectrum. The Europeans and Japanese reluctantly concluded that only CDMA seemed capable of using available spectrum efficiently enough to achieve the target data speeds. They knew this was a gamble because their manufacturers specialized in TDMA-based technologies, but they expected that in a competition based on traditional criteria of economies of scale, marketing, and systems integration for carriers, they could eventually surpass their U.S. rivals. This calculation overlooked an essential difference in the CDMA market. To a degree not initially appreciated, a single U.S. company, Qualcomm, controlled the key intellectual property for CDMA.[28]

Qualcomm's control of the IP platform severely undercut the typical arrangements for telecom networks in global standards bodies. The formal ITU rules about licensing are artfully ambiguous about expected terms for licensing, but no standard can emerge without the consent of all significant IP holders.[29] In this case Qualcomm controlled most of the key IP, which was its main competitive asset. Qualcomm could not give its IP away and survive because it was too new and too small to fight it out in a competition hinging on traditional criteria. It simply was not a traditional, vertically integrated supplier of telecommunications equipment. Therefore, Qualcomm insisted on collecting royalties. In addition, although Qualcomm was not a traditional leader in standards processes and had virtually no profile in Europe's ETSI, it insisted on a significant role in designing the 3G architecture.

Key players slowly realized the implications of Qualcomm's claims. European and Japanese suppliers resented Qualcomm's claim that it knew the best way to design a global wireless network. Incumbents viewed Qualcomm as an arrogant upstart with a cavalier attitude toward the global standards-setting processes. They certainly did not want to pay significant royalties to Qualcomm. So, Europe and Japan proposed a series of design features that would modify CDMA's performance for 3G by incorporating some features of GSM. They called this package W-CDMA, for Wideband-Code Division Multiple Access. These features also would have created new intellectual property that would weaken Qualcomm's control or provide Europeans with IP bargaining chips to force better licensing terms from Qualcomm.[30]

Qualcomm considered these design features as arbitrary or technically inferior. It worried (correctly, it turned out) that the W-CDMA design would have many teething problems that might jaundice carriers about 3G. It also worried about the implications for its IP holdings and suspected that the main purpose was to complicate and slow the seamless transition from 2G CDMA to 3G CDMA, thus strengthening GSM sales of 2G systems.[31] Qualcomm believed that if the transition from 2G CDMA was smooth this strengthened the case for buying CDMA at once. If the transition to 3G CDMA was likely to be complex, regardless of the choice of 2G technology, then there was less of a downside in selecting GSM.[32]

Qualcomm also recognized that in many countries with multiple technologies in 2G, CDMA was the choice of a newer entrant. This led the dominant incumbent to favor W-CDMA. NTT DoCoMo, for example, had a strong interest in urging the ITU to choose W-CDMA as the only 3G option because its technical specifications would make the 2G network of its rival, DDI (now KDDI), much less valuable for the third generation.[33] Similar stories, each with their own national nuances, appeared in India and China as they introduced greater competition.[34]

The traditional ITU players maneuvered to have W-CDMA adopted as the only ITU standard for 3G. Qualcomm responded by refusing to license its IP to the proposed ITU standard. Under ITU rules this refusal theoretically made it nearly impossible to set a global standard.[35] To guard against any possible loophole to its rights in the standards process, Qualcomm then won the support of a few key governments to back it in the ITU consensus system. The United States, of course, was essential. Qualcomm worked intensively with Lucent and U.S. carriers committed to CDMA to rally support in Washington. They triumphed, despite objections from GSM and TDMA carriers.

The political key was that Qualcomm and CDMA had become a showcase of how spectrum auctions could induce new technological successes. The Clinton administration worried that the global standards process might undermine the success of this "showcase" of the reform process. It justified its intervention in the fracas among American firms by relying on the established U.S. position that standard setting and licensing for 3G should be technologically neutral. Accordingly, the U.S. government vigorously pushed the ITU to adopt either a single standard acceptable to Qualcomm or simply endorse multiple standards. The United States intervened with Europe and Japan at the highest political levels.[36]

The initial positions of developing countries depended on their technological infrastructure for 2G. In practice, in 1999 most of Europe and Africa, large parts of Asia, and some South American countries relied on GSM.[37] However, important CDMA networks existed in Korea, Argentina, Brazil, Mexico, and Venezuela. In addition to strong support from Korea, Qualcomm courted a solid commitment to its ITU position by major operators in the Americas.[38] For example, Canada had a technology neutral policy, but CDMA was the choice of a powerful market leader. Even large operators such as Telefonica and Bell South, which did not use CDMA in their home markets, embraced CDMA in several South American properties where they were market leaders. By May 2007 there were 215 CDMA2000 operators in 105 countries offering commercial 3G services and 36 more planning to launch services by the end of 2007 using CDMA2000 technologies to deliver 3G services (CDMA Development Group 2007).

The ITU system has a strong regional component to its decision-making process. The CDMA camp in the Americas meant that North and South America insisted on policies that made it difficult for the ITU to take any decision on standards (or spectrum) that would undermine the Qualcomm position. The W-CDMA camp could not paint this as an issue of the United States versus the world.

Ultimately, there was a compromise. The major suppliers recognized Qualcomm's IP while Ericsson, the last major company to license from Qualcomm, purchased Qualcomm's network supply business to shore up its CDMA position.[39] Only then did Qualcomm compromise on its 3G design to allow the GSM camp to build in some special features for one version of 3G that Qualcomm had previously rejected. This horse-trading meant that, contrary to the ITU's original 3G plan, *three versions of 3G were initially sanctioned*. The first, cdma2000 1X, was a direct descendent of Qualcomm's 2G cdmaOne technology. The second, W-CDMA (also called Universal Mobile Telecommunications System, or UMTS), drew more from GSM and incorporated some features that Qualcomm had resisted. The third, TD-SCDMA (Time Division-Synchronous Code Division Multiple Access), is an idiosyncratic blend of CDMA and TDMA that will drop from the marketplace unless China continues to champion it.[40]

Regional strategies had intersected with global institutional dynamics to thwart a single technical design for 3G. It also created an intense industrial rivalry between the two main "flavors" of 3G. Although large players planned to sell into both camps, each side had a clear preference. cdma2000 enjoyed a head start in deployment because the transition to it from existing CDMA systems was straightforward. With South Korea leading the way in network deployment, the standard was fully specified and the chip sets shipped in considerable commercial numbers by the first half of 2002. Japan's KDDI and several carriers in the Americas and Asia soon followed. In contrast, in mid-2002 W-CDMA standards were not yet fully specified in Europe and much of Asia (Japan's DoCoMo was the exception) and therefore no commercially viable handsets and chip sets were available. Substantial shipping of W-CDMA slipped to 2004.

The significant delay in 3G buildout plans has profound consequences for the economics and performance of 3G. Although the numbers should be viewed with caution because they come from Qualcomm, figures on data speed and costs show that all 3G systems have better performance than 1G or 2G networks.[41] A new system, 2.5G, emerged in 2001 as a transition offering because it could be deployed on 2G networks as an upgrade. Inevitably, even this limited upgrade became mixed up in politics. GSM carriers in countries like India preferred to upgrade their infrastructure to 2.5G and focus on keeping handset costs low in order to build out the market for lower-income users. This approach also was used to justify an argument that the Indian government should give priority to expanding spectrum for existing GSM networks rather than upgrade to 3G where the commercial battleground might look very different.[42]

Spectrum Management and the Assignment of Licenses

The ITU process coordinated 3G spectrum planning but the bargaining positions emerged out of regional dynamics with different legacies from 2G. European suppliers and carriers began the 3G process with the goal of creating a uniform global band and a homogenous network environment (W-CDMA) (Commission of European Communities 1998; Council of the European Union 1999). Given the dominance of GSM in Asia, Asian band allocations approximated those of the EU. Therefore many European and Asian carriers systematically considered building a global footprint from the start.

North America was the largest stumbling block for band harmonization because the United States had never fully committed to 3G. Indeed, at the 1992 WRC the U.S. position favored a commitment to facilitating mobile services, without giving special priority to 3G over 2G or mobile satellite services. The United States did not begin clearing the spectrum designated elsewhere for 3G until 2003 (*Telecommunications Reports International*, June 1, 2001).[43] Even then, the United States declared that 2G spectrum could be used for 3G, thereby creating diversity in the global spectrum band. As a result, economies of scale in equipment are hurt. This may impair performance for consumers because even phones on the same standard often must contain chips designed to work on two sets of frequencies to allow global roaming.[44]

By 1998 most industrial countries had competition in mobile services that went beyond duopoly. The policy for licensing 3G spectrum depended on the political economy and institutional processes in each regional market. For example, the weakness of EU institutional capabilities drove significant aspects of the auctions in Europe. An unexpected consequence of the intersection of the politics of licensing systems with the pursuit of global networks by large carriers was the emergence of a consolidated set of megacarriers, not the expected radical expansion of market participants.[45]

Europe Insiders and journalists were obsessed about the cost of the 3G auctions in Europe (exceeding $100 billion), especially the United Kingdom and Germany auctions, and frequently attributed the early failure of 3G to auctions. This emphasis missed the three pillars of the politics that set the strategic context of the auctions: spectrum allocation, standards setting, and institutional processes.

The EU initially decided that, for reasons of technology and industrial policy, incumbents could not use their 2G networks to deliver 3G services. Reallocating spectrum is politically difficult and the EU governments eased the task by reframing it as a major coup for industrial policy. Accordingly, EU governments decided that 2G spectrum was already too crowded, and 3G would benefit from having substantial capacity on "virgin" spectrum. They wanted all of the new equipment and service providers operating on exactly the same European bandwidth. Reinforcing the decision to require separate licenses for 3G was the implicit decision to restrict licenses only to W-CDMA technology. Although not illegal, any licensee using another technology was at risk if it wished to create a pan-European network. Thus, Europe implicitly tied licensing to a technology standard (Cave 2002, 216–217). Uniformity meant that European suppliers could maximize their economies of scale and the natural public relations advantage of early continent-wide deployment of a single new 3G network. The auctions for licenses also were intended to force carriers to roll out networks quickly in order to create revenue streams to pay off licensing costs. This quick deployment was intended to guarantee European leadership in 3G, allowing duplication of the GSM successes.[46] It is essential to note that these two market-steering policies—the uniform virgin band for 3G and implicit compulsory standards—had nothing to do with auctions. However, combined with regulatory institutional processes, they explain a significant part of the auction story.

Requiring a license for virgin spectrum meant that the major incumbents in each market had to win the 3G auction or forfeit the 3G market. Moreover, the lack of clear institutional power for the European Union over spectrum licensing made it implausible to create a single European-wide auction (as happened in the continental U.S. market for 2G). Buying a place at the table became especially expensive because the German and British auctions took place early in the European auction cycle when large players believed that they needed to win in both of these two key countries or forfeit development of a pan-European network.[47] Therefore, there was a strong temptation to pay a premium not to lose the pan-European option. This was an inherent risk of issuing national spectrum licenses in sequence, as opposed to simultaneous European-wide auctions. When auction bids skyrocketed, so did the debt burdens of the winning bidders. When teething problems for W-CDMA technology delayed delivery of the equipment, the stock market soured on telecom carriers and problems mounted because carriers were caught with 3G licenses containing technology and rollout requirements.

In addition, 3G licensing also induced less new entry in Europe than originally predicted.[48] This was partly because the variation in national auction designs caused some countries to have less competitive entry than others (Klemperer 2001). France and a few other countries dispensed with auctions. Overall, although some new local industrial firms entered, they only selectively altered the complexion of the European and major global markets. Each region was mostly dominated by existing licensees, many drawn from the ranks of old wireline networks, which often were part of a small pool of emerging global supercarriers. In Europe, for example, a few traditional incumbents like British Telecom, Deutsche Telekom, France Telecom, and Telefonica along with two newer supercarriers, Vodafone and Hutchinson, initially commanded a large share of the key auctions. (However, DoCoMo was a key minority investor in Hutchinson.) Alliances of Scandinavian incumbents also played key roles in Northern Europe.[49] (The slow deployment of 3G later forced reorganization of these strategies, especially the retreat of British Telecom.)

Japan Japan had diverse technology standards due to the legacy of U.S.-Japan trade negotiations. But the government continued to manage the market to ensure the stability of incumbent carriers and the ability of NTI to assist equipment suppliers. Therefore, it used a beauty contest to award 3G licenses to the three incumbent wireless carriers.[50] The KDDI group, the beneficiary of the Motorola trade war, adopted the cdmaOne and cdma2000 standards. DoCoMo, NTT'S mobile wireless group, built around the W-CDMA standard. So did J-Phone, an affiliate of Vodafone descended from consolidation of three other carriers. Although DoCoMo was the first to roll out 3G service, KDDI grew more quickly, in part by selecting less expensive and more reliable handsets made possible by seamless compatibility with CDMA's 2G technology (Nakamoto 2003, 17)[51] Meanwhile, DoCoMo experienced severe, early technical performance problems on its reported $10 billion network plan. Still, fuelled by revenues and stock valuation made possible by its success in 2G instant short-messaging services (i-mode), DoCoMo invested heavily in minority shares in AT&T and European carriers to leverage a package of i-mode and W-CDMA. This strategy also assisted its traditional group of Japanese equipment suppliers. Through 2006 the mobile market had three characteristics. The first was high prices as other carriers operated under the price umbrella of DoCoMo. The second was extensive technological innovation because

KDDI had successfully created a technology race with DoCoMo while not upsetting pricing. The third was a reorganization of competition in the market as the two most aggressive entrants in wired broadband networking, Softbank and eAccess, entered the wireless market.

Korea Korea tied licensing to technology and export promotion goals. The Korean government, which continues to play a strong but less than transparent role in the selection of technologies, hedged its bets. It insisted that all three carriers use the preliminary version of cdma2000, 1XRTT, in the short run but hedged on long-term choices. Knowing Korea's determination to be an early adopter of 3G, all major equipment suppliers put money into Korean carriers during the height of the Asian financial crisis in order to influence technology choices.[52] The Europeans and Japanese stressed the benefits of building experience with W-CDMA at home to gear up for export. In addition, Korea Telecom, the dominant incumbent for wireline, was second in the wireless market, and saw W-CDMA as a way to differentiate itself.[53] In the end, the government effectively required one carrier to provide cdma2000 and two others to provide W-CDMA.[54] Later, the government focused on introducing a Korean-designed version of a new wireless technology (the WiBro variant of WiMax technology) into the Korean market as an effort to increase the control over intellectual property of Korean manufacturers. The success of this strategy was still very uncertain as of 2006.

China and India The crucial question in Asia was what will happen with China and India. GSM dominated in both countries and the governments openly designated technical standards for services. Following the precedent of Hong Kong and hoping to develop export technology markets on the Korean model, China opted for technology diversity once it had commitments by major equipment suppliers to license CDMA and GSM exports. China licensed Unicom to use both GSM and cdmaOne for 2G mobile. It has charted a course to have all three versions of 3G deployed. India accommodated the entry demands of its only industrial giant without a wireless play, Reliance, by granting it a license for fixed (limited mobility) services for CDMA.[55] As a result, customers probably will have a choice between both flavors of 3G in the largest Asian markets even if W-CDMA predominates. But the timing of 3G expansion is still in question because 2G incumbents argue that a poverty alleviation strategy should give priority to 2G network expansion, which will have lower costs for customers.

United States Technology neutrality and (limited) service neutrality in licenses meant U.S. carriers could convert their 2G networks to 3G when and how they chose. However, it took until mid-2003 for the United States to begin allocating additional spectrum for new auctions.

This policy mix resulted in a mixture of carrier strategies for upgrading to 2.5G or 3G. The CDMA carriers (Alltel, Sprint, and Verizon) focused on the large North American market because CDMA coverage was so spotty elsewhere in 2G. Markets beyond North America were a bonus, but could not be counted on.[56] However, these carriers pressed to win advantages from first deployment of new services because it was easier for them to upgrade from CDMA for 3G. By contrast, the U.S. GSM/TDMA carriers (AT&T Wireless, Cingular, and VoiceStream) faced larger technology challenges on W-CDMA because they had to replace their core network equipment. Ultimately, AT&T and Cingular merged because the 3G network deployment favored those with larger economies of scale. Until Deutsche Telekom's contentious purchase of VoiceStream was approved, foreign carriers were cautious about U.S. entry, slowing effective global alliances entering the United States, but T-Mobile is now one of the four major U.S. national carriers.

What Next?

3G ran into trouble because it was an unusually ambitious effort to coordinate global technology planning. It began in an era dominated by monopoly but had to evolve in a more competitive milieu. The politics of introducing competition meant that most major wireless carriers were offshoots of the traditional wired network carriers. These carriers spread into territories of their rivals and began to widen their pool of the equipment suppliers: But the key to 3G remained the symbiotic relationship of the small pool of carriers and equipment suppliers, a swan song of a less competitive era.

The politics of introducing competition, discussed earlier, eventually doomed 3G planning because they blocked timely achievement of all three of its premises. Qualcomm, an upstart with a strong IP position and different business incentives, disrupted global standardization institutions and forced a diversity of standards. A uniform global band plan only emerged slowly and imperfectly because national and regional incentives for managing spectrum worked against global strategies. And the licensing of 3G was bedeviled by problems because technology promotion and other goals effectively hobbled market flexibility, thereby hindering the ability of carriers to adapt during the telecom downturn.

The 3G implosion in 2001 shook established and newer carriers. As a politically prominent reform that generated investment and jobs stumbled, government leaders scrambled to provide relief, even at the cost of undercutting market efficiency. Pressure increased especially in Western European countries where governments feared deep job cuts or bankruptcies. Many chose to revisit their licensing strategies because plunging stock values for heavily debt-burdened carriers impaired their financing capabilities just when they needed to incur the substantial cost of network buildout. The policy question is what to learn from efforts to address these problems.

Policies for Financial Assistance

One way Europe tried to help its carriers was using pedestrian tinkering to forge financial relief. Quite direct financial relief was undertaken in the Netherlands and France (Andrews 2002; Tagliabue 2002).[57] Another strategy was to change the licensing terms to provide financial relief. Thus, France extended the term of 3G licenses and reduced license fees (*Telecommunication Reports International*, October 26, 2001, 1–2).[58] Both approaches have all the usual flaws of industrial subsidy packages. Relaxing regulation to allow carriers in Germany and the U.K. to share the buildout of certain network infrastructure also provided relief (*Telecommunication Reports International*, September 28, 2001, 12 and 20).[59] The verdict is still out on the competitive effects of this strategy.

Financial relief can also occur through inaction, such as when a government moves slowly to address market conditions that yield large profits for hard-pressed carriers. Although regulation should address competition problems, not oppose high profits, analysts worry that some wireless profits arise from the exercise of market power. For example, mobile operators in most countries other than the United States profit handsomely from high fees they charge to terminate calls that originate on the terrestrial network. There were trade complaints about DoCoMo's manipulation of such fees in Japan (*Telecommunication Reports International*, April 3, 2001, 2–3). This issue came under discussion within Europe in 2003, but a price cap on roaming fees did not emerge until 2007.[60] Similarly, termination fees are even higher for international calls made when customers use their home country services in another country or receive international calls on their mobile phones (Noam 1993, 46–47).

New Approaches to Spectrum and Licensing

More fundamental and promising reforms were on the table. Some of the largest European players began to advocate strengthened property rights on spectrum licenses, thus granting more market flexibility to deal with adverse circumstances. The EU agreed that starting in July 2003 3G licensees may trade spectrum and licenses to provide financial relief, not just to deal with the awkward problem of direct subsidies for existing licensees (*Wall Street Journal Online*, 2002a). Even more significant are two developments.[61] One is to allow greater flexibility on the choice of technology and services on a spectrum band. While the precise amount of flexibility to be allowed on the already allocated 3G bands was still uncertain in early 2007, the movement to more flexibility was clear (Cave, Prosperetti, and Doyle 2006; Cave and Nakamura 2006). The other was the opening up of a diversity of spectrum bands for the same generation of services, such as the deployment of 2G and 3G on the 450 MHz band in some European countries. Finally, the difficulties of deploying new wireless technology to a unified European market are driving the EU toward a common spectrum policy agency.

Recommendations: A Different Approach to Global Innovation

Even as 3G plays out, many urge a vigorous push toward 4G that would introduce an integrated model of wireless technologies, especially on unlicensed bands (such as 802.11b, known more generally as Wi-Fi, and its successors), to permit much higher speeds and other capabilities. While the technological contenders are of many flavors, WiMax (heavily promoted by companies like Intel and Samsung) is an especially prominent candidate. Although a fixed wireless technology, it promises much higher speeds than the original 3G design and has a plan to achieve mobile capabilities. Many see it as the start of a 4G infrastructure. There are many international planning groups for 4G now in existence.

Planning for 4G could easily become an example of not learning from experience. Its premise is that 3G was the right idea, but flawed either by bad timing (prematurely pushing for high-speed wireless before better technologies were available) or poor execution (including the corporate battles over rollouts). This misses the point. 3G assumed that massive global coordination of standards, spectrum, and licensing policies was possible in a timely manner. But the stakeholders in wireless communications, even in the insiders' community, have diversified significantly while the coordination mechanisms remain relatively weak.

The goal of 4G also assumes that the shape of the future is known. This severely taxes the ability to forecast in any technologically innovative, competitive market. For example, as a result of the market pressure from WiMax, 3G has entered into a much more rapid path of technological change, including the incorporation of elements of technologies that many associate with 4G. Our interviews in 2007 showed a sharp divergence of opinion among carriers about which technology approach would generate the fastest average speed, the greatest network capacity, and the "right" cost structure. From the viewpoint of public policy, this technological race is good for consumer welfare. And a collective planning process to select technologies threatens to slow innovation.

A better model for standards and IP resembles the modal type of the information industry. Collective efforts on standardization of technologies and supporting business processes embrace a pluralistic view of the future. There are competing models of the future and various collective efforts to advance these visions. Although markets, technology communities like the Internet Society, or even governments may evolve a single standard for key parts of the landscape, the goal is ***not*** to develop a single consensus model of the future. The capabilities associated with 4G can be nurtured through much more vigorous test-bed processes and narrow, specialized standards setting. Emphasizing interoperable interfaces so that different technologies could interconnect smoothly would be a desirable goal, and public R&D could further encourage it. The IP process broke down in the standards process for 3G precisely because a monolithic design raised the costs for the players.

Spectrum allocation would improve if it embraced "spectrum flexibility," using more flexible tools to manage spectrum. Private coordination mechanisms facilitated by market incentives should supplement or replace government coordination or both. Although it will be politically contentious, this idea of spectrum flexibility needs to be introduced into ITU processes where a U.S.-EU coalition could give it considerable traction.

We endorse three recommendations of the special report on spectrum submitted to the British government. (1) Do not harmonize spectrum globally in the absence of large cost-benefit advantages. (2) If harmonizing, rely on broad service categories such as mobile wireless, not particular technology descriptions such as 3G. In other words, use the minimum number of parameters to describe the harmonization. (3) Harmonize only for a limited period of time (Cave 2002). To this we would add, (4) Encourage regional experimentation, especially in the higher frequency bands (5 GHz or higher perhaps) and parts of existing television spectrum. More broadly, we conclude that top-down planning of future technology is unlikely to work well. Instead of picking winners, governments should allow new technologies to emerge and succeed organically by emphasizing requirements that whenever possible the goal should be to minimize interference rather than place restrictions on the use of band.

Developing Countries and Reform

As of late 2006, the lessons derived from the 3G experience have significant implications for most developing countries. Wireless networks are far more significant for the general communications infrastructure of these countries than for wealthy nations. They are also becoming the locus for market growth as the industrial countries mature.

Except for the distinct minority of countries already embracing CDMA in 2G, most developing countries did not have 3G systems licensed in early 2003. In general, they also have fewer competitors for mobile services than in industrial countries (*Wall Street Journal Online*, 2002b; Ramakrishnan 2002; Mitchell 2002). Thus, they had an opportunity to examine the merits of spectrum reform, technology flexibility, and competition policy before replicating an approach to wireless policy that has underperformed and run into great difficulty.

By 2006 many countries were beginning to deploy wireless broadband. But there were many of the familiar problems of traditional wireless policies. Some countries, such as Colombia, license one wireless technology but not the others as they try to balance out competitive equities among firms in markets where government still tries to micromanage competition.[62] Others, such as India, consider using spectrum policy to favor technological platforms out of the hope of managing economies of scale to lower costs. In India, this took the form of a heated debate over the merits of more spectrum for 2G rather than 3G. But such debates assume that broadband data is a service for the wealthy in the same way that there used to be policy debates over telephone and short message service (SMS) connectivity for the poor. The demand may be less intense among the poor, but they are not totally isolated from global society and its markets. And the cost differentials between generations of services rapidly diminish because the fundamentals of declining costs of electronics are relentless drivers of lower prices. The real lesson of the mobile revolution is an extension of the general lessons of the digital revolution—policy should encourage competition and experimentation. Old practices are not good predictors of service innovations that serve society well. It is time to put out some warning signs—"copy only if facing congestion"—on many traditional tools for managing radio frequencies. Too much of the global advisory process conveys the opposite message.

A second assumption should be that there will not be a single, neat technology or market model for 3G (or its rivals). This provides an opportunity for at least some relative commercial newcomers to compete. The success of Hutchinson of Hong Kong is indicative because it had the advantages of local experience when deciding how to build 3G services in the fast-growing, but now fragmented, developing markets. Many of the most aggressive operators in poor countries are emerging out of other developing markets that draw on the management and operation experiences of their home markets. Technological variety allows fine-tuning of service and business models. Indeed, 3G is morphing into even more variants as it becomes a technological hybrid. Variety will not matter because innovation on the handset can allow interoperable networks to become a reality.

Even more important is the success of South Korean companies specializing in CDMA. Market diversity opened the way to a commercial breakthrough. Such specialized entry is more, not less, likely in a world where policy induces less uniformity. The search for profits to sustain 3G may drive megacarriers like Vodafone and Orange to turn to specialized suppliers of equipment, applications, and network software upgrades (*Business Week Online*, 2002). In a sense Qualcomm is an early version of this stripped-down, specialized supplier. Its business model allows it to partner in creating new equipment suppliers in key markets because it is not in the general equipment business. At least for developing economies that are nurturing advanced centers for innovation, the growth of suppliers with this kind of strategy may open future opportunities. This is particularly true because much of the service innovation will require specialized component and software technology on the handset in the future. A look into the future is the LG mobile terminal introduced in 2006 that is optimized to do remote diabetes testing for patients. As this product illustrates, the mobile terminal is now a platform for specialized solutions designers, and many of these will come from developing countries.[63]

The more general lesson for developing countries is simple. The industrial countries, out of painful experience, will have to re-engineer their spectrum allocation, licensing, and standards policies. Developing countries relying more heavily on wireless networks need to move even faster and more radically to adapt their policy approaches.

Conclusion

The creation and implementation of 3G wireless networks is a story of technological innovation in a marketplace undergoing structural transformation and a policy system lagging behind the pace of innovation. The third-generation effort was both ambitious and flawed for the same reason. 3G was supposed to create the new pool of high-margin revenues that would ensure the growth of longstanding dominant players while accommodating some new entrants. However, interests diverged as the number of entrants grew, especially as different world market centers adopted different compromises between incumbents and entrants. The global decision-making process for setting standards and coordinating spectrum could not reconcile the clashes. As a result, third-generation networks have more varied technology and spectrum plans than originally envisioned. Furthermore, commercial strategies—such as those in Europe—based on a quick deployment of the networks stumbled. The lesson of 3G is simple—major shifts in wireless technology in the future need to emerge out of a different policy process, one more attuned to the consequences of competition. For their part, in the future developing countries should adopt regulatory strategies that anticipate significantly different paths to technology innovation, and better consumer welfare.

Notes

1. Despite the investment bubble, procompetitive reform boosted efficiency and improved consumer welfare in the global communications market. Wireless communications expanded the availability of communications in developing countries and boosted connectivity rates in mature markets.

2. Where network externalities exist, networks grow more valuable to individual users as more people use or are connected to them.

3. Wireless networks are somewhat path dependent. Sunk costs in current network equipment means that new technologies must provide returns sufficient to abandon existing technology infrastructures.

4. Equipment vendors can reap large advantages if they "lock in" customers to a more specialized technology platform. Once a carrier installs a supplier's network equipment, the carrier is locked in and is unlikely to switch equipment vendors. Global carriers prefer suppliers with global

support capabilities, so this limits entry for both network and handset equipment. (Interviews with European and Asian suppliers, November 2002 and December 2002.)

5. Krasner emphasizes the role of power in determining which approach to coordination wins out. He argues that the spectrum problem typifies elements of what game theorists call "the battle of the sexes." We believe that power matters in the context of political processes that shape the preferences of countries and the way in which power is applied to decision making as described in Austin and Milner 2001.

6. Property rights are assignments of the ability to control and use an economic resource. They typically include a mix of rights (e.g., the ability to make a profit and resell the resource) and responsibilities (e.g., liability responsibilities for damages) for owners of the rights.

7. The new alliance brought together large corporate users that constituted a large percentage of long-distance traffic, equipment suppliers outside of the traditional vendors to telephone companies, and carriers that had identified potentially profitable entry strategies in the market.

8. Flawed property rights are difficult to fix. This makes it difficult to use commercial side payments as an alternative to continued regulatory micromanagement.

9. The structure of government institutions, the nature of electoral systems, or the form of executive power (e.g., parliamentary or presidential) influences how these strategies play out in different countries. Our analysis of global markets handles these factors on an ad hoc basis.

10. Transparency and due process make regulatory commitments to protect private property more credible, but also alter the balance of influence among stakeholders by rewarding those with resources to participate intensively and who do not need fast decisions.

11. Governments were heavily involved in the standards-setting process for telecommunications because, in most countries, they owned the telephone companies.

12. The ITU was created in 1865. At the end of 2002 there were 189 member states and over 650 sector members.

13. The laws of physics make bands differ in their radio propagation characteristics, so spectrum is not equally tractable for all tasks. For example, spectrum bands over 100 MHz permit straight-line transmissions that can be power efficient.

14. The absence of private property rights for spectrum partly reflects high transaction costs in assigning and monitoring individual property rights in the early days of radio. It emerged from a tradition of state-building that reserved commons for government ownership. Government control also satisfied the large demands for spectrum of military and police services (about 30 percent of the spectrum) that few political leaders wanted to oppose.

15. The arcane regulatory process is fiercely contested. Advocates debate what would constitute a threat of interference and how to reallocate different pieces of spectrum to different uses. These proceedings raise enormous informational problems for government decision makers. The glacial process cumulatively favors incumbents. Political leaders could change the system but so far have been content to allow institutional dynamics to slow the pace of change.

16. In addition, member governments have committed to work within ITU allocations. Therefore, national bargaining positions must take these ITU dynamics into account.

17. Countries also viewed commercial services as local, which served as a self-fulfilling prophecy.

18. Governments subsidized carriers by not charging them for using valuable resources. The rents created by this choice were shared with labor and equipment suppliers.

19. Given weak property rights, commercial compromises among companies may not emerge without a credible enforceable guarantee. Political decision-making processes shape possible trade-offs.

20. In May 1998, 80 million subscribers still used one of the three major families of 1G analog systems. There were 125 million digital 2G subscribers (70 million used GSM systems, 26 million Japan's PDC system, 15 million on CDMA and 13 million mainly split among technologies idiosyncratic to the United States, the EU, or Japan. The market estimates were provided in internal documents of one market supplier in May 1998.

21. National and regional standard-setting processes varied. Usually voting procedures, to the extent they were specified, favored larger incumbents. Effective participation required both a significant commercial presence and the ability to fund staffers who could dedicate extensive time to the standards process. Voting, if used, often was weighted according to market revenues and required supermajorities.

22. When additional bands in a higher frequency opened for 2G, the EU still required use of GSM.

23. If incumbent wireline operators had not controlled major wireless firms this probably would not have been politically, viable.

24. For example, the first competitive British license for wireless went to Cellnet, owned by Racal, a British equipment company. Later 2G licenses all went to U.K. firms (Mercury, Vodafone, and Orange). The United States also limited the pool of potential entrants using restrictions on foreign investment rights. Although subject to waiver, Until 1997 the FCC limited foreign investment in wireless carriers to 20 percent. Even then, the FCC's true intent was not irrevocably clear to foreign investors until its approval of Deutsche Telekom's purchase of VoiceStream in 2000.

25. There was a bipartisan political consensus made possible by the diversity of U.S. industry. The FCC declared technology neutrality, agreeing that government could not usually select the right mandatory technology even if there were cases when it might be hypothetically advantageous to do so.

26. For example, as the second generation matured, the national Japanese wireline carriers evolved into three groups with wireless subsidiaries—NTT DoCoMo, J-Phone, and KDDI. DoCoMo was part of the NTT business group, the traditional domestic incumbent. The other two represented the consolidated carriers from the former international monopoly (KDD) and the three long-distance entrants licensed in the 1980s. Most European countries in this period did not allow long-distance competition and awarded licenses to only one or two new competitors in wireless. This pattern did not change until the late 1990s.

27. The size of the market of your "flavor" of 3G influences the total cost structure for the technology. Within that cost envelope, any individual carrier's buying power depends on factors such as the size of its potential purchases.

28. A series of patent suits brought mainly by Motorola and Ericsson did not weaken Qualcomm's supremacy. They were settled in 1999.

29. Traditionally some standards-setting organizations, including the ITU, demanded "royalty-free licensing." Many others now require "reasonable and nondiscriminatory" licensing. (This discussion relies on Patterson 2002.) In 2000 the ITU Telecommunication Standardization Bureau stated: "The patent holder is not prepared to waive his rights but would be willing to negotiate licenses with parties on a nondiscriminatory basis on reasonable terms and conditions." The bureau does not set precise criteria for these conditions and leaves it to negotiations among the parties. But, the relevant factors for setting royalties include costs for development and manufacturing plus profits (Patterson 2002, 1053–1054 and note 40).

30. Even in 2003 other vendors commonly complained of the "Qualcomm tax," the royalty rate charged by Qualcomm for its IP.

31. This description is based on materials provided to the authors by Qualcomm.

32. Concern over 2G sales explains why neither side followed the economic logic of compromise to grow the market size that is set out in Shapiro and Varian, 1999, 237–242.

33. The key event producing the W-CDMA initiative was a successful negotiation on common interests among the largest expected winners in Europe and Japan—DoCoMo, Nokia, and Ericsson. Lightman with Rojas 2002, 90–94, point out that if the ITU had standardized only around W-CDMA specifications, the chip rate in the system would have been incompatible with seamless upgrading from second-generation CDMA systems.

34. The United States had no comparably dominant wireless incumbent. AT&T was a TDMA carrier as were the wireless groups of several large Bell operating companies. Verizon and Sprint ran the flagship CDMA networks. So, the carriers quarreled bitterly over the U.S. position in the ITU on standardization.

35. Qualcomm notified the standards bodies involved in 3G that it held patents that were essential to all proposed versions of 3G. It offered to license, on reasonable and nondiscriminatory terms, to an ITU standard either based on Qualcomm's proposed standard or a single converged ITU standard for 3G (an acceptable hybrid of W-CDMA and Qualcomm's proposed standard). It declared that it would not license to other versions of 3G, such as the EU's W-CDMA standard. (Qualcomm press release, June 2, 1998.)

36. For example, on October 13, 1999, Secretary of Commerce William Daley, U.S. Trade Representative Charlene Barshefsky, and FCC Chairman William E. Kennard released a letter to EU Commissioner Erkki Liikanen protesting EU policy on 3G.

37. Most low-income developing markets rely more on European suppliers of network equipment than they do on North American suppliers. This partly reflects the legacy of colonialism that led these European companies into a far earlier drive toward serving these markets.

38. South Korea was a particularly strong Qualcomm backer. Early on it supported CDMA through heavy investment in its networks and equipment in the hope of building a significant export equipment market. This calculation proved correct. In 2002, CDMA handsets were reputedly the largest single export item in the Korean high-tech sector.

39. As part of the deal, Ericsson also concluded its patent suits with Qualcomm.

40. According to Kynge (2002), in November 2002, China set aside large amounts of 3G spectrum for this blended technology standard. As of 2007, it appears that China will deploy multiple standards.

41. A separate debate rages over the top end for performance of the 3G "flavors." Qualcomm, of course, argues that cdma2000 can evolve into a much higher-speed, lower-cost network solely for data than can W-CDMA.

42. Most agree that the cost of transmitting data is radically lower over 3G networks compared to over 2 or 2.5G networks. So, 2.5G networks have much smaller data potential and still emphasize voice. The discussion of India is based on author interviews in New Delhi in July 2006.

43. Pleas for urgent action by carriers endorsing W-CDMA were countered by the military and public safety agencies that held the desired spectrum. Officially, cdma2000 carriers endorsed reallocation, but their real preferences were unclear because they could launch 3G without new spectrum. Opponents included the politically powerful UHF television broadcasters.

44. Some phones will be able to handle both 3G modes, to be both dual band and dual mode. This increases costs for production in a market where consumers demand low prices.

45. However, horizontal cross-entry by the large supercarriers clearly invigorated competition.

46. This logic is exactly opposite of the reasoning of auction critics like the economist Martin Cave who believe auctions drain potential investment capital.

47. Recall that weak spectrum property rights meant that these players could not confidently turn to the resale market to purchase another license.

48. Auctions in Britain and Germany yielded the most licensees (five and six, respectively). Italy had only five final bidders for five licenses, later reduced to four. Spain and France allowed in fewer new competitors initially.

49. The Scandinavian carriers, such as Sonera, sought Scandinavia-wide footprints and selective entry into major roaming markets for their customers. They ran up large debt burdens even though Sweden decided to distribute its 3G licenses in a beauty contest.

50. See www.itu.in/itunews/issue/2001/08/licensing3g.html.

51. The problems of handling asynchronous data transfer on mobile handsets caused short battery lives and overheating of early W-CDMA handsets (cdma2000 used synchronous data transfer). (Interview data, January 2003.) DoCoMo bailed out its equipment suppliers on development costs.

52. Qualcomm led funding for one group. A Hong Kong consortium, with European supplier backing, funded a second carrier. DoCoMo put money into a third. (Interview data, January 2003.)

53. The incumbent carrier favors W-CDMA because it trails in the mobile wireless market and hopes to use the technology to create brand differentiation. (Interviews, Seoul, December 2002.)

54. Samsung, the largest Korean equipment supplier, was required to supply phones for both standards. (Interviews, Seoul, December 2002.)

55. China's Unicom runs a GSM network for the mass market and a cdmaOne network for business customers. A small cdmaOne carrier in Hong Kong completes the China footprint for CDMA. Qualcomm invested $200 million in 2002 in the Reliance Group of India, the country's largest firm, to demonstrate its support for the Reliance CDMA plan. Later, Qualcomm and Reliance had a falling-out over licensing terms and the Tata Group, the traditional giant of Indian commerce, aligned with Qualcomm and CDMA.

56. Verizon, the largest CDMA carrier, also had substantial minority ownership by Vodafone, which limited its own overseas activities.

57. Debts were high for major European carriers. This included 65 billion euros for Deutsche Telekom, and 64 billion euros for France Telecom. In 2001 the Dutch government assisted KPN, the traditional carrier, in a new financial offering to allow it to refinance debt. In 2002 France provided a direct financial subsidy to France Telecom.

58. The term went from fifteen to twenty years in 2001 while fees went from a 5 billion euro fee to 619 million euros plus an annual royalty payment to be based on earnings. Bouyugue Telecom, which had dropped out of the auction because of the high price, was quietly promised a license on the same terms.

59. Skeptics suggest that real savings will amount to 5 to 15 percent (*Telecommunication Reports International*, April 27, 2001, 4).

60. In 2003, British Telecommunications (BT) decided to shed its major wireless carrier and successfully persuaded the British regulator to cut termination fees from wired to wireless networks, thus providing financial relief to BT at the expense of Vodafone. We thank Chris Madsen for this point.

61. This would allow new technology into 2G bands. However, there is sharp opposition in many EU quarters. (Interviews with EU and U.S. equipment suppliers, January 2003.)

62. In Colombia the government worried that companies licensed for long-distance service were suffering. So, it decided to give them a head start in wireless broadband by licensing them to deliver WiMax. (Based on author interviews, October 2006.)

63. One of the authors sat on an international prize committee in 2007 for the design of innovative mobile service "solutions." The grand prize winner was from India. Two of the prizes went to Indonesia and China. Several of the runners-up were from low-income countries.

Bibliography

Andrews, Edmund L. 2002. "Lower Goals for Telecom in Europe." *New York Times*, March 20: W1.

Austin, Marc, and Helen Milner. 2001. "Product Standards and International and Regional Competition." *Journal of European Public Policy* 8, no. 3 (June): 411–431.

Besen, Stanley, and Joseph Farrell. 1991. "The Role of the ITU in Standardization, Pre-eminence, Impotence or Rubber Stamp?" Santa Monica, CA: Rand Corporation.

Brock, Gerald. 1981. *The Telecommunications Industry*. Cambridge, MA: Harvard University Press.

Bruce, Robert, and Jeffrey Cunard. 1994. "Restructuring the Telecommunications Sector in Asia: An Overview of Approaches and Options." In *Implementing Reforms in the Telecommunications Sector: Lessons from Experience*, ed. Bjorn Wellenius and Peter Stern, 197–231. Washington, DC: The World Bank.

Business Week Online, 2002. "Eavesdropping at Europe's Wireless Bash," March 6. Available at www.businessweek.com/bwdeaily/dnflash/mar2002/nf2002036_7136.html.

Cave, Martin (chairman). 2002. *Review of Radio Spectrum Management* (February). Reported to the Chancellor of the Exchequer and the Secretary of State for Trade and Industry.

Cave, Martin, and Kiyoshi Nakamura. 2006. *Digital television: Policy and Practice in the Americas, Europe and Japan*. Camberley, UK: Edward Elgar.

Cave, Martin, Luigi Prosperetti, and Chris Doyle. 2006. "Where Are We Going? Technologies, Markets and Long-range Public Policy Issues in European Communications." *Information Economics and Policy* 18: 242–255.

CDMA Development Group. 2007. "More Than 250 Operators in 105 Countries Worldwide Have Selected CDMA2000." May 15, Available at http://www.primenewswire.com/newsroom/news.html?d=119427.

Chadran, Achmad. 1994. "Examining the Latest Regulatory Developments in Asia." Presentation, September 12, to IIR Global Telecoms. London.

Commission of the European Communities (CEC). 1998. "Amended Proposal for a European Parliament and Council Decision on the Coordinated Introduction of Mobile and Wireless Communications in the Community, July 27, Brussels. COM (98): 496.

Council of the European Union. 1999. *Council Decision on the Coordinated Introduction of Third Generation Mobile and Wireless Communications System (UMTS) in the Community*. 128/1999/EEC.

Cowhey, Peter F. 1990a. "The International Telecommunications Regime: The Political Roots of High Technology Regimes." *International Organization* 44 (Spring): 169–199.

Cowhey, Peter F. 1990b. "Telecommunications." In *Europe 1992: An American Perspective*, ed. Gary Hufbauer, Washington, DC: The Brookings Institution, pp. 159–224.

Cowhey, Peter. 1993. "Domestic Institutions and the Credibility of International Commitments: The Cases of Japan and the United States." *International Organization* 47: 299–326.

Cowhey, Peter, and Mikhail Klimenko. 1999. "The WTO Negotiations and Telecommunications Reform." World Bank Trade Policy Support Papers (September).

Cox, Gary, and Mathew D. McCubbins. 1993. *Legislative Leviathan.* Berkeley, CA: University of California Press.

Drake, William J., and Kalypso Nicolaidis. 1992. "Ideas, Interests, and Institutionalization: 'Trade in Services' and the Uruguay Round." *International Organization* 46 (Winter): 37–100.

Farrell, Joseph, and Paul Klemperer. 2007. "Coordination and Lock-In: Competition with Switching Costs and Network Effects." *Handbook of Industrial Organization* 3. M. Armstrong and R. Porter, eds. Amsterdam: Elsevier B.V., 1967–2072.

Gandal, Neil, David Salant, and Leonard Waverman. 2003. "Standardas in Wireless Telephone Networks." NERA and CEPR Working Paper (February).

Greenstein, Shane. 1994. "Invisible Hand versus Invisible Advisors: Coordination Mechanisms in Economic Networks." Columbia Institute of Tele-Information Working Paper, Columbia University (August).

Hazlett, Thomas. 2001. "The Wireless Craze, The Unlimited Bandwidth Myth, the Spectrum Auction Faux Pas, and the Punch Line to Ronald Coase's 'Big Joker.'" Working Paper No. 0101, AEI-Brookings Joint Center for Regulatory Studies (January).

Hjelm, Bjorn. 2000. "Standards and Intellectual Property Rights in the Age of Global Communication—A Review of the International Standardization of Third-Generation Mobile System." Paper presented at the Fifth IEEE Symposium, Computers and Communications, Antibes-Juan Les Pins, France (July 3–6).

Hudson, Heather. 1997. *Global Connections.* New York: Van Nostrand.

"ITU Secretary General Utsumi Presses US on 3G Allocations." 2001. *Telecommunications Reports International* (June 1): 1.

Jagannathan, Shankar, Stanislav Kura, and Michael J. Wilshire. 2003. "A Help Line for European Telcos." *McKinsey Quarterly* no. 1. Available at www.mckinseyquarterly.com.

Johnson, Chalmers. 1989. "MITI, MPT, and the Telecom Wars: How Japan Makes Policy for High Technology." In *Politics and Productivity: How Japan's Development Strategy Works*, ed. Chalmers Johnson, Laura Tyson, and John Zysman, 177–240. New York: Harper Business.

Klemperer, Paul. 2001. "How (Not) to Run Auctions: The European 3G Telecom Auctions." Draft, November. Available at www.paulklemperer.org.

Krasner, Stephen. 1991. "Global Communications and National Power: Life on the Pareto Frontier." *World Politics* (April): 336–366.

Kynge, James. 2002. "China Backs Homegrown Software Market." *Financial Times* (November 11). *China Business Briefing* (November 25–December 1). Available at www.swisschinacham.org/news/businessbriefl21.html.

Lee, Kwan Min. 2002. "Modeling Regional Differences in the 3G Mobile Standardization Process: The Entrepreneur, the Committee, and the Investor." *Communications & Strategies* 47 (3d quarter): 27.

Lehr, William, and Lee W. McKnight. 2003. "Wireless Internet Access: 3G vs. WiFi." *Telecommunications Policy* 27: 351–370.

Lightman, Alex, with William Rojas. 2002. *Brave New Unwired World: The Digital Big Bang and The Infinite Internet.* New York: John Wiley.

MacAfee, E. Preston, and John McMillan. 1996. "Analyzing the Airwaves Auctions." *Journal of Economic Perspectives* 10 (Winter): 159–176.

Mitchell, Sarah, 2002. "Central Europe: The Transition to Information Society-Mobile Growth and the Prospects for 3G." *World Marker Research Centre* (December 16).

Nakamoto, Michiyo. 2003. "DoCoMo Plans to Subsidise 3G Handset Makers." *Financial Times* (January 17).

Neuchterlein, Jonathan E., and Phillip J. Weiser. 2005. *Digital Crossroads: American Telecommunications Policy in the Internet Age.* Cambridge, MA: The MIT Press, p. 230.

Noam, Eli. 1993. *Telecommunications in Europe.* New York: Oxford University Press.

Noll, Roger, and Frances Rosenbluth. 1995. "Telecommumcations Policy: Structure, Policy, Outcomes." In *Structure and Policy in Japan and the United States,* ed. Peter Cowhey and Mathew McCubbins, 119–176. New York: Cambridge University Press.

Office of Technology Assessment (OTA) (United States Congress). 1993. *The 1992 World Administrative Radio Conference: Technology and Policy Implications.* Washington, DC: Government Printing Office (May): 77–94.

Olson, Mancur. 1971. *The Logic of Collective Action: Public Goods and the Theory of Groups.* Cambridge, MA: Harvard University Press.

Owen, Bruce, and Gregory Rosston. 2001. "Spectrum Allocation and the Internet." SIEPER discussion paper (December): 1–9.

Patterson, Mark R. 2002. "Invention, Industry Standards, and IP." *Berkeley Technology Law Journal* 17 (Summer): 1043–1083.

Pelkmans, Jacques. 1998. "The GSM-Standard, Explaining a Success Story." Paper presented to the Political Economy of Standard Setting conference, European University Institute, Florence (June 4–5).

"Qualcomm supports converged standard for IMT-2000." 1998. CDMA Development Group, press release (June 2). Available at www.cdg.org/news/jun_98.asp.

Ramakrishnan, Jessica. 2002. "Malaysia: Telekom's Opportune Mobile Challenge," June 28. World Market Research Centre, www.worldmarketsanalysis.com.

Red Herring. 2002. "Wireless Briefing" (March): 68–69.

Richards, John. 1999. "Toward a Positive Theory of International Institutions: Regulating International Aviation Markets." *International Organization* 53, no. 1 (Winter): 1–37.

Sandholz, Wayne, and John Zysman. 1989. "Recasting the European Bargain?" *World Politics* 42, no. 1 (October): 1–30.

Schmidt, Susanne K., and Raymund Werle. 1998. *Coordinating Technology: Studies in the International Standardization of Telecommunications*. Cambridge, MA: The MIT Press.

Schoppa, Leonard. 1997. *Bargaining with Japan: What American Pressure Can and Cannot Do*. New York: Columbia University Press.

Shapiro, Carl, and Hal Varian. 1999. *Information Rules*. Cambridge, MA: Harvard Business School Press, 237–253.

Sugrue, Tom J. 2001. Opening Remarks, Sixth Annual Commercial Mobile Radio Services (CMRS) Competition Report, Federal Communications Commission, June 20.

Tagliabue, John, 2002. "France Struggles to Ease Debt of Phone Company." *New York Times*, September 14, 2002. Available at www.nytimes.com.

Telecommunication Reports International. Various dates.

Tsbelis, George. 2002. *Veto Players: How Political Institutions Work*. Washington, DC: Princeton, NJ: Princeton University Press.

Wall Street Journal Online. 2002a. "EU Telecom Ministers Agree State Aid Inappropriate" (December 5). Available at www.online.wsj.com.

Wall Street Journal Online. 2002b. "Thai Government Helps Telecoms Company Beat Competition" (December 11). Available at www.online.wsj.com.

Wellenius, Bjorn. 1994. "Telecommunications Restructuring in Latin America." In *Implementing Reforms in the Telecommunications Sector: Lessons from Experience*. Washington, DC: The World Bank, ed. Bjorn Wellenius and Peter A. Stern, 113–144.

5 The GATS Agreement on Basic Telecommunications: A Developing Country Perspective

Boutheina Guermazi

On February 15, 1997, after many arduous years of discussion on conceptual and political issues, basic telecommunications services were brought under the umbrella of the multilateral trading system. The successful conclusion of the Basic Telecommunications Agreement (BTA) was depicted as a landmark achievement of the World Trade Organization (WTO).[1] It marks a remarkable shift in the global governance regime for telecommunications, from a heavily regulated regime subject to cooperative arrangements under the aegis of the International Telecommunications Union (ITU) to a globally traded service under the WTO.[2]

Although international trade in telecommunications services has taken place since the inception of international telecommunications,[3] countries and telecom administrations originally viewed international telecommunications not as a trade issue, but rather as a cooperative endeavor between sovereign entities. From the beginning, international telecommunications have been treated as jointly provided services between monopolistic telecommunications operators in two countries. The success of the regime of joint provision was made possible with key instruments developed by the ITU with regard to standardization, spectrum management, and an accounting regime that is as old as ITU itself. While the new WTO-based regime does not supplant or supersede the ITU-based regime, the global shift toward a trade framework has triggered an immense strain on key instruments of the ITU-based telecommunications regime, notably with respect to accounting and settlements for cross-border traffic.

The BTA is not a stand-alone agreement with a distinct preamble and final provisions. Rather, it is a set of negotiated schedules annexed to the General Agreement Trade in Services (GATS) as its Fourth Protocol. The GATS governs this package of commitments and provides the general legal framework within which the liberalization of basic telecommunications services is to take effect. The GATS is the first multilateral agreement for the global liberalization of trade in services. Since its inception, the multilateral trading system was only concerned with trade in goods on the assumption that services do not obey the trade rule of comparative advantage.[4] In the 1980s and under increased pressure exerted by large multinational corporations, the issue of service liberalization emerged as the new U.S. priority. However, the proposal to extend General

Agreement on Tariffs and Trade (GATT) rules to services was fought with increased resistance from developing countries, most of which ran deficits in this area. This initial rejection was softened slightly when the ministerial meeting that launched the Uruguay Round embraced the issue of development as a central objective of the GATS framework.

Taking into account the situation of developing countries, the GATS preamble stipulates that the fundamental objective of the new regime for free trade in services is the economic growth of all trading partners and the economic development of developing countries. The preamble puts emphasis on increasing the participation of developing countries in trade in services and expanding their exports inter alia by strengthening their domestic markets and enhancing their efficiency and competitiveness. Measured against the statement of purposes listed in the GATS preamble, the success of the new regime in promoting international trade in telecommunications services must be viewed as mixed.

This chapter proposes to elucidate how power dynamics and policy considerations have shaped the current legal regime governing international trade in telecommunications services under the GATS BTA and have resulted in a regime that only partially reflects developing country concerns. It will scrutinize the agreement from the perspective of developing countries and consider the extent to which it provides them with an opportunity to "catch up" with the rest of the world, or is, alternatively, a threat that could hinder their efforts.

Studying the regime of international trade in telecommunications services from a developing country's perspective is a worthwhile endeavor in at least two respects. First, when negotiations of trade in services were put on the Uruguay Round agenda, an intense North–South controversy erupted. Many studies were conducted to determine the impact of a new regime for developing countries.[5] All studies were reluctant to assess the impact of the GATS in general. As stated by Sauvé, the GATS is an "empty shell" and is therefore difficult to judge in *abstracto*.[6] The BTA represents the first occasion on which sector-specific negotiations were undertaken under the GATS, and it thus deserves separate and focused analysis. Second, under article XIX of the GATS, WTO members committed themselves to the progressive liberalization of services. A new program of services liberalization was launched in early 2000. The GATS 2000 initiative was then incorporated into a wider negotiating round instituted during the 2001 Doha ministerial meeting. The Doha Round was launched with the promise of special attention to the needs of developing countries.

This chapter surveys those areas of the BTA and the GATS that need to be taken into account in the ongoing negotiations to ensure that the multilateral trading system responds to the needs of developing countries, and plays its promised role in promoting their economic growth and development. This approach will also take into account the current, fervent debate on the opportunities and risks of globalization, and the role of the rule of law in ensuring an even distribution of the benefits of globalization. Such an investigation is rendered all the more urgent by heightened controversy over the legitimacy of the multilateral trading process, especially in light of protests against globalization and the Doha Round.

The study of the trade regime governing telecommunications services reveals that the negotiations and outcomes of the new arrangement were largely shaped by power dynamics and conflicting priorities between developed and developing countries. The resulting compromise contains numerous areas that might be propitious to developing countries. It also fails in many respects to take into account developing country needs and concerns. The first section of this chapter provides background on the current international mechanism governing international trade in telecommunications. It also describes the BTA agreement and examines how its commitments fit within the wider GATS context. The second section considers the impact of power dynamics in shaping agenda setting, negotiations, and outcomes of the GATS and the BTA. It focuses mainly on the areas that proved controversial during the negotiations. It highlights the conflict between the attitudes of developed and developing countries, and the impact of the compromise on both parties. The third section addresses the importance of wider market and corporate dynamics in shaping the current regime. It sheds light on two areas of paramount importance for developing countries: the issue of accounting rates and the liberalization of foreign direct investment. Wider market dynamics and corporate interests in both areas could prove more important than the current governance regime. Although the WTO regime fell short of addressing many concerns, the fourth section argues that opting out of the system is not an option for developing countries. Instead, this section calls for developing countries to use the empowerment possibilities embodied in the GATS to maximize the benefits of liberalization and minimize its risks. Finally, the fifth section puts forward a set of recommendations for further advancing telecommunications service liberalization to benefit all trading partners and the information society as a whole.

The Global Governance of Trade in Telecommunications Services

To have a comprehensive understanding of the Basic Telecommunications Agreement and its impact on developing countries, it is important to explain first how the telecommunications commitments fit within the structure and rules of the GATS.

The General Agreement on Trade in Services

The adoption of the GATS was one of the major accomplishments of the Uruguay Round, the eighth multilateral round of trade negotiations. Launched at Punta Del Este, Uruguay, in September 1986, the round was the most comprehensive negotiation ever undertaken by the GATT contracting parties. In addition to the traditional topic of tariff reduction, the round was successful in broadening the scope of the multilateral trading system to cover new topics, such as trade in services, intellectual property, and investment measures. In addition, the round created a new institutional framework, the WTO, a full-fledged international organization responsible for administering the trade regimes.

The GATS contains rules, principles, and procedures for the liberalization of international transactions in services. It is important to highlight that one of the earliest challenges arising from the introduction of services in the negotiating agenda of the Uruguay Round was how to define trade in services. Yet a concise definition was viewed as crucial in determining the scope of the agreement.[7] However, negotiators finally agreed to not define "services" but to focus only on the definition of "trade in services."[8] The GATS adopted the same four-part typology of trade in services developed in economic literature. Mode 1 refers to cross-border trade, by which services are provided by a service provider in country "A" to a consumer in country "B" without any physical proximity. The GATS defines the second mode of supply as "the supply of a service in the territory of a Member to the service consumer of any other Member." Mode 3 refers to services supplied by a service supplier of one member through the commercial presence in the territory of any other member, and mode 4 refers to the temporary presence of natural persons in the territory of the consumer.

This typology resulted in a wider reach of the GATS disciplines compared to those of the GATT. Trade in services encompasses, in addition to traditional cross-border trade, factor mobility and labor mobility. The inclusion of commercial presence encountered difficulties during the GATS negotiations. Developing countries initially rejected the extension of trade disciplines to investment-related matters. They finally accepted this development in exchange for the promise that labor mobility would also be included in the new framework.

The GATS edifice rests on three pillars: a framework agreement, a set of sectoral annexes, and sector-specific commitments. The framework agreement contains general rules and principles of multilateral trade liberalization. It lays down the legal framework for specific commitments and provides for progressive liberalization of trade in services. The framework rules are binding on each member regardless of specific commitments. The GATS general obligations and disciplines resemble to a large extent the rules embodied in the GATT for trade in goods. Examples of these obligations include the most favored nation (MFN) obligation, transparency, and rules on domestic regulation. In addition, the GATS contains several other disciplines with no counterparts in the GATT, which were introduced into the GATS to respond to the characteristics of international trade in services.

The GATS contains several annexes setting out specific rules for specific sectors, which constitute the second pillar.[9] The Telecommunications Annex deals with measures affecting access to and use of public telecommunications transport networks and services. It addresses the telecommunications sector not as a service sector per se, but as a facilitator of the other services, such as financial services and professional services. The annex requires that access to and use of public networks and services be provided for foreign service providers on a reasonable and nondiscriminatory basis.

Finally, the third pillar of the GATS consists of the actual liberalization of services in the form of schedules of negotiated commitments that are unique to each member. Specific commitments consist of market access and national treatment guarantees under each mode of supply that constitute the definition of trade in services, as well as additional commitments that a member may undertake for specific sectors.

Article XVI of the GATS dealing with market access requires that with respect to each mode of supply, each party must accord foreign providers access no less favorable than that provided in its schedule. Article XVI includes a list of measures that restrict market access and are prohibited unless specifically maintained in the national schedule. This means that any or all measures can continue to be applied to any sector and mode of supply as long as they are scheduled.[10] As in the case of market access, members can subject national treatment commitments to further limitations and qualifications. However, unlike article XVI on market access, article XVII on national treatment does not provide an exhaustive list of measures inconsistent with national treatment.

This brief overview of the GATS scheduling mechanism reveals how the GATS is built around a hybrid approach to liberalization. A "positive list" approach was taken for sectors subject to liberalization, meaning that only sectors listed are subject to liberalization. However as mentioned by Hoekman, within each sector and in each mode of supply, liberalization is based on a negative list approach for policies that either condition or limit market access and national treatment.[11]

The Basic Telecommunications Agreement

From the outset of the Uruguay Round, telecommunications services occupied a central place in the negotiations of trade in services. Despite the growing importance of the telecommunications sector, the path toward the liberalization of international trade in telecommunications services proved to be problematic and awkward. The process began with the liberalization of enhanced telecommunications services and the GATS Telecommunications Annex before gradually embarking on the liberalization of basic telecommunications services. The enhanced services were easier to liberalize because they were already provided in a competitive and deregulated environment in an overwhelming number of countries. By contrast, economic and social interests combined to justify monopoly treatments for basic telecommunications services.

At the conclusion of the Uruguay Round, only a few countries made commitments on basic telecommunications services; furthermore, some countries expressed their desire to exempt basic telecommunications services from the MFN or nondiscrimination principle. In order to ensure that these services were not excluded from GATS, the trade ministers agreed to extend the period for negotiation of trade in basic services without withholding the entry into force of the GATS. Meanwhile, and until new offers entered into effect, members suspended the application of MFN on offers already made in the sector. The one-off possibility to take additional MFN exceptions on basic telecommunications offers was also extended until the new offers entered into force.

The negotiations, which were scheduled to end in April 30, 1996, failed to conclude successfully the basic telecom agreement. The main reason for this failure was that the scheduled commitments did not constitute a "critical mass of offers" acceptable to major trading partners and especially to the United States.[12]

On April 30, 1996, the Council for Trade in Services adopted a decision that authorized the continuation of the discussions and allowed parties to make adjustments to their offers. Between November 1996 and February 1997, twenty-three new offers were submitted and thirty-two schedules were revised. The package was open to acceptance until November 30, 1997, and entered into force on February 5, 1998.

The schedules contained specific commitments in the field of basic telecommunications from countries representing over 90 percent of the world's basic telecommunications revenues.[13] In addition to market access and national treatment commitments, the package contained a negotiated set of procompetitive regulatory principles contained in a Reference Paper on basic telecommunications. The agreement covers basic telecommunications services whether local, long distance, or international; includes services for public and nonpublic use provided on an infrastructure basis as well as through resale; and covers telecommunication services provided by any means of technology including cable, wireless, and satellites.[14] The commitments are annexed to the GATS and form an integral part of GATS in application of article XX:3.

In making commitments, members under the Fourth Protocol are bound by their specified level of liberalization. In other words, they undertake to not impose any new measures that would restrict market access or national treatment beyond the level entered in their schedules. The schedules are complex documents, which indicate for each telecommunications subsector as defined in the classification list document developed during the Uruguay Round,[15] the level of market access and national treatment allowed by a particular country. The commitments and limitations are in every case entered with respect to each of the four modes of supply that constitute the definition of trade in services.

The level of commitments varies from total liberalization to no commitments at all. In the first case, the member inscribes "none" under the market access and the national treatment columns; in the second case, the word "unbound" appears in both columns. In most cases, however, schedules reflect a combination of full commitments, unbound entries, and limitations with regard to each subsector, in each mode of supply, and under both market access and national treatment.

Developing Country Commitments under the Fourth Protocol

Focusing on how developing countries participated in the liberalization process, it is important to note that almost 60 percent of the participants in the BTA are developing countries. This high percentage is misleading and does not imply a high level of participation of developing countries for fewer were even WTO members at the time. Indeed, taking into account the big ITU family, the level of participation of developing countries is only around 22 percent, which leaves about 80 percent of developing countries outside the liberalized regime. However, one of the most revealing aspects of developing countries' participation in the basic telecom liberalization is the fact that, since the entry into force of the basic telecommunications agreement, many additional developing countries have joined with unilateral commitments on basic telecommunications.[16] This marks a precedent in international economic relations, as the first time that commitments from developing countries have been made outside the context of wider negotiations. The WTO general director views this spontaneous reaction as evidence of a general conviction of the importance of liberalization for the development of the telecom markets. This conviction was once again tested when most newly admitted members to the WTO included basic telecommunications offers in their accession package.[17]

Recognizing the importance of autonomous liberalization is one of the items inscribed in the ongoing round of services negotiations. On March 6, 2003, negotiators agreed on the modalities for treatment of liberalizations measures undertaken unilaterally by WTO members outside multilateral negotiations in application of article XIX:3 of GATS and paragraph 15 of the Doha ministerial declaration. Under these modalities, countries that undertake autonomous liberalization measures are entitled to seek credit from their trading partners. They have to enter into bilateral negotiations to prove that their measures have a trading value and to negotiate concessions from their partners.[18] Bilateral negotiations might, however, discourage further unilateral liberalization, because most developing countries lack the required leverage to extract concessions bilaterally from more powerful trading partners.

The study of developing country's schedules demonstrates a reserved approach to liberalization under mode 1 compared to the other modes of supply. Few countries entered "unbound," meaning they could undertake no commitments of liberalization with regard to this mode.[19] Most countries committed to cross-border trade with various kinds of limitations, mainly including the prohibition of bypass or the obligation to route traffic through the incumbent telecommunications carrier.[20] Liberalization of international trade in telecommunications services under this mode of supply means that participating countries undertake to eliminate relevant impediments, primarily restrictions on network access. Restrictions can range from a total prohibition on access to the network to discriminatory access and non-cost-based access.

Whereas cross-border trade is a priority for developed countries because they are well equipped to reap the benefits of unrestricted cross-border trade in telecommunications services, through the use of international simple resale and through full circuit provision, it is less so for developing countries. For developing countries with underdeveloped infrastructure and lack of access to technology, the economic benefits of free cross-border trade are less prominent than the immediate benefits of foreign direct investment in the telecommunications sector.

By undertaking commitments under mode 3, participants agreed to liberalize their foreign investment regime to allow foreign participation and investment in their telecommunications industries. There are numerous possibilities through which telecommunications operators have been approaching foreign markets. The most popular mechanisms are investment in a privatized incumbent, funding infrastructure development through BTO regimes, joint ventures, and mergers and acquisitions. The study of national schedules reveals that mode 3 is the mode with the least number of unbound entries.[21] It is important to note that although developing countries were initially opposed to the inclusion of foreign investment in the definition of trade in services under GATS, their schedules in the basic telecommunications services as well as other sectors have reflected a higher level of commitment to commercial presence than on mode 1.

Under mode 3, three types of limitations with regard to market access and national treatment were most common: limitation on the type of legal entity,[22] limitation on the number of suppliers,[23] and limitation on the participation of foreign capital.[24] In addition, numerous other limitations were registered in the country schedules, which do not easily fit into any of the six limitations developed in the GATS. Such limitations include restrictions on bypass of monopoly facilities or requirement to use monopoly facilities, as well as restrictions on resale of excess capacity.

Consumption abroad is of growing importance in the telecommunications sector because of developments in mobile communications technology. Mobile technology allows customers to move to foreign countries with their mobile handsets and consume telecommunications services outside of their countries of residence. Another example of consumption abroad is the use of calling cards in foreign countries. Consumption abroad recorded the highest level of full commitments by sectors and by countries. Few unbound entries were recorded. Even those limitations that were introduced are confined to the prohibition of callback. With developments of mobile technology and mobile applications, the relevance of consumption abroad will likely rise in the near future.

In the telecommunications sector, liberalization under mode 4 received a very low level of commitments. Although basic telecommunications were subject to sectoral negotiations, most countries entered unbound with regard to mode 4 unless otherwise provided in their horizontal commitments applying to all sectors. Only a few low-income countries made full commitments under mode 4.[25] These results reflect once again the resistance of countries to liberalizing labor and the lack of balance in the treatment of labor and capital in the services liberalization context.

A major negotiating goal for the ongoing round of services negotiations as revealed from country proposals is to have more countries join the liberalization wave with new and improved commitments. The proposals also highlight the need for the old members to revise their schedules and decrease the number and scope of limitations to market access and national treatment.

Power Dynamics in Agenda Setting, Negotiations, and Outcomes

Power dynamics and policy considerations have had a major impact on shaping the global governance regime for telecommunications services. This section starts by illustrating how these dynamics have impacted the rules of the GATS and the BTA. It attempts to go beyond legalistic analysis of the rules to show how they resulted from the power relationship between developed and developing countries during the GATS negotiations. The section then sheds light on areas reflecting longstanding conflicts in attitude between developed and developing countries, which were deliberately omitted from the current trade framework. The silence of the GATS and the BTA on these development-sensitive areas is another indication of how power dynamics have shaped the outcome of the negotiations.

Tensions and Compromises in the GATS Framework and the BTA

From the launching of the Uruguay Round, developing countries expressed skepticism over the idea of a GATT-like agreement for services. Although developing countries led by Brazil and India at first rejected the extension of trade rules to services, they finally accepted the proposal after developed countries paired progress in services with progress in areas of special concern to developing countries.

An overview of the GATS negotiating history clearly shows a constant tension between developed and developing countries over almost every aspect of the GATS. In the developments to be covered here we focus on the issue of MFN, the special rules on developing countries, and the Reference Paper on regulatory principles as those areas of the GATS framework and the BTA that best illustrate the mounting tension between developed and developing countries.

Unconditional Most Favored Nation Treatment: A Triumph for Developing Countries?

Under article II of the GATS, each member is to accord immediately and unconditionally to services and service suppliers of any other member treatment that is no less favorable than what the member state accords to like services and service suppliers of any other country. The MFN principle is the backbone of trade liberalization in the

WTO scheme, and is included in the GATT, the GATS, and the TRIPS agreements. The MFN principle is a general obligation under the GATS, which means that it applies even in the absence of specific commitments. The principle of unconditionality implies that these commitments are extended to third countries regardless of whether they made similar commitments under the BTA or not.

Although long considered and proved to be the most important pillar for liberalization of trade in goods, the most favored nation principle was one of the most difficult items to negotiate during the GATS discussions. Countries with different levels of domestic liberalization were divided about whether the GATS should embody a conditional or unconditional MFN principle. Disagreement over the formulation of the MFN principle threatened on many occasions to collapse the GATS negotiations. The United States was concerned about the free-rider impact of an unconditional MFN. Having one of the most liberalized markets for services, the United States feared that unconditional MFN would allow other countries to benefit from the U.S. market without having to present similar concessions to U.S. firms. Accordingly, the initial U.S. position was that the MFN principle should not govern the GATS as a whole but rather should be negotiated on a sectoral basis and granted upon satisfactory market-opening commitments from interested parties. On the other hand, most developing countries and a group of smaller OECD countries pressed the case for unconditional MFN treatment. The compromise reached during the negotiations was that the GATS should be based on unconditional MFN, but with the possibility for countries to take MFN exceptions.

However, a closer analysis of the circumstances surrounding the negotiations of the BTA reveals that in this area, the MFN principle was not genuinely unconditional after all. Concerns over the application of unconditional MFN in the basic telecom area explain to a large extent how this sector ended up in a separate set of negotiations. In addition, the difficult negotiating history of the BTA testifies to the fact that in the telecom sector, the prospects of the MFN principle were conditioned by minimum acceptable market-opening commitments, summarized by the famous U.S. slogan of the need for a "critical mass" of offers.

From a developing country's perspective, the introduction of unconditional MFN should be acclaimed as the first development-friendly provision in the GATS, at least theoretically. Developing countries can benefit from the unconditional extension of basic telecommunications commitments in at least two ways. First, the application of unconditional MFN to the market-access commitments of developing countries creates competition between foreign investors interested in the particular market of a developing country, especially those markets characterized by high demand and high return. In this case competition might prove to be very beneficial to network growth in these target markets, as foreign investors will try to offer additional benefits to gain market access. Hoffman and Hobday cited the case of the fierce competition between foreign suppliers to win an Indian contract for switching systems. France's Alcatel won the contract after the intervention of the French government to ensure additional benefits to the Indian industry in the form of favorable loans and a commitment to favorable transfer of technology. This strategy is widely used today by different suppliers to win contracts in large developing countries.[26]

The other benefit of unconditional MFN for developing countries is that far-reaching market-opening commitments of developed countries are extended on an MFN basis and without condition to all developing countries. In principle, unconditional MFN ensures that countries with weaker economic leverage can benefit from the best trading conditions negotiated among the stronger powers. It should be mentioned, however, that this legally sanctioned possibility could be meaningless in practical terms absent specific measures to allow developing countries to benefit from market-access opportunities in developed countries. For most developing countries, the possibility to export their services to developed countries, especially under mode 3 of supply, is remote if not unforeseeable. Since the entry into force of the agreement, the analysis of trade flows in basic telecommunications services between developed and developing countries shows that developing countries are in the importing end of the spectrum, through improved foreign presence of the firms of developed countries.

GATS-Specific Rules on Developing Countries

At the outset of the Uruguay Round negotiations, a tension arose between developed and developing countries as to how a new agreement on services could be formulated to attain the objectives set out in the ministerial declaration. While the United States argued that development is best attained through immediate and progressive liberalization of trade in services, other countries defended the idea that the agreement should contain specific measures to increase participation of developing countries in trade in services, including inter alia transfer of technology, training, and preferential access to distribution and information channels.[27]

The GATS position toward developing countries differs fundamentally from that of the GATT. Whereas the GATT treats developing countries as a different class of members with special rules based on the principle of preferential and more favorable treatment, the GATS treats the development component as an integral part of the agreement. To a large extent, GATS treatment of developing countries reflects the changing conception of which trade policies are best suited to meet the development objectives. Like other agreements resulting from the Uruguay Round of trade negotiations, the GATS embraced a new conception of preferential treatment for developing countries.

GATS-specific rules on developing countries are spread throughout the agreement. Essentially, these provisions embody two approaches to preferential treatment: the first aims to increase trade opportunities for developing countries, including technical cooperation; the second consists of giving developing countries flexibility in their commitments.

Article IV of the GATS identifies the need to increase trade opportunities for developing countries through many different vehicles, including specific negotiated commitments relating to strengthening their domestic services' capacity, efficiency, and competitiveness, as well as access to technology on a commercial basis. Article IV also emphasizes the importance of liberalization of market access in sectors and modes of delivery of export of interest to developing countries. Despite its importance, article IV does not exempt developing countries from any obligation under the GATS, but rather is an invitation to enter into specific commitments on the bilateral level. The article does not guarantee that the negotiations will actually lead to the strengthening of a developing country's domestic service capacity, efficiency, and effectiveness. But the idea of access to technology is particularly important in this context. The main constraint for developing countries, impeding their active participation in trade in services in general and trade in telecommunications services in particular, is the lack of technological capability. In this regard, article 6(d) of the Telecommunications Annex supplements article IV of the GATS by setting out in greater details measures to increase the participation of the least-developed countries (LDCs) in trade in telecommunications services. It mentions that members shall encourage foreign suppliers of telecommunications services to assist in the transfer of technology and training as well as other activities that could boost LDCs' telecommunications services trade.

The second category of preferential treatment in the GATS relates to flexibility offered to developing countries in their commitments. An overview of provisions on special and differential treatment as embodied in trade agreements resulting from the Uruguay Round reveals that flexibility is by far the most widespread item of special and differential treatment accorded to developing countries.[28] The flexibility built into the GATS is of two kinds. The first is flexibility resulting directly from the structure of GATS, under which market access and national treatment disciplines are elaborated through specific commitments rather than general obligations. The issue of national treatment was an area of controversy between developed and developing countries during the negotiations. Developed countries led by the United States defended the idea of a GATT-like national treatment obligation that would apply to all services and all modes subject to possible exceptions. Such a full application of national treatment would amount to a total elimination of all domestic measures that favored or protected particular industries, a proposition that was not welcomed by developing countries.[29] Ultimately a compromise was reached whereby market access and national treatment were treated as specific commitments subject to further limitations and specifications as inscribed in each country schedule.

With regard to the second kind of flexibility, many other articles of the GATS offer additional policy space to developing countries. Most important, article XIX of the

GATS emphasizes the importance of appropriate flexibility for developing countries in allowing them to open fewer sectors, liberalize fewer types of transactions, and progressively extend market access in line with their development objectives. This possibility has been widely criticized by many economists, who argued that flexibility discourages developing countries from reaping the benefits of liberalization, since it allows them to liberalize at a different speed than their partners.[30] The BTA exemplifies how developing countries have made extensive use of this possibility in the first round of negotiations. Indeed, developing countries were more interested in liberalizing commercial presence than in liberalizing cross-border trade in basic telecommunications services because they perceived more benefits under mode 3 and more risks under mode 1. In addition, the review of the kinds of limitations introduced in their schedules reveals their desire to have more domestic participation and less bypass of their networks.

Ensuring an increased participation of developing countries in international trade in services would normally be a priority item in the negotiations within the Doha "Development" Round. In this context, the Council for Trade in Services has recently adopted a set of guidelines and modalities for the special treatment of LDC members in the Doha negotiations. The adoption of these modalities is mandated by article XIX of the GATS. Their objective is to ensure that LDC enjoy maximum flexibility in the negotiations and overcome institutional deficiencies to analyze and respond to offers and requests. The document simply gathers in a single place the development provision items, with special emphasis on the situation of LDC, that were scattered throughout the GATS framework. What is worth mentioning here, however, is that paragraph 9 recognizes that mode 4 is one of the most important means for LDC to supply services internationally and invites WTO members to "the extent possible" to consider undertaking commitments to provide access in mode 4. In addition, paragraph 10 invites WTO members to grant LDC special credit for autonomous liberalization and refrain from requesting credit from them.

It is important to note that despite the importance of the GATS disciplines for developing countries, its provisions accord developing countries including LDC only limited preferential treatment, contrary to the provisions of the GATT. Most of the provisions are broadly drafted, and are not very likely to provide adequate means to fulfill the aspirations for development embodied in the GATS preamble.

Tensions over Regulatory Autonomy and the Reference Paper Compromise

From the outset of the GATS negotiations, developing and developed countries alike expressed their concerns about national regulatory autonomy. The concerns were that the extension of free trade principles to services would amount to international interference with domestic economic policies and thereby undermine their sovereign right to determine national regulations according to their national priorities.[31] To alleviate these concerns, the ministerial declaration launching the Uruguay Round solemnly declared that any multilateral regime for trade in services should respect the policy objectives and national laws and regulations in the service sector.

During the negotiations of the BTA, it became clear that the GATS disciplines on regulatory matters were too weak to ensure competitive market conditions in the telecommunications sector, and that market access and national treatment commitments could be nullified in the absence of additional rules to govern domestic barriers. However, domestic regulatory barriers are difficult to control because they are rooted in domestic policy choices and reflect national priorities. Conscious of this dilemma, the negotiators of the agreement on telecommunications decided to craft a legal mechanism that responded to the prerequisite of liberalization of trade in telecommunications services without undermining, at least theoretically, the principle of regulatory autonomy affirmed in the GATS preamble. In this respect, the additional commitment mechanism is a novel approach to an old trade dilemma.[32]

Under article XVIII of the GATS, parties are allowed to schedule "additional commitments" in addition to market access and national treatment commitments.[33] In principle, those additional commitments are binding on the countries that make them and are enforceable through the WTO dispute settlement procedures. The Reference Paper on basic telecommunications represents the regulatory component of the BTA. It is a set of common guidelines for a regulatory framework to guarantee effective market access and foreign investment commitments. The Reference Paper outlines six regulatory principles for competitive safeguards, interconnection, universal service, licensing, allocation and use of scarce resources, and creation of an independent regulator.

The Reference Paper constitutes an original international legal instrument at least on two fronts: it is the first international instrument to introduce enforceable competition law principles in a trade framework; and the first international document that embodies concepts and elements of telecommunications policy and regulations. It provides a road map of the basic guidelines for modern, procompetitive regulation.

The Reference Paper contains a common set of rules agreed upon multilaterally, but gives countries flexibility to choose from these rules according to their national needs. The results of the BTA already demonstrate this flexibility. Many participants adopted selected elements of the Reference Paper. Other countries drafted their own wording to additional commitments in regulation. Finally, only Tunisia and Ecuador decided not to undertake any regulatory commitments.

Theoretically, no member is obliged to adopt the Reference Paper. However, in practical terms countries that want to participate in liberalization would be faced by a de facto obligation to adopt these regulatory principles, as a minimum guarantee to attract foreign operators. In addition the Reference Paper presents important advantages for developing countries engaging in regulatory reform. It embodies and compiles in one short document the result of many years of regulatory practices of other countries.

For many countries, adherence to the Reference Paper's principles will not affect their national rules. For the United States for instance, the principles of the paper already mirror the principles of the U.S. Telecommunications Act of 1996. For other countries, adherence to these new principles translates into extensive legal and regulatory revisions of their telecommunications framework. However, without additional international commitments to guide developing countries and coach them in this difficult exercise, the Reference Paper could become a heavy burden for such countries.

There are many areas where regulators in developing countries would probably experience difficulties. Interconnection is a complex issue requiring expertise from regulators. Most of the challenges would be related to the price structure of interconnection. This problem is acute even in developed countries with a long tradition in market regulation. Despite the insistence on cost-based interconnection charges, the related provisions of the Reference Paper are limited. It is likely that developing countries will encounter numerous obstacles to setting acceptable criteria for determining such costs. In addition to the difficulties in implementing interconnection disciplines, developing countries will also face difficulties in controlling cross-subsidization and other anticompetitive practices. This is due, for the most part, to the lack of a competitive tradition. But perhaps the most daunting task for developing countries in complying with the Reference Paper is the creation of independent regulatory agencies. Although the paper's proposal appears relatively straightforward, the experience of many countries demonstrates that creating an independent authority is a very delicate task.[34] Developing countries could also suffer from a lack of resources and a scarcity of well-trained staff that could potentially risk the effectiveness of these institutions. Although it is difficult to establish institutions, it is even more difficult to maintain efficient and effective ones. Hence, "It is not surprising that, with few exceptions, these new regulatory agencies have been slow to get off the ground and perform poorly."[35]

Among all aspects of the BTA, the Reference Paper on regulatory principles is perhaps the most significant and groundbreaking. The paper's legacy reaches far beyond its attempt to create a global regulatory framework to guarantee that market access and national treatment commitments are not impaired by domestic measures or omissions. The paper offers the international community an approach to paradoxical issues with which policy makers and analysts have been struggling for many years. It offers an example of how to reconcile national regulatory autonomy with the prerequisites of free trade and how to address competition elements in a trade framework.

The Reference Paper is at the center of attention in the ongoing round of services negotiations. Numerous proposals highlight the importance of adoption of the paper by all WTO members. Other proposals invite adopting a similar solution to other service sectors. For example, the EU proposal for postal services called for agreeing upon a reference paper in the postal/courier sector, which would cover elements of importance such as "universal service" and "competitive safeguards."[36] Similar invitations to adopt a reference paper are proposed by the United States for the energy sector[37] and by some Latin American countries for the tourism sector.[38]

Central to developing countries' concerns in the whole debate over the fate of the Reference Paper in the basic telecommunications sector is whether its disciplines should be made applicable to Internet delivery service. This is an area of mounting controversy among WTO members, yet a positive position could have far-reaching implications for developing countries. At the heart of the controversy is an ongoing definitional debate on whether Internet delivery is a basic or value-added telecommunication service. If Internet delivery services were declared basic telecommunication services, the reference paper disciplines, especially interconnection, would apply to international Internet charging arrangements. Those disciplines translate into an obligation, subject to the WTO dispute settlement mechanism, of major backbone service providers to provide fair, nondiscriminatory terms and cost-based charging to smaller ISPs. Such an approach would prove beneficial to developing countries.

Loopholes in the Current Trade Regime

Assessing the basic telecommunications agreement from a developing country's perspective requires that one goes beyond the disciplines sanctioned in the agreement to shed light on those areas of concern to developing countries, which were either postponed until future negotiations, or completely omitted from the liberalization framework. Safeguard measures and concerns over restrictive business practices are two areas reflecting a longstanding conflict in attitude between developed and developing countries. The silence of the GATS and the BTA on these development-sensitive issues is another indication of how power and policy considerations have had a major impact on shaping the global governance regime for telecommunications services.

Safeguard Measures

During the negotiation of the GATS, several items proved difficult to negotiate and were deferred. One was safeguard measures,[39] protective measures that could be used by all parties to a trade agreement, under defined conditions, with a view to temporarily withdrawing or modifying their normal trade obligations under the agreement. The philosophical underpinning of safeguard measures is that they provide governments with safety mechanisms to react to injury or threat of injury resulting from liberalization of trade. Although, theoretically, trade liberalization is supposed to yield positive results to all participating countries, it is sometimes imperative to adjust to a surge of imports in a particular industry.

Article X of the GATS, entitled "Emergency Safeguards Measures," invited negotiations on safeguard rules and their entry into force by 1998. Negotiations on safeguards proved a difficult matter. Many countries clearly oppose the introduction of safeguard measures in a service agreement. Despite the ongoing work and debate within the WTO on the issue of emergency safeguards, the prospects for a successful conclusion of the negotiations are not very promising. The main areas of disagreement include the desirability, feasibility, and possible form of an emergency safeguard mechanism in the GATS. When dealing with safeguards in services, negotiators, policy makers, and commentators have recourse to the institution as applied in the trade in the goods context.[40]

The legal regime of emergency safeguards under the GATT is very complicated, with tests that are difficult to meet and complex remedial actions. Many commentators feel that a similar regime could not be transplanted in the GATS. The lack of such disciplines marks one of the most salient weaknesses in the legal regime of trade in services liberalization, especially as far as developing countries are concerned. For them, the elaboration of rules on emergency safeguards is a very important element in a sensitive area like telecommunications services. Because of the characteristics of the telecommunications sector, the possibility exists that competition from foreign operators can cause injury to domestic suppliers. Here it is important to stress that actions by foreign suppliers need not be unfair. Even in cases of fair terms of trade, an escape action can be taken to shield the industry from competition for a certain period of time. Although protective in nature, safeguard measures can be seen as a procedure to encourage trade liberalization. Given the absence of disciplines on subsidies, dumping, and restrictive business practices by foreign service suppliers, an escape clause is very much needed for weaker parties in a service liberalization scheme.

Today, the only possibility for the disposition of injured parties is contained in paragraph 2 of article X, which allows parties to modify or withdraw a specific commitment one year after the commitment enters into force. This regime is transitional in nature and is supposed to be applied until the results on negotiations on safeguards enter into force. This paragraph also constitutes an exception to the regime of withdrawal and modification of schedules as contained in article XXI, which does not allow withdrawal of concessions before three years from any concession's entry into force.

By allowing parties to invoke withdrawal procedures a year after the entry into force of the commitment, the GATS negotiators tried to take into account the urgency of the need to react to injury. It should be emphasized, however, that in areas like telecommunications one year is already a long period to wait before a party can react to injury. In addition to this first shortcoming, the procedure for withdrawing concessions is very cumbersome and might constitute a reason for parties to prefer not to use it. The procedure requires notification of the council for trade in services and the negotiation of compensatory adjustments. Should no agreement be reached, the affected parties are allowed to take the matter to arbitration.

Free Trade and Fair Competition: Concerns over Restrictive Business Practices of Multinational Firms

During the negotiations of the GATS, major concerns were raised in regard to the interaction between free trade and fair competition. The GATS and the Reference Paper on regulatory principles stressed the need for competition rules to ensure that foreign service providers enjoy effective market access in the face of a powerful incumbent. The competition disciplines did not extend to the possible anticompetitive behaviors of foreign suppliers in the domestic market.

Developing countries expressed their concern that free trade in telecommunications services will enhance the present competitive advantage of multinational corporations and lead to abuse of dominant position. The possibilities of such harmful effects on the domestic market of developing countries are numerous and the lack of international disciplines to address these threats constitutes one of the major weaknesses of the GATS as it applies to the telecommunications sector. An overview of the competition laws and trends in developed countries reveals that the application of laws to ensure competition in domestic markets has not been accompanied by similar attention to the activities of their firms in foreign markets. This area has always been left to competition laws of the host countries. During the negotiations of the UNCTAD set of Multilaterally Agreed Equitable Principles and Rules for the Control of Restrictive Business Practices, the issue of control of anticompetitive practices of firms in foreign markets was raised by developing countries. Developed countries, however, strongly argued that this set of rules should not extend to the behavior of enterprises in foreign markets.[41]

For countries without a legal tradition of competition, it is very difficult even in the presence of young competition laws to detect the anticompetitive practices of large corporations. In the case of the telecommunications sector, a new wave of global diversification of carriers is taking place through mergers, acquisitions, joint ventures, and alliances. Global carriers, because of their technological sophistication, can enjoy cost advantages over incumbent carriers in the countries where they are present. They can have recourse to techniques of cross-subsidization between their different affiliates as well as to internal transfer pricing and to profit shifting between affiliates to maximize their profits.[42] The problem with such practices is that the reallocation of funds is likely to reduce funds otherwise available for reinvestment in the domestic network and the quality of services may suffer as a result. Those techniques are not easily discernible by the host country; effective monitoring of operations between foreign and local affiliates is also difficult on a daily basis. Although it is difficult to quantify the loss that could occur, the risk is present and needs to be addressed on a multilateral level.

The question of extending trade rules and disciplines to private anticompetitive practices of national firms has been a prevalent issue for many years. Lately, the issue received renewed attention with the first ministerial meeting of the WTO's working group on competition and trade. It is inscribed as one of the most important topics to be addressed in the Doha Round of trade negotiations. However, the debate has proved contentious.[43] Controversies over competition matters proved too complicated during the Cancun ministerial meeting and were a major reason for the collapse of the negotiations.

Importance and Impact of Wider Market and Corporate Dynamics

In order to appreciate the impact of the new governance regime on developing countries, it is important to go beyond theoretical analysis of the intergovernmental rules and see how wider market and corporate dynamics have a marked impact on how these rules are translated in practice. In this context this section proposes to focus on two areas of particular importance for developing countries: foreign direct investment and international accounting rates.

The Distribution of Benefits under Mode 3

In light of the huge need for capital to upgrade their infrastructure, and faced with the insufficiency of internal funding as well as the inadequacy of international borrowing and aid mechanisms, developing countries are likely to find that inviting foreign investors into their market is the most convenient way to develop their telecommunications sector. Some authors even argue that when entering into basic telecommunications negotiations, developing countries knew well that the real benefits from the regime are derived from opening their markets to imports rather than opening up the opportunity for their providers to serve foreign markets as service exporters.[44]

Due to the shrinking of public sector financing since the 1980s, the private sector has been called upon to assume the responsibility for funding infrastructure development. Private participation in infrastructure development has undergone tremendous growth both in developed and developing countries. Data gathered by the World Bank's project database on private participation in infrastructure clearly shows that telecommunications is the leading sector in this area.[45] Between 1990 and 2001, more than $300 billion in private investment have been committed to over 650 telecommunications projects around the world. The figure covers both greenfield projects and divestitures. In addition to providing funding for infrastructure development, foreign direct investment (FDI) provides other benefits in the transfer of technology and expertise. Access to technology is increasingly recognized as an important factor for economic development. Developing countries, which remain almost exclusively technology importers,[46] have a vital interest in accessing modern technology. This technology could be used to upgrade their telecommunications capacity and to gain a competitive strength in the information-based economy. Although transfer of technology can be channeled through various means,[47] FDI is considered the most significant channel for dissemination of technology to host countries. In the case of telecommunications services, the potential for transfer of technology extends both to hard and soft technology. By encouraging FDI in telecommunications services, developing countries would not only import more efficient technology but also generate technological spillovers for their firms. Indeed, because telecommunications is a capital-intensive sector that is very sensitive to technological development, developing countries hosting foreign investors in this sector expect to have access to state-of-the-art technology in both fixed and mobile telecommunications. While increased foreign investment in the telecommunications sector is both necessary and desirable as a strategy for telecommunications development, the prospect that increased FDI could bridge the telecommunications gap soon fades away if one considers the dramatically uneven distribution of private capital among developing countries. A study by the South Centre documented that only twelve developing countries account for 80 percent of private flows to developing countries, with China alone accounting for over 25 percent of all FDI flows to developing countries.[48] This leaves the majority of developing countries with little access to private capital. The lack of access is particularly acute for poorer countries. As reported by the UNCTAD in the 2001 World Investment Report, developing countries' share of global FDI flows declined in 2001 to 19 percent compared with the much higher level of 41 percent in 1999.[49] The share of least developed countries has further deteriorated to 0.3 percent of world inflows compared to 1 percent in 1996.[50]

This pattern of unequal distribution is particularly apparent in the telecom sector where foreign operators compete fiercely for investment opportunities in large lucrative markets. The problem of geographical disparity is further accentuated by a "sector disparity" phenomenon, where foreign presence is mainly recorded in the most lucrative sectors of the host telecom market, as typically proven in the case of Latin America.[51] For instance, some foreign investors regard Brazil, with its large population and low teledensity, as "the world's largest potential phone market."[52] This explains to a large extent the fierce competition between numerous bidders to operate two cellular networks in the two largest cities of Brazil. Similar interest of foreign investors has been expressed in the Chinese telecom market, which is also considered to be a major lucrative market mainly because of its size and the huge demand for new connection—this despite the numerous regulatory hurdles encountered by foreign investors in China.[53] With this concentration on big markets, only 2 percent of total telecom FDI reached Africa. Even in this case, most investment was concentrated in South Africa.[54] For other African countries, foreign investment is mostly restricted to the lucrative mobile sector.[55]

This situation should come as no surprise, because the private sector seeks to invest where conditions are most favorable. FDI decisions in particular are subject to detailed calculations of risks and profits. A study of foreign direct investment flows to low-income countries over the last two decades shows a high concentration in a few countries, which all enjoy a large market size, low labor costs, and highly available natural resources.[56] Other elements enter into account as determinants of FDI flows such as the degree of political stability in the host countries and the availability of legal and regulatory regimes to guarantee foreign investment against different risks.

It should be emphasized also that although a considerable amount of private capital is available for telecommunications projects, attracting such capital is an extremely competitive endeavor. Poor countries have to compete with the rest of the world for investment opportunities, while foreign investors are being offered expanding investment possibilities that go far beyond the traditional telecommunications market. Foreign investors are increasingly focusing on the newer markets of broadband deployment, converged services, wireless Internet, dot.coms, and so on.

The uneven sectoral distribution of foreign direct investment is particularly clear in foreign investors' preferences for the most lucrative sector of the market.[57] This explains to a large extent why the cellular market is attracting foreign investors even in the poorest countries of the world while very little foreign presence is recorded in the fixed rural market. The cellular market has in most cases benefited from a competitive environment in contrast with the predominately monopolistic environment of the fixed-line network. Indeed, competition in the mobile market is almost the rule in both developed and developing countries, with multiple suppliers even in countries with low per-capita income, like Bangladesh and Uganda, and countries with low density, like Botswana and Côte d'Ivoire.[58] However, it is unlikely, even with the liberalization of trade in basic telecom and the lifting of restrictions on foreign investment in fixed networks, that similar levels of FDI presence will be recorded in the fixed network market, especially in the rural sector, at least in the near future.

It follows that only a handful of developing countries are best equipped to reap the benefits offered by the new regime. Global market failures and the concentration of private financial flows in a small number of countries and in lucrative parts of the telecom market are likely to exacerbate the digital divide and reinforce the marginalization of many smaller developing countries, especially LDCs.

The Market-Driven Solution to the Accounting Rate Debacle

The tension between developed and developing countries in the global governance regime for international telecommunications is perhaps nowhere more evident than in the area of international accounting rates. Interestingly, solutions to the accounting rate debacle seem to be emerging more from the wider market and corporate dynamics than the ITU's slow efforts to bring accounting rates closer to cost, or the BTA's unbroken silence on the issue.

As previously noted, the accounting rate regime is a key instrument that has been developed by the ITU. It is described in the International Telecommunications Regulations (ITRs)[59] and expanded in the ITU-T D-Series Recommendations, notably recommendation D.140. The accounting rate regime is to a large extent the cornerstone of the cooperative telecommunications regime that prevailed for decades before the WTO-based competitive regime. The cooperative regime is based on the notion of the half-circuit regime by which an operator in the country of origination conducts the call to a certain point in the international gateway. The call then is taken over from the interior point by another operator in the country of destination and delivered to the recipient.

Under the traditional cooperative mode, carriers in the originating countries compensate carriers in the country of destination for matching the international circuits and providing switching capabilities and domestic routing to deliver calls to their recipients. The compensation is based on an agreed rate per unit of traffic, negotiated bilaterally between the two operators. In case there is imbalance in the volume of incoming and outgoing traffic, the operator that originates more traffic pays a settlement rate (usually half the accounting rate) to compensate the terminating operator.

Developing countries have long benefited from the traditional accounting mechanism because they were (and continue to be) net receivers of international telecom minutes and hence net recipients of settlement payments. In 1998, when the BTA entered into force, developing countries received significant settlement payments each year. Indeed, according to ITU calculations, developing countries receive an average total of $10 billion of settlement payments per year.[60]

Any discussion about international trade in telecommunications services would necessarily bring the issue of accounting rates to the forefront. The accounting rates problem was discussed during the Uruguay Round[61] and more specifically during the negotiations of the BTA. The debate was stirred by a mounting concern about the fate of a regime that was developed to govern telecommunications in a fundamentally different era. The debate revealed a disagreement on the need to include the issue of accounting rates in a new agreement, and whether it would be feasible. The contentiousness of the issue revolved mainly around a North-South controversy. On the one hand, developed countries suffering from deficit payments are eager to reform the accounting rates regime and align it with cost. For example, the United States in 1997 alone paid out $5 billion more than it received in international settlement (almost 5 percent of its trade deficit). On the other hand, developing countries, which largely benefit from the current regime, are reluctant to consider any change to it. Developing countries that derive an important portion of their telecommunications revenues from settlement receipts were concerned that any change in the traditional accounting mechanism would result in a drain in their telecom revenues and affect their network buildout programs.

The conflict in attitudes was resolved during the basic telecommunications negotiations by an understanding among members to waive their rights to WTO dispute settlement procedures to challenge accounting rates, subject to a sunset clause. The BTA's silence on the issue of accounting rates has been criticized for concealing a significant discrepancy and anomaly in the new liberalized regime for international telecommunications services under the WTO. The accounting rate regime could be argued to contradict the GATS principles and disciplines.[62] Under the current accounting rules and practices, discriminatory treatment of the same service based on its national origin is a common practice. Termination of international calls are charged at very different rates, which are negotiated bilaterally, in violation of the MFN principle under GATS.

In addition, the traditional accounting rate system based on confidential bilateral arrangements does not stand the transparency test enshrined in GATS. The most striking problem with the system is that the rate levels for international services have not necessarily been dictated by effective competition or even the underlying cost structures. The discrepancy between accounting rates and the costs of conducting international telecommunications has grown extremely wide in recent years. Studies show that accounting rates are sometimes ten times higher than the actual cost of delivering international service.[63] To the extent that accounting rates exceed the cost of terminating an international call, high rates can prevent consumers from enjoying cheaper telecommunications services made possible by technological developments, the unleashing of market forces, and liberalization efforts at the multilateral level.

Although the BTA preserved the status quo by not subjecting the accounting rate system to dispute settlement procedures, developing countries are not likely to continue to benefit from the system for a longer time. The silence of the GATS cannot be interpreted as a triumph for developing countries. Here also wider market and corporate dynamics seem to have a marked impact on the solution to the accounting rate problem. Market and corporate dynamics are pushing toward new approaches that bypass the accounting rate system. In this case, developing countries that continue to resist the winds of change will only see an increased portion of their traffic settled outside the accounting rate system.

These dynamics were largely behind the U.S. Federal Communications Commission's (FCC) benchmark order, the development of alternative calling procedures, and the increasing portion of international traffic handled over IP-based networks. The accounting rate system can also be made obsolete by specific commitments under modes 1 and 3 of supply.

It is the dissociation between international accounting rates and the cost of terminating international services that caused the dissatisfaction of large U.S. firms and increased pressure on the FCC to unilaterally bring accounting rates closer to cost. According to the FCC, above-cost termination services constitute a subsidy paid by U.S. consumers to foreign operators. In 1997, the FCC adopted a new order, which established low benchmarks for settlement rates that U.S. carriers are allowed to pay to foreign carriers for terminating calls originating in the United States.[64]

The U.S. position has triggered immense controversy and mounting dissatisfaction among foreign carriers, many of which joined forces to challenge the order in the U.S. Court of Appeals. The court held that the FCC order was a valid exercise of its regulatory authority under the Communications Act of America.[65] The level of implementation of the benchmark order can be deduced from data gathered by the FCC International Bureau on accounting rates between the United States and over 250 international points.[66] Data shows that for many destinations accounting rates have dropped closer to the benchmarks. The immediate impact of the entry into force of the order for the overall U.S. settlement deficit is already apparent. In 1999, U.S. net settlement payments declined by one billion dollars compared to 1997 despite the growing imbalance of traffic in 1999 (18 billion minutes in 1999, up from 13.4 billion in 1997). In 2001, net settlement decreased by a further 10 percent to stand at $3.5 billion.[67]

The second market-driven solution to the accounting rate problem is increasingly brought about by the evolution of new technologies bypassing the national network. Examples of alternative calling practices include call back and refile. These techniques are growing in importance as a share of global international traffic.[68] According to the ITU, the portion of international traffic settled outside accounting rates is as high as 30 percent.[69] This portion will grow even faster with the exponential growth of voice traveling over Internet-based protocols in lieu of the traditional public switched telephone network (PSTN). In 2000, 4 billion minutes traveled over IP networks. As mentioned in the ITU's Internet Reports, with the economies of scale enjoyed by IP networks and the possibility to conduct international calls at a fraction of the price, it is very likely that IP networks will grow to provide an attractive alternative to the PSTN for international traffic.[70] Traffic traveling over IP networks is not settled through traditional accounting, but rather is based on peering and transit arrangements.

Finally, it is important to stress the impact of liberalization under GATS on the issue of accounting rate. According to the ITU, the impact of GATS is likely to be manifold: first, liberalization will drive accounting rates closer to the cost of delivering international services.[71] Second, an increasing portion of international traffic can be conducted outside the accounting rate regime through cross-border interconnection and under the full-circuit regime. Under cross-border interconnection the foreign operators can interconnect to the incumbent operator's network on cost-oriented nondiscriminatory terms and offer end-to-end services. The foreign operator can either establish a switch in the foreign country or use leased lines. Under the full-circuit regime, the foreign operator self-terminates its traffic through direct access to the foreign operator's PSTN. Third, with liberalization under mode 3, operators with point of presence in foreign countries would be able to terminate their own traffic in the country of destination using their own network. And finally, with increased global alliances handling international traffic, the accounting rate regime becomes irrelevant for traffic between members of the alliance.

The forces challenging the accounting system do seem irreversible and the impact of change is extending to all countries, even those that started with a clear approach to the whole issue back in 1997. The case of India is worth noting here. India's revenues from telecommunications services constitute a significant portion of the national economy and therefore, India's immediate interests consisted in preserving the status quo. In its schedule, India did not commit to any level of liberalization in cross-border trade, with unbound entries for cross-border trade in all telecom subsectors.[72] Despite the lack of commitments under mode 1 and India's original eagerness to preserve the status quo, it does not seem to have succeeded in resisting the winds of change. The year 2002 seems to be the year of doing things differently in India, which succeeded in breaking up its national monopoly two years ahead of its scheduled commitment in the BTA. In addition, in April 2002, Internet service providers were allowed to offer voice telephony over the Internet. In late 2002, India reported a 30 percent fall in settlement rates between 2001–2002, as well as a tremendous cut in international collection tariffs by up to 40 percent in the U.S./India route and 25 percent in India/Europe, which amounted to a total revenue decline of 10.72 percent.[73]

It is not clear how countries would like to reopen the debate on the moratorium on accounting rates disputes in the ongoing round. What is clearer, however, is that pressure toward reducing the level of accounting rates, and more generally replacing the accounting rate regime with an interconnection regime, will continue to build. The WTO's first dispute settlement case on trade in services between Mexico and the United States will likely encourage this change. According to the panel report, international accounting rates are subject to cost-based interconnection discipline. In other words, while countries can continue to use the accounting rate regime as a commercial modality to compensate companies for jointly providing international service, adoption of the Reference Paper requires that such payments to major suppliers be cost-oriented. While the U.S./Mexico case does not have the value of a binding precedent for future disputes, the panel's analysis and conclusions are expected to have a significant influence on how countries will decide the fate of the accounting rate regime in the context of global trade.[74]

Using the GATS's Empowerment Possibilities

The GATS contains numerous empowerment possibilities that developing countries should be aware of and use to maximize the benefits of liberalization and minimize the risks that result from the lacunae of the current order. This section focuses on the GATS architectural design and built-in flexibility, the precommitment mechanism, and the possibility of maintaining performance requirements along with commitments under mode 3. It should be stressed from the outset that while the first two possibilities are straightforward empowerment mechanisms, the last mechanism is deduced from a development-oriented interpretation of various provisions of the GATS.

The GATS Architecture

The real tangible benefit for developing countries will prove to be the structure of the GATS itself and the built-in flexibility of the whole framework. As outlined earlier, market access and national treatment are treated in the GATS as individual rather than general obligations. Developing countries therefore are free to choose their own pace of liberalization according to their national priorities. Indeed, nothing in the GATS obliges members to schedule commitments in particular sectors. If a country for economic, political, or other reasons decides to shield a particular sector from competition it can do so by not listing this sector. The technique of positive approach embodied in the GATS offers countries a large degree of discretion. Although this technique was subject to criticism from eminent economists, who preferred a negative approach,[75] the built-in flexibility of the GATS undoubtedly is an important asset for developing countries. Before the launching of GATS 2000, there were rumors that several developed countries, unsatisfied with the current GATS architecture, called for its review. The proposal was soon withdrawn due to developing countries' objections to the notion of reviewing the GATS structure and considering a negative approach to liberalization in the current round. While developing countries have succeeded in preserving the GATS design, they should be careful in addressing the so-called cluster approach.

The cluster approach consists of bundling together interrelated services for the purpose of negotiations. It is based on the fact that commercial linkages between sectors are not adequately reflected in the GATS services classification. The approach first appeared in a proposal from developing countries to treat the tourism sector as a separate annex of the GATS. The proposal is motivated by the fact that commitments to liberalize the tourism sector would be meaningless if no commitments are made in "supporting" services such as air transport, computer reservation systems, and financial services.[76] The cluster idea has been supported by the United States, the EU, and Australia as a negotiating tool in the ongoing round of negotiations in addition to the offer-request technique. There is a well-founded concern among developing countries that the cluster approach may indirectly undermine the positive listing approach of the GATS. This risk should be carefully watched in the ongoing negotiations.

The Precommitment Mechanism

During the negotiations of the GATS, developing countries requested that infant industry considerations be enshrined in the framework. The main argument was that such considerations would give them an opportunity to develop their service sector to become competitive in the world market.[77] Defenders of infant industry protections argue that the specific characteristics of the service industry in most developing countries are such that they could be treated as infant industries and thereby benefit from an exceptional treatment under the GATS. Telecommunications is one case of an industry that can be considered an infant both in terms of its age and its degree of technological development in developing countries. It is also important to stress that the telecommunication industry is characterized by huge capital investment and by economies of scale. The notion of learning by doing, which characterizes most service industries, including telecommunications, also legitimizes the treatment of the telecommunications industry as an infant industry.

Despite the above arguments, the GATS did not incorporate infant industry considerations, opting instead for flexibility in the commitments themselves. According to many analysts, the lack of such protection does not constitute a weakness in the GATS or a development-unfriendly gesture.

Although the infant industry argument in theory is motivated by development goals, some analysts argue that recourse to such a policy is likely to be harmful to developing countries, as it will delay their participation in the multilateral trading system. In this context, the GATS offers an important mechanism that satisfies developing countries' infant industry concerns without suffering "infant industry" results. Under the GATS, countries could defer market-access and national treatment commitments to a future date. The possibility of phased-in commitments allows countries certain flexibility so they can buy time and amend laws and regulations before opening their telecommunications sector to competition. At the same time, it creates a legal obligation to honor the offers by the time specified. In the basic telecommunications agreement, numerous schedules included specific commitments with regard to specific subsectors to liberalize by a specific future date.[78] As such, phased-in commitments strike a compromise between the need to liberalize trade in services and concerns about whether a specific industry is ready for liberalization.

Performance Requirements

Finally, maintaining performance requirements, deduced from the GATS rules and disciplines, could benefit the development cause in the telecommunications sector. Such requirements could be imposed on foreign investors by host governments as a condition for opening markets. Performance requirements could be incorporated in the host country's domestic laws or negotiated bilaterally between parties. This issue continues to be one of the main areas of policy controversy for developing countries and a major source of contention in economic literature.[79]

In the case of telecommunications services, performance requirements could be an essential development policy for developing countries. Examples of development-oriented performance requirements in the telecommunications sector include tying market access to infrastructure upgrade, stipulating the use of joint ventures, requiring technology transfer to the host country, and requiring foreign investors to hire nationals.

However, the legal regime for performance requirements is puzzling. While such requirements are curtailed for trade in goods under the Trade-Related Investment Measures (TRIMs) agreement,[80] they are not explicitly prohibited in the context of trade in services. As outlined earlier, market access under the GATS is not a general obligation. It only applies to sectors and modes inscribed in a country schedule. Even when specifically scheduled under the positive list approach, market access is subjected to conditions and limitations explicitly listed. Article XVI of the GATS includes a list of measures that restrict market access and are prohibited unless specifically maintained in the national schedule. This means that any or all measures could continue to be applied for any sector and mode of supply as long as they are scheduled. These limitations can be grouped into three different categories: measures related to admission and establishment, measures related to ownership and control, and measures related to operations of a foreign supplier. The list provided in article XVI does not include all of the measures that restrict market access. This means that members can adopt other restrictive measures without an obligation to schedule the limitation. In this case the limitation should only be tested with regard to the national treatment obligation, to the general disciplines on domestic regulations, and to the MFN obligation.[81]

A possible application of performance requirement is clearly stipulated in article XIX:2 dealing with the negotiation of specific commitments. This article provides an example of how developing countries can use negotiated commitments to achieve the development objectives of article IV. When opening their markets to foreign providers, developing countries are allowed to make access contingent on conditions aimed at strengthening their own service capacity, effectiveness, and competitiveness. This provision is of paramount importance. It introduces a new dimension of preferential treatment. The possibility for developing countries to attach conditions for foreign service providers as a quid pro quo for opening their markets can play an important role in basic telecommunications trade. Given that the biggest handicaps facing developing countries are the lack of adequate infrastructure and low teledensity, developing countries might have recourse to a conditionality approach under which they can require network enhancement as a condition for market access. This technique has already been used in many cases and might prove to be a good tool to achieve universal service objectives in the basic telecommunications area. The Reference Paper's disciplines on universal service only attach the most general conditions to such an approach. It should be noted, however, that the use of performance requirements is much more complicated than the textual reading of the GATS would suggest. From a legal perspective, performance requirements would increase dispute settlement cases based on nonviolation complaints. If members of the GATS believe that the expected benefits are being nullified or impaired as a result of any measure that is not in conflict with the GATS, they can still have recourse to dispute settlement as indicated in GATS article XXIII:3. The nonviolation cases under GATT gave rise to difficulties in defining the nullification and impairment of benefits.[82] Invoking the same ground in the context of trade in services would prove even more difficult. This ground for dispute settlement will be resorted to with increasing frequency as the issue of performance requirements continues to be challenged by developed countries. Developed countries led a campaign against performance requirement during negotiations of TRIMS.[83] They further defended their position toward this subject during the negotiations of the OECD draft

Multilateral Agreement on Investment (MAI).[84] As a result, a list of performance requirements including transfer of technology, joint ventures, and domestic equity participation were prohibited in the draft text.

From a policy point of view, developing countries should make cautious use of performance requirements. Although they can be used as a development policy tool, developing countries should be aware of the fact that onerous requirements may effectively deter foreign direct investment. For example, in the computer industry companies investing abroad have often refused to invest in countries that impose transfer of proprietary technology as a performance requirement.[85]

Recommendations

Given the crises in the negotiations during 2003 and 2006, it is not clear whether and how the long-awaited Doha Development Round will conclude. In line with the single undertaking approach embraced in Doha, progress in services negotiations is intrinsically linked with progress in any other area, including those that caused the deadlock in Cancun.[86] Developed countries consider the breakdown of the talks a deplorable missed opportunity for both developed and developing countries. Developing countries celebrate what happened in Cancun as a demonstration of their long-lost unity.

This last chapter section looks at telecommunications services liberalization in the wider context of the Doha Development Round. It puts forward a set of recommendations to ensure that further liberalization of international trade in telecommunications services responds to the development objectives of progressive liberalization as stipulated in article XIX of the GATS; that is, progressive liberalization with a view to "promoting the interests of all participants on a mutually advantageous basis and to securing an overall balance of rights and obligations."

Preserving the Development-Friendly Aspects of the GATS

As previously mentioned, the GATS contains many provisions that might be propitious to developing countries. The first strategy for developing countries is to fully understand these development-friendly approaches and to make sure they are not altered in the ongoing negotiating round. Examples of such items are the very architecture of the GATS and the fact that market access and national treatment are negotiated commitments subject to the "positive list" approach and not on the basis of opening up all service sectors, except those specifically excluded, through "the negative list" approach.

The choice of a particular negotiating technique can have far-reaching implications on these elements even if the GATS architecture is not technically altered.

Widening and Deepening Commitments under Mode 4

During the early days of service negotiations, developing countries stressed the inherent weakness of their service suppliers to effectively compete in the world market for services. They argued that to be acceptable, a framework for trade in services should guarantee preferential access for their service providers to the market of developed countries. Both the argument and request were rejected and the GATS did not embody the concept of preferential access to export markets as a component of special and differential treatment for developing countries.

The argument advanced by the economists in this context was that the request for privileged access is based on an erroneous conception that developing countries do not enjoy a comparative advantage in trade in services. According to available statistics, developing countries enjoy a comparative advantage in sectors like tourism, construction, and labor-intensive services.[87] It follows that, theoretically, as long as developing countries could secure liberalization in sectors and modes of interest to them, a framework for trade in services would be beneficial for them. This idea is based on the principle defended by trade economists that trade is not a zero-sum gain and that a gain in market access for one country does not mean a loss for the importing country because it will be compensated by access in other sectors and under other modes.

The whole debate over the parity of treatment of labor and capital movement, which underlined the launching of the GATS negotiations in the late 1980s, is still relevant today, taking into account the level of actual commitment to liberalization under mode 4 in all services. The outcome of the current negotiations needs to produce a significant improvement of liberalization under mode 4, the mode through which developing countries can supply service internationally. Allowing the rule of comparative advantage to work to the benefit of all members requires a more balanced treatment of labor and capital. One possible approach is to revisit the annex on movement of natural persons and ensure effective market access through mode 4.[88]

Finishing the "Unfinished Business"

As previously outlined, many areas have been left outside the GATS agreement and subjected to future negotiations. Negotiations of these areas started even before the GATS 2000 talks were launched. However, little tangible progress has been made. The issue of emergency safeguard is particularly important as a security measure that will help developing countries overcome their fear and join with improved trade offers. The need to tackle this issue is particularly important in light of the absence of global competition rules, and the difficulty of developing such rules in the near future, especially after competition proved a stumbling block in the Cancun meeting.

Another area of unfinished business, though one that was not inscribed in the original GATS agenda but added on later, is the issue of e-commerce. In recent years, e-commerce has emerged as a rapidly growing market sector. Because of the significance of e-commerce for international trade, the WTO adopted a work program on the e-commerce issue, which consists of highlighting trade-related issues of electronic commerce to be studied by different WTO organs.[89] The items cover a wide range of areas like market access, customs duties, protection and enforcement of copyright, the development implications of e-commerce, as well as the relation of e-commerce to trade in other services.[90] Discussions on GATS and e-commerce cover many complex issues, most of which will probably not be settled in the ongoing round. However, taking into account developing country needs and concerns, it is important to stress that any approach to e-commerce can have far-reaching implications on developing countries.

E-commerce has changed the way business is conducted at the international level and is becoming a driving force behind the growth of the global economy.[91] It presents an important potential for development,[92] promising increased productivity and increased access to the global markets. For developing countries, access to global markets by small entrepreneurs for product marketing increases their export performance and revenue generation. In turn, this access increases competitiveness of developing countries in the global economy. Many success stories from developing countries are already evident in sectors affected by e-commerce, such as tourism,[93] entertainment, and retail services.[94] Ensuring that any solution or framework on e-commerce takes into account the development potential of e-commerce should form a negotiating priority for developing countries in the Doha Development Round.

Revising the Reference Paper to Reflect Technological Developments

The ongoing round of trade negotiations is unfolding against the backdrop of spectacular technological developments in the telecommunications industry. Technological convergence is blurring the traditional distinctions between services, which use voice, text, video, and data. It is also blurring the traditional line between content and conduit. Convergence not only promotes the development of new services but also creates new challenges to preexistent regulatory definitions, classifications, and requirements.

As far as developing countries are concerned, pushing for a clear position on the issue of Internet delivery services is one area to focus on in the short term to address the current imbalance. In the longer term a more complex exercise needs to be undertaken, not within the confined area of trade in telecommunications services but within the larger WTO mandate on e-commerce.[95] Resolving the issue to the benefit of developing countries would, however, prove very controversial.

The issue of imbalance in Internet delivery and the risks for developing countries has become a hot policy question at the international level. The Asia-Pacific Economic Cooperation (APEC) is the first forum to call for the study and development of "compatible and sustainable international charging arrangements for Internet services."[96]

Defending a development-oriented approach to the issue of Internet delivery through extending the reference paper disciplines to Internet delivery services would tip the balance in favor of developing countries and encourage their participation in the online age.

Helping Developing Countries Reap the Benefits of Liberalization

An important negotiating objective for the telecommunications sector as deduced from different countries' proposals is to ensure that most countries adopt the reference paper on regulatory principles. The need for an effective and efficient regulatory framework is recognized as a priority item for developing countries if they intend to reap the benefits of liberalization. While the importance of this element has been recognized since 1997 with the adoption of the Reference Paper, it is now more apparent than ever. With the sector going through difficult times since 2001,[97] the importance of effective regulation becomes not only important but also key to investors' confidence.

For many developing countries, poorly planned regulatory reform will lead to negative outcomes. Failure to adopt effective frameworks that foster competition translates into scaring away potential investors and exacerbating the digital divide. One possible approach would be to single out regulatory assistance programs as a mechanism that fits under article IV on increasing participation of developing countries.

Conclusion

In addition to the terms and conditions inscribed in the schedules of each member, with regard to market access and national treatment, liberalization of basic telecommunications services is subject to the general rules and obligations of the GATS framework. To determine the importance and impact of the BTA on developing countries, it was necessary to go beyond the study of the content of the schedules annexed to the Fourth Protocol and to analyze the legal framework that conditions the application of the commitments.

Our analysis of the GATS framework has highlighted numerous areas that might be propitious to developing countries and contribute to bridging the digital divide. Most important, the structure of the GATS, its built-in agenda, the unconditional MFN, as well as the attempt of the GATS framework to respect domestic regulatory autonomy, are all areas where tangible benefits could accrue to developing countries. The attempts to guarantee some form of preferential treatment developing countries were also discussed. Despite these attempts, the chapter demonstrated that the whole regime contains many lacunae and that the development cause was not handled well. These lacunae will be very heavily felt as the liberalization of trade in telecommunications services proceeds in developing countries. While the GATS and the BTA failed in many respects to address developing country needs and concerns, opting out of the liberalization system is not a viable option. Instead, developing countries should seek to take advantage of the potential benefits that a liberalized international environment of telecommunications services offers and work in concert to ensure that their needs and concerns are better reflected in the ongoing Doha Round. Developing countries should approach the current and future negotiations with a clear understanding that a country's inability to provide adequate telecommunications services can be a costly handicap for their entire economics.

Notes

The chapter was written when the author was a doctoral candidate at McGill University and assistant professor on the faculte des sciences juridiques, politiques, et sociales at the University of Tunisia. Currently the author is a regulatory specialist in the World Bank group. The ideas expressed in the paper are the author's and do not represent the institutions with which the author was or is currently associated.

1. The WTO deal on basic telecommunications has attracted the attention of many analysts. See, for example, Tuthill 1997, 783–798; Drake and Noam 1997, 799–818; Krein and Freytag 1997, 477–491; Cowhey and Klimenko, 1999.

2. Drake 1999.

3. Cowhey 1990.

4. For economic analyses of the tradeability of services including telecommunications, see Feketekuty 1988; Nicholaidis 1989; and Benz 1985.

5. Srinivasan 1998; Chakravarti 1990; Sapir 1985.

6. Sauvé 1994.

7. Renya 1993.

8. Abu-Akeel 1999, 189.

9. There are eight annexes dealing with article II MFN exemptions: the movement of natural persons supplying services, air transportation services, financial services (two annexes), negotiations on maritime transport services, telecommunications, and negotiations on basic telecommunications.

10. The list comprises measures affecting (1) the number of service suppliers allowed; (2) the value of transactions or assets; (3) the total quantity of service output; (4) the number of natural persons that may be allowed; (5) the type of legal entity through which a service supplier is allowed to supply a service; and (6) the participation of foreign capital as a maximum-percentage limit of foreign shareholding or the absolute value of foreign investment.

11. Hoekman 1995.

12. The United States concluded that among the forty-eight schedules, only twelve represented acceptable market-opening commitments. According to the statement of Ambassador Charlene Barshefsky, over 40 percent of world telecom revenue and over 34 percent of global international traffic were not subjected to acceptable offers.

13. Initially, the number of participants in the BTA was sixty-nine members including the EU. Today the number has reached eighty-nine with additional members undertaking unilateral commitments or making commitments as part of their accession package for the WTO. The list of telecommunications commitments as well as individual schedules are available at http://www.wto.org/english/tratop_e/serv_e/telecom_e/telecom_commit_exempt_list_e.htm.

14. Note by the chairman of the Group on Basic Telecommunications for scheduling basic telecommunications commitments. S/GBT/W/2 Rev. 1, January 16, 1997.

15. The list consists of voice telephone services, including packet-switched data transmission, circuit-switched data transmission, telex, telegraph, facsimile, private leased circuits, electronic mail, voice mail, online information and database retrieval, electronic data interchange, enhanced/value-added facsimile services including store and forward, store and retrieve, code and protocol conversion, online information and/or data processing (including transaction processing), and others.

The negotiators of the BTA discussed additional services to be liberalized. These are considered basic telecommunications services and are classified under the "other" category. These services include: cellular whether analog or digital, mobile data, paging, personal communications, satellite-based mobile, fixed satellite, gateway earth station, teleconferencing, and trunked radio system.

16. There are many countries that chose to offer commitments unilaterally. Examples include Barbados, Dominica, Egypt, Estonia, Guatemala, Lithuania, Moldova, the Philippines, Suriname, Uganda, Papua New Guinea, and the separate customs territories of Taiwan, Penghu, Kinmena, and Matsu. An updated list is available at the WTO services gateway at http://www.wto.org/english/tratop_e/serv_e/telecom_e/telecom_commit_exempt_list_e.htm, visited September 2003. See Henderson 2005.

17. Examples of members whose accession packages include offers on BTA are Albania, China, Croatia, Oman, Jordan, Kyrgyz Republic, Latvia, and Mongolia. Schedules are available at http://www.wto.org/english/tratop_e/serv_e/telecom_e/telecom_commit_exempt_list_e.htm.

18. See modalities for treatment of autonomous liberalization adopted by a special session of the council for trade in services, March 6, 2003, available at http://www.wto.org/english/news_e/pres03_e/pr335_e.htm.

19. Examples include Cyprus, India (with respect to data, cellular, voice, fax, and private leased networks), Philippines, Dominican Republic, and Ecuador.

20. The fact that a few developing countries committed to no limitation for cross-border trade in most subsectors is also significant. Examples include Chile, Columbia, Côte d'Ivoire, El Salvador, EU, Hong Kong, Iceland, Japan, New Zealand, Norway, and the United States.

21. There are seven countries that did not allow any form of commercial presence into their markets—four in the Americas region (Antigua and Barbuda, Dominica, Grenada, and Trinidad and Tobago) and three in Asia (Pacific Bangladesh, Brunei Darussalam, and Papua New Guinea), while other regions like Africa, the Arab States, and Europe have recorded no case of total prohibition of commercial presence.

22. See, for example, the schedules of Ghana and Indonesia, GATS/SC/35/supp.1, April 11, 1997, and GATS/SC/43/Suppl.2.

23. See, for example, the schedules of India, GATS/SC/42/Suppl.3 and Tunisia, GATS/SC/87/Supp.1, April 11, 1997.

24. This is the most widely used limitation in schedules of both developed and developing countries. Examples include Japan, Canada, India (foreign equity must not exceed 25 percent), Hungary (foreign investment limited to 25 percent of Hungarian interests plus one vote for MATAVRT, GATS/SC/40/supl.2, April 11, 1997), Indonesia (foreign equity participation limited to 35 percent, GATS/SC/43/Suppl.2, April 11, 1997), Malaysia (foreign shareholding up to 30 percent, GATS/SC/52/supp.2, April 11, 1997), and Israel (no more than 80 percent of shares may be owned by a foreign entity), GATS/SC/44/sup1, April 11, 1997.

25. Côte d'Ivoire schedule, GATS/SC/23/Suppl.1. Trinidad and Tobago, GATS/SC/86/Suppl.1.

26. Hoffman and Hobday have documented that the same strategy has been used by Alcatel, Ericsson, and Cable and Wireless in Singapore, Chile, Argentina, Brazil, Venezuela, China, and Indonesia. Hoffman and Hobday 1989, 242.

27. For example, the United States argued for development through liberalization in a study submitted to the GATT in the early 1980s.

28. Michapoulos 1998.

29. Renya 1993.

30. Mattoo 2000.

31. Mattoo and Sauvé 2004.

32. The tension between respect for national regulatory autonomy on the one hand and the requirement of an international trading system to minimize obstacles to trade on the other hand is as old as trade itself. In the context of trade in goods, the issue was first debated in the Tokyo Round of trade negotiations, which highlights the need to control intricate barriers arising from regulation on the domestic level.

33. Sherman 1998, 61.

34. Jacob 1994.

35. Smith and Wellenius 1999.

36. Communication from the European Communities and their Member States, 2001.

37. See U.S. proposal on trade on energy services, S/CSS/W/24, December 18, 2000. A reference paper on energy to ensure nondiscriminatory third-party access to and interconnection with energy networks and grids, where they are dominated by government entities or dominant suppliers.

38. Dominican Republic, Honduras, and El Salvador propose that a reference paper be applied in the tourism sector.

39. The other two items are government procurement and subsidies.

40. Currently, emergency safeguard measures are governed by the safeguards agreement concluded under the Uruguay Round. The agreement codifies to a large extent the same principles and rules under article XIX of the GATT.

41. A multilateral voluntary code on restrictive business practices in the areas of goods and services was developed by UNCTAD IV in Nairobi 1976 and approved by the United Nations General Assembly in resolution 35/63 of 1980.

42. Capithorne 1971; Rugman and Eden 1985.

43. Hoekman 1995.

44. Mattoo 2000.

45. Roger 1999.

46. Maskus 1998.

47. The different methods of transfer of technology include imports of capital goods and components, foreign technology licensing agreements, and foreign education and training.

48. South Centre 1999.

49. UNCTAD 2001.

50. UNCTAD 2000.

51. Luxner 1997a.

52. Luxner 1997b.

53. Ryan 2001.

54. Noam 1999, 6.

55. See Izaguirre 1999.

56. See UNCTAD 2001.

57. Izaguirre 1999.

58. ITU 2003.

59. Under article 4 of the ITU Constitution, the legal instruments of the Union are the Constitution, the Convention, and the Administrative Regulations. While the Constitution and Conven-

tion provide a general legal framework for the operation of the union, the regulations contain provisions regulating specific uses of telecommunications. The Administrative Regulations are further subdivided into the Radio Regulations (RRs) and the International Telecommunications Regulations (ITRs). The current ITRs were approved in the final acts of the World Administrative Telegraph and Telephone Conference (WATTC) held in Melbourne in 1988.

60. The sums at stake are very high if we consider, for instance, that the cash flow from settlement payments for one year is more than twice as much as the annual amount of telecom investment in all of Africa. It is also estimated that the net settlement that flowed to developing countries from 1992 to 1998 was sufficient to fund forty-five million new lines. See ITU 1999a, 73.

61. The issue of accounting rates came up during the drafting of the telecommunications annex under the GATS. The annex applies to access to and use of public telecommunications transport networks and services by service providers. The first draft of the telecommunications annex contained a provision that access to and use of public telecommunications transport networks and services (PTTNS) should be cost-oriented. This provision triggered a controversy between those who argued that above-cost access is a barrier to trade and those who considered that pricing is a commercial matter and should not be subject to GATS rules. The pricing clause was completely deleted from the final version of the annex.

62. Guermazi 1999.

63. In the benchmark order, the United States estimates the actual cost of terminating a call to be not higher that 8 cents a minute (settlement rate). Many developing countries charge as much as 80 cents. See also ITU 1999, ch. 3.

64. FCC, "In the Matter of International Settlement Rates: IB Docket No 96-261. Report and Order Adopted August 7, 1997." Released August 18, 1997. Available at FCC web site, http://www.fcc.gov/bureaus/international/orders/1997/fcc97280.html. The benchmarks proposed by the United States have three different rates: 15.4 cents per minute for upper-income countries; 19.1 cents per minute for middle-income countries; and 23.4 cents per minute for lower-income countries. The regime took effect in January 1998 and was scheduled to be in operation in 1999 for upper-income countries, 2000 for upper-middle income countries, 2001 for lower-middle income countries, 2002 for low-income countries, and 2003 for countries with a teledensity of less than one.

65. United States Court of Appeals for the District of Columbia "Circuit, Cable and Wireless PLC vs. FCC, January 12, 1999." The court's opinion is available at http://www.fcc.gov/ogc/documrnts/opinion/1999/cable.html.

66. Current U.S. accounting rates as well as historical data from 1985–2002 are published by the FCC (International Bureau). Available at http://www.fcc.gov/ib/td/account.htnnl.

67. This FCC data is available at http://www.fcc.gov/Bureaus/Common_Carrier/Reports/FCC-State_Link/Intl/4361-f01.pdf.

68. For a brief description of each of these forms, see ITU 1997.

69. ITU and Telegeography 1999.

70. It is reported that voice-over Internet (VOIP) doubles every hundred days. ITU 2000.

71. Data published by the FCC on traffic between the United States and selected countries reveals that in many cases the rate can drop more than 50 percent following the introduction of competition. For example, for the U.S./Hong Kong route, it was reported that one year before the introduction of competition in Hong Kong, the accounting rate between both ends was at $0.79 per each minute of traffic. One year after competition was introduced, the rate dropped sharply to $0.13 (a reduction of 83 percent). The same trend is also evident in relations between the United States and developing countries as reported by the FCC. In the case of Indonesia, accounting rate reduction after the introduction of competition is around 12 percent and 17 percent in the case of Chile. FCC, Section 43.61 of, "International Traffic Data, 2000 Report." Available at http://www.fcc.gov/Bureau/Common_carrier/Reports/FCC-state_Link/int'l.html.

72. India's Commitments, GATS/SC/42/Suppl.3, April 11, 1997.

73. This data is from the annual reports of the Indian government's Department of Telecommunications and of Videsh Sanchav Nigram Ltd. (VSNL). VSNL annual report and DOT annual reports.

74. Wellenius, Galarza, and Guermazi 2005.

75. The negative approach for trade liberalization in services supposes that all sectors are subject to liberalization unless specifically excluded by members. This approach is used in NAFTA and is supposed to have a larger impact on liberalization than the positive approach of GATS. The positive approach could slow the liberalization process. For a criticism of positive approach, see Snape and Bosworth 1999, 185–204.

76. Proposal by the Dominican Republic, El Salvador, and Honduras.

77. Renya 1993.

78. For example, Argentina committed to full liberalization by November 8, 2000. Chile maintained limitation for long-distance competition for four years, starting on August 1994. India committed to reviewing its policy on international competition in 2004.

79. See for example UNCTAD 1991.

80. The TRIMs agreement deals with trade-related investment measures. It outlaws investment measures that are inconsistent with the GATT national treatment obligation or the prohibition of the use of quantitative restrictions measures. The TRIMs provides a list of prohibited measures that include local content requirements, trade balancing requirements, foreign exchange balancing requirements, as well as domestic sales requirements. M. Ariff 1989, 349.

81. Mattoo 1997 cites the example of fiscal measures as the most significant in this case. According to the author, members are allowed to maintain nondiscriminatory high taxes in certain services without the obligation to schedule such measures even if the impact of the tax can severely limit market access.

82. For an overview of the nonviolation nullification or impairment clause, see Rossler 1997.

83. For a description of the position of developed countries toward performance requirements during the Uruguay Round, see, for example, Fatourous 1989, 201.

84. The draft text as well as a commentary as of April 24, 1998, and an overview of each country's position is available on the OECD site at http://www.oecd.org//daf/invest/fdi/reports.htm.

85. UNCTAD and World Bank 1994, 75.

86. The issue of services negotiations was not a hot subject during Cancun. What caused the deadlock was a mounting disagreement between developed and developing countries over the Singapore trade issues (investment, competition, government procurement, and customs procedures) and mainly the controversial issue of subsidies in agriculture.

87. For an explanation of the theory of comparative advantage in services, see, for example, Feketuty 1988.

88. For a comprehensive overview of the issue of mode 4 and the challenge, difficulties, and merits of wider liberalization, see Self and Zutshi 2002.

89. See WTO, "Work Program on Electronic Commerce adopted by the General Council on September 25, 1998." The WTO bodies dealing with the e-commerce project are the Council for Trade in Goods for issues related to market access, valuation, import licensing procedures and standards, rules of origin, and classification issues and customs duties; the Council for Trade in Services for issues related to modes of supply, transparency, MFN, increasing participation of developing countries, domestic regulation, standards and recognition, competition, market access, national treatment, and issues related to access and use of public telecommunications transport networks. In addition, the Council for TRIPS is required to address three items: copyrights, trademarks, and access to technology as related to e-commerce. Finally, the Committee on Trade and Development is required to study five items related to e-commerce and development: the impact of e-commerce on small- and medium-sized enterprises, how to enhance participation of developing countries in e-commerce, the impact of e-commerce on delivery of physical goods, financial implications of e-commerce for developing countries, and the use of information technology in the integration of developing countries in the multilateral trading system. For further information visit the WTO e-commerce site at http://www.wto.org/english/tratop_e/ecom_e/ecom_e.htm.

90. On the treatment of e-commerce under GATS, see Drake and Nicolaidis 2000.

91. See the WTO publication "Electronic Commerce and the Role of the WTO." Available at http://www.wto.org/english/res_e/booksp_e/special_study_2_e.pdf.

92. There is a vast, growing literature on the development dimension of e-commerce. See, for example, Goldstein and O'Connor, "E-commerce for Development: Prospects and Policy Issues." Paper of the OECD Development Center available at http://www.oecd.org/dev/ENGLISH/New/documents/tokyo2.pdf. See also OECD, *The Economic and Social Impact of Electronic Commerce: Preliminary Findings and Research Agenda* (Paris: OECD, 1999).

93. The UNCTAD Expert Meeting on Electronic Commerce and Tourism was held in Geneva on September, 18–20, 2000. See background document "Electronic Commerce and Tourism: New

Perspective and Challenges for Developing Countries." Available at UNCTAD e-commerce site at http://www.unctad.org/ecommerce/index.html.

94. See ITU 1999b, chapter 3, paragraph 3, dealing with Internet commerce in selected industries.

95. See Drake and Nicolaidis 2000.

96. For international Internet interconnection charging studies and findings, visit http://www.apii.or.kr/apec/atwg/pritgtgid.html. In the Cancun Ministerial Meeting, the ministers reaffirmed the importance of governmental intervention to promote competition in the case of dominant players of de facto monopolies. The declaration also highlights the importance of further studies on the issue. See Cancun Declaration, May 24–26, 2000.

97. Since the end of 2000, the telecom industry has experienced a significant decline. While some argue that this is a bad sign for the industry, others who are more optimistic argue that this is a "gale of creative destruction" after which the telecom industry will be reinvented. See ITU 2001.

References

Abu-Akeel, A. 1999. "Definition of Trade in Services under the GATS: Legal Implications." *Georges Washington Journal of International Law and Economy* no. 32: 189–216.

Ariff, M. 1989. "TRIMS: A North-South Divide or a Non-Issue?" *World Economy* 12, no. 3: 347–360.

Benz, S. F. 1985. "Trade Liberalization and the Global Service Economy." *Journal of World Trade Law* 19: 95–120.

Capithorne, L. W. 1971. "International Corporate Transfer Prices and Government Policy." *Canadian Journal of Economics* (August): 324–341.

Chakravarti, R. 1990. *Recolonization, GATT, the Uruguay Round and the Third World.* London: Zed Books.

Communication from the European Communities and their Member States. 2001. "GATS 2000: Postal/Courier Services, Council for Trade in Services." S/CSS/W/6123, March 2001. See also "The Doha Development Agenda and the EC Initial Offer on Services: The European Postal Operators Mind the GATS," June 25, 2003. Available at http://www.posteurop.org/eng/print.asp?ne#id=524.

Cowhey, P. 1990. "The International Telecommunications Regime: The Political Roots of Regime for High Technology." *International Organizations* 44, no. 2: 196–199.

Cowhey, P., and M. Klimenko. 1999. "The WTO Agreement and the Telecommunications Policy Reform." Washington, DC: The World Bank, Mimeo.

Drake, W. J. 1999. *Toward Sustainable Competition in Global Communications: From Principle to Practice—Summary Report of the Third Aspen Institute Roundtable on International Telecommunications.* Washington, DC: Aspen Institute.

Drake, W. J. 2000. "The Rise and Decline of the International Telecommunications Regime." In *Regulating the Global Information Society*, ed. C. T. Marsden, 124–177. London: Routledge.

Drake, W. J., and E. Noam. 1997. "The WTO Deal on Basic Telecommunications: Big Bang or Little Whimper?" *Telecommunications Policy* 21, no. 9/10: 799–818.

Drake, W. J., and Kalypso Nicolaïdis. 2000. "Global Electronic Commerce and the General Agreement on Trade in Services: The 'Millennium Round' and Beyond." In *GATS 2000: New Directions in Services Trade Liberalization*, ed. Pievre Sauve and Robert M. Stern, 399–437. Washington, DC: The Brookings Institution.

Fatourous, A. 1989. "Possible Standards for Trade-Related Investment Measures." In *United Nations, Technology, Trade Policy and the Uruguay Round*," Papers and proceedings of a round table held in Delhi, Greece, April 22–24, 1989, 201.

Feketekuty, G. 1988. *International Trade in Services: An Overview and Blueprint of Negotiations*. Cambridge, MA: Ballinger Publishing Co.

Feketekuty, G. 2000. "Assessing and Improving the Architecture of GATS." In *GATS 2000: New Directions in Services Trade Liberalization*, ed. P. Sauvé and Peter Stern, 85–112. Washington, DC: Brookings Institution Press.

Guermazi, B. 1999. "International Accounting Rates, Developing Countries and the WTO: The Dilemma and a Possible Solution." *International Journal of Communications Law and Policy*, no. 3, (Summer).

Guermazi, B. 2002. "Bridging the Digital Divide: Beyond the Basic Telecom Agreement, Towards a Global Universal Service Regime." Doctoral dissertation, Institute of Air and Space Law, McGill University, Montreal, Canada.

Henderson, A., I. Gentle, and E. Ball. 2005. "WTO Principles and Telecommunications in Developing Nations: Challenges and Consequences of Access." *Telecommunications Policy* 29: 205–221.

Hoekman, B. 1993. "Safeguard Provisions and International Trade Agreements Involving Services." *World Economy* 16, no. 1: 29–49.

Hoekman, B. 1995. "Tentative First Step: An Assessment of the Uruguay Round Agreement on Trade in Services." Paper presented at the World Bank Conference on the Uruguay Round and the Developing Economies. Washington, DC.

Hoffman, K., and M. Hobday. 1989. "The Third World and US Telecommunications Policy." In *New Directions in Telecommunications Policy*, volume 2: Information Policy and Economic Policy, ed. Paula R. Newberg, 231–265. Durham, NC: Duke University Press.

International Telecommunications Union (ITU). 1997. *World Telecommunications Development Report: Trade in Telecommunications*. Geneva: ITU.

International Telecommunications Union. 1999a. *World Telecommunications Development Report: Mobile Communications*. Geneva: ITU.

International Telecommunications Union. 1999b. *Challenges to the Network: Internet for Development*. Geneva: ITU.

International Telecommunications Union. 2000a. *Internet Reports: IP Telephony*. Geneva: ITU.

International Telecommunications Union. 2000b. *Trends in Telecommunications Reform: Interconnection*. Geneva: ITU.

International Telecommunications Union. 2001. *World Telecommunications Development Report: Reinventing Telecoms*. Geneva: ITU.

International Telecommunications Union. 2003. "Mobile Overtakes Fixed." September 2003, available at http://www.itu.int/osg/spu/ni/mobileovertakes/Resources/Mobileovertakes_Paper.pdf.

International Telecommunications Union and Telegeography Inc. 1999. *Direction of Traffic: Trading Telecom Minute*. Geneva: ITU.

Izaguirre, A. K. 1999. "Private Participation in Telecommunications: Recent Trends." Public Policy for the Private Sector Series, Note No. 204. Washington, DC: World Bank.

Jacob, H. S. 1994. "Building Regulatory Institutions in Central and Eastern Europe." Paper prepared for the OECD/World Bank Conference on Competition and Regulation in Network Infrastructure Industries, Budapest, June 28 to July 1, 1994.

Krein, M. F., and A. Freytag. 1997. "Telecommunications and WTO Discipline: An Assessment of the WTO Agreement on Telecommunications Services." *Telecommunications Policy* 21, no. 6: 477–491.

Lahouel, M., and K. Maskus. 1999. "Competition Policy and Intellectual Property Rights in Developing Countries: Interests in Unilateral Initiatives and a WTO Agreement." Paper prepared for The WTO/World Bank Conference on Developing Countries in a Millennium Round. WTO Secretariat, Center William Rappard, Geneva, September 20–21, 1999.

Low, P., and A. Subramanian. 1995. "TRIMS in the Uruguay Round: An Unfinished Business," in *The Uruguay Round and the Developing Economies*, ed. Will Martin and Alan Winters. Washington, DC: World Bank.

Luxner, L. 1997a. "The Dream of Mobility: Cellular Meets Pent-up Demand and Sparks Explosive Growth in Latin America." *Global Telephony* (December 1, 1997), available at www.luxner.com.

Luxner, L. 1997b. "US Carriers Head South: Latin America Proves a Sunny Climate for Formerly Provincial Telecos." *Telephony* (June 2, 1997), available at www.luxner.com.

Marr, A. 1997. "Foreign Direct Investment Flows to Low-Income Countries: A Review of the Evidence." Briefing paper. London: Overseas Development Institute.

Maskus, K. S. 1998. "The Role of Intellectual Property Rights in Encouraging Foreign Direct Investment and Technology Transfer." *Duke Journal of Comparative and International Law* 9: 109.

Mattoo, A. 1997. "National Treatment in the GATS: Corner-Stone or Pandora's Box." *Journal of World Trade Law* 31, no. 1: 5.

Mattoo, A. 2000. "Is There a Better Way: Alternative Approaches to Liberalize under GATS." Research paper prepared for the World Bank. Available at http://www1worldbank.org/wbiep/trade/papers_2000/Bpgats.pdf.

Mattoo, A., and P. Sauvé, eds. 2004. *Domestic Regulation and Services Trade Liberalization.* New York: Oxford University Press for the World Bank.

Michapoulos, C. 1998. "Differential and More Favorable Treatment of Developing Countries in the WTO: A Conceptual Classification." Research paper. Geneva: WTO.

Murkherjee, N. 1999. "GATS and the Millennium Round of Multilateral Negotiations." *Journal of World Trade* 33, no. 4: 87–102.

Nicholaidis, P. 1989. "Economic Aspects of Services: Implications for a GATT Agreement." *Journal of World Trade Law* 23, no. 4: 65–67.

Noam, E. N., ed. 1999. *Telecommunications in Africa.* New York: Oxford University Press.

Prica, D., and B. Christy. 1996. "Agreement on Trade Related Investment Measures (TRIMS): Limitations and Prospects for the Future." In *The Multilateral Trade Framework for the 21st Century and US Implementing Legislation,* ed. Terrence P. Stewart, 439–445. Washington, DC: American Bar Association.

Renya, K. 1993. "The General Agreement on Trade in Services." In GATT Uruguay Round: A Negotiating History, ed. Terence P. Stewart. London: Kluwer Law International.

Roger, N. 1999. "Recent Trends in Private Participation in Infrastructure." Public policy for the private sector. Note No. 196. Washington, DC: World Bank.

Rossler, F. 1997. "The Concept of Nullification and Impairment in the GATT/WTO Dispute Settlement System." In *The GATT/WTO Dispute Settlement System,* ed. Ernst-Ulrich Petersman, 142–149. London: Kluwer Law International.

Rugman, A. M., and L. Eden. 1985. *Multinationals and Transfer Pricing.* New York: St Martin's.

Ryan, R. "Golden Opportunities, Hidden Difficulties." *Telephony* (December 3, 2001).

Sapir, A. 1985. "North-South Issues in Trade in Services." *World Economy* 8, no. 2: 27–42.

Sauvé, P. 1994. "Liberalizing Trade in Services." Discussion Paper 243. Washington, DC: World Bank.

Saunders, R. J., J. J. Warford, and B. Wellenius. 1994. *Telecommunications and Economic Development.* Baltimore and London: Johns Hopkins University Press.

Self, R., and B. K. Zutshi. 2002. "Temporary Entry of Natural Persons as Services Providers: Issues and Challenges in Future Liberalization under the Current GATS Negotiations." A paper presented in the joint WTO-World Bank symposium on movement of natural persons under the GATS. WTO, Geneva, April 10–12, 2002. See http://www.wto.org/english/tratop_e/serv_e/symp_apr_02_zutshi_self_e.doc.

Sherman, L. 1998. "Wildly Enthusiastic about the First Multilateral Agreement on Trade in Telecommunications Services." *Federal Communications Law Journal* 51 (December).

Smith, P. L., and B. Wellenius. 1999. "Strategies for Successful Telecommunications Regulation in Weak Governance Environment." Memorandum. Washington, DC: World Bank.

Snape, R. H., and M. Bosworth. 1999. "Advancing Services Negotiations." In *The World Trading System: Challenges Ahead*, ed. Jeffery J. Schott, 5–204. Washington, DC: Institute for International Economics.

South Centre. 1999. *Financing Development: Issues for a South Agenda* (April 1999). Geneva: South Institute. Available at http://www.southcentre.org/publications/financing.

Spata, J. 2002. "Common Carrier Approach to Internet Interconnection." *Federal Communications Law Journal* 54 (March): 225–280.

Srinivasan, T. 1998. *Developing Countries and the Multilateral Trading System from the GATT to the Uruguay Round*. Boulder, CO: Westview Press.

Tuthill, Lee. 1997. "The GATS and New Rules for Regulators." *Telecommunications Policy* 21, no. 9/10: 783–798.

UNCTAD. 1991. *The Impact of Trade-Related Investment Measures on Trade and Development: Theory, Evidence and Policy Implications*. New York: United Nations.

UNCTAD. 2000. *The Least Developed Countries: 2000 Report: AID, Private Capital Flows and External Debt: The Challenge of Financing Development in the LDC*. UNCTAD: Geneva.

UNCTAD. 2001. *World Investment Report: Promoting Linkages*. UNCTAD: Geneva.

UNCTAD and World Bank. 1994. *Liberalization of International Transactions in Services: A Handbook*, 75. United Nations: New York and Geneva.

Willenius, B., J. Galarza, and B. Guermazi. 2005. "Telecommunications and the WTO: The Case of Mexico." Policy Research Working Paper.

II The Global Governance of Networked Information, Communication, and Commerce

6 Trade Barriers or Cultural Diversity? The Audiovisual Sector on Fire

Byung-il Choi

In the uncharacteristically warm weather of mid-November 2002, in Pusan, the second-largest city and the busiest seaport in Korea, the opposition party presidential candidate Lee Hoi-chang was compaigning. The election was just a month away and the race was dead locked: one poll showed that the margin of the lead Lee enjoyed was rapidly evaporating and the race had become too close to call. In a desperate effort to not lose the advantage, Lee promised to protect the Korean film industry to the audience gathered at the Pusan International Film Festival (PIFF). The Korean film industry people had flexed their muscles to turn the PIFF, the largest and the most successful film festival in Asia, into a political pork barrel. They wanted to secure protection of their industry from the candidates, and they got what they wanted: both candidates from the ruling party and the opposition party made pledges to maintain the screen quota requiring that theater owners show Korean movies for 146 days a year (40 percent of the screening days).

The screen quota has been a hot potato ever since then-outgoing President Kim Dae-jung proposed a bilateral investment treaty (BIT) with the United States in 1998. Kim intended to utilize the Korea-U.S. BIT as the means and signal to lure foreign investors to Korea, which was undergoing an unprecedented financial crisis. In the light of the magnitude of the crisis unfolding, he was quite confident that the BIT would be quickly agreed upon. As it turned out, the Korean film industry was visibly upset by the idea of the BIT, because it would spell the end of the screen quota. To block the negotiations, the film industry forged a solid coalition with various and disparate NGOs such as environmental groups, labor unions, and teachers' organizations. The coalition mounted a series of rallies, arguing that the BIT would trade the Korean culture for measly economic gains. They chanted in unison, "Film should not be judged only by the market principle." Several Korean actors, clad in black, staged a mock funeral marking the death of the fledgling Korean film industry. Some shaved their heads to express their protest. Kim's administration, founded on the notion of populism, neither persuaded those angry people nor found any compromise. By the end of his tenure Kim could not conclude the BIT due in large part to the stonewalling of the powerful local film industry. Kim is now long gone, but the controversy over the screen quota continues to incite the nation.

The battle in Korea mirrors the ongoing global conflict concerning the cultural sector in the era of globalization.[1] A rising tide of globalization is now reaching to the sector in which principal players and policy makers have been operating without having to fathom the trade implications of their practices and policies. The cultural sector in general, and the audiovisual sector in particular, is a case in point. Despite the obvious dual characteristics of the audiovisual sector (comprised of film and video production, distribution and projection, television programs and broadcasting, and music) both as a cultural asset and as an industry that provides jobs and helps to grow the global economy, in many parts of the world the economic rationale has taken a back seat in the modus operandi of the sector. Many countries have organized their audiovisual sector as if they were immune to the economic law of supply and demand. Under the name of cultural promotion, public institutions were built, taxpayers' money was earmarked, and fat subsidy programs were designed to create and sustain some cultural products about which many consumers were not enthusiastic.

The economic principle having been shoved to the background, the political economy of regulations took the center stage. Politicians designed a labyrinth of regulations which discouraged new entry and favored the incumbents. There was not much competition to speak of. Innovative ideas and technological breakthroughs were kept on the drawing board and had to confront manmade entry barriers if they were associated with actors outside the sector. Time and again, people who happened to be there at the genesis thrived under the policy of industry protection and promotion.

Across the world, the cultural sector, especially the audiovisual sector, is politically well connected and influential. In this regard, it is not surprising to find the audiovisual sector claiming a disproportionately large share of policy attention and budgets relative to other sectors. Politicians directed their favors to the audiovisual sector, and in return the audiovisual sector granted much-coveted media attention to them. In the process, a cartel has been forged between regulators and regulated. Over time, through the reciprocal exchange of favors, both legally and illegally, the cartel has become more and more consolidated to the extent that it is turning into a fortress.[2]

Now, the rising tide of globalization is threatening to overflow into this fortress, throwing residents into bewilderment. It is the United States that is pushing the longstanding audiovisual cartel to the brink. Over the past decades, audiovisual products have shown stellar economic performance in international trade. Driven by commercial interests, U.S. companies, spearheaded by Hollywood, are pressing hard on the U.S. trade negotiators to break the regulatory cartel in the audiovisual sector. As an integral part of the global strategy of keeping communism at bay, and also mindful of the colossally disastrous consequences of the "beggar-thy-neighbor" policy of competitively erecting protective barriers in the late 1920s and early 1930s on the eve of the

Great Depression, the United States has been vigorously promoting trade liberalization. U.S. leadership has been instrumental in sustaining the momentum of trade liberalization through multilateral negotiations at the General Agreement on Tariffs and Trade (GATT). After four decades of pushing trade liberalization in the field of goods, the focus has now shifted to the field of services and agriculture. Two factors have contributed to this shift: first, through continuous multilateral negotiations, the average rate of tariffs on most industrial products has been pushed to a low level.[3] Second, over time, the service sector has gained in importance in the world economy. Hence, the stage was set for a close encounter between culture and trade.

The Uruguay Round, the multilateral trade negotiations started in 1986 under the auspices of the GATT, aimed to create new multilateral trade rules in services. In the seven-year negotiations, differences and gaps among the major players in dealing with the audiovisual sector proved to be regular deal breakers. The Uruguay Round witnessed two parallel arguments for and against trade liberalization of the audiovisual sector. One view, which we can call the "trade perspective," argued that the audiovisual sector be placed in the stream of progressive trade liberalization like any other sector. The other view, which we can call the "cultural perspective," argued that the audiovisual sector should not be subject to trade negotiations. The United States was the principal architect of the trade perspective, while the EU, led by France and Canada, were the staunch proponents of the cultural perspective.

This trans-Atlantic rift almost derailed the Uruguay Round at the eleventh hour. A deal was reached only after the United States conceded to allow the EU to exclude their audiovisual from MFN (most-favored-nation) treatment, and the EU conceded that the audiovisual sector as such would not be taken outside of the new General Agreement on Trade in Services (GATS).[4] In practical terms, the EU could continue to maintain the preferential policy of discriminating against non-EU contents in its policy on audiovisual products such as TV programming and film. At the same time, the United States had a trophy to show to its constituents, in ensuring that the audiovisual sector would not be carved out of the service agreement.

More than a decade has lapsed since the conclusion of the Uruguay Round. In the meantime, the world has witnessed the dazzling advances of technology, mainly the emergence of digital technology, the blossoming of the Internet, broadband access, and the convergence of media, thereby fundamentally reshaping the landscape for producing, distributing, and consuming audiovisual services. These technological breakthroughs and immense room for innovative commercial applications in a fashion unimaginable only a few years ago gave rise to new concerns and unforeseen challenges for stakeholders.

The DVD (digital video disc or digital versatile disc) became an important medium of storing visual contents, and more and more people are watching movies not at the cinema but at the place of their own choosing, such as at home. The Internet has become the most popular medium of delivery, both for legal downloading of films and illegal circulation of pirated films. Digital technology overcame capacity constraints, while the Internet conquered the tyranny of distance. The new audiovisual environment means more diverse and easy access to the unprecedented variety of products for consumers, as well as more delivery channels for suppliers. Even though the cinema still holds its importance as the first outlet for a film, other competing delivery mediums such as DVD, Internet downloading, cable television, and satellite have fundamentally changed how audiovisual contents are manufactured, delivered, and consumed. New media, such as cable, direct-to-home satellite, and digital networks that distribute content locally and also internationally cast challenges to the conventional policy of promote and protect, dating back to the days of analogue, with cinema and over-the-air channel broadcasting in mind.

All these new developments serve as a warning signal of growing tensions and new friction between the trade perspective and the cultural perspective. For major content exporters, the goal is to take advantage of the more integrated global market and to ensure their products against piracy, increasingly common due to digital technology and the Internet. As for the defenders of cultural identity, they fear the increasing dominance of exporters and the eventual extinction of their local products in the market. Against this backdrop, the new multilateral trade negotiations have been underway at the WTO. To counter this movement, in 2005 proponents of the cultural perspective won a new international Convention on Cultural Diversity in the United Nations Educational, Scientific and Cultural Organization (UNESCO) in an attempt to insulate the cultural sector from trade disciplines. At the same time, in contrast to the multilateral approach, the United States has shown its strong appetite for the bilateral approach, pursuing a series of bilateral investment treaties (BIT) and free-trade agreements (FTA).

This chapter attempts to understand the ongoing dispute over trade liberalization in the cultural sector through the lens of international negotiations. To make this task manageable, the chapter deals with the audiovisual services as the main focus. It is organized as follows. Section one presents the audiovisual industry's situation in the global market. Section two traces trade disputes in the audiovisual sector, and section three discusses the current international trading regime for the cultural industry. Section four presents the status of the film industry in Canada, Mexico, and Korea. All three countries have negotiated with the United States, including on their cultural sectors, but with remarkably different outcomes. Insights gained from the case studies of each will be instrumental in understanding the ongoing friction at the interface of trade and culture from two levels: domestic and international. In-depth analysis of the international dimension of this ongoing battle is offered in section five. Section six offers conjecture on the future of the ongoing clash, and then the chapter closes with recommendations and conclusions.

What Is at Stake?

A rough sketch of the global audiovisual services (film, TV broadcasting, and music) goes as follows. The combined market of the three major economies—the United States, EU (fifteen member states), and Japan—is assessed as 202 billion euro as of 1999, and experts forecast annual growth of 10–15 percent over the next decade. The United States has 49 percent of the overall market, followed by the EU with 34 percent, and Japan with 16 percent. In terms of per capita consumption, a U.S. consumer spends 364 euro, the EU 185 euro, and Japan 266 euro. Within the EU, a UK consumer tops by spending 302 euro.

Among the subsectors of film, TV broadcasting, and music, the largest market is TV broadcasting, with a share of more than 60 percent of the audiovisual sector. In the EU the share of TV broadcasting is almost 70 percent. On the other hand, film takes 25.2 percent of the share in the United States, far exceeding the corresponding share of film in the EU and Japan (figure 6.1).

According to the OECD statistics, the United States is the largest trading country in the audiovisual service sector as illustrated in table 6.1. Most EU countries record a trade deficit, with the notable exception of the United Kingdom. During 1990–1999, the EU trade deficit more than doubled, whereas the U.S. trade surplus more than tripled.

A closer look into the trans-Atlantic trade reveals more stylized facts. The balance of trade with the United States is deeply to the disadvantage of the European audiovisual industries—and the deficit is rising, from a little more than $2 billion in 1989 to more than $7 billion in 1999 (table 6.2).

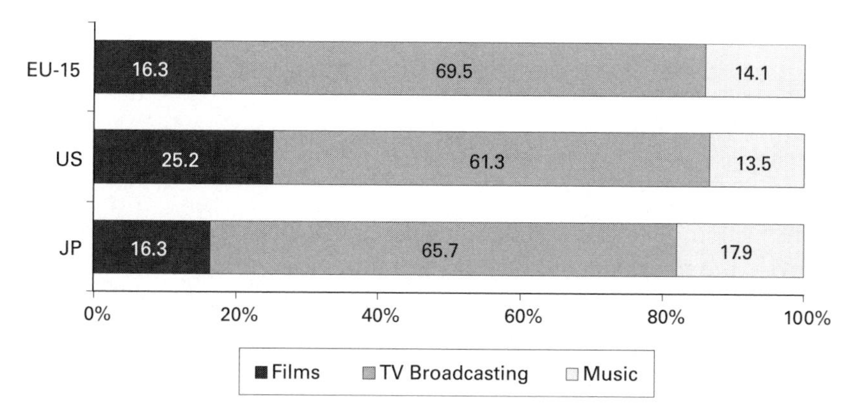

Figure 6.1
Composition of the Audiovisual Market (1999)
Source: European Commission (2001), *Statistics on Audiovisual Services Data 1980–1999*.

Table 6.1
Trade in audiovisual service sector (in million USD)

	Export		Import		Trade balance	
	1990	1999	1990	1999	1990	1999
United States	1,459	5,186	65	232	1,394	4,954
United Kingdom	868	1,023	840	689	28	333
Canada	298	1,055	608	1,158	–309	–104
Australia	—	99	323	410	–323	–311
France	632 [1]	819	900 [1]	1,332	–268 [1]	–513
Italy	234 [1]	196	716 [1]	690	–482 [1]	–494
Germany	126	128	864	3,314	–738	–3,187
EU	2,744 [2]	3,299	4,852 [2]	8,520	–2,108 [2]	–5,221
Korea	0	8	12	80	–12	–72

Source: OECD 2000, *OECD Statistics on International Trade in Services 1990–1999.*
[1] Data for 1992.
[2] EU total for 1992 does not include the Netherlands, Denmark, Ireland, Greece, Portugal, and Austria.

Table 6.2
Trade balance in audiovisual works, United States-European Union (in million USD)

	1989	1990	1991	1992	1993	1994	1995	1996	1997	1998	1999
United States to European Union	3,133	3,280	3,947	4,106	4,642	4,886	5,331	6,262	6,645	7,313	8,117
European Union to United States	404	464	279	300	429	566	517	613	668	706	853
Balance	—	—	—	—	—	—	—	—	—	—	—
European Union	2,729	2,816	3,668	3,806	4,213	4,320	4,814	5,649	5,977	6,607	7,264

Source: European Audiovisual Observatory.

The Film Industry

Due to the construction spree in multiplex cinemas and some successful Hollywood blockbuster movies, the global film industry has been in an expansion phase ever since the mid-1990s. The United States continues to dominate as the largest film exporter in the world, followed by Hong Kong. In terms of production, India is the biggest film producer (producing mainly for its domestic market), followed by the United States and Hong Kong. Europe is the principal destination of U.S. entertainment products: in 1995 about 70 percent of all U.S. exports of audiovisual services went to Europe. The American share of the European market rose from 56 to 78 percent over the 1990s. During this period, the market share of European films in their home turf

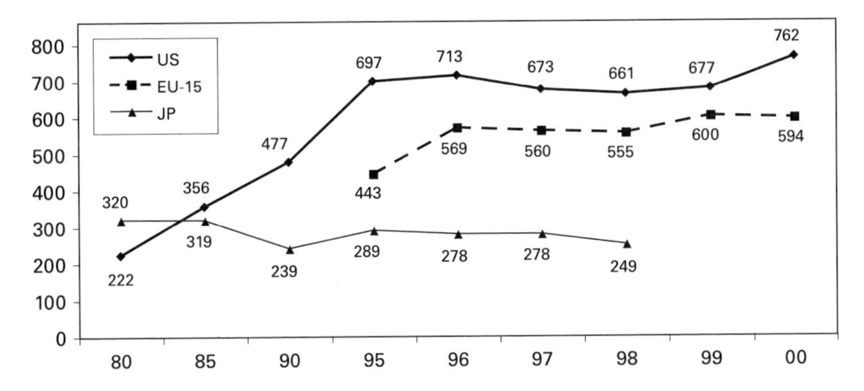

Figure 6.2
Film production trends in major regions (1980–1999)
Source: European Commission (2001), *Statistics on Audiovisual Services Data 1980–1999.*

diminished from 19 to 10 percent. As of 2000, the U.S. film industry enjoys a 73.7 percent share in the EU. A notable exception is France, where French films claim 36 percent of its market and EU films 33 percent in year 1999 (figure 6.2).

A major competitive advantage for Hollywood films derives from the huge U.S. domestic market. Having recouped high fixed costs in the home market, it is very easy for a U.S. film to successfully enter foreign markets at a marginal cost.[5] Economies of scale work to the benefit of the U.S. film vis-à-vis local films with smaller domestic markets. Besides, the U.S. film has been helped by its global marketing and distribution channels.

While the audiovisual sector in the EU countries is struggling to deal with fragmented distribution structures, the U.S. entertainment industry is investing vast amounts in the promotion and distribution of its products throughout Europe, coordinated by its powerful lobby group the Motion Picture Association of America (MPAA). All these factors gave rise to the situation where a dozen U.S. distributors take the majority of box-office receipts in Europe, while over a thousand European distributors compete for the rest of the market.

Taken together, all data show a clear-cut and decisive edge for the U.S. audiovisual sector in the world. The U.S. dominance in the cinema field is an old story. It goes back at least to the 1920s. An equally, or perhaps more important part of the story is the phenomenal growth of U.S.-based commercial television, comparitively free from regulations. The audiovisual sector is one of the few industrial sectors where the United States actually maintains a trade surplus.[6] Forging ahead and deepening this entrenched position is naturally not only a matter of private interest on the part of U.S. companies in this sector: it may also be a grave matter on the national agenda of the U.S. government in the post–Cold War era, when "soft power" through images and symbols are becoming more influential in international relations.

If television was the moving force behind the rapid increase in the U.S. trade surplus in the audiovisual sector over the last decade, some worry a new technological leap forward may carry it even farther and deeper. Convergence of media and digital compression techniques are, such observers conjecture, likely to create networks with the capacity to transmit content that will far surpass the availability of content. On the other hand, such a dramatic capacity increase would spell more opportunities for content suppliers from small countries to find their niche market at home and abroad. It is not just the major Hollywood studios that are jockeying for the lead. Publishing, utilities, and telecommunications companies are making inroads to audiovisual sectors, taking advantage of new technological convergence and mergers and acquisitions. Non-U.S.-originated companies like Bertelsmann, Vivendi, BT, Telefonica, and NTT are visible in this regard.

In all likelihood, the future of the audiovisual industry will be characterized by coexistence, although not peaceful, of domineering giants and pioneering small players. Nonetheless, the fear of a "conquest of the world by the big bad giant" could drive the cultural sector of the rest of the world to a more protectionist stance.

Blowing Trade Wind in the Cultural Sector

Evolution

Trade friction in the film industry between the U.S. and European countries traces back to the early twentieth century. During the inter-war years, the American motion picture industry pressed the U.S. Department of State to deal with the European screen quotas. Screen quotas, which require mandatory showing of local films at the expense of foreign ones, were deemed to be trade barriers. More significantly, screen quotas run afoul of the cornerstone trade principles of nondiscrimination: most favored nation (MFN), which requires that countries be treated equally, and national treatment (NT), which requires that foreign and domestic products, services, or nationals be treated equally.

This trans-Atlantic dispute eventually was resolved by establishing an exception to the national treatment principle in the case of film—in the GATT, which was negotiated in 1947. Article IV of the GATT specifically authorizes screen quotas, thereby removing them from normal GATT disciplines; in practice their application in the immediate postwar era was of limited duration due to a decline in U.S. film exports matched by a strong European film industry.

Conflict over the cultural sector was manifest in several international forums, including the United Nations Committee on the Uses of Outer Space in the 1960s, the UNESCO debate on the New International Information Order in the 1970s, and the GATT. During the Tokyo Round negotiations (1973–1979), the United States complained about the subsidization of cinema and television by no fewer than twenty-one countries. The United States also sought unsuccessfully to challenge European Community restrictions on the televised showing of non-European films in 1991, but the EC viewed television broadcasting as a service and considered that it could *not* be covered by the goods regime of the GATT.

The Uruguay Round witnessed a major clash of two opposing views on applying trade disciplines to the cultural sector. Encouraged by its strong performance in audiovisual products, the United States claimed that film and television products were marketable commodities, and therefore should be subject to the same trade rules like any other products. The EU, Canada, and others countered that trade should not interfere with culture, and criticized the U.S. request as infringement on the sovereign right of national cultural expression and diversity. The United States responded that cultural diversity is only a smokescreen for protection, and argued that its opponents were just trying to preserve the conventional way of organizing their cultural sector, depriving consumers of more choices.

The underlying tension between these two opposing views led to the idea of a "cultural exception" (*exception culturelle*). At the Montreal ministerial meeting in December 1988, the EU and Canada sought to secure specific language in the draft negotiating text on services so that the audiovisual sector would not be covered by the negotiations. They argued that audiovisual works were neither goods (as covered by the original GATT concept) nor services (as applied in the new GATS), but rather were "cultural goods" and could therefore not be covered in the negotiations. This demand for the so-called cultural exception led to a stalemate in negotiations in December 1993. To salvage the Uruguay Round negotiations—which by then had been in progress for seven years and, more significantly, had missed their self-imposed deadlines of 1990 and 1991, to the frustration of political leaders and business people—a great compromise was made:[7] the audiovisual sector was to be ruled under the GATS, but MFN would not be applicable for the time being. The clash among countries in the Uruguay Round was only the beginning of the discord to come.

Right after the conclusion of the Uruguay Round, trade in the cultural industry proved another deal breaker in the negotiations for the Multilateral Agreement on Investment (MAI). Intended as a free-investment instrument under the aegis of the OECD, the MAI only made it to the drawing-board, but it was a close call. The idea was to establish an international treaty that would facilitate and protect foreign investment. When the talks were begun in 1995, the world economy had been in recession. Policy makers saw foreign investment as an impetus toward recovery and rebuilding of the global economy. The MAI was building on the concepts of nondiscrimination (MFN and NT) in investment. After three years of drafting and negotiations the MAI treaty was ready for signature in April 1998. By that time an impromptu alliance of unrelated NGOs abruptly stole the show by attacking the MAI deal as the end of cultural and environmental sovereignty. France and Canada were adamant in insisting on securing some sort of safeguards in opening the cultural sector to foreign investment.

The MAI talks in the OECD collapsed. It was withdrawn from the OECD agenda and passed on to the WTO for possible inclusion in the so-called Millennium Round. In tandem with these developments in the international arena, the cultural sector constituents became more cohesive. To resist market-opening pressure from the United States, cultural sector policy makers, suppliers, and activists from the EU, Canada, and a number of developing countries forged a coalition on an international scale under the name of cultural diversity.

International Governance of Trade in Culture

Various multilateral, plurilateral, regional, and bilateral trade agreements cover some aspects of trade in culture, as summarized in table 6.3. The existing international rules on the circulation of cultural goods and services can be put into one of two categories: (1) those that establish or leave room for exceptional treatment for cultural goods and services in trade agreements; and (2) those designed to facilitate the circulation of the goods and services in question by eliminating trade barriers. The second group of legal texts makes it clear that the right to protect threatened national cultural production is not something to be taken for granted, reflecting the trade perspective. On the other hand, the legal texts in the first category support a vision of cultural diversity that protects and promotes national culture as well as encourages openness to the cultural production of others and cultural diversity in general, reflecting the cultural perspective.

Many bilateral or plurilateral free trade agreements have obvious built-in pressures to open the cultural sector. Negotiations over whether or not to include the cultural sector in these trade agreements proved stormy in some cases. As evidenced in the case of Korea-U.S. BIT talks, the screen quota issue has held the negotiations at bay. Their concern about U.S. dominance propelled the Canadians to negotiate the so-called cultural exception in the Canada-U.S. free trade agreement (FTA) and the North American Free Trade Agreement (NAFTA) talks. The OECD, considered an exclusive group of advanced industrial economies, set higher standards for international investment in its Code of Liberalization of Current Invisible Operations. Within this code restricting barriers to international investment, exceptions concerning screen quotas have been negotiated by countries such as Italy, Spain, and Korea.

For both sides of the conflict over trade and culture, the focal point is the WTO. The WTO offers the baseline for assessing the current status of trade in culture and negotiating for a freer flow of cultural products. For policy makers, industry people, and activists striving to protect the cultural sector, it is their strategic imperative to stop the process at the WTO. The GATT and the GATS are both relevant to trade in cultural products, though most of the negotiations would be under the GATS. The WTO dispute settlement mechanism provides another significant venue for resolving trade-related conflicts.[8]

Table 6.3
International rules on the circulation of cultural goods and services

1. *Article IV of GATT 1994*
Screen quota

2. *Articles II, XVI, and XVII of GATS*
Most-favored-nation treatment for all services and service suppliers of all members, irrespective of whether commitments have been made; market access; or national treatment.

3. *NAFTA Article 2106 and Annex 2106, Canada-United States and Canada-Mexico*
Exemptions related to measures concerning cultural industries; annexes I and II for Mexico (specifically concerning audiovisual)
Cultural industries exempt from provisions of this agreement, except as specified.

4. *OECD Code of Liberalization of Current Invisible Operations*
For cultural reasons, support for the production of printed films for cinema may be maintained provided that it does not significantly distort international competition in export markets.

5. *Agence de la Francophonie, Final Declaration of the Moncton Summit (1999)*
Affirms the right of states and governments to freely define their cultural policies.

6. *Agreement for Facilitating the International Circulation of Visual and Auditory Materials of an Educational, Scientific and Cultural Character (UNESCO, Beirut, 1948)*
Customs agreement regarding imports that covers the following categories of materials: films, filmstrip, microfilms, sound recordings, glass slides, models both static and moving, wall charts, maps, and posters.

7. *Agreement on the Importation of Educational, Scientific and Cultural Materials (UNESCO, Florence, 1950)*
Designed to remove customs tariffs and other obstacles that impede exchanges not only of visual and auditory material, but also of several other categories of material; it also provides for duty-free entry of very diverse items.

8. *Protocol to the Florence Agreement, adopted in Nairobi in 1976*
Protocol extends exemption from customs duties to various groups of material not covered by the agreement.

9. *Recommendation Concerning the International Exchange of Cultural Property (UNESCO, Nairobi, 1976)*
Facilitating the legal circulation of collectors' objects among museums and other cultural institutions by exchanges, loans, etc.

10. *European Convention on Transfrontier Television (ETS no. 132, 1989, in force 1993) and Protocol Amending the Convention on Transfrontier Television (ETS no. 171)*
Legal framework for the free circulation of transfrontier television programs in Europe, through minimum common rules.

11. *Resolution of the European Council and of the Representatives of the Governments of the Member States, Meeting within the Council of 25, January 1999, Concerning Public Service Broadcasting Official Journal no. C 030 of 05/02/1999.*
Supporting the role and funding of public service broadcasting; broad public access; the benefits of new audiovisual and information services and technologies, and quality public programming and services.

12. *European Council Resolution of February 8, 1999, on fixed book prices in homogeneous cross-border linguistic areas (1999/C 42/02)*
Emphasizing the importance of a balanced assessment of the cultural and economic aspects of books, the resolution promotes cultural development and diversity in Europe, and cultural benefits to the consumer.

13. *Bilateral cultural cooperation agreements*
Cultural and technical cooperation is exempted from customs duties.

The GATT

Prior to the establishment of the GATS, friction in the cultural sector—trade versus protection—was not in the limelight, mainly due to the fact that countries could hide under the internationally recognized protection that had been specially designed for the film industry. Contrary to its founding philosophy of nondiscrimination, the GATT provided the basis for favoring local films at the expense of foreign ones. Under article III, the GATT provides one of its central principles, "non-discriminatory national treatment to all imported goods." Yet article III also provides an explicit exemption to national treatment for motion pictures. Moreover, article IV's "Special provisions relating to cinematograph films" allows signatory nations to establish screen quotas and give preference to films from preferred countries. Even thought article IV does not elaborate by what standards the national origin of films should be determined, or how to apply screen quotas, this provision functions as an exception to article I's most-favored-nation treatment and article III's national treatment provision.

Article IV of the GATT ("Special provisions relating to cinematograph films") stipulates as follows:

> If any contracting party establishes or maintains internal quantitative regulations relating to exposed cinematograph films, such regulations shall take the form of screen quotas which shall conform to the following requirements:
> (a) Screen Quotas may require the exhibition of cinematograph films of national origin during a specified minimum proportion of the total screen time actually utilized, over a specified period of not less than one year, in the commercial exhibition of all films of whatever origin, and shall be composed on the basis of screen time per theatre per year or the equivalent thereof;
> (b) With the exception of screen time reserved for films of national origin under a screen quota, screen time including that released by administrative action from screen time reserved of national origin, shall not be allocated formally or in effect among sources of supply;
> (c) Notwithstanding the provisions of sub-paragraph (b) of this Article, any contracting party may maintain screen quotas conforming to the requirements of sub-paragraph (a) of this Article which reserve a minimum proportion of screen quota for films of a specified origin other than that of the contracting party imposing such screen quotas; *provided* that no such minimum proportion of screen time shall be increased above the level in effect on April 10, 1947;
> (d) Screen quotas shall be subject to negotiation for their limitation, liberalization or elimination.

Since its establishment, members have tried to clarify the meaning of article IV. In the early 1960s, the United States challenged restrictions against showing foreign television programs as a violation of article III: 4. The GATT working party was convened, but its members were unable to come to any agreements on the subject. In 1991, while negotiators at the Uruguay Round were dissecting the audiovisual sector, the United States tried to corner the EU, blaming EU measures for restricting the showing of non-European films on television. The EU responded that the question of broadcasting, whether by television or by any other means, belonged essentially to the area of services, where the negotiations were still underway.

In summary, the GATT provides the legal basis for protecting the local films through the screen quota measure, although it is inconsistent with the GATT principle of non-discrimination. However, as inscribed in article IVd, the screen quota is not blanket protection from trade discipline: it is subject to negotiations.

The GATS

Creating the GATS achieved two important victories for trade constituents: first, it put the audiovisual sector on the center stage; second, the burden of proof shifted to the cultural sector in demonstrating the necessity of protection. Now, the ball is in the court of those countries that have trade-restrictive measures in place.

Negotiators agreed that the coverage of the GATS would be universal, meaning that as a matter of principle no service sector would be excluded from the multilateral trade discipline of the GATS. As discussed in the previous section, the universal coverage of the GATS was seriously resisted. During the Uruguay Round, where the negotiations on services were conducted for the first time in the history of the multilateral trading regime, and eventually created the GATS, the EU argued that culture in general, and the audiovisual sector in particular, should be outside the scope of multilateral trade disciplines, reflecting the discriminatory policy on non-EU contents for the audiovisual services.

The final outcome agreed at the conclusion of the Uruguay Round was as follows:

- coverage of the GATS is universal, meaning that there is no cultural exception for the audiovisual sector;[9]
- however, market access and national treatment of every service sector is to be negotiated, not automatically granted;
- and the EU obtained an exemption to MFN treatment for its audiovisual sector.

The universal coverage does not imply that all service sectors should automatically be opened up to foreign services and service suppliers.[10] What can a country do, then, if it is determined to protect its audiovisual sector? Mainly, two options exist: an MFN exemption and making no or few specific commitments. Article II of the GATS ensures MFN treatment of all the WTO members. Article II:1 states that, with respect to any measure covered by the GATS, each member shall accord immediately and unconditionally to services and service suppliers of any other member treatment no less favorable than it accords to like services and service suppliers of any other country. However, it has been agreed in the Uruguay Round that particular measures inconsistent with the MFN obligation can be maintained—in principle for not more than ten years and subject to review after not more than five years. Such measures must have been specified by members in a list of MFN exemptions submitted by the end of the Uruguay Round or by the conclusion of extended negotiations on certain sectors for which the delayed submission of related exceptions was expressly authorized. Subsequently, requests for exemptions from article II (MFN) can only be granted under the waiver procedures of the Marrakesh agreement. The EU invoked this exception clause of the MFN for its audiovisual sector. By the time the Doha Round was launched in late 2001, forty-nine countries had listed MFN exemption.

These countries secured an international legal right to preserve their ability to discriminate against foreign audiovisual service suppliers on the basis of origin of supply. Under this exemption, a specific foreign country can be targeted for trade restrictions. While the MFN exemption was granted only at the beginning of the GATS as a temporary measure, in principle lasting no more than ten years, there is still room for discriminating between foreign and local suppliers. Mainly, two options are possible: to make no "specific commitment" or a less specific commitment.

Any country that would allow the foreign suppliers access to its domestic audiovisual market would make a specific commitment through negotiations with other WTO members. Major negotiating points are market access and national treatment. The negotiations on terms and conditions of market access (article XVI of the GATS) and national treatment (article XVIII of the GATS) would determine the nature, scope, and speed of a country's liberalization of its service sector. Once inscribed in the national schedule under the GATS, such specific commitments would apply to all the WTO members on equal terms, pursuant to the MFN principle. The exact details of a specific commitment are subject to negotiations and may vary by countries. For instance, China made a specific commitment covering only some audiovisual services, with certain limitations on market access and national treatment on those offered services, such as requiring a joint venture of a foreign service supplier with a local partner for distribution of audiovisual services, except motion pictures, which are subject to censorship by the Chinese authority.

If countries could avoid substantive market access and national treatment commitments, then why did they fight so hard on the issue of coverage? The answer is that when a sector is covered by the GATS, it is subject to the MFN exemption and the progressive liberalization, inter alia.

The United States has long sought to introduce the free trade principle into the cross-border exchange of cultural products. Despite intense diplomatic and commercial efforts, the outcome has not been an unqualified success. Only a distinct minority of the WTO members has agreed to schedule bound liberalization commitments in audiovisual services under the GATS: by the time the Doha Round was launched in late 2001, only twenty-six countries had made specific commitments in the sector of audiovisual services.[11]

Although more limited, the commitments by India (the world's largest film producer), Hong Kong, and Japan are meaningful, because some countries with large production and influential cultures deem to consider the issue of liberalization in audiovisual services with an "open mind."[12] The remaining WTO members, led by the EU, have severely limited access to their markets by not making any specific commitments and making MFN exemptions. Salient aspects of these commitments are easily noticeable.

Table 6.4
WTO/GATS audiovisual service commitments

Service subsector	Countries that made specific commitments	Total number
Films and video production/ distribution	Albania, Central African Republic, Gambia, Georgia, Hong Kong, India, Israel, Japan, Jordan, Kenya, Korea, Kyrgyz, Lesotho, Malaysia, Mexico, New Zealand, Nicaragua, Oman, Panama, Singapore, Thailand, United States	22
Film projection services	Albania, Central African Republic, Gambia, Georgia, Japan, Jordan, Kyrgyz, Kenya, Lesotho, Mexico, New Zealand, Nicaragua, Oman, Panama, United States	15
Radio and TV production services	Albania, Central African Republic, Gambia, Georgia, Jordan, Kyrgyz, Lesotho, New Zealand, Panama, Thailand, United States	11
Radio and TV transmission services	Albania, Central African Republic, Dominican Republic, El Salvador, Gambia, Kyrgyz, Lesotho, Malaysia, New Zealand, United States	10
Sound recording	Albania, Central African Republic, Georgia, Hong Kong, Japan, Korea, Kyrgyz, Panama, Singapore, United States	10
Other	Albania, Central African Republic, Dominican Republic, El Salvador, Hong Kong, New Zealand, United States	7

Source: WTO Secretariat, November 2001.
Note: China and Chinese Taipei are omitted for lack of details.

First, most commitments were not concerned with further liberalization but rather were intended to bind the existing level of market access and national treatment. Most countries deem it preferable, as a matter of trade policy, to refrain from taking on legally binding GATS obligations in order to pursue their cultural policies. Some negotiators, however, claim that this sort of "standstill" commitment was still meaningful because the pace of liberalization was established and at least reneging on the existing level of liberalization would not take place or would be very costly.

Second, the EU, Canada, and Australia did not make any specific commitments. Only three countries, Albania, the United States, and Central African Republic, made some form of commitment, though not for complete liberalization, in all of the audiovisual service sector (table 6.4).

David against Goliath: The Film Industries of Canada, Mexico, and Korea versus the United States[13]

In this section, the film industries of three countries—Canada, Mexico, and Korea—are analyzed and compared. Each presents a vivid case of pitting the argument of cultural exception against trade liberalization. In each of the three economies, the country's strong trade and investment relation with the United States was the driving force behind free trade agreements (in the cases of Canada and Mexico) and the BIT negotiations (in the case of Korea).

For all three economies, the United States offers its top export market, and has demanded more access to the products of the Canadian, Mexican, and Korean markets by removing trade and investment barriers. Canada and Mexico established NAFTA with the United States, but with an interesting contrast in the cultural sector: Mexico did not ask for any special and differential treatment of its cultural industry in the course of the negotiations with the United States, whereas Canada claimed the cultural exception.

In the NAFTA negotiations, the United States, as a leading audiovisual producer and exporter, pushed Canada and Mexico to cooperate in eliminating all barriers to the flow of trade and investment in their cultural industries. Canada and Mexico expressed quite different stances on this issue. As a result, a double standard was applied: cultural industries are exempted from the agreement between the United States and Canada, whereas Mexico allowed cultural industries to be governed by NAFTA provisions with minor exemptions.

There exists a salient disparity in the production capacity and market strength of U.S., Canadian, and Mexican film producers. U.S. producers could take advantage of economies of scale, first-mover advantages, a competitive environment that favors production for audience maximization, and the world's wealthiest English-speaking audiences. Compared with the United States, both Canada and Mexico have rather small domestic markets. But Mexico was not defensive in negotiating NAFTA with the United States. Why did Canada and Mexico respond differently to the U.S. film industry at the NAFTA negotiations?

The concept of cultural distance may be helpful in this context.[14] With 80 percent of the Canadian population living within 100 miles of the United States, no language barriers (except in French-speaking Quebec), and similar cultural parameters (in language, genre preferences, and viewing habits), U.S. cultural products travel easily across the U.S.-Canadian border. Because of this "short" cultural distance from the United States, Canada built a high fortress behind which its content providers could feel safe. However, cultural distance alone fails to account for what is unfolding in Korea. Despite its "far" cultural distance from the United States, the Korean film sector feels uneasy about dismantling protectionist measures and the government is under the powerful influence of local film people.

Local film industry resistance is a prime example of a "collective action" problem. Compared with the Hollywood film industry, the local film industries of Canada, Mexico, and Korea are small. If existing trade and investment barriers (or, from the viewpoint of the local film people, policy measures to promote "cultural diversity") are to be abolished, it would be more beneficial to the local economies in general, if one believes in the principle of comparative advantage and, as history has shown, that trade expansion is mutually beneficial. The problem is that this removal of barriers gives rise to losers as well as winners.[15] The groups that had been receiving fat protections stand to lose and will protest. If the calculus of cost-benefit prevails, however, such resistance should not stand in the way of tearing down barriers.

Yet their voices and protests do matter. More often than not, the losers get the center stage and succeed in their sabotage. How does this happen? Typically, trade benefits are diffused over the public in general, whose identity tends to be anonymous. On the other hand, costs are concentrated on a few players, in a relative sense, who are easily identifiable. Hence, the per capita benefit is small compared with the per capita loss: each loser has a lot at stake, while the benefits for each winner are marginal. Moreover, trade benefits tend to be vague and do not materialize in the short term, whereas losses are specific and immediate. All these considerations lead to a situation in which a small group of losers becomes strongly consolidated to exercise its political influence so that the barriers remain, whereas the voices of the winners are seldom heard. The fact that there are so many winners but their per capita benefits are rather small produces a free-rider problem. Even though it is beneficial to society to have the barrier removed, competition in the political arena is tilted in favor of the losers. A litmus test for a society seeking to advance its prosperity is whether it can overcome this collective action problem. In this regard, political leadership (or political entrepreneurship) is what separates Mexico and Korea.

Canada

Given its geographic proximity, historical ties, and common language with the United States, Canada has been particularly concerned about the "Hollywood Juggernaut." For instance, in 1997, 96 percent of all films shown in Canada were of foreign origin and mostly from the United States. Three quarters of the music heard on the radio is not Canadian, and neither is 80 percent of its newsstand magazines or 60 percent of its books. Common language, common cultural and institutional origin, and the geographic proximity of Canada to the United States, which has a huge domestic market in terms of both absolute purchasing power and relative size vis-à-vis Canada, have long pressed the Canadian government to develop defensive and protective cultural policies.

Canada's small domestic market is an obstacle to Canadian firms achieving cost efficiencies, thereby causing significant inefficiency in competing with the U.S. firms. The film industry is the case in point. Annually, Canada produces thirty to thirty-five feature films, with a single-digit national market share. For instance, in 1999 Canadian films earned only 2.1 percent of all box office receipts, which is the lowest performance among comparable film-making countries: in 1999 homegrown films earned 37 percent of box office receipts in Japan, 30 percent in France, and 14 percent each in Italy and Spain. The poor performance of Canadian films is attributed to their low budgets for production and marketing compared to those of the foreign films.[16]

In the past, the Canadian government tended to rely on subsidies to support the cultural industries and to achieve the country's cultural objectives. Over time, government support has evolved in the form of tax and investment measures coupled with regulatory measures in the television, film, music, and book publishing industries. Border measures such as tariffs were used in the past, but are gradually being phased out.

Canada makes direct grants and investments through an organization called Telefilm Canada, whose mission is to foster and promote the development of the nation's feature film industry. Telefilm Canada's support covers a wide range of areas, including investments, loans and loan guarantees, advances, lines of credit, credits, national and international marketing and distribution, exportation, and accelerated tax depreciation for qualified productions. Every year Telefilm supports some thirty new feature films, and for the past thirty years it has contributed to the production of over six hundred feature films.

Foreign investments in Canada's cultural sector are regulated. The Canadian government allowed foreign ownership in the audiovisual sector in return for those foreign investors accepting obligations to provide more local content, or to abide by performance requirements.[17] Canada also has special regulations to protect its French heritage. The province of Quebec has established a special quota to protect its dubbing industry. Under legislation passed in 1987, feature motion pictures must be released in French versions within forty-five days of the release of the English version.

The Canadian government defines what constitutes Canadian content. Such content is defined differently for different types of cultural products. For television programs and feature films, Canadian content is based on a point system. For example, programs can earn two points for using a Canadian director and a point for each leading Canadian actor. Programs must be produced by a Canadian and earn at least six points overall to be considered Canadian. To qualify for financial assistance from the Canadian Television Fund, a production must attain a minimum of ten points. Canadian content rules can be flexible in some cases, however. To encourage the production of Canadian programming and increase the cultural industries' access to capital and export markets, the Canadian government has signed coproduction agreements with more than thirty countries. Under these agreements, productions with as little as 20 percent Canadian participation may meet the requirements for Canadian content.

Film distributors based in the United States have traditionally been able to purchase the rights to distribute films for both the United States and Canada. This practice deprives Canadian distribution firms of the opportunity to distribute major films in the Canadian market, and of the revenue that those distribution rights would generate, which could then be invested in developing new Canadian films. In 1987, Canada considered establishing an import licensing system that would limit foreign firms in distributing their own films or films for which they had the world rights. Foreign firms operating in Canada objected and the legislation did not proceed.

This longstanding protectionist policy in the cultural sector and the collective action problem aggravated by the lack of political entrepreneurship propelled the Canadian negotiators to work on a cultural exception in the Canada-U.S. FTA. In return for accepting such an exception, the Canada-U.S. FTA gives the U.S. government a right to retaliate. The United States and Canada agreed on a cultural exception as follows:

- the tariffs are removed for goods used as "input" in production by the cultural industries;
- the requirement that magazines be typeset and printed in Canada in order for companies to deduct advertising expenses for corporate tax calculations is removed; and
- copyright payments for cable redistribution of distant signals are imposed, regardless of nationality.

Mexico

Mexico's film industry has undergone a dramatic change in recent years. During 1988–1990, about 100 Mexican films were produced annually. Since 1991, the production of local films has plunged to the level of thirty to forty annually. As a measure to prevent the further decline of Mexico's film industry, the Mexican government introduced a screen quota as a federal law covering cinematography on January 1, 1993. The law earmarked 30 percent of screening days for Mexican films in 1993. However, the government of Mexico soon began to phase out the screen quota.

Removing the screen quota in the midst of the local industry's lackluster performance would risk the outrage of the local film industry. Mexican political leaders knew this, yet bit the bullet. Why? First of all, in terms of cultural distance, Mexican cultural producers are better positioned than their Canadian counterparts due to intrinsic barriers of language, content, and genre. Less inundated by foreign cultural products, particularly in the broadcasting sector, the Mexican audiovisual sector is comparatively competitive with well-positioned domestic producers that are content exporters as well. Also thirty million Hispanics residing in the United States, who maintain cultural ties to their countries of origin and are far wealthier than Mexican audiences, can be a crucial target for Mexican investors and content providers.

Moreover, the ruling Mexican party of Salinas wanted to seize the momentum of NAFTA to fuel its ongoing economic reform toward deregulation, privatization, liberalization, and competition. Policy makers pursued "neo-liberal" communication policies, which are based on three pillars: privatization of national networks, deregulation of audiovisual markets, and diminished public intervention in the production and distribution of audiovisual products. Despite resistance from the entrenched domestic cultural sector, they came to conclude that the cultural exemption was unnecessary and the local audiovisual sector could even benefit from trade and investment liberalization.

After NAFTA went into effect on January 1, 1994, the screen quota lost its effectiveness in promoting local film. In 1994, local industry produced forty-six local films. Ever since, the number of Mexican films produced yearly has consistently remained below twenty. In 1998, production hit the lowest ever: ten films. The market share of the U.S. film industry in Mexican theaters soared to 90 percent. It was a free-fall for the Mexican film industry. Having sensed the risk of collapse of the domestic film industry, in 1998 the Mexican government introduced a ten-percent screen quota, but this has had limited effects.

Some blame the downfall of the Mexican film industry on the elimination of the screen quota. A close scrutiny reveals a more complicated scenario. Prior to the economic reforms of Salinas, the government set the entrance fee to movie theaters at an artificially low rate, which made the production of movies unprofitable. As a result, most local movies were low-budget and the quality was far below the demands of consumers. Some talented movie makers and actors started to avoid investing in Mexican movies, propelling further deterioration in their quality. Although many Mexican movies were made until 1990, the number is misleading and largely unrelated to local moviegoers' preferences and the films' commercial success. In 1992, President Salinas lifted the price cap on entrance fees, privatized the government-owned theaters, and opened the movie market to foreign investors and films. The deregulation, privatization, and liberalization policies of Salinas brought about the construction of movie theaters with modern facilities catering to the demands of movie goers and the increased import of foreign movies, in particular from the United States.

The decline of the local film industry has had little to do with weak protectionist policies. The screen quotas may have induced local film makers to produce more movies. However, as long as consumers do not choose to view Mexican films, protectionism limits profitability and diminishes the financial resources for investing in new films.

While the number of local films has remained below twenty in recent years, some commercially successful domestic films have started to appear, capturing the attention of young talented people and investors. Even though local film production is still at a low level, its market share is on the rise. Important factors behind this trend are the rise in theater entrance fees and the government subsidy, which is funded by a special tax on TV broadcasting and box-office sale revenues (1 percent of the entrance fee is allocated to the film promotion fund).

Korea

At present, the screen quota is the only trade restrictive measure on the film industry in Korea. In the past, the Korean government had restricted the importation of foreign films and the distribution of film, but no longer. In the late 1980s, the Korean government took a series of liberalization measures when faced with U.S. trade retaliation.[18] Lifting restrictions on the distribution of foreign films was one consequence. Now, as discussed earlier, the screen quota has emerged as a deal breaker in the process of Korean-U.S. negotiations on a BIT.

The goal of BIT negotiations is to provide foreign investment with "treatment no less favorable" than that accorded to domestic investments. Many BITs, which the United States negotiated, require that foreign investments should not be mandated "to achieve a particular level or percentage of local content." Under these provisions, U.S. investment projects would not have to comply with certain domestic regulations. Therefore, if the United States were to establish its own theaters in Korea, they would not have to comply with the screen quota mandate. At the first round of negotiations for the Korea-U.S. BIT in Washington, DC, in July 1998, the United States claimed that the screen quota requires theater owners in Korea to screen Korean films and was not in line with the standard model of the BIT. Responding to the U.S. request, at the third round of talks in November 1998, the Korean side offered to limit the mandatory screening period to ninety-two days; the United States rejected the proposal, demanding complete elimination.

In retrospect, this offer was the maximum Korea could afford.[19] As the talks moved on, it became clear that the Korean domestic side could not agree on the matter. The local film industry wasted no time in mobilizing its resources to forge a powerful political coalition with diverse civic groups unrelated to the film industry to resist any change to the screen quota. They organized mass rallies and chanted anti-U.S. and anti-Hollywood slogans. And they pushed the Korean negotiators into the corner by labeling them traitors. Facing stonewalling from the film industry, the Ministry of Culture and Tourism switched its prior position by taking sides with them. The ministry stated publicly that the screen quota would be maintained until the Korean film industry achieved 40 percent of the market share, which was 25.1 percent in that year. When the BIT talks were launched, negotiators were optimistic about the completion of the work within a year; most importantly Korea wanted the BIT to be blessed by President Kim. The government was not prepared to deal with the resistance of the local film industry. Faced with diverging domestic views on the issue, Korean politicians claimed their stakes in the game. In January 1999, the Korean National Assembly passed a resolution that the screen quota should remain as it was until the local film industry obtained 40 percent of the market.

With the groundswell of domestic resistance, the Korean negotiators proposed to their U.S. counterparts to take the screen quota off the negotiating agenda. The United States turned this down. Early in the talks, the U.S. private sector representatives proposed investing $500 million in building Korean multiplex cinemas if the screen quota was removed. The Korean government came up with the idea of building theaters exclusively for local films and proposed a massive subsidy. These ideas were greeted

Table 6.5
Growth of Korean film industry

	1998	1999	2000	2001	2002	2003	2004
Number of screens	517	588	720	818	977	1132	1351
Korean films produced	43	49	61	57	82	80	82
Market share of Korean film (%)	25.1	39.7	35.2	50.1	48.3	53.5	59.3
Number of foreign films imported	296	348	404	339	262	271	285

Source: The Korean Film Commission.

by a cold response. To the eyes of the Korean film industry, the negotiations amounted to a conspiracy to kill them. The situation took an interesting twist in early 2002. After recording a 39.7 percent market share in 1999 and 35.2 percent in 2000, the Korean film industry achieved 50.1 percent in year 2001, surpassing the threshold of 40 percent. Now, the Korean Ministry of Finance and Economy struck back. Pointing out the resolution of the National Assembly and the public statement of the Ministry of Culture and Tourism, they argued that it was time to reduce the screen quota in order to make a BIT deal. The business sectors of the United States and Korea rallied behind. Initial reaction from the local film industry was as if it could not believe its own success; members claimed it was premature to make any decision since the target of 40 percent had just been achieved for the first time. They went on to charge that the BIT would not bring any substantial economic gains.

Evidently, the Korean film industry has achieved tremendous success since the late 1990s. The market share of local films has been staying close to the level of 50 percent, more screens are being built, and more local films are being manufactured (table 6.5). Notwithstanding this success, the Korean film industry continues its practice of sticking to the screen quota no matter what. The minute the screen quota is removed, they claim, the local film industry will begin to dwindle. Is it the case?

The screen quota system in Korea was first adopted in 1966 to protect the Korean film industry from foreign films. The lifting of restrictions on the direct distribution by foreigners in 1987 dealt a major blow. With the other protections gone, the screen quota was the only buffer shielding local films from foreign competition. Notably, the local film industry created its own civic organization to make sure theater owners were abiding by the mandatory screening days.[20] Allegedly, many theaters had not been sticking to the screen quota, due in large measure to the lack of consumer interest in the local movies. To the frustration of the local film industry, enforcement had been lax as well. Now, the local film sector assumed the role of law enforcement officer.

As it turns out, allowing the U.S. film distributors to freely distribute their own films was a blessing in disguise. Its survival threatened, the Korean film industry was pushed onto the ropes. And then it sprang into action. Relaxing censorship triggered the flood of a new generation of Korean movies. Local filmmakers hit the theaters with a series of blockbusters. Encouraged by the success of some local films, more investment and human resources were channeled to local productions. On the distribution side, many multiplexes were built and old theaters were renovated to attract more moviegoers. The purchase of movie tickets became easier due to reservation services using the Internet and telephone. In addition, the massive and innovative marketing strategies promoting local films played a role.

In this remarkable turnaround of the Korean film industry, what role has the screen quota played? In all likelihood, the presence of the screen quota gave some assurance to local film makers that their movies would reach the screens and the moviegoers. But this psychological buffer zone has been in place for a long time, since the 1970s and 1980s; the screen quota is no new, magic formula causing local filmmakers' growing success. Apparently, the Korean film industry members refuse to accept these facts. They allege, or they want to believe, that their current success is only possible because of the screen quota, which is the mirror image of accepting no concessions on the screen quota whatsoever.

The New Battle: The WTO Doha Round versus the UNESCO Cultural Convention

Using the WTO as a battleground, the United States, the principal author of trade rules in cultural products, is looking for the second round of a major battle with cultural protectionists. At the same time, these activists have been groping for the magic sword that can save their pristine world from the imperialistic ambition of the United States.

The WTO Doha Round

Pursuant to the GATS principle of progressive liberalization, as stipulated in article XIX, stating that "members shall enter into successive rounds of negotiations, beginning not later than five years from the date of entry into force of the WTO agreement and periodically thereafter," a new round of multilateral talks has been underway at the WTO.

Progressive liberalization is likely to be a step-by-step process over consecutive rounds of negotiations similar to the GATT talks. In the field of goods, the multilateral process of liberalization started in 1947 and carried over eight rounds of negotiations that led to the substantial opening of world trade. In the light of the growing importance of audiovisual services and the strong interest of the American entertainment industry in opening foreign markets, it is only natural to presume that the audiovisual sector will be treated rather seriously.

This new round, which miserably failed to launch at the WTO Seattle Ministerial Conference in December 1999, got off the ground in November 2001 at Doha, Quatar. The new round, officially named the Doha Development Agenda (DDA) negotiations, in an effort to bring on board many marginalized and less developed countries, has the goal of making a grand trade deal over the vast range of complicated issues from agriculture to trade facilitation. Achieving such a goal with more than 140 countries at different stages of economic development is a tall order. Deadlock, delay, and deferred decisions are familiar features of the multilateral negotiations. DDA negotiations have already suffered a severe deadlock at the Cancun Ministerial Meeting in September 2003, failing to produce a compromise to make progress. To save the DDA negotiations, the WTO members agreed to push its original deadline of January 2005 to December 2005.[21] A subsequent extension yielded a similar fate, with the talks being suspended in July 2006.

In the process leading to the DDA round, three countries, the United States, Switzerland, and Brazil circulated a paper expressing their views on pending audiovisual services negotiations.[22] Highlighting how significantly different the landscape of the audiovisual sector is from that of the Uruguay Round period when "negotiations focused primarily on film production, film distribution, and terrestrial broadcasting of audiovisual goods and services," the United States claimed the negotiations should provide predictable and transparent trading rules to maximize the opportunities for new technology in the global market. The United States called the negotiating approach of the Uruguay Round an "all-or-nothing-game"—liberalize or carve out from the trade rule—and argued that new service negotiations should not be black and white. In particular, the United States proposed reclassification of audiovisual services in light of the increasing inadequacy of the current classification due to technological changes, more market access commitments from the members, and development of an agreement concerning disciplines for subsidies in the area of audiovisual services. Such an agreement, according to the United States, "will respect each nation's need to foster its cultural identity by creating an environment to nurture local culture."[23]

Japan sought improvement in market access in the audiovisual sector. Specifically, it called for removing MFN exemptions in the audiovisual sector, lowering quantity restrictions such as quotas, and elimination of limitations on national treatment.[24] Switzerland was also of the view that the cultural exception would not be possible and expressed its wish to see discussions "reveal to what extent the GATS has the necessary flexibility to offer appropriate solutions to the specificity of the audiovisual sector, and to sufficiently take into account the cultural, social and democratic objectives of Members."[25] Switzerland listed several items for review and negotiation, from safeguards for cultural diversity to regulation against monopolistic market behavior by dominant players.

While echoing the inappropriateness of the cultural exception, Brazil presented the needs and aspirations of developing countries. On the one hand Brazil favored a liberalization of trade in—and preferential treatment within the WTO rules for—audiovisual products from developing countries, while on the other hand it argued for safeguards against dumping and other "crowding-out" strategies by foreign players in the audiovisual marketplaces of developing countries.[26] Brazil furthermore placed strong emphasis on the autonomy of the nation-state to regulate in order to promote cultural policy objectives.

Among these submissions, the U.S. proposal was the most noteworthy: it showed a remarkable change in the approach to the negotiations by the major player that spearheaded trade negotiations in the sector. When the United States was dubbing the audiovisual negotiations at the Uruguay Round as "all-or-nothing," it was not just accusing the EU and others of animosity and stalemate, but was also putting blame on itself. Citing the case of financial service negotiations and basic telecommunications negotiations where sector-specific concerns were addressed by creating an annex or reference paper, and also pointing out GATT article IV as a flexible way of making a specific commitment, general exception, and subsidy (complete details yet to be negotiated) at the GATS, the United States called for establishing trade rules in the audiovisual sector compatible with the realities of quotas, subsidies, ownership restrictions, and content regulations.

It is interesting to note that Brazil, the self-proclaimed champion of the developing countries, is not retreating behind the veil of a cultural exception. There is a deep silence on the part of the EU and Canada, which made the MFN exemption "indefinitely" so that they could continue to maintain discriminatory measures favoring their local cultural content. Apparently, they are happy as long as they can maintain their MFN exemption. However, time is ticking toward the end of the maximum time allowed for such exemptions.

The GATS negotiations on specific commitments proceed on a request-offer basis. Countries make their requests to other countries of interest, and these countries respond by tabling offers. The gap between request and offer, sometimes significant, is a prevailing aspect of negotiations, which are all about narrowing the gap. Deadlines were set for submissions of initial requests by June 30, 2002, and initial offers by March 31, 2003. Between March 31, 2003, and October 30, 2003, thirty-nine members submitted initial offers, including Australia, Canada, China, EU, Japan, Korea, and the United States.[27] Most initial offers on the table did not seem to provide any significant improvement of commitments in audiovisual services made at the Uruguay Round.[28] In the case of the United States, it reclassified some audiovisual services: radio and television broadcast transmission services were moved to communications services, and cinema theater services, including cinema projection services, were moved to entertainment services. The United States also created a new category of home video entertainment services, including, but not limited to, video tapes and optical discs.[29] Other than such reclassification, no major commitment was made by the United States in its initial offer. As expected, the EU and Canada did not make any initial offers in audiovisual services.

It is only natural and may be fair that many countries were targeting the EU's audiovisual sector. According to the EU:

Half of the requests received so far address audio-visual services. . . . Most requests are more targeted, asking for the binding of the current level of market opening or for commitments in some specific subheadings, usually film production, distribution and projection services, sometimes sound recording. Radio and television services are less targeted, but the elimination of quotas, the scheduling of discriminatory measures and the reduction of discriminatory subsidies is requested. One request asks for foreign service suppliers to be allowed to produce motion pictures and radio and TV programs in co-operation with domestic producers. Another one asks for the elimination of discriminatory withholding tax treatment applied to films in some Member States.[30]

Based on the principle of a maximum ten-year duration for MFN exemptions, some countries are calling for the removal of the MFN exemption. They lambasted major trading members for their lack of leadership by taking "indefinite" MFN exemptions as the rule rather than the exemption. Highlighting that 98 percent of the exemptions have unspecified length despite the ten-year stipulation, they claim that "this 10 year duration should not be viewed as a minimum period of exemptions, but should be viewed as a maximum period of transition for the Members."[31] Essentially, these countries were requesting that "all registered MFN exemptions should be eliminated by the end of 2004 or the conclusion of the current negotiations, whichever comes earlier."[32]

FTA as Strategic Option

Remarkably, the United States is not insisting upon the elimination of the MFN exemptions in the audiovisual sector made by the EU and Canada. Apparently, the United States does not believe that the Doha Round would achieve a great deal in the liberalization of the audiovisual sector. Under its Republican leadership, the U.S. government has been vigorously pursuing bilateral investment treaties and free trade agreements. Starting with the FTA it negotiated with Chile, concluded in December 2002, and Singapore, concluded in February 2003, the United States has been pursuing a new negotiating strategy, departing from its Uruguay Round approach, which addressed audiovisual products like any other commodity. Now, the United States seems willing to make concessions to countries that want to require local content or ownership in the "traditional" audiovisual sector, that is, film and radio; however, no such requirements will be allowed in the category of "electronic commerce." This negotiating stance reflects the view shared by the U.S. government and U.S. industry that all cultural content (films, television, music) will be transmitted digitally in the near future.[33] The U.S.-Australia FTA that was concluded in February 2004 and entered into force in January 2005 is an important achievement for the United States for two reasons: first, digital products receive nondiscriminatory treatment and are subject to customs duties; and second, Australia did not make any commitment in the WTO and sometimes had been viewed as part of the "rest of the world" opposing the U.S.

In its *2005 National Trade Estimate Report on Foreign Trade Barriers*, the Office of the U.S. Trade Representative (USTR), the body that negotiates international trade agreements for the U.S. government, released its annual listing of measures of foreign government that it considers significant trade barriers to U.S. exports. National content quotas for radio, film, and television along with foreign ownership limits for national publishing companies, broadcasters, and cable and satellite companies figure prominently on this list. In this report, France's national content requirements for television broadcasters are singled out as being a significant barrier to access for U.S. programs in the French market.

The *2005 National Trade Estimate Report* says: "France continues to apply its more restrictive version of the EU Broadcast Directive. In implementing the Directive, France chose to specify a percentage of European programming (60 percent) and French programming (40 percent), which exceeded the requirements of the Broadcast Directive. Moreover, these quotas apply to both the 24-hour day and prime time slots, and the definition of prime time differs network to network." It went on to claim: "The prime time rules are a significant barrier to access of US programs to the French market." The same report also argues: "the US continues to be concerned that broadcasts of American music are limited by radio broadcast quotas (40 percent of songs on almost all French private and public radio stations must be Francophone)."

France is hardly alone in having quotas for radio, television, or film, however. Similar quotas in Australia, Brazil, Canada, Colombia, Italy, Korea, Malaysia, Spain, and Sri Lanka—to name just a sampling of countries—are also cited as barriers in the USTR report. The same holds true for limits on foreign ownership of cultural enterprises. Countries including Brazil, Canada, Italy, Turkey, Venezuela, and Vietnam are cited. Those policy measures, assailed as trade barriers by the United States, are considered by the cultural sector constituents as essential linchpins for promoting and safeguarding cultural diversity in the era of globalization. Table 6.6 provides an illustrative list of the most commonly observed measures to promote domestic cultural industry.

UNESCO Convention on Cultural Diversity

It is not difficult to find international documents that stress the importance of cultural expression and diversity. The Universal Declaration of Human Rights; the United Nations International Covenant on Economic, Social and Cultural Rights; and declarations of UNESCO and the International Organization of the Francophonie are prominent examples. However, all these are declarations which do not give rise to any meaningful international rights and obligations among governments when it comes to culture. On the other hand, the WTO has preempted the debate on trade implications of the cultural sector. The WTO system coupled with rapid technological changes

Table 6.6
Domestic support measures used in cultural industry

1. *Subsidies, including grants and loans,* for the production of cultural works, most notably audiovisual products. For example, Eurimages, an initiative by the Council of Europe, provides subsidies for the coproduction of European audiovisual works. The Media II program of the European Communities, while excluding the support of production, focuses on training for professionals, development of attractive projects, and transnational distribution of audiovisual programs and films. National programs providing subsidies to the domestic film industry exist in France, Germany, the United Kingdom, Canada, the United States, and Switzerland. Canada also subsidizes its publishing industry through grants, marketing assistance, interest-free loans, and postal subsidies.
2. *Local content rules,* especially measures regulating radio and television broadcasting content. For example, the European Union, the Council of Europe, Australia, Canada, and France use domestic broadcast content to control access to their television broadcast and film markets.
3. *Market access restrictions,* in particular measures that control access to film markets, including screen quotas for cinemas (as in France, Mexico, Korea, and Spain); rebates on box office taxes for cinemas that show national films (Italy); prohibitions of dubbing of foreign films (Mexico); and dubbing licenses (e.g., in Spain, film distributors can only receive a dubbing license for foreign films when they contract to distribute a certain number of national films).
4. *Regulator restrictions,* especially measures that control access to radio or television broadcasting through regulatory or licensing restrictions.
5. *Tax measures* France, for example, uses taxes on box office revenues, on receipts of broadcasters, and on video recordings to support local film production.
6. *Foreign investment and ownership restrictions.*
7. *Border measures,* which may include tariffs or quantitative restrictions, as in India, which used to restrict the import of film titles to 100 per year.
8. *Film coproduction agreements* A leading example of which is the Council of Europe, which has established a convention on film coproduction.

Source: Footer and Graber 2000.

threaten to overwhelm the capacity of many countries to maintain their own cultural industries. Driven by this fear and mindful of the shortcomings of existing international arrangements on culture, these countries are confronting the serious work of creating an international treaty on which they can rely.

Their major forum is UNESCO. Following the "Our Creative Diversity" (1995) project and the Stockholm Action Plan (1998), UNESCO held events such as a symposium on pluralism (January 1999), a conference on cultural diversity and trade (June 1999), and a roundtable of ministers of culture, on "Culture and Creativity in a Globalized World" (November 1999), to explore major cultural diversity issues. In November 2001, UNESCO adopted the Universal Declaration on Cultural Diversity, which declared that cultural diversity is the common heritage of humanity and should be recognized and affirmed for the benefit of present and future generations. This declaration inspired the UNESCO member states to adopt a new Convention for Safeguarding of

Intangible Cultural Heritage, in 2003. Despite its moral force as a milestone for international cooperation, the declaration was regarded by many members as an inadequate response to specific threats to cultural diversity in the era of globalization.

The cultural sector players from the "rest of the world" beyond the United States have been working with a view to establishing a legally binding international instrument on cultural diversity of which the prime objective is to dissociate the cultural sector from trade liberalization. To achieve this goal, they have forged an international coalition of cultural sector policy makers, content suppliers, and activists, calling their group the International Network for Cultural Diversity (INCD) and the International Network on Cultural Policy (INCP). The INCD represents the NGO side of the battle, while the INCP represents the government side. The INCD draws members from various sectors of the cultural industry in about seventy countries, ranging from new media artists to traditional artisans. The INCP is an informal group of over forty culture ministers that has been meeting annually since 1998.

At its meeting in Lucerne in 2001, the INCP directed a working group on cultural diversity and globalization to prepare a draft of a new international instrument on cultural diversity. Since their respective meetings in Lucerne, the works of the INCD and INCP have proceeded in tandem. Thus in May 2002, the first draft of the INCD proposal for a convention on cultural diversity was presented to the ministers' working group at its meeting in Johannesburg. In July 2003 the INCP published its own draft convention. Parallel with these developments, in October 2003 UNESCO agreed to work on a convention on cultural diversity. After a series of drafting sessions, the final draft proposal for the convention was introduced to the UNESCO general conference in October 2005. Only two countries, the United States and Israel, voted no, on the grounds that the final draft was inconsistent with the principle of trade liberalization. On a yes vote of 148 countries, the convention was adopted.

The Convention on the Protection and Promotion of the Diversity of Cultural Expressions (commonly referred to as the Convention on Cultural Diversity) consists of seven chapters: (1) Objectives and Guiding Principles; (2) Scope of Application; (3) Definitions; (4) Rights and Obligations of Parties; (5) Relationship to Other Instruments; (6) Organs of the Convention; and (7) Final Clauses. There is also an Annex on Conciliation Procedure. The scope of the convention is ambitious: article 3 states, "This Convention shall apply to the policies and measures adopted by the Parties related to the protection and promotion of the diversity of cultural expressions." According to article 4, "cultural expressions" are those expressions that result from the creativity of individuals, groups, and societies, and that have cultural content; "cultural activities, goods and services" refers to those activities, goods, and services that are considered as a specific cultural attribute, use, or purpose, and embody or convey cultural expressions, irrespective of the commercial value they may have; and "cultural industries" refer to industries producing and distributing cultural goods or services as previously defined. Furthermore, "cultural policies and measures" refer to those policies and measures related to culture, whether at the local, regional, national, or international level, which are either focused on culture as such or are designed to have a direct effect on cultural expressions of an individual, group, or society, including on the creation, production, dissemination, distribution of, and access to cultural activities, goods, and services.

The convention declares: "Each Party may adopt measures aimed at protecting and promoting the diversity of cultural expressions within its territory." Such measures may include those that provide opportunities for domestic cultural activities, as well as goods and services for the creation, production, dissemination, distribution, and enjoyment of such activities, including provisions relating to the language used for them.

The convention reflects keen awareness of the significance of cultural disparities among the countries. Article 16 calls for preferential treatment for developing countries: "Developed countries shall facilitate cultural exchanges with developing countries by granting, through the appropriate institutional and legal frameworks, preferential treatment to artists and other cultural professionals and practitioners, as well as cultural goods and services from developing countries." Article 18 inscribes the establishment of the International Fund for Cultural Diversity.

The relationship to other international regimes was the most controversial issue. An early draft of February 2005 offered two options, reflecting the obvious disagreement among the parties. Option A stated, "(1) Nothing in this Convention may be interpreted as affecting the rights and obligations of the States Parties under any existing international instrument relating to intellectual property rights to which they are parties. (2) The provisions of this Convention shall not affect the rights and obligations of any State Party deriving from any existing international instrument, except where the exercise of those rights and obligations would cause serious damage or threat to the diversity of cultural expressions." Option B contained one sentence: "Nothing in this Convention shall affect the rights and obligations of the States Parties under any other existing international instruments."

This draft drew sharp criticism from both sides of the debate. Cultural activists wanted to have a complete and uninterrupted degree of freedom in designing and implementing cultural policies.[34] In order to achieve such a goal, the convention should be not only legally binding but also superior to any international trade disciplines. More specifically, their goal was to override any existing trade commitment in the audiovisual sector, whenever and wherever possible. Exactly for this reason, they wanted to have a stronger version of option A. On the other hand, countries sympathetic to trade expansion in the cultural sector believed the convention interfered with the trade system by creating too much space going beyond the legitimate realm for the cultural policy, and clearly preferred option B.

The end result of this debate was rather bizarre. Article 20 of the adopted convention states as follows:

> 1. Parties recognize that they shall perform in good faith their obligations under this Convention and all other treaties to which they are parties. Accordingly, without subordinating this Convention to any other treaty,
> (a) they shall foster mutual supportiveness between this Convention and the other treaties to which they are parties; and
> (b) when interpreting and applying the other treaties to which they are parties or when entering into other international obligations, Parties shall take into account the relevant provisions of this Convention.
> 2. Nothing in this Convention shall be interpreted as modifying rights and obligations of the Parties under any other treaties to which they are parties.

Both options A and B found their way into the final convention, hence creating legal fuzziness and uncertainty. Apparently, paragraph 1 and 2 in article 20 are mutually contradictory: while paragraph 1 opens the door wide for making discretionary cultural policy that may override a country's international commitment, paragraph 2 says such a policy would not be possible. It is precisely because of this ambiguity that so many countries cast yes votes for the convention, interpreting the legal text to their own advantage. Such a decision may be an easy political compromise, but a recipe for chaos.

In order to ensure the legal effectiveness of the convention, the INCD has developed a strategy of maintaining the status quo both at the DDA negotiations and at FTA negotiations. Basically, the strategy calls for 'neither request nor offer' in the audiovisual service in any negotiating forums. INCD members recognize that the WTO is just leaving the starting block in the race for trade liberalization in the cultural sector. Each area of investment, competition, policy, and procurement, where the WTO has failed to launch negotiations, encompasses the frequent conflicts of domestic policy guided by trade liberalization, with policy necessary to protect the indigenous cultural sector. By maintaining the status quo, the INCD calculates that it can preserve various options to protect the diversity and integrity of national cultures, as would be allowed under the convention.

The convention allows ample room for creating tensions with the WTO. It is perfectly likely that a member government, belonging to UNESCO and the WTO, may create a situation where it asserts the withdrawal of its existing trade commitments under the GATS by citing the convention. In such an eventuality, what would be a proper venue to deal with conflict? The WTO or UNESCO? What happens if the WTO and UNESCO arrive at different conclusions? What if the complaining country brings its case to the WTO and gets a favorable ruling, whereas the defending country brings its case to UNESCO and manages to get a sympathetic decision? If the convention can be used to derogate any existing obligation of a sovereign government under the WTO, it would wreak havoc on the stability and predictability of international affairs. Welcome to the brave new world!

Sneak Preview of the Future: Coming to a Theater near You

A sovereign nation brings its domestically negotiated agenda to international negotiations. At the same time, negotiations between governments are affected by the likelihood of winning domestic support and ratification at the end of negotiations. For these reasons, international negotiations take place at two levels, domestic and intergovernmental, with constant interactions between them.[35] As the case of Korea exemplifies, failure to reach a domestic consensus would dash any hope of making an international trade deal.

For those in favor of special treatment of the audiovisual sector, the playing field is tilted to their disadvantage. They may admit that it is a big world and there may be lots of room for all sorts of films and all sorts of film philosophies.[36] Nonetheless, their fear of U.S. dominance is unabated, because the United States can "inspire the dreams and desires of others, thanks to the mastery of global images through film and television."[37] This notion needs to be scrutinized.

Since the late 1990s, the foreign market has replaced the U.S. domestic market as the major source of revenue for Hollywood films. Major Hollywood studios have been mapping out the strategy of "going global," trying to attract the biggest possible international audience. Curiously, this strategy of global reach gives rise to the unintended consequence of "losing the voice of America."[38] To satisfy the preferences of foreigners, Hollywood movies pursue more and more stories involving sci-fi or animation, which are completely unrelated to the U.S. landscape, history, and culture. To reach the non-English-speaking audience, U.S. filmmakers rely more on action and violence than subtle dialogue and psychology. It is not uncommon to see that some Hollywood movies are even selling and exploiting anti-Americanism. While other countries regard globalism as a chance to reveal their national psyches and circumstances through film, the American film is becoming more and more un-American. Hence, it is not clear what the commercial success of some un-American Hollywood films spells for U.S. dominance in the sphere of audiovisual culture.

Other than the argument of cultural diversity, one may invoke the strategic industry argument for the protection of the cultural sector. The argument would go as follows: In the light of rapid technological innovations such as digitalization, compression, and the Internet, the cultural sector has enormous potential for contributing to the gross domestic product (GDP), creating jobs, and boosting some regional economies. However, for many countries with small domestic markets, it is difficult for the local suppliers to go down the cost curve so that their production cost is economized. If foreign cultural products were freely allowed to cross their borders, the agony of local cultural suppliers would be increased. Seen in this light, protecting local markets from foreign competition and promoting them through various measures, even though they may distort trade, would be the logical conclusion.

Sound as the case may be, such logic invites more questions. Since any protection and promotion causes efficiency loss and transfer of money from the more productive sector to the less productive sector, economic principles would dictate providing the least possible protection and promotion. But if protection and promotion were to be maintained, it should be done according to clear national priorities. When it comes to the priority list, it is questionable how compelling a case can be made for the cultural sector. Curiously, the strategic industry case for the cultural sector may be construed to imply that the sector may not be strategic enough to deserve policy attention from the perspective of the overall national economy.

Even if policy makers decide that the local cultural sector is a strategic industry, that does not finish the story. If it is indeed a strategic sector, one would argue, protection and promotion should make the domestic cultural sector sufficiently productive some time in the future. When such a time comes, the local cultural products should be able to compete with foreign ones, without the institutional help from protection and promotion. Would any local cultural sector players accept this determination? The answer is no: they think they deserve permanent protection and promotion. While the cultural diversity argument calls for a blanket protection, the strategic industry argument for the cultural sector has interesting ramifications. In either case, the ground is set for the battle among domestic players to figure out how much and how long protection would be needed. This domestic competition, or negotiation, determines in large measure what a country can take to the international negotiations table. The concept of a "win-set" is derived from this aspect of negotiations.

What can we expect for the audiovisual sector in the context of the DDA negotiations? From the earlier analysis, one can identify a range of preferred outcomes for countries other than the United States, which is very much skewed to the status quo or a slow phase-out of trade-restrictive measures over the long term. This range defines in turn the win-set for the non-US negotiator in the audiovisual sector when he encounters his counterpart at the negotiations table, since any agreement outside of this range would not be ratified at home.[39]

As long as the stalemate in the negotiations continues, the win-set at the DDA negotiations would be illustrated as shown in figure 6.3. The preferred outcome for the rest of the world (the outcome that wins unanimous approval from both losers and winners of the trade deal) is indicated as R_m. R_1 represents the point associated with the minimal vote necessary for ratification by the opponents of trade liberalization, and the area between R_m and R_1 represents their win-set. Likewise, the area between U_m and U_1 represents the win-set for the United States. As shown in figure 6.3, the two

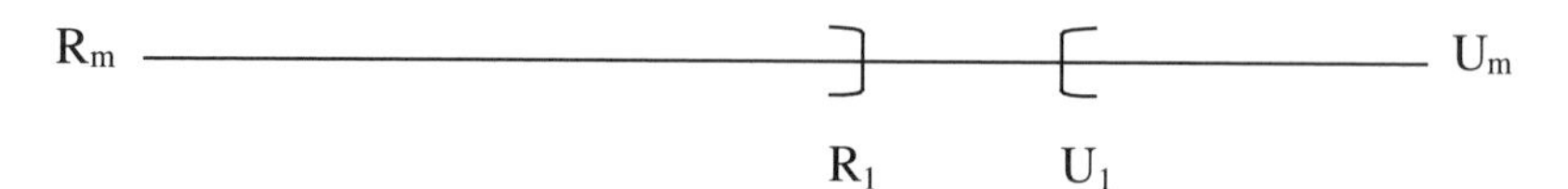

Figure 6.3
Win-set for a trade negotiation

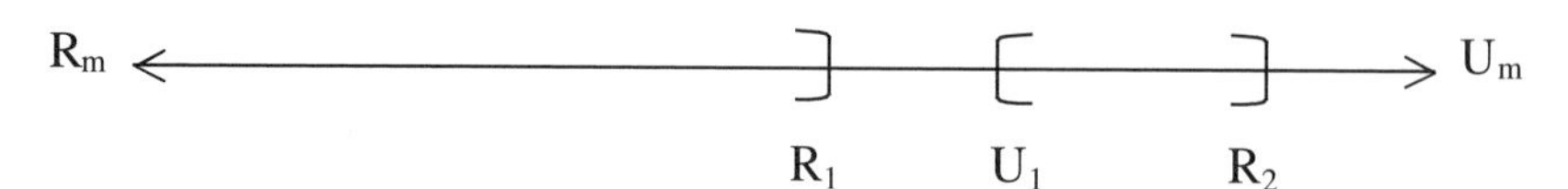

Figure 6.4
Expanding win-set

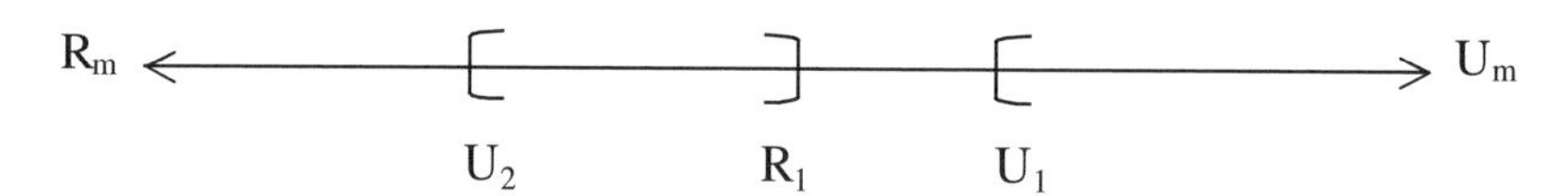

Figure 6.5
Expanding win-set

win-sets do not overlap, implying no "zone of possible agreement" (ZOPA) between the United States and other countries.

To reach an agreement, at least one side needs to broaden its win-set so that both sides can find a ZOPA, as shown in figure 6.4 and figure 6.5. If neither party is willing to broaden its win-set to make it more favorable to its counterpart, the negotiations will stall and no agreement is possible. Which party is more likely to broaden its win-set depends on, among other things, the cost of disagreement. Other factors being equal, the higher a party's cost of disagreement the likelier it will extend its win-set so as to obtain an agreement.

If negotiations were to collapse, both parties would end up seeking their own alternatives outside of the possible agreement. If party A has a relatively stronger alternative than party B, then disagreement is more costly for party B. In this case, one may expect party B to extend its win-set. Such a move would involve domestic negotiations, including persuasion or compensation or both for the losers. At the ongoing DDA negotiations, from the U.S. viewpoint, the alternative if no agreement is reached is to make a series of bilateral trade agreements. The alternative for other countries is UNESCO's Convention on Cultural Diversity. Both alternatives look equally strong. Therefore, each side is likely to push the other side hard at the DDA negotiations. Despite the decision to defer the deadline of the DDA talks, the road ahead is bumpy. Are trade protagonists and cultural activists on a collision course? Is there any way to avoid this? Here are some policy recommendations that may help to avert a global clash between the respective proponents of the two perspectives.

Recommendations

Remove the MFN Exemption

MFN is the mainstay of the multilateral trading regime. When the MFN exemption was negotiated at the Uruguay Round, there was the understanding that during the exemption's ten-year duration that members were not securely exempted from obligations without doing anything, but rather should actively seek ways to bring their MFN inconsistent measures in conformity with the general principles of the GATS. Otherwise, the annex on MFN exemption would not have been worded as it is. Even after the ten years, the stubborn insistence on the MFN exemption by major trading partners does not stand up to this responsibility. Either they should seek a formal waiver or remove the MFN exemption.

Establish Safeguards Specific to the Audiovisual Sector

Some of the concerns raised by cultural activists should not be taken lightly. Otherwise, unnecessary and unproductive confrontation for the sake of confrontation will continue, while the world community fails to seize upon the unprecedented opportunities offered by the expanding global network and market. Taking as precedent the Reference Paper on basic telecommunications, which addressed regulatory concerns in the case of liberalization without prejudicing negotiations in other undeveloped areas of the GATS, negotiations should be launched with a view to addressing the contingency of failure of competition and of safeguarding local cultural content.

Negotiate Rules on Competition and Subsidies for a Level Playing Field

Competition is the mirror image of protection. A subsidy scheme promoting local content should be allowed as long as it does not significantly change the terms and conditions of competition among local and foreign content and content suppliers. The existing GATS is insufficient and limited in addressing anticompetitive behavior and the discipline on subsidy is yet to be developed.

UNESCO Convention Should Respect Existing International Commitments

While the establishment of the UNESCO Convention on Cultural Diversity is a monumental milestone, the parties to the convention should respect their international commitments at other forums. In any event, the convention should not be used as a back-door excuse in the name of cultural exception, overriding any prior international commitments and creating legal chaos and unpredictability; such an eventuality would not be in anyone's interests.

Assure Technology-Neutral Delivery of Content

The application of new media to the audiovisual sector may only be at the beginning stage. No one can foresee precisely the future direction and pace of technological development. Unreasonable regulations and restrictions favoring one particular technology over others may stifle the untapped potential of the audiovisual sector. One significant measure in this context is to ensure the delivery of content using any feasible medium. Such an assurance of technology-neutral delivery of content would promote the plethora of delivery mediums and diversity of cultural expression. Any multilateral forum may sponsor the negotiations of such an assurance.

Conclusion

In February 2006, the Korean filmmakers and actors took to the street again. But there was something unusual about their rally this time. In a dramatic turnaround, the Korean government had just announced that it would reduce the screen quota from 146 days to 73 days starting in July 2006. The announcement took the film industry by surprise. It was such a bold decision, and the timing was hard to believe. For the first three years of the Roh Moon-hyun government, there had been much talk but no action. Every time the government hinted at some reduction of the quota, the film industry mobilized and succeeded in maintaining the quota. The film industry firmly believed that the government was on its side. Now, the decision to kick off the Korea-U.S. FTA talks reshaped the whole debate. The decision to launch the FTA talk was on Korea's initiative: The FTA is bigger than the BIT in scope and significance, and the Korean government wanted to leapfrog over the stalled BIT talks. The U.S. side was stubborn in requesting that substantial progress on the screen quota be made before the parties could sit at the FTA negotiating table. After some hesitation, the Korean government bit the bullet and swallowed hard: it cut the screen quota in half.

The Korean film industry players resorted to their familiar argument: "Don't trade cultural sovereignty for the dubious FTA." Now, they accused the United States of being an imperialistic bully and the empire of evil. In the battle of trade and culture, cultural activists had preempted the symbols,[40] portraying themselves as the good David against the brutish Goliath. They label their domestic opponents as puppets of the United States.

While the Korean film sector stages demonstrations day after day, the Korean people seemed unsympathetic to their cause. Even though the Korean film industry claims it faces an uphill battle with Hollywood, Korean moviegoers tend to view movies not so much as a matter of national contest, but as a matter of choice for how to spend their money. To them, the Korean film industry is no longer fragile. Korean movies perform better than foreign ones. The infrastructure of the domestic film industry has been bolstered. Hollywood film distributors no longer dominate the market; it is home-grown distributors who get the lion's share of the market. The export of Korean films is steadily on the rise. The phenomenal success of Korean films is making the case for protection awkward.

In this age of the digital global village where myriad international coalitions among local civic organizations are being forged and carry a big stick, it is still the power of a sovereign government that counts more, as the saga of the screen quota in Korea amply demonstrates. The success of the cultural activists is only possible if they find the government on their side. As it turns out, many governments in this global village consider the issue of economic sovereignty more significant than that of cultural sovereignty. After all, cultural diversity alone would not mean very much to the people in the street. If cultural activists are indeed visionaries, then they should be clever enough to embrace economic logic without fearing it too much and seek solutions that include this economic dimension.

Notes

1. Garnham used the term *cultural industries* to describe "those institutions in our society which employ the characteristic modes of production and organization of industrial corporations to produce and disseminate symbols in the forms of cultural goods and services, generally, although not exclusively, as commodities" (Garnham, 1990, 155–156). According to UNESCO, cultural industries are those "that combine the creation, production and commercialization of contents which are intangible and cultural in nature. These contents are typically protected by copyright and they can take the form of goods or services," which include "printing, publishing and multimedia, audio-visual, phonographic and cinematographic productions, as well as crafts and design. For some countries, this concept also embraces architecture, visual and performing arts, sports, manufacturing of musical instruments, advertising and cultural tourism."

2. Even the economically struggling countries of Central and Eastern Europe throw strong public support toward film production.

3. In many advanced industrial economies, the average tariff rate is below 4 percent.

4. In principle, MFN exemptions should not exceed a period of ten years. Furthermore, in any event, they shall be subject to negotiation in subsequent trade-liberalizing rounds. Refer to the GATS, Annex on Article II (MFN) Exemptions.

5. In 1997, the average production cost of a Hollywood film was 12.9 million euro, whereas countries with small film industries such as Mexico and Argentina had an average cost per film of 1.5 million euro or less. (Data from the European Audiovisual Observatory.)

6. In testimony to the U.S. Congress in 2002, Jack Valenti, then chairman of the MPAA, said "Copyright industries (consisting of movies, TV programs, home videos, books, music, video games and computer software) are responsible for some five percent of the GDP of the nation. They gather in more international revenues than automobiles and auto parts, more than aircraft, more than agriculture. They are creating new jobs at three times the rate of the rest of the economy. The movie industry alone has a surplus balance of trade with every single country in the world. No other American enterprise can make that statement. And all this at a time when the U.S. is bleeding from some $400 billion in deficit balance of trade."

7. It would be inaccurate to argue that a grand compromise over the audiovisual sector alone saved the Uruguay Round from failure. Since the UR was designed as a "single-undertaking," all sectors needed to reach agreement to call the round a success, including the most controversial sector, agriculture.

8. There is a precedent in the cultural sector of using the WTO dispute settlement mechanism. The United States challenged the Canadian policy on periodicals at the WTO; the United States won the dispute.

9. Twelve sectors were identified to facilitate the negotiations at the GATS: business; communication; construction and engineering; distribution; education; environment; financial; health; transport; tourism and travel; recreation, cultural, and sporting; as well as "other." Each sector contains subsectors. Audiovisual service is classified as a subsector of the communication sector.

10. As it turns out, this is one of the most widespread misunderstandings concerning the GATS. Many NGOs and activists mislead people to believe that the GATS is an open invitation to the complete market without any safeguards at all, and is therefore a "conspiracy" of the unbridled multinational corporations.

11. Most of these twenty-six countries made their commitments at the end of the Uruguay Round (1993), while a few members, like China and Chinese Taipei, made their commitments as part of their WTO accession negotiations.

12. Messerlin, Siwek, and Cocq 2004.

13. This section draws on Choi 2002a.

14. The concept of "cultural distance" is used to establish the cultural proximity of two societies. Variables such as language, content, and genre preference are considered in measuring cultural distance. Galperin 1999 is a prime example.

15. Consumers and export industry are "winners."

16. Over the 1980s and 1990s, average production budgets of Canadian feature films have decreased from $3 million to $2.7 million for English-speaking films and from $3.1 million to $2.2 million for French-speaking films. Films produced in other, comparable countries have increased their production budgets over the same period: by 13 percent in the UK, 21 percent in Australia, and 42 percent in France. In recent years, nearly two-thirds of Canadian feature films with government support had average marketing budgets of less than $150,000. By comparison, the

Hollywood-based studio productions had average expenditures of $37 million for marketing in 1999.

17. When Canada approved PolyGram N.V.'s (a UK-based entertainment multinational owned 75 percent by Philips Electronics of the Netherlands) application to establish a new film production and distribution company in Canada, PolyGram was required to limit its distribution activity to only proprietary films. The company also agreed to spend a certain portion of its Canadian revenues on cultural production in Canada.

18. During those years, Korea was enjoying an unprecedented economic boom, running a huge trade surplus with the United States.

19. An in-depth analysis of the interests, power, and constraint of stakeholders in the Korean screen quota controversy is offered in Choi 2005.

20. This is an interesting case of "private enforcement of public authority." It was fitting that the new organization was named the Screen Quota Watch Group.

21. Pushing deadlines back is now a familiar facet of the multilateral trade negotiations. The previous round of multilateral trade negotiations missed its original deadline of 1990 before declaring a deal in December 1993.

22. Refer to United States 2000; Switzerland 2001; and Brazil 2001. Japan (2000) set its negotiating goal in the audiovisual service in the broad context of the service negotiations.

23. United States 2000.

24. Japan 2000.

25. Switzerland 2001.

26. Brazil 2001.

27. See the WTO web site.

28. Since some of these requests and offers are exchanged among the negotiating countries on the basis of nondisclosure to the public, it is difficult for an outsider to the negotiations to have the full picture of offers. However, some members make their offer publicly available, and they are available from the WTO web site.

29. United States 2003.

30. See EU web site.

31. Korea 2001.

32. Japan 2000.

33. Bernier 2004.

34. At their annual meeting in October 2003 at Croatia, cultural activists made it explicit that they wanted to be free from any obligation from the WTO.

35. The metaphor of a "two-level" game, which was pioneered by Putnam 1988, succinctly describes this aspect of international negotiations.

36. "While it is true that, on a given Friday, most of the world's multiplexes will be playing franchise products from American studios, it is not hard to imagine a future in which an American suburban marquee will boast a Chinese martial-arts picture, a Korean action thriller, a Mexican cop drama and a French romantic comedy" (Scott 2004).

37. Vedrine 2001.

38. Hirschberg 2004.

39. To be more precise, the dichotomy should be understood as the United States and like-minded countries vs. the EU and like-minded countries.

40. For the symbolic aspect of protection, see Choi and Park 2002.

References

Bernier, Ivan. 2004. "The Recent Free Trade Agreements of the United States as Illustration of Their New Strategy Regarding the Audiovisual Sector." Available at www.bilaterals.org.

Brazil. 2001. "Communication from Brazil: Audiovisual Services." S/CSS/W/99, July 9.

Choi, Byung-il. 2002a. *Culture and Trade in the APEC: Case of Film Industry in Canada, Mexico and Korea*. APEC Study Series 02-01. Seoul: Korea Institute for International Economic Policy.

Choi, Byung-il. 2002b. "When Culture Meets Trade—Screen Quota in Korea." *Global Economic Review* 31, no. 4 (December): 1–19.

Choi, Byung-il. 2005. "When Culture Meets Trade—Screen Quota Conundrum." In *Korea and International Conflicts: Case Studies*, vol. 1, ed. B. Choi, 129–170. Seoul: Institute for International Trade and Cooperation, Ewha Womans University.

Choi, Byung-il, and Seah Park. 2002. "Symbols Do Matter—the Screen Quota Dispute between the US and Korea." *The US-Korea Tomorrow* no. 4 (October).

European Commission. 2001. *Statistics on Audiovisual Services Data 1980–1999*. Brussels.

Footer, Mary E., and Christopher Beat Graber. 2000. "Trade Liberalization and Cultural Policy." *Journal of International Economic Law* 3, no. 1: 115–144.

Galperin, Hernan. 1999. "Cultural Industries Policy in Regional Trade Agreements: The Cases of NAFTA, the European Union and MERCOSUR." *Media, Culture & Society* 21: 627–648.

Garnham, Nicholas. 1990. *Capitalism and Communication*. London: Sage Books.

Germann, Christopher. 2001. "Cultural Diversity and Free Trade in Audiovisual Services: Is There an Abuse of Dominant Market Positions?" Presented at the conference on "Cultural Industries and New Information Technologies," Strasbourg, November 19–20.

Graber, Christopher Beat. 2003. "Audiovisual Policy: The Stumbling Block of Trade Liberalization?" In *The WTO and Global Convergence in Telecommunications and Audiovisual Services*, ed. Damien Geradin and David Luff, 165–214. Oxford: Oxford University Press.

Hirschberg, Lynn. 2004. "What Is an American Movie Now?" *The New York Times*, November 14.

Japan. 2000. "Communication from Japan: The Negotiations on Trade in Services." S/CSS/W/42, December 22.

Korea. 2001. "Communication from the Republic of Korea: A Thought on the Alternatives to MFN Exemptions in the Annex on Article II Exemptions." S/CSS/W/127, November 30.

Messerlin, Patrick A., Stephen E. Siwek, and Emmanuel Cocq. 2004. *The Audiovisual Services Sector in the GATS Negotiations*. Washington, DC: The American Enterprise Institute.

Office of United States Trade Representative. 2004. *The 2004 National Trade Estimate Report on Foreign Trade Barriers*. Washington, DC.

Putnam, Robert D. 1988. "Diplomacy and Domestic Politics: The Logic of Two-Level Games," *International Organization*, no. 42 (Summer): 427–460.

Scott, A. O. 2004. "What Is a Foreign Movie Now?" *The New York Times*, November 14.

Switzerland. 2001. "Communication from Switzerland—GATS 2000: Audiovisual Services." S/CSS/W/74, May 4.

United States. 2000. "Communication from the United States: Audiovisual and Related Services." S/CSS/W/21, December 18.

United States. 2003. "Communication from the United States: Initial Offer." TN/S/O/USA, April 9.

Vedrine, Hubert. 2001. *France in an Age of Globalization*. Washington, DC: Brookings Institution Press.

Web Sites

Coalitions for Cultural Diversity at http://www.cdc-ccd.org.

EU at http://europa.eu.int.

International Network on Cultural Policy at http://www.incp-ripc.org.

International Network on Cultural Diversity at http://www.incd.net.

Korean Coalition for Diversity in Moving Images at http://www.screenquota.org.

Media Trade Monitor at http://www.mediatrademonitor.org.

Motion Picture Association of America at http://www.mpaa.org.

UNESCO at http://www.unesco.org/culture/industries/trade.

USTR at http://www.ustr.gov.

WTO at http://www.wto.org.

7 The Global Governance of Mass Media Content

Cees J. Hamelink

Much of the debate about international media focuses on the messages it carries across borders. Those messages that are packaged as audiovisual (film, television, video) or printed commodities represent such categories as news, entertainment, education, political propaganda, public relations, and advertising. Together they can be referred to as "content." The core business of the international media conglomerates is content. Several of their recent mergers are motivated by the desire to gain control over rights to content, such as content invested in film libraries or in collections of musical recordings. German media tycoon Leo Kirch made much of his enormous fortune through the acquisition of rights to content. Kirch bought from United Artists/Warner Brothers the rights to some hundred successful American TV programs; and he acquired the rights to over 50,000 hours of such TV series as *Baywatch* and copyrights to 16,000 film titles that he bought from Columbia Pictures, Paramount, and Universal. In a similar way the British music company EMI purchased the copyrights to over one million of the world's most popular songs.

The trading of media content has become global business in an expanding and profitable market. From the year 2000 the world market for media content exceeded the $3 trillion mark. The expansion of this market is to a large extent due to the concurrent worldwide processes of the deregulation of broadcasting and the commercialization of media institutions. These developments are reinforced by the process of economic globalization and the prevailing neoliberal trade policies as pursued by institutions like the World Trade Organization (WTO). There is a clear trend worldwide towards an increasing demand for audiovisual content, especially American-brand entertainment. A remarkable feature of this trend is that Europeans and Japanese are buying into this successful export commodity. The big European and Japanese media companies have outgrown their saturated home markets and the logical way towards further growth is transborder expansion. Joint ventures have been initiated and foreign investors have acquired traditional U.S. entertainment companies. An important feature of the trend toward globalization is that the trading by the megacompanies is shifting from the international exchange of local products to production for global markets. Consequently, content has become an item on the international trade agenda.

In 1865 the first multilateral treaty to deal with world communication was signed. In this International Telegraph Convention the freedom (in the sense of "secrecy") of correspondence across national borders was secured. At the same time, however, governments reserved the right to interfere with any message they considered dangerous for state security or in violation of national laws, public order, or morality. This tension between freedom and interference remained over the years a much debated topic among politicians, regulators, content carriers, and users. Since the late nineteenth century, media content in particular became an issue of international concern. The ambiguity of the freedom of such flows versus the need to interfere with this freedom posed a challenge for attempts at global governance. On the side of the freedom of content one finds classical civil and political rights arguments in favor of "free speech." On the side of interference there are arguments about "national sovereignty" and about the responsibility of speech vis-à-vis the rights and reputations of others. The "free speech" argument promotes an unhindered flow of messages into and out of countries. The sovereignty argument provides for protective measures against flows of messages that may impede the autonomous control over social and cultural development. The "responsible speech" argument claims the right to protection against the harmful effects of such free flows.

The fundamental human rights provisions on free speech (as in the Universal Declaration of Human Rights (UDHR) and the International Covenant on Civil and Political Rights) imply the need to deal with this ambiguity. There are explicit human rights claims (in the articles 19 in both instruments) to the freedom to hold opinions, to the freedom to express opinions, and to the freedoms to seek, receive, and impart information and ideas. However, only for the freedom to hold opinions is there the peremptory norm that this should be "without interference." For all other forms of information traffic interference is not excluded. This is inherently problematic as any interference would obviously imply the threat of erosion of the claim to freedom.

Over the years the international community and individual national governments have repeatedly tried to establish governance mechanisms (rules and institutions) to deal with the freedom versus interference concern. This chapter will describe some of these attempts. The conclusion is that no satisfactory arrangement has been produced. No satisfactory rules for balancing freedom versus interference have been developed and no relevant global institutions have emerged to deal with this.

Freedom versus Interference: The "Responsible Speech" Argument

Evolution of the Debate

In the early twentieth century the League of Nations addressed the problems of false news and propaganda, but it did not address the protection of freedom of expression. The International Federation of Journalists, however, stated the concern about freedom of information in the first article of its statute in 1926: "Its purpose is, notably, to safeguard in all possible ways the liberty of the Press and of journalists which it will endeavour to have guaranteed by law." During World War II the professional community would continue to express its concern about the freedom of information. At the second congress of the International Federation of Journalists of Allied or Free Countries (IFJAFC) in 1942 a resolution was adopted on the Charter of the Free Press that in part read, "The Congress recognises, however, that the freedom of the Press is no greater and no less than the freedom of the individual and carries with it the obligations, which the journalist worthy of the name will freely accept and observe, not to publish false or distorted news, not to do anything likely to discredit his profession, his paper, or himself. To this end, the Congress urges the universal adoption of a Code of Professional Conduct as has been done by several of its affiliated organizations."[1]

Debates on freedom of information have always had an association with reflections and viewpoints on the social responsibility of the media of mass communication. The key normative provisions on freedom of information permit freedom of expression "without fetters," but also bind this to other human rights standards. The clear recognition of the right to freedom of information as a basic human right in the Universal Declaration of Human Rights was positioned in a standards-setting instrument that also asked for the existence of an international order in which the rights of the individual can be fully realized (article 28 of the UDHR). This implies that the right to freedom of information is linked with the concern for a responsible use of international media. This linkage laid the basis for a controversy in which one position emphasized the free flow principle, whereas another position stressed the social duty principle.

The constitution of the United Nations Economic, Social, and Cultural Organization (UNESCO), adopted in 1945, contained the tension between the two approaches. It accepted the principle of a free exchange of ideas and knowledge, but it also stressed the need to develop and use the means of communication toward a mutual understanding among nations and to create an improved factual knowledge of each other. This could also be seen in the postwar development of the professional field: the time IFJAFC convened a world congress of journalists held in Copenhagen, June 3–9, 1946. This congress was attended by some 165 delegates from 21 countries. In the invitation letter the executive committee of the IFJAFC listed among the purposes of the congress, "to discuss methods of assuring the freedom of the press."[2] The discussions largely focused on the establishment of a new international professional organization, a provisional constitution was unanimously adopted, and the International Organisation of Journalists (IOJ) was created. Special attention was given to the debate on the liberty of the press and at the end of the Congress a statement of principle on the freedom of the press was adopted: "The International Congress of Journalists affirms that freedom of the press is a fundamental principle of democracy and can function only if channels of information and the means of dissemination of news are made available to all."[3]

The statement stressed "the responsibility of every working journalist to assist by every means in his power the development of international friendship and understanding" and instructed the Executive Committee to "examine the various codes of professional ethics adopted by national bodies, particularly in respect of any journalist deliberately and knowingly spreading—whether by press or radio or news agencies—false information designed to poison the good relations between countries and peoples."

The social duty dimension was even stronger in the resolution on press and peace that stated that "this congress considers the cementing of lasting international peace and security the paramount aim of humanity, and calls upon all the 130,000 members of the IOJ to do their utmost in support of the work of international understanding and co-operation entrusted to the United Nations."

As the Cold War was already growing, the social duty principle and the free flow principle collided in the early UN debates, largely in East/West ideological confrontations. The Yugoslav delegation, for example, proposed in the UN General Assembly in 1947 legislation to "restrict false and tendentious reports calculated to aggravate relations between nations, provoke conflicts and incite to war." This was unacceptable to the Western delegations, and eventually a compromise text (proposed by France) was adopted that recommended the study of measures "to combat, within the limits of constitutional procedures, the publication of false or distorted reports likely to injure friendly relations between states" (UNGA Res. 127[II]).

After 1948 the principles of freedom of information and social responsibility largely followed separate paths and could be found as separate provisions in standards-setting instruments that were adopted since 1948. Some instruments tried to link the two principles, such as the International Covenant on Civil and Political Rights (1966) and the UNESCO Mass Media Declaration (1978).

On freedom of information the essential standard was set by the Universal Declaration of Human Rights in its article 19, which provides the following: "Everyone has the right to freedom of opinion and expression; this right includes freedom to hold opinions without interference and to seek, receive and impart information and ideas through any media and regardless of frontiers." An important observation is that the authors of the article constructed freedom of information in five components. The first is the classical defense of the freedom of expression. The second is the freedom to hold opinions. This provision was formulated on the American insistence on protection against brainwashing, the forced acceptance of a political conviction. The third is the freedom to gather information. This reflected the interests of the U.S. news agencies to secure freedom for foreign correspondents. The fourth is the freedom of reception. This has to be understood as a response to the prohibition of foreign broadcasts reception during the war. The fifth is the right to impart information and ideas. This is a recognition of the freedom of distribution in addition to the freedom of expression.

The article 19 formulation became important guidance when later international documents articulated the freedom of information concern. Important illustrations are the European Convention for the Protection of Human Rights and Fundamental Freedoms (1950), the International Covenant on Civil and Political Rights (1966), the American Convention on Human Rights (1969), and the African Charter on Human and Peoples' Rights (1981).

This free speech standard like other human rights norms is not absolute and its exercise can be subjected to limitations. This obviously implies the risk of abuse by those actors (and particularly governments) who are intent on curbing free speech. Limitations could easily erode the significance of a normative standard. For this reason a threefold test has been developed in international law to assess the permissibility of limitations. These must be provided by law. They must serve purposes expressly stated in international agreements and they must be shown to be necessary in a democratic society. The UN special rapporteur on freedom of information has at times expressed concern about the tendency of governments to invoke article 4 of the International Covenant on Civil and Political Rights in justification of the suspension of free speech. This article lists the human rights provisions that are nonderogable. This means that under no circumstance, not even in times of war, can they be suspended. The right to freedom of expression is not listed in article 4. However, the Human Rights Committee, in its General Comment No. 29 (CCPR/C/21/Rev.1/Add.11), has identified the conditions to be met for a state to invoke article 4 (1) of the International Covenant on Civil and Political Rights to limit certain rights enshrined in its provisions, including the right to freedom of opinion and expression. Inter alia, the measures must be strictly limited in time, provided for in a law, necessary for public safety or public order, serve a legitimate purpose, not impair the essence of the right, and conform with the principle of proportionality. It is the view of the special rapporteur that in many of the cases brought to his attention over the years, all or some of these conditions are not being met, and that the argument of the fight against terrorism is used by governments as an illegitimate justification for the restriction of human rights and fundamental freedoms in general, and the right to freedom of opinion and expression in particular. There are cases where the feeling of insecurity caused by recent terrorist attacks has provided states with an opportunity to adopt such measures that had long been on the authorities' agenda, cases where the argument of national security is used to cover direct attacks against free media, investigative journalism, political dissent, and human rights monitoring and reporting. In this context, the special rapporteur does observe, however, that in practice it is quite difficult to monitor the legitimacy, necessity, and proportionality of anti-terrorism measures in the absence of a universally accepted, comprehensive, and authoritative definition of terrorism. This, on the one hand, leaves ample space for abusive restrictions based more on varying definitions of terrorism that respond to individual states' interests than on a universal concept of what a terrorist act is, and, on the other hand, makes it all the more difficult to monitor and evaluate the necessity and proportionality of such restrictions.

Propaganda

The technical possibility to directly reach out to vast audiences made the world community in the early twentieth century aware of both the opportunities and risks this entailed. International radio broadcasting raised in particular the issue of the right of sovereign states to prevent the reception of signals they consider harmful to their interests. This issue came up very strongly in the context of hostile political propaganda. Although propaganda has been exercised through newspapers, magazines, and movies, radio broadcasting became the essential medium: "the invention of radio in the late nineteenth century totally altered for all time the practice of propaganda (Jowett and O'Donnell 1986)."[4] In spite of the availability of more advanced communication technologies, radio broadcast propaganda is still today a vibrant activity engaging vast sums of money around the world. There is an estimated audience between 100 and 200 million listeners worldwide. Radio programs are transmitted abroad for a variety of reasons. Among them are the contacts with nationals in foreign countries; the international dissemination of news; the distribution of information about a country's politics, economy, and culture; and hostile propaganda. During World War I an extensive use was made of the means of propaganda. This psychological warfare continued after the war had ended and international short-wave radio began its proliferation. Germany began with international radio broadcasting in 1915 and was followed in 1917 by the Soviet Union. During the 1920s there was a rapidly growing interest in international radio broadcasting and more nations became involved. By 1939 some twenty-five countries were actively engaged in broadcasting to foreign countries in a multitude of languages. Many broadcast programs in this period contained propaganda and most of it was hostile. Programs defamed other governments or their leaders, made attempts to subvert foreign leadership, or incited to war.

From its establishment, the League of Nations was concerned about the role of the mass media in international relations and addressed such problems as the contribution of the press to peace. In 1931 the League of Nations had decided to ask the Institute for Intellectual Cooperation (the predecessor of UNESCO) to conduct a study on all questions raised by the use of radio for good international relations. In 1933 the study was published ("Broadcasting and Peace") and it recommended the drafting of a binding multilateral treaty. This treaty was indeed drafted and concluded on September 23, 1936, with the signatures from twenty-eight states. The fascist states did not participate. The International Convention Concerning the Use of Broadcasting the Cause of Peace entered into force on April 2, 1938, after ratification or accession by nine countries (Brazil, the United Kingdom, Denmark, France, India, Luxembourg, New Zealand, the Union of South Africa, and Australia). Basic to the provisions of the convention was the recognition of the need to prevent, through rules established by common agreement, broadcasting from being used in a manner prejudicial to good international understanding. These agreed upon rules included the prohibition of transmissions that could incite the population of any territory "to acts incompatible with the internal order or security of contracting parties" or were likely to harm good international understanding by incorrect statements. The contracting parties also agreed to ensure "that any transmission likely to harm good international understanding by incorrect statements shall be rectified at the earliest possible moment."

World War II led to an enormous expansion of radio propaganda and this continued in the years after the war. "As the dynamics of world politics were being played out, international radio broadcasting became a prominent weapon in the arsenal of propaganda"[5] (Jowett and O'Donnell 1986, 86). Immediately after World War II the issue of harmful broadcasting was discussed at the United Nations Conference on Freedom of Information in 1948. From this conference evolved the Convention on the International Right of Correction (adopted by UNGA Res. 630 [VII], on December 16, 1952). The convention desired "to protect mankind from the scourge of war . . . and to combat all propaganda which is either designed or likely to provoke or encourage any threat to the peace, breach of the peace, or act of aggression."

The United Nations General Assembly adopted a series of resolutions addressing the problem of war propaganda and the diffusion of false or distorted reports "likely to injure friendly relations between States" (UNGA Res. 127[II] of 1947). The concern about propaganda was codified in article 20 of the International Covenant on Civil and Political Rights with the provision "Any propaganda for war shall be prohibited by law." Also the UNESCO General Conference adopted a series of resolutions on the role of UNESCO in generating a climate of public opinion conducive to the halting of the arms race and to the transition to disarmament. For example, Resolution 1878 (twentieth session) recommended members "to pay particular attention to the role which information, including the mass media, can play in generating a climate of confidence and understanding between nations and countries as well as increasing public awareness of ideas, objectives and action in the field of disarmament."

Following the distinction proposed by Whitton,[6] there is subversive propaganda, defamatory propaganda, and war propaganda. The category of subversive propaganda that incites to revolt has been recognized in various legal instruments as an illegal act. It has been seen as a form of illegal intervention and as a form of aggression. Several multilateral treaties urge states to refrain from this type of propaganda. The category of defamatory propaganda in which foreign leaders are vilified has also been ruled illegal and a danger to international peace. A special problem in this context has been "the question of the publication of defamatory—especially false—news by the independent press and radio of a country." Although there is general agreement in the world community that false news should be prohibited, "there is wide discord over the best method to accomplish this."[7] There is no unequivocal collectively adopted agreement on this type of propaganda.

In the early twentieth century, war propaganda came to be seen as a serious danger to international peace by the League of Nations and later by the United Nations. The preliminary draft of the General Convention to Improve the Means of Preventing War by the League of Nations stated that aggressive propaganda could "take such offensive forms and assume such a threatening character as to constitute a real danger to peace." In various instruments the world community expressed its concern about war propaganda, but it never managed to establish a robust multilateral accord on the issue. A peculiar problem in this area has been the fact that much hostile communication originates with private actors and opinions have been very divided over the extent of state responsibility in such cases. Moreover, only a limited number of states have adopted the prohibition of war propaganda in their domestic legislation. It remains a matter of dispute whether a state could be held responsible if independent media under its jurisdiction were guilty of warmongering. There remains also a lack of clarity about the precise definition of propaganda for war.

Discrimination

An important provision in international law that affects the content of media concerns discrimination. Article 20 of the International Covenant on Civil and Political Rights contains next to its provision on propaganda, a paragraph that states, "Any advocacy of national, racial or religious hatred that constitutes incitement to discrimination, hostility or violence shall be prohibited by law." This was even more strongly formulated in article 4 of the International Convention on the Elimination of All Forms of Racial Discrimination (adopted as UNGA Res. 2106 A[XX], December 21, 1965). Here "all dissemination of ideas based upon racial superiority or hatred, incitement to racial discrimination, as well as acts of violence or incitement to such acts against any race or group of persons of another colour or ethnic origin" was declared a criminal offense. Essential to this provision is that states are also required (in article 4b) to declare illegal organizations that promote and incite racial discrimination. The UNESCO 1978 Declaration on Race and Racial Prejudice mentioned in this context the mass media specifically in article 5. The declaration provides a strong prescription for media conduct as it urges "the mass media and those who control or serve them" to "promote understanding, tolerance and friendship among individuals and groups and to contribute to the eradication of racism, racial stereotyped, partial, unilateral or tendentious pictures of individuals and of various human groups." The mass media are also told to be "freely receptive to ideas of individuals and groups which facilitate communication between racial and ethnic groups."

The issue of racial discrimination was specifically addressed in the most extensive regulatory instrument to address the concern for social responsibility in the mass media: the UNESCO Mass Media Declaration.[8] Its full title is Declaration on Fundamental Principles concerning the Contribution of the Mass Media to Strengthening Peace and International Understanding, to the Promotion of Human Rights and to Countering Racialism, Apartheid and Incitement to War. This declaration originated from the 19th General Conference of UNESCO (in 1979 at Nairobi), at which convened the first multilateral discussion about a draft declaration on fundamental principles governing the use of the mass media in strengthening peace and international understanding and in combating war propaganda, racism, and apartheid. The heart of the controversy at that meeting was a proposed article XII of the draft declaration that stated, "States are responsible for the activities in the international sphere of all mass media under their jurisdiction." For many member states this reference to state responsibility suggested the possibility of state control over the mass media. In particular for Western member states the text of the declaration might imply that standards would be set for media content by their users and among those users states could play a prominent and potentially dangerous role. The draft declaration that was proposed at the nineteenth General Conference had been adopted in 1975 during an experts conference. The Group of Nine EC countries (the Federal Republic of Germany, Denmark, Ireland, Britain, France, Italy, Belgium, Luxembourg, and the Netherlands) had walked away in protest from this meeting. For them the text was unacceptable because of its reference to the United Nations General Assembly Resolution 3379 on Zionism as a form of racism and because of the fact that the state was accorded an important role with regard to the mass media. During the General Conference of 1976 there were three positions. For the Western countries, Japan, and some Latin American countries the draft declaration was totally unacceptable. Several delegations from Eastern Europe pushed for the adoption of the draft declaration. A group of African and Asian countries supported the draft declaration but did plead for postponement in order to design a new text with better chances for consensus.

The decision taken in Nairobi referred the draft to the UN director general and the General Conference invited him "to hold further broad consultations with experts with a view to preparing a final draft declaration which could meet the largest possible measure of agreement [as well as to] request the Director General to submit such a draft declaration to member states at the end of 1977 or early in 1978." ... It was also decided to establish a commission for "a comprehensive study on the problems of communication in the modern world." This became the International Commission for the Study of Communication Problems, usually called the MacBride Commission after its chairman Sean MacBride.

During the twentieth General Conference of UNESCO in 1978 at Paris the amended draft declaration was unanimously adopted. The conference also adopted a resolution aiming at practical recommendations concerning the mass media declaration. The resolution proposed the holding of a congress to discuss the application of the declaration and was accepted with sixty-one votes in favor (the socialist and developing countries), one vote against (Switzerland), and twenty-six abstainers (all of them Western countries).

During the 1980s and 1990s there was no followup to the declaration. An attempt was made jointly by the leading journalists' bodies, the International Federation of Journalists and the International Organization of Journalists, to arrange an international congress to discuss the declaration. Largely due to the unwillingness to cooperate expressed by the Federation of International Editors of Journals (FIEJ) and the International Press Institute (representing editors-in-chief), the meeting never took place.

Crimes against Humanity

At the dramatic core of many contemporary ethnic conflicts is the grand-scale perpetration of crimes against humanity. As the term suggests these are criminal acts that render their perpetrators enemies of the human species. Crimes against humanity transgress taboos that apply in most cultures, such as the murder or torture of defenseless men and women, and the killing of children. Although crimes can be committed without apparent motivation, the exercise of gross violence on a grand scale—as in crimes against humanity—needs motivating beliefs. In order to get people to commit such crimes, they need to believe that the violent acts are right. In situations where crimes against humanity are committed one usually finds a systematic distribution of hate propaganda and disinformation. The purpose of this is to promote and justify the social and/or physical elimination of certain social groups. Members of such groups are often first targeted as "socially undesirable"; they are publicly ridiculed, insulted, and provoked (often in the media); and when the harassments are put into acts the victims are finally beaten up and killed. In the propagation of "elimination beliefs," the "other" are dehumanized, whereas the superiority of one's own group is emphasized. The propagandists convincingly suggest to their audiences that the "others" pose fundamental threats to the security and well-being of society and that the only effective means of escaping this threat is the elimination of this great danger. The use of violence in this process is presented as inevitable and thus not only acceptable, but also absolutely necessary. The elimination beliefs that motivate people to kill each other are not part of the human genetic constitution. They are social constructs, which need social institutions for their dissemination. Such institutions include religious communities, schools, families, and the mass media. Because crimes against humanity are unthinkable without elimination beliefs, the institutional carriers of such beliefs should be seen as enemies of the human species. This implies that all those who propagate beliefs in support of genocide, through whatever media, have to be treated as perpetrators of crimes against humanity.

Among the crimes against humanity—as defined by international law—are murder and extermination of civilian populations, genocide, and apartheid. The Charter of the

International Military Tribunal of Nuremberg (1946) defines (in article 6) crimes against humanity as follows: "murder, extermination, enslavement, deportation, and other inhuman acts committed against any civilian population, before or during the war, or persecutions on political, racial, or religious grounds in execution of or in connection with any crime within the jurisdiction of the Tribunal, whether or not in violation of the domestic laws of the country where perpetrated." An important feature of this definition is the recognition that states can commit such crimes against their own populations: "against any civilian population." It is also important that states cannot refer to domestic legislation as justification: "whether or not in violation of domestic law."

The Convention on the Non-Applicability of "statutory limitations" (by which the General Assembly of the United Nations establishes that war crimes and crimes against humanity cannot be precluded by the lapse of time) explicitly mentions apartheid and genocide in addition to the crimes defined by the Nuremberg Tribunal. It is moreover provided that it is not relevant whether these crimes are committed in war time or in peace.

Although impunity is characteristic of the treatment of those who commit violations of human rights, under international law there is an obligation to prosecute crimes against humanity. War crimes and crimes against humanity, including the incitement to these criminal acts wherever committed, must be punished. In 1996 the international community began—finally—to take this matter seriously and the UN General Assembly decided on a concrete agenda for the establishment of an international criminal court. In July 1998 an international diplomatic conference that was convened by the United Nations (in Rome) produced a treaty establishing the permanent international criminal court (ICC). The ICC will deal with war crimes and crimes against humanity. In accordance with existing treaties the court will have the mandate to prosecute those who incite to genocide by propagating elimination beliefs.

The Convention on the Prevention and Punishment of the Crime of Genocide (1948) proposes in article III that "direct and public incitement to genocide" shall be punishable. The convention also applies this criminal responsibility to "private individuals" (article IV). The Draft Declaration on the Rights of Indigenous Peoples also refers to incitement when it states in article 7 that "Indigenous peoples have the collective and individual right not to be subject to ethnocide and cultural genocide" and refers in this context to the prevention of and redress for "any form of propaganda directed against them."

Protection of Minors

Special concerns have been raised in the international political arena in connection with media content and its impact upon children. When the United Nations concluded in 1989 the Convention on the Rights of the Child these concerns were expressed through article 17 of the Convention:

States Parties recognize the important function performed by the mass media and shall ensure that the child has access to information and material from a diversity of national and international sources, especially those aimed at the promotion of his of her social, spiritual and moral well-being and physical and mental health. To this end, States Parties shall:

(a) Encourage the mass media to disseminate information and material of social and cultural benefit to the child and in accordance with the spirit of article 29;
(b) Encourage international co-operation in the production, exchange and dissemination of such information and material from a diversity of cultural, national and international sources;
(c) Encourage the production and dissemination of children's books;
(d) Encourage the mass media to have particular regard to the linguistic needs of the child who belongs to a minority group or who is indigenous;
(e) Encourage the development of appropriate guidelines for the protection of the child from information and material injurious to his or her well-being, bearing in mind the provisions of articles 13 and 18.

The interesting implication of this provision is that rather than trying to restrict certain types of content, an effort is made to stimulate the production and dissemination of content suited for children; it will be interesting to see if the Convention attracts enough ratifications to come into force, and if the stipulations of article 17 then will be taken seriously.

Freedom versus Interference: the "National Sovereignty" Argument

The main regulatory attempts that addressed the tension between the free speech argument and the national sovereignty argument focused on the implications of direct broadcasting by satellites and the demand for a new international information order.

Direct Satellite Broadcasting

The possibility of International Direct Broadcasting by Satellite (IDTBS) raised the hot political issue of prior consent. The world community has been divided on the question of whether direct satellite transmissions of media content need the advance consent of receiving states.

IDTBS raised difficult questions on the impact of international telecommunication. Can satellite TV content imply damage to a state's political, commercial, and cultural interests? If so, who is to be held responsible? Are states responsible for TV transmissions originating in their territories with private institutions? Are recipient states responsible for the protection of the integrity of their territories and as such responsible for incoming TV signals?

The IDTBS issue first emerged in a meeting of the United Nations Committee on the Peaceful Uses of Outer Space (hereinafter cited as COPUOS) in November 1963. This meeting was preparing the draft Declaration of Legal Principles Governing the Activities of States in the Exploration and Use of Outer Space. In connection with the draft outer space declaration, the delegate from Brazil stated, "The Declaration should also incorporate a ban on the utilization of a communication system based on satellites for purposes of encouraging national, racial or class rivalries and a reference to some international scrutiny of global satellite communication."[9] Also, the Soviet delegate proposed a provision regarding the content of space communication: "The use of outer space for propagating war, national or racial hatred or enmity between nations shall be prohibited."[10] These interventions articulated a demand to impose restrictions on program content in IDTBS. The outer space declaration did not incorporate the demand when it was passed by the General Assembly on December 13, 1963 (Res. 1962 [XVIII]). The resolution, however, made a reference to the earlier UNGA Resolution 110 (II) of November 3, 1947, which "condemned propaganda designed or likely to provoke or encourage any threat to the peace, breach of peace, or act of aggression." The resolution also recognized that the anti-propaganda provision was applicable to outer space. It was, however, not explicitly stated that this standard on program content was to be imposed on space communication. As a matter of fact, the Outer Space Treaty (1965) that was to follow from the declaration (and that became the leading legal document on the use of space resources) does not mention space communication. There is a reference (in the eighth preambular paragraph) to UNGA Resolution 110 (II), but IDTBS is not mentioned. The space communication issue had been raised however and would soon be put on the agenda of the Legal Sub-Committee of COPUOS. The UNGA Resolution 2222 of December 19, 1966, to which the Outer Space Treaty was annexed contained a request to COPUOS "to begin ... the study of questions relative to the definition of outer space and the utilization of outer space and celestial bodies, including the various implications of space communications." In response to this request, COPUOS reported in 1964 to the General Assembly that it had decided "to consider questions relating to the use of satellites for transmitting radio and television programmes for direct reception by the general public." Four years later, in 1968, the General Assembly unanimously adopted on December 20 Resolution 2453 (XXIII) by which the COPUOS Working Group on Direct Broadcast Satellites was created. In the report of its first session in 1969, the working group recognized both the positive effects ("the promise of unprecedented progress in communications and understanding between peoples and cultures, e.g. improved health and education and a greater flow of news and information") and the adverse effects ("as resulting from the receipt of television programs in the territories of states whose governments do not wish to have them received.... Programs aimed at achieving certain political objectives, e.g. propaganda, incitement and interference; Programs that are objectionable because they use materials or techniques outlawed in the receiving states, e.g. violence, libel, obscenity; Programs that are unwanted mostly because they are foreign and have the potential to undermine national integrity, e.g. news or public information programs").

At this first session all members of the working group, with the exception of the United States, were in favor of some form of standards-setting for direct satellite broadcasting. The USSR had already submitted a draft of provisions for the utilization and exploitation of satellites for broadcasting. However, no consensus was achieved save the minimal acceptance (at the working group sessions in 1969 and 1970) of the major sources of international law that would be applicable to IDTBS. These were the UN Charter, the Outer Space Treaty, the ITU Convention and Radio Regulations, the Universal Declaration of Human Rights, and relevant resolutions of the UN General Assembly.

This was endorsed by the UN General Assembly in Resolution 2733. The resolution also pointed to the issue of responsibility for IDTBS as it recommended in article V "that the COPUOS should study through its Legal Sub-Committee, giving priority to the Liability Convention, the work carried out by the Working Group on Direct Broadcasting Satellites under the item on the implications of space communications."

In the first session of the Working Group it was also proposed that a distinction be made between the deliberate broadcasting into the territory of another state and unintentional "spill-over" into the territory of another state. In the second session (1969) the issue of program content was discussed under the headings of political programs, cultural and social programs, and commercial programs. The dominant theme in discussing these three categories was the two-sided assessment of the impact of IDTBS, potential benefits versus adverse effects.

In 1970 the French delegation to the COPUOS presented a Code of Good Conduct relative to space communication. This code, while accepting the right of states to broadcast from space, stated a number of restrictions that would all need the prior consent of the receiving state. Programs that would be considered illegal were identified in the code. These were programs with propaganda detrimental to international peace; programs undermining individual dignity, cultures, religions, or traditions; and programs that would interfere with domestic or foreign policies of other states. The code also proposed restrictions on advertising and recommended that advertisements should demonstrate an artistic or educational interest and contain truth. The French proposal was not accepted. It did bring clearly to the surface, however, the essential issues involved in IDTBS negotiations: a controversy about the principle of prior consent, and a controversy about the question of state responsibility.

In the ensuing debates on IDTBS three different positions emerged: A position that emphasized prior consent, sovereignty, and noninterference (held by France, the Soviet Union, and developing countries); a position that opposed prior consent and defended the human right to freedom of information as prohibitive of any interference with direct broadcasting (especially the United States); a position that advanced the possibility of a compromise agreement, largely based on bilateral arrangements between sending and receiving countries (Canada and Sweden).

In 1972 the Soviet Union submitted a draft Convention on Principles Governing the Use by States of Artificial Earth Satellites for Direct Television Broadcasting. Article 5 of the proposed convention read that satellite broadcasts should be contingent upon "the express consent" of the receiving states. The Soviet proposal also promoted the control over program content. In the debates that followed, a Soviet resolution calling for a binding treaty on IDTBS was amended to Resolution 2916 (adopted by the General Assembly in 1972) on Preparation of an International Convention on Principles Governing the Use by States of Artificial Earth Satellites for Direct Television Broadcasting. The only dissenting vote came from the United States. In 1972 Canada and Sweden proposed a compromise to overcome the IDTBS controversy between those states that wanted to regulate space communication strictly, and those that abhorred such regulation. The proposal adopted the principle of consent, but as a form of licensing for international IDTBS operations (similar to national licensing schemes) and with participation by and in close cooperation between broadcasting and receiving states. It was suggested by the initiators of the compromise solution that this position would make a provision on program content superfluous.

The hope was that the provision on "consent, participation, and cooperation," would satisfy the pro-regulation countries and that the absence of a program content provision would satisfy the anti-regulation countries. The compromise remained just that. It was still too restrictive and impractical for the free flow protagonists, and unsatisfactory for those who felt that the principle of consent would still need to be expanded with standards on program content since international cooperation among states would not be a guarantee against "unlawful" broadcasts.

In 1972 UNESCO adopted a declaration that reflected largely the position of the countries that felt that there should be express obligations for the operation of international IDTBS. The Declaration of Guiding Principles on the Uses of Satellite Broadcasting for the Free Flow of Information, the Spread of Education, and Greater Cultural Exchange did prescribe a set of standards for IDTBS. Its preamble made references to article 19 of the Universal Declaration of Human Rights as well as to the anti-propaganda resolution of 1947. Article V stated that the objective of satellite broadcasting for the free flow of information is to ensure the widest possible dissemination, among the peoples of the world, of news of all countries, developed and developing alike. While stressing the freedom of information principle, the declaration also recognized the prior consent principle in article IX: "In order to further the objectives set out in preceding articles, it is necessary that States, taking into account the principle of freedom of information, reach or promote prior agreements concerning direct satellite broadcasting to the population of countries other than the country of origin of the transmission."

The prior consent principle was again stressed in a 1974 proposal made by Argentina for a Draft International Convention on Direct Broadcasting by Satellite. This proposal attempted to accommodate both the prior consent and the free flow principle and stressed the need for regional agreements (UNGA document A/Ac.105/134, July 5, 1974).

A different position was taken in the draft convention presented by the United States in the same year. Article 3 of the draft convention stated that all states had the right to emit international broadcasts by means of IDTBS. The proposal promoted the free flow of information principle. In the COPUOS debates the U.S. delegate explained that according to his country's view international law did not contain any obligation of prior consent.

In 1976 the UN General Assembly adopted a resolution by which the COPUOS was requested to draft principles for IDTBS as a matter of priority. A more technical approach to IDTBS was taken within the International Telecommunication Union (ITU) through the World Administrative Radio Conference for the Planning of the Broadcasting Satellite Service (WARC 1977). The conference drew up a detailed plan for space communication and confirmed the consultation procedures that had been established at the 1971 WARC. These procedures implied that the transmission of signals to the territories of other countries should be reduced unless previous agreement had been reached between the countries concerned. The pertinent Radio Regulation 7, section 428A, reads: "In devising the characteristics of a space station in the television broadcasting satellite service, all technical means available shall be used to reduce, to the maximum extent practicable, the radiation over the territory of other countries, unless agreement has been previously reached with such countries." At the WARC 77 the United States had no objection to this form of prior consultation. There were different views on whether the ITU form of prior consultation made the provision of prior consent in the draft principles on IDTBS superfluous. Some countries held the view that the technical agreements that had been reached under the ITU plans made further agreements unnecessary. For the U.S. delegation these provisions were seen to provide adequate protection. Another claim (by the USSR delegation) was that coordination agreements under the ITU regulations did cover exclusively technical matters and that they in no way concerned regulatory questions on the political and legal level. Given the claim by the majority of states that they did not wish to accept foreign radio and television interference through space communication unless previously agreed upon, the ITU technical standards were not sufficient to satisfy this claim and further work on guiding international legal principles for IDTBS activities remained necessary.[11]

An important argument by those who felt the ITU arrangements were sufficient was that the 1977 WARC defined satellite broadcasting largely as a national operation with limited international ramifications. The plans for IDTBS adopted in the WARC 1977 Final Acts were for national satellite broadcasting (and a few regional arrangements) and provided for permission for a small number of states to engage in state-to-state broadcasting. For most countries in the world IDTBS would in fact be national broadcasting and would need no special protection.

Contrary to these expectations satellite television developed into a global arena. Since the late 1980s it was widely used across Europe and it began to grow in the United States during the 1990s. Television transmission by satellite began to use digital technology and the compression of signals to require less bandwidth. This has considerably increased the number of satellite television services available today. It has also made it possible to encrypt television signals so that transmitted content can only be unscrambled by those audiences who pay for decoding equipment. Concerns about undesirable content continue to be raised. Asian countries in particular have expressed reservations about Anglo-American-dominated flows of news and entertainment that are beamed into their countries. Although such reservations tend to be couched in terms of the erosion of local cultures, they usually refer to concerns about the domestic political effects of foreign broadcasts. A multilateral consensus on IDTBS has not been reached, and with the advent of the Internet, the use of digital technologies, and the increasing internationalization of broadcasting, this seems by now a highly unlikely project. In relation to satellite television, international regulation now focuses more on technical and economic questions rather than on content issues.

A New International Information Order

In the 1970s the movement of nonaligned countries initiated an attempt to define rules and practices for the international provision of news. At the core of this initiative was the recognition of a crucial standard in international law: the right to self-determination.

This right—articulated in several international conventions—provides that peoples should freely pursue their economic, social, and cultural development. Developing countries that were at the receiving end of international information flows experienced the free flow of cultural products into their countries as a new form of colonialism that threatened their cultural autonomy. The key agenda issue for this debate was the demand for a new international information order. This demand expressed the Third World concern about disparity in communication capacity along three lines.

First, there was the concern about the impact of the skewed communication relations between North and South on the independent cultural development of the Third World nations. Actually, the first summit of nonaligned states, the Asian-African Conference in Bandung, Indonesia, in 1955, already referred to the impact of colonialism on culture. "The existence of colonialism in many parts of Asia and Africa, in whatever form it may be, not only prevents cultural cooperation but also suppresses the national cultures of the peoples.... Some colonial powers have denied their dependent peoples basic rights in the sphere of education and culture." At the 1973 summit of nonaligned states at Algiers, members expressed their concern about cultural colonialism as the effective successor to the earlier territorial modes of colonialism.

Second was the concern about the largely one-sided exports from the North to the countries of the Third World and the often distorted or totally absent reporting in the media of the North about developments in the South. The Algiers summit called for the "reorganization of existing communication channels, which are a legacy of the colonial past and which have hampered free, direct and fast communication between developing countries." This disequilibrium in the exchange of information between the North and the South controlled by a few Western transnational information companies began to be criticized by the nonaligned movement as an instrument of cultural colonialism.

Proceedings from the Tunis symposium of 1976 stated: "Since information in the world shows a disequilibrium favouring some and ignoring others, it is the duty of the nonaligned countries and other developing countries to change this situation and obtain the decolonization of information and initiate a new international order of information." The New Delhi Declaration on Decolonization of Information stated that the establishment of a New International Order for Information is as necessary as the New International Economic Order.

The third line of concern addressed the transfer of media technology. On balance it was concluded in the early 1970s that precious little technology had been transferred and that by and large only technical end-products had been exported from the industrial nations. This was often done under disadvantageous conditions so that in the end the technical and financial dependence of the receiving countries had only increased. Proceeding from its Algiers summit in 1973, the nonaligned movement continuously articulated its position of strong support for the emancipation and development of media in the developing nations.

Already in 1970 the minutes of the UNESCO General Conference read: "Delegates from a number of developing countries stressed the need to ensure that the free flow of information and international exchanges should be a two-way operation. They asserted that the programme must continue to emphasize the rights of less privileged nations, to preserve their own culture."

In its first phase (1970–1976) the debate was characterized by the effort to "decolonize." In this period political and academic projects evolved that fundamentally criticized the existing international information order and developed proposals for decisive changes. Several years of declarations, resolutions, recommendations, and studies converged into the demand for a New International Information Order (NIIO).

The concept surfaced at the Tunis information symposium in March 1976. With this concept (formally recognized by nonaligned heads of state in August 1976 at Sri Lanka), a clear linkage was established to the proposal for a fundamental restructuring of the international economy that was put forward in 1974 (the New International Economic Order, NIEO).

Although the precise meaning of the NIIO was not defined, it was evident that its key notions were national sovereignty and cultural autonomy. The NIIO reflected the nonaligned states' aspiration to an international information exchange in which states that develop their cultural systems in an autonomous way and with complete sovereign control of resources fully and effectively participate as independent members of the international community. From 1976 onward, the Western news media began to take a critical attitude toward the demand for an NIIO. The majority of international mass media organizations expressed their opposition against the nonaligned movement's initiative. Among the essential objections was the suspicion that the proposal for a restructuring of the international information order would mainly serve the interests of authoritarian states and would seriously undermine the standard of freedom of information.

After much political commotion, the twentieth General Conference of UNESCO (in 1978 at Paris) adopted a request for a special international commission (the so-called MacBride Commission) to propose measures that could lead "to the establishment of a more just and effective world information order." In fact, this conference was a turning point in the debate insofar as at this meeting the hostile opposition toward the idea of a new order was softened. There began to be almost unanimous acceptance that Third World countries had justifiable complaints and that concessions must be made by the industrialized states. The original formula coined by the nonaligned movement, NIIO, was replaced by the proposal for a "new, more just and effective world information and communication order," called NWICO. According to the interpretation of United States Ambassador John Reinhardt at the 1978 General Conference, this new order required "a more effective program of action, both public and private, to suitable identified centers of professional education and training in broadcasting and journalism in the developing world ... [and] ... a major effort to apply the benefits of advanced communications technology ... to economic and social needs in the rural areas of developing nations." The new order (NWICO) that was now acceptable to all UNESCO member states was mainly interpreted as a program for the transfer of knowledge, finances, and technical equipment. The problem of the international information structure was being reduced to mere technical proportions. In response to this an intergovernmental program to support the development of communication was launched as a Western initiative in 1980.

The twenty-first General Conference in 1980 at Belgrade adopted by consensus a resolution concerning the establishment of the International Programme for the Development of Communication (IPDC). Also a Venezuelan resolution was adopted (without consensus and against mainly Western opposition) that proposed study into the new information order. The resolution aimed "to initiate studies necessary for the elaboration of principles related to a New World Information and Communication Order." The Venezuelan resolution was adopted with fifty-three votes in favor, six against (among them the United States, UK, and Switzerland), and twenty-six abstainers.

During the UNESCO general conferences of 1976, 1978, and 1980, the Western minority managed to achieve most of its policy objectives against the expressed preference of the majority of member states. In the end, the debate did not yield the results demanded by the developing countries. Their criticism of the past failures of technical assistance programs was answered by the creation of yet another such program: the International Programme for the Development of Communication. It was seen by many Third World delegates as the instrument to implement the standards of the NWICO. The UNESCO General Conference of 1980 had stated (Res. 4.19) that among these standards were the elimination of the imbalances and inequalities that characterize the present situation, to increase the capacity of the developing countries to achieve improvement of their own situation, notably by providing infrastructure and by making their information and communication means suitable to their needs and aspirations and to the sincere will of developed countries to help them. The IPDC was not going to meet these expectations. Apart from the inherent difficulty in that the IPDC did represent a definition of world communication problems that had in the past not worked to the benefit of Third World nations, the program would also suffer a chronic lack of resources from the outset.

Moreover, in 1981 there were clear indications that the opponents of the Third World demand for a new information order had not yet been satisfied. They continued to see dangers to the liberties of the Western mass media. At Talloires, France (May 15–17, 1981), a "Voices of Freedom Conference of Independent News Media" was held. The conference was attended mostly by representatives of publishers' interests (from twenty-one countries). There were no representatives of the international journalists' federations. The conference participants adopted the Magna Carta of the Free Press, which stated:

> We believe the time has come within UNESCO and other intergovernmental bodies to abandon attempts to regulate news content and formulate rules for the press. Efforts should be directed instead to finding practical solutions to the problems before us, such as improving technological progress, increasing professional interchanges and equipment transfers, reducing communication tariffs, producing cheaper newsprint and eliminating other barriers to the development of news media capabilities.... We are deeply concerned by a growing tendency in many countries and in international bodies to put government interests above those of the individual, particularly in regard to information. We believe that the ultimate definition of a free press lies not in the actions of governments or international bodies, but rather in the professionalism, vigor and courage of individual journalists.

The International Federation of the Periodical Press (FIPP) reinforced this position at its meeting in Washington, DC, on May 20, 1981. In the same year the first signs of the later (1984) U.S. withdrawal from UNESCO (in 1985 followed by the UK) emerged and included references to the Talloires meeting. A letter by President Reagan to the House of Representatives (September 17, 1981) said: "We strongly support—and commend to the attention of all nations—the declaration issued by independent media leaders of twenty-one nations at the Voices of Freedom Conference, which met at Talloires,

France, in May of this year. We do not feel we can continue to support a UNESCO that turns its back on the high purposes this organization was originally intended to serve." The U.S. withdrawal was subsequently officially announced on December 29, 1983, in a letter by Secretary of State George Schultz. According to the letter, "Unesco has extraneously politicized virtually every subject it deals with, has exhibited hostility toward the basic institutions of a free society, especially the free market and the free press, and has demonstrated unrestrained budgetary expansion."

In the effort to maintain a fragile international consensus, the UNESCO General Conference began to erode the original aspirations of the NIIO proponents and shifted from the need to establish a regulatory structure to the new order as "an evolving and continuous process." (Paris, general conference, 1982, Res. 3.3).

In September 1983, UNESCO convened a round table on a "New World Information and Communication Order" in Igls, Austria. The meeting focused strongly on assistance to developing countries for the development of their communication infrastructures. A second, and last, round table on the NWICO took place in 1986 (April) in Copenhagen, Denmark.

Although the concept of a new order remained central for some years to come, nothing was done with regard to its implementation. Gradually UNESCO withdrew support from all research, documentation, or conference activities intended to contribute to the establishment of a new information order and moved toward a "new strategy" in its Medium-Term Plan for 1990–1995 with emphasis on the free flow of information and the freedom and independence of the media, the priority of operational activities, and the importance of information technology. Whereas the nonaligned summit in Belgrade (September 1989) reiterated its support for the New International Information Order (NIICO), the UNESCO General Conference strove hard to reach consensus on formulations that represent conventional freedom of the press, pluralism of the media, freedom of expression, and free flow of information positions. According to its director general (in 1989), plans for a new information order no longer existed in UNESCO.

In 1978 the United Nations General Assembly had expressed its support for the establishment of a new world information and communication order in G. A. Resolution 33/115. The General Assembly concluded that it "affirmed the need to establish a new, more just and more effective world information and communication order, intended to strengthen peace and international understanding and based on the free circulation and wider and better-balanced dissemination of information." In part C of the same resolution, the General Assembly had established the Committee to Review United Nations Public Information Policies and Activities. In the report of this committee submitted to the 1979 meeting of the General Assembly, it was stated that the review of United Nations information policy should be "a part of the wider undertaking involving the general evolution of the question of a new world information order." In the same year the General Assembly adopted G. A. Resolution 34/182 (of December 18,

1979) by which the Committee to Review was renamed the United Nations Committee on Information and in which the committee mandate was formulated. Pursuant to the resolution, the committee would continue "to examine United Nations public information policies and activities, in the light of the evolution of international relations, particularly during the past two decades, and of the imperatives of the establishment of the new international economic order and of a new world information and economic order." The committee was also "to promote the establishment of a new, more just and effective world information and communication order intended to strengthen peace and international understanding and based on the free circulation and wider and better balanced dissemination of information and to make recommendations thereon to the General Assembly."

Since adopting Resolutions 33/115 and 34/182 (of 1978), the General Assembly has each year confirmed the importance of the effort to establish a new, more just, and more effective world information and communication order. Support for the effort was also expressed by the European Economic Community: its representative to the 1985 meeting of the Committee on Information stated that the ten members of the EEC "reaffirmed their commitment to a more just and more effective world information and communication order, which should emerge through an evolutionary process." In 1990 the resolution forwarded by the UN Committee on Information to the General Assembly shifted its language to a recognition of the call for a new world information and communication order as an evolving and continuous process.

The Right to Communicate

In 1969, Jean d'Arcy introduced the concept of a right to communicate by writing, "the time will come when the Universal Declaration of Human Rights will have to encompass a more extensive right than man's right to information.... This is the right of men to communicate."[12] Communication needs to be understood as an interactive process. The criticism was also that adopted rules focus too much on the content of the process. "It is the information itself which is protected."[13] Also: "The earlier statements of communications freedoms... implied that freedom of information was a one-way right from a higher to a lower plane."[14] There is an increasing need for participation, for "more and more people can read, write and use broadcasting equipment and can no longer, therefore, be denied access to and participation in media processes for lack of communication and handling skills."[15]

The right to communicate is perceived by the protagonists as more fundamental than the information rights as accorded by current international law. The essence of the right would be based on the observation that communication is a fundamental social process, a basic human need, and the foundation of all social organization. Since 1974 the notion has been included in UNESCO's program. At the eighteenth session of the UNESCO General Conference a resolution was adopted (Res. 4.121) that affirmed "that all individuals should have equal opportunities to participate actively in the means of communication and to benefit from such means while preserving the right to protection against their abuses."

In its final report the MacBride Commission concluded that the recognition of this new right "promises to advance the democratization of communication."[16] The commission stated, "Communication needs in a democratic society should be met by the extension of specific rights such as the right to be informed, the right to inform, the right to privacy, the right to participate in public communication—all elements of a new concept, the right to communicate. In developing what might be called a new era of social rights, . . . the implications of the right to communicate should be further explored."[17]

The commission also observed that "Freedom of speech, of the press, of information and of assembly are vital for the realization of human rights. Extension of these communication freedoms to a broader individual and collective right to communicate is an evolving principle in the democratization process."[18] According to the commission, "The concept of the "right to communicate" has yet to receive its final form and its full content . . . it is still at the stage of being thought through in all its implications and gradually enriched."[19]

The 1980 UNESCO General Conference in Belgrade confirmed the concept of a right to communicate in terms of "respect for the right of the public, of ethnic and social groups and of individuals to have access to information sources and to participate actively in the communication process" (Res. 4.19, 14[xi]).

The UNESCO General Conference in Paris of 1983 adopted a resolution on the right to communicate stating, "Recalling that the aim is not to substitute the notion of the right to communicate for any rights already recognized by the international community, but to increase their scope with regard to individuals and the groups they form, particularly in view of the new possibilities of active communication and dialogue between cultures that are opened up by advances in the media" (Res. 3.2). The twenty-third UNESCO General Conference in 1985 at Sofia requested the director general to develop activities for the realization of the right to communicate. In the early 1990s the right to communicate had practically disappeared from UNESCO's agenda. In the Medium-Term Plan for 1990–1995 it was no longer a crucial concept. The right to communicate was mentioned but not translated into operational action.

In 1992, Pekka Tarjanne, Secretary General of the International Telecommunication Union, took up the issue of the right to communicate and stated "I have suggested to my colleagues that the Universal Declaration of Human Rights should be amended to recognize the right to communicate as a fundamental human right."[20]

During the preparations for the United Nations World Summit on the Information Society (WSIS) held in 2003 in Geneva and in 2005 in Tunis, the discussion on the right to communicate was revitalized. This was particularly due to the activities of the

Communication Rights in the Information Society (CRIS) campaign during the preparatory committee meetings (in July 2002 and February 2003). It is especially significant that the UN secretary general in his public message on World Telecommunication Day (May 17, 2003) reminded the international community "that millions of people in the poorest countries are still excluded from the 'right to communicate,' increasingly seen as a fundamental human right."

Nevertheless, many individuals and institutions object to the adoption of this right as they suspect that its protagonists are intent on restricting the freedom of expression and the freedom of the press. They reject any attempt to link the standard of freedom with notions of responsibility and accountability. As such, the concept has not been fully embraced at the global level.

The debate on the right to communicate could have produced a new and productive approach to the freedom versus interference controversy. Unfortunately, so far it has mainly reproduced a fruitless debate between those who defend this addition to international law provisions and those who contest it on the grounds of its potential abuse.

Recommendations

In the near future, a satisfactory approach to global governance of content will require the establishment of an international court of human rights. For the prosecution of gross violations of human rights (crimes against humanity) the international community has managed to establish the permanent International Criminal Court. Despite all its shortcomings, this global institution is an encouraging development that should inspire the effort to found a similar supranational body to address the violations of human rights standards (civil, political, social, economic, and cultural) that the international community has formally endorsed. More than fifty years ago, the international community adopted a set of universal moral standards for the conduct of its members, and it is bizarre that the implementation of these standards has not yet been made its foremost concern.

With the growing significance of information and communication in the twenty-first century, the proposed international court of human rights may need a special chamber to address violations of communication and information rights. This chamber would receive complaints about violations from governmental and nongovernmental parties, from both institutions and individuals, and such complaints would pertain to infringements by states, by corporations, and by individuals.

Conclusion

The global governance of content represents a classical case of the need to balance fundamental standards in international law. The freedom to transport content across borders is explicitly recognized as a standard in international human rights law. However, this freedom is not provided as a peremptory norm. Interference with the freedom of content is an acceptable legal option. States have used this option whenever they defined certain types of content as illegal by law and their transport as criminally punishable. The international community has also accepted that—irrespective of national legislation—some content, such as child pornography and incitement to racial hatred, violates human dignity and should be ruled illegal. Moreover, in many countries one finds that some content is considered morally impermissible and thus "harmful." This often pertains to materials that may be harmful to minors and "sensitive people" but that are permissible for adults.

Since notions as illegal and harmful tend to have elastic interpretations, the difficult question arises of when does interference begin to unduly erode freedom? The international community has not managed to develop a satisfactory answer to this question. Striking a balance between the standard of freedom of information and the standards of responsible speech and national sovereignty has turned out to be too difficult a challenge!

Governance of content has been mainly a matter for national regulators. In many national legislations one finds rules that protect the state against so-called subversive information and that protect citizens against libel, slander, or hate speech and other forms of prohibited speech. In several countries there are also mechanisms such as professional codes of conduct and self-regulatory bodies such as press councils that address the freedom versus interference tension through forms of voluntarily adopted rules and practices.

The issue of governance of mass media content is on the national level a complex and a strongly contested domain. This becomes even more complicated when there are bilateral or multilateral dimensions at stake. The bilateral dimension arises when country A produces and exports media content to country B, which finds this content illegal or harmful or both. The multilateral dimension arises when the international community concludes that internationally available materials are illegal in terms of international law or harmful in terms of universally shared moral standards or both. It is difficult to identify universal standards to judge illegality and harmfulness. Yet, any form of global governance would have to take into account cultural differences in the assessment of what is permissible and what is not. International initiatives would also have to take into account different legal and moral standards in different countries and identify rules and practices that protect people against illegal and harmful materials without undermining the free speech standard.

The complexity of this challenge is not made easier by the ambiguous attitudes and practices of the key participants in the international debate. In the governance arena we find on the one hand national governments and on the other hand the industrial producers and distributors of content. There is obviously also the third party of a variety of small users, consumers, and citizens that—although increasingly organized in such coalitions as the CRIS campaign—have not yet been able to contribute as a major player in issues of global governance. This is partly due to the refusal of many United Nations member states to take citizens movements seriously as a negotiating partner, and partly to the heterogeneous and often conflicting interests that these movements represent.

National governments are not necessarily the most reliable partners when free speech is at stake! Interference by national governments is often motivated by very questionable purposes and often operates through limits on information freedom that are defined with considerable elasticity. The post-9/11 information restrictions in many countries are an illustrative case in point. Moreover, around the world many governments are engaged in efforts to censor contents available through the Internet (see for more detail the chapter in this book by Peng Hwa Ang).

An ironic observation in this context is that governments may worry about free flows of content and at the same time support liberalization and privatization policies that make effective regulation of media content increasingly a marginal affair. The content industries that are in favor of information freedom would appear to have more acceptable credentials than national governments. Yet their most vocal representatives are part of a global media business that conceives of freedom primarily in terms of the unhindered access to markets and consumers. This implies for example that media business strongly promotes total trade freedom for all goods and services, including those in the cultural domain. The WTO-led 'commodification of culture' has important consequences for cultural policies that are based upon policy measures such as import quotas for foreign content and subsidies for domestic cultural content. Such measures do indeed constitute trade barriers and WTO member states have agreed upon a basic commitment to remove such barriers to the global liberalization process. This process is steered by economic interests and not by cultural considerations. Any interference with content is perceived as a violation of international trade law. The representatives of the international content business argue strongly that only forms of governance that consist of modalities of self-regulation are acceptable to the marketplace. They suggest that in the end one can trust the marketplace to find the necessary balance between freedom and interference (for more detail, see chapter 6).

The problem with this proposition is, however, that reliance on the mechanisms of the free market does not imply there will be no interference. Like governments that have specific interest in interfering with the content of message flows, commercial operators also have specific interests in steering such flows. They will interfere whenever content is likely to bring them in conflict with stockholders, advertisers, audiences, or governments.

For more than a century, the politics of world communication has confronted the intellectual and political challenge of establishing a balance between 'no limits on the freedom of media content' and 'legitimate interference with such content.' Among the reasons why the international community has not made any real progress on the issue of content governance is the absence of an international institution that over the years could propose best-governance practices through its case law and the resulting jurisprudence. In the European region a unique balancing act between freedom of content and interference with content has been performed through the opinions of the European Court of Human Rights. A similar institution is urgently needed in the global arena.

This is evidently a major challenge for the international community. As societies around the world become "information societies," there are however urgent reasons to come to terms with the freedom versus interference controversy. Among those reasons are the rise of political, commercial, and cultural censorship around the globe; the increase in modes of speech that incite to discrimination, hate, and genocide; emerging threats to the freedom and secrecy of private correspondence; the proliferation of criminal and terrorist uses of new media; the withering away of public spaces for freedom of expression; and the corporate monopolization of the production and distribution of content. The WSIS offered a unique opportunity for the international community to address the need for such a global governance institution in the domain of information and communication.

The WSIS was not well equipped, however, to deal with content issues. Its orientation was largely toward infrastructural and funding issues and its coordinating UN agency—the ITU—was the international organization with adequate expertise in this field. The WSIS was strongly influenced by the technocentric discourse that is characteristic of current discussions and negotiations regarding the so-called information society.

It took considerable lobbying force from civil society participants in the Geneva meeting to give—as a minimum recognition of international human rights—the reference to article 19 of the Universal Declaration of Human Rights in the final summit declaration. Evidently, this article that provides for the freedom of content is an essential normative position that fortunately was not ignored. However, the article is part and parcel of a much larger construction that also touches on grounds of limits to this freedom, as in the case where content is used to incite to genocide. In the light of recent events around the world, the summit could at least have discussed this. And even had it decided to restrict its commitment to article 19, it could have produced some serious proposals on governance structures that could implement this commitment.

Unfortunately, since the international community chose to let the opportunity pass, other moments will have to be created to deal with the persistent problem of content governance.

Notes

1. Kubka and Nordenstreng 1986, 83.

2. Ibid., 10.

3. Ibid., 115.

4. Jowett and O'Donnell 1986, 82.

5. Ibid., 86.

6. Whitton 1979, 217–229.

7. Ibid., 223.

8. Nordenstreng 1984.

9. COPUOS 1963.

10. Ibid.

11. Matte 1982, 191.

12. D'Arcy 1969.

13. Fisher and Harms 1983, 8.

14. Ibid., 9.

15. Ibid.

16. International Commission 1980, 173.

17. Ibid., 265.

18. Ibid.

19. Ibid., 173.

20. Tarjanne 1992, 45.

References

Committee on the Peaceful Uses of Outer Space. 1963. *Fifth Session Report.* A/5549/Add.1. New York: United Nations.

D'Arcy, Jean. 1969. "Direct Broadcasting Satellites and the Right to Communicate." *EBU Review* 118: 14–18.

Fisher, Desmond, and L. S. Harms, eds. 1983. *The Right to Communicate: A New Human Right?* Dublin: Boole.

International Commission for the Study of Communication Problems. 1980. *Many Voices, One World.* Paris: UNESCO.

Jowett, Gareth, and V. O'Donnell. 1986. *Propaganda and Persuasion.* London: Sage.

Kubka, Jiri, and Kaarle Nordenstreng. 1986. *Useful Recollections. Part I and Part II.* Prague: International Organization of Journalists.

Matte, Nicolas. 1982. *Aerospace Law: Telecommunications Satellites*. London: Butterworth.

Nordenstreng, Kaarle. 1984. *The Mass Media Declaration of UNESCO*. Norwood, NJ: Ablex Publishing.

Tarjanne, Pekka. 1992. "Telecom: Bridge to the 21st century." *Transnational Data and Communications Report* 15, no. 4: 42–45.

Whitton, John B. 1979. "Hostile International Propaganda and International Law." In *National Sovereignty and International Communication*, ed. K. Nordenstreng and H. I. Schiller, 217–229. Norwood, NJ: Ablex Publishing.

8 International Regulation of Internet Content: Possibilities and Limits

Peng Hwa Ang

The most striking feature of the Internet must be the seemingly easy access to free-wheeling content. All kinds of content can be found online—from the strange and wonderful to the criminal and dreadful. To the lay user clicking away on the mouse, the Internet does appear to be beyond the reach of government. Without a central governing authority and with an architecture designed to route around blockages, there seems to be little that governments can do to police the Internet. The operative word, however, is "seems." For since at least the mid-1990s, governments the world over have attempted to regulate Internet content with varying degrees of success.

There was little need to police content when the Internet began in universities and research institutions; the community of academics and their respective institutions could do it themselves. The potential for damage to one's reputation through posting something unseemly online was a check on the community. But when the Internet diffused to the public, such checks on self-restraint diminished to the vanishing point.

This chapter traces some of these efforts and outlines the likely outcome of international attempts to regulate Internet content. The chapter defines Internet content regulations broadly: at the core are regulations that directly affect free expression; at the periphery, privacy and copyright are included in the discussion as regulation of these areas impinges on content.

The chapter argues first that the Internet will be increasingly regulated, though not necessarily in a manner that will be censorial. Much of the impetus for regulation in fact is coming from the more developed countries; it is only natural that less developed countries follow apace. Second, much of the process of developing and enforcing regulations will and should involve national governments. The days of not respecting kings and governments are over in part because of the difficulty in excluding them from having any regulatory role whatsoever, and in part because of the failure of the Internet community to self-regulate. Third, because laws promulgated by national governments bear their respective cultural imprint, it follows that Internet regulations would also bear the quirks of the governments and countries. Here, it is not just a case of China attempting to censor the Internet the way that it censors newspapers, but also the case of the United States attempting to prevent its citizens from indulging in online gambling.

National Regulatory Responses

Early Days

Since the diffusion of the Internet to the public, there have been three overlapping and parallel "waves" of attempts to regulate its content. The waves reflect the attempts of national governments and international agencies to wrestle with the Internet. In the first wave, roughly from 1994 to 1997, which were the early days of the mass-access Internet, national governments were the first to step into the regulatory void. Often, the approach was to try to fit the Internet into the same mold as traditional mass media. That is, governments simply used legislation and existent executive powers and resources that most closely resembled that part of the Internet to regulate it. If that part of the Internet functioned like a newspaper, then newspaper laws applied; if another part functioned as a broadcasting station, then broadcast rules applied. This "functional approach" to regulating the Internet[1] is understandable because lawyers and judges are trained to use precedents and analogies in their arguments.

As an example, U.S. Judge Stewart Dalzell in his opinion striking down the Communications Decency Act and declaring the "democratizing" effects of Internet communication compared messages posted on the Internet to the pamphlets of yore: "Federalists and Anti-Federalists may debate the structure of their government nightly, but these debates occur in newsgroups or chat rooms rather than in pamphlets. Modern-day Luthers still post their theses, but to electronic bulletin boards rather than the door of the Wittenberg Schlosskirche."[2]

One advantage of using precedents is that the law will then tend to change incrementally, making it easier for citizenry to adjust. With a new medium such as the Internet, however, there are limits to the application of precedents and analogies to resolve legal puzzles. Many attempts to draw similarities between the Internet and traditional mass media will therefore be limited, if not forced. The legal landscape of the Internet today is littered with laws purporting to regulate it that have either been struck down or simply abandoned in practice through lack of enforcement. Examples abound from both the East and the West.

Asian governments seem to have taken a more cautious approach before making the Internet publicly available. They seem to have considered the potential impact of the Internet on the traditional media. For example, Singapore was not the first country in Southeast Asia to allow public Internet access. Its relatively less technologically savvy neighbor, Malaysia, was the first country in the region to do so. China too appears to have proceeded cautiously, deliberately making the Internet available only in universities first (when many countries by then had made it publicly available) before gradually allowing its diffusion to the wider public. Vietnam at first declared that it would not allow the Internet into the country in April 1996 before changing its mind.[3] Not surprisingly, in each of these countries, Internet regulations were passed soon after public access was possible.

It was Singapore that promulgated the world's first code of practice for web site owners in 1996. It drew worldwide attention because it was the first Internet-specific legislation aimed at regulating content. The code licensed web site owners, in line with Singapore rules that require licensing of all who publish any periodicals. The license would be revoked if the terms for them were breached. In a nod that recognized the uniqueness of the Internet, licenses were automatically issued—that is, one was given a license when one operated a web site.[4] Why was this fiction of an automatic "class license" created when it is well-nigh possible simply to have a law that criminalized the breach of the rules in the code? Well, functionally, the Singapore regulators were treating the Internet as a traditional mass medium: anyone who ran a mass medium had to have a license. However, for more than six years after the passage of the code and even at the height of the local dot-com boom, not one Singaporean has been found to have breached it.

In South Korea, the offline rules pertaining to national security were imported to apply to the Internet.[5] Under the National Security Law, any publication that praised North Korea would be banned. So when one page of the massive Geocities web site praised the North Korean leader Kim Il Sung, South Korea blocked access to the entire Geocities site.[6] After the hailstorm of international criticism that followed that incident, South Korea removed the block and has not attempted a similar move.

In contrast to the more interventionist and cautious approach of several Asian countries, Western countries in general began with a more hands-off approach in the adoption of the Internet. But as the type of content available online became better known, there was pressure on politicians to intervene and "do something." In 1996, the French passed a telecommunications law that empowered the Conseil Supérieur de la Télématique to rule on what types of content were permissible—only to have its Constitutional Council reject its provisions.[7] Across the Atlantic, in 1997, the U.S. Supreme Court struck down the most significant part of the Communications Decency Act dealing with content that was deemed "indecent," "lewd," or "harmful to minors."[8] The Communications Decency Act does seem to have been passed out of apprehension and confusion rather than malice.[9]

In traditional content regulation, a distinction is drawn between negative regulation, which discourages certain types of content, and positive regulation, which encourages certain types of content. There was an initial attempt at positive content regulation by Canada. It explored the possibility of applying to the Internet a quota system that had successfully increased the amount of Canadian-made content on television that was available on the Internet.[10] In this case, the aim was to increase French content on the Internet using the television content rules as a model. The attempt, however, went nowhere. No matter how morally—or socially—uplifting the content, it was just not possible to legislate it on the Internet.

In short, between 1995 and 1997, it looked as if the Internet could never be regulated. Countries that were trying to regulate it were either failing before the courts or failing in practice. Laws in the United States and France had been struck down. Singapore was giving regulation a symbolic shot; South Korea ended up backpedaling on its attempt to block access.

In the conceptual space, scholars such as David Johnson and David Post argued that the Internet could not be governed by existing legal regimes because it crossed borders and that therefore the Internet needed a new legal regime.[11] The U.S. government under President Bill Clinton declared a hands-off approach to Internet communication and e-commerce because of its "decentralized nature and . . . tradition of bottom-up governance."[12] John Perry Barlow's famous "Cyberspace Independence Declaration" denying sovereignty of cyberspace to governments, made at a World Economic Forum meeting no less, may be seen as the high-water mark of this belief.[13]

Second Wave

While governments were stumbling in their attempts to legislate, law enforcement agencies, particularly those from the West, were certainly not adopting the hands-off approach. This second wave of regulation, roughly commencing from 1995, started with law enforcement, which began to police the Internet and with some success. It should be noted that law enforcement for the most part is territory-based. That is, with some exceptions, such as airplane hijacking or the doctrine of hot pursuit, one enforces the law only in one's geographic borders. With the Internet, however, enforcement agencies recognized the futility of attempting that and instead began cooperating internationally. Such international action requires countries first to have similar laws that recognize an act to be criminal and then to cooperate to stop the offenders. The two main areas where international sweeps have been used have been child pornography and consumer fraud.[14]

The first international Internet sweep was conducted by United Kingdom police in July 1995 in Operation Starburst. That resulted in thirty-seven arrests across a few countries.[15] In March 1996, Operation Ripcord run by the New York Attorney General's Office resulted in seventy-five search warrants issued for New York State alone and more than 220 referrals internationally.[16] Then in 1999, twelve countries in the Organization for Economic Cooperation and Development (OECD) banded together in Operation Cathedral to scour the Internet for child pornography sites. The sweep resulted in 180 arrests and the breakup of several child-porn rings.[17] Since then, there have been other similar sweeps. In 1999 in Operation Avalanche, several U.S. agencies cooperated to shut down a site that had a list of more than 250,000 subscribers.[18] Of that list, more than 7,000 names were handed over to UK police.[19] Today, police in several countries regularly patrol the Net for child pornography.

The international sweep against consumer fraud on the Internet was started by Australia in 1997.[20] No new Internet laws were passed for the sweep. Instead, the Australian Fair Trading Office relied on existent consumer protection laws. Since 2002, Australia coordinates the annual sweep with some thirty countries, members of the International Marketing Supervision Network, an association of trade practices law enforcement authorities that are mostly from the West.[21,22]

That such action is possible indicates that given the political will, the "borderless" nature of the Internet is no barrier to law enforcement. It would, however, require a convergence of law, coordination, and cooperation as well as sufficient technical know-how. It is conceivable that sweeps could be used to guard against the use of the Internet for terrorist networks, drug smuggling, and other acts that a sizeable group of countries deem to be criminal.

Self-Regulatory Responses

The limits of international cooperation, however, are all too obvious.[23] Such coordinated actions demand so many resources that they require the heavy involvement of governments and even then are suitable only for extremely socially harmful acts. On the Internet, as with life, there are irritations and annoyances for which such crackdowns would be deemed overkill. For example, mild intrusions of privacy and the odd pornographic spam would be considered irritations and annoyances by many Internet users. In any event, the Internet community avoids government intervention as far as possible, adopting instead the stance that self-regulation of the Internet is the best if not the only way to go. Industry clearly hopes to keep legislation at bay while attempting to regulate itself with a light hand. If there is a time that marked the shift toward self-regulation, it would probably be the late 1990s. This move is a significant departure from the original no-laws-are-applicable paradigm, as the Internet community began to realize that the position was not tenable.

Conceptually, there are factors that work in favor of self-regulation of the Internet. First, for a fast-moving industry such as the Internet, fewer formal processes of self-regulation make it more flexible in adapting rules and therefore less likely to stifle innovation or excessively limit consumer choice. Second, in so new a business realm, it is industry that has access to the best information to ensure the efficacy of potential courses of action. For all their good intentions, lawmakers and judges have made rules that undermine the workings of the Internet. For example, when the Bavarian judge stopped CompuServe from serving up a newsgroup that he considered pornographic, he blithely ignored the impact this had on the operations of CompuServe in the rest of Europe.[24] Third, because industry bears the cost of regulation, it has incentives to keep enforcement and compliance costs down.

Intuitively at least, given the difficulty though not impossibility of regulating and policing the Internet, self-regulation is an attractive alternative. It should be noted at this juncture that self-regulation in this context is a technical term with a specific meaning that has been misunderstood and misused. In the medical arena, self-regulation is often used in conjunction with biofeedback. On the Internet, posts to online discussion groups suggest that some users, and even industry players, erroneously believe self-regulation to mean the person ("self") controlling ("regulating") himself or herself. So when a parent installs filtering software on the home computer to screen out undesirable sites from their child's use, that is sometimes cited as a successful model of self-regulation. This confuses self-help with self-regulation. For any regulation to be effective there must be some form of sanction. However, it is hard to imagine individuals or entities punishing themselves for violating their own rules.

The traditional meaning of self-regulation is industry regulating industry. In fact, this meaning was so well understood that a 1993 PhD thesis at Oxford University titled "Theories of Self-Regulation" offered no definition of the word.[25] Self-regulation in this technical sense works best when the following elements are present: a motivated and competitive industry, a small number of large players, maturity in the market, clearly defined problems that do not pose large social risks to consumers, an active and cohesive industry association that covers much of the industry, and a government regulatory backstop.[26] The Internet industry as it stands does not meet most of the criteria, the most significant of which is that the industry has not sufficiently matured.

Self-regulation is most successful when there is government legislation to back it up. In many developed countries, a significant amount of advertisements are self-regulated. That is, the industries involved set the rules, police them, and sanction the violators. The enforcement mechanism is the media companies themselves: they will not carry advertisements that violate the advertising code. Pure industry self-regulation, where only the industry sanctions the offender, may be effective, but only up to a point. Not all offending advertisements are caught: owners who distribute flyers to advertise their businesses would not be caught by the self-regulatory rules because they are both the advertiser as well as the media. Recalcitrant offenders, for example a media company that publishes violative advertisements, cannot be sanctioned through such a self-regulatory process; government legislation is necessary in such extreme cases. However, once government is involved, the Internet community considers this to be no longer self-regulation but rather the industry regulating on behalf of the government. And so what used to be called self-regulation is now called co-regulation. The distinction is probably necessary given the clearer meaning and the avoidance of misunderstanding. In fact, the two countries leading in the area of self-regulation—the United Kingdom and Australia—have drawn the distinction as well.

The role of the government in self- or co-regulation cannot be underestimated. Industry and lobby groups in the United States are "allergic" to self-regulation apparently because there has been a history in which, when self-regulatory rules work, the agencies in question "harden" the rules so they become law.[27] Hence while industry talks of self-regulation on the one hand, there is also wariness. On the other hand, when self-regulation does not work, the government steps in anyway. Therefore, the U.S. Federal Trade Commission, after extending several one-year deadlines for the industry to regulate the privacy of children using the Internet, finally passed the Children Online Privacy Protection Act in 2000.[28] A different approach has been taken in Australia where the Internet Industry Association emerged with a code of practice after five years and only when the federal government passed a law that, among other things, *compelled* the development of the code.[29] Europe has expressed greater interest in self-regulation. There, Internet service providers' associations have often developed codes to minimize their own liability as well as to self-regulate some aspects of business. Painting with a broad stroke, it is probably accurate to say that Asia lacks cohesive industry associations and has tended to go with law. Given the nature of the Internet, the move towards increasing use of self- and co-regulation is inevitable.

However, it should be noted that there are also significant downsides to self-regulation. First, regulation is perceived as never going against the interests of the self. The perception may not be borne out in practice, as this author has personally witnessed in self-regulation of advertising in Singapore, where business people have voted against their interests. In any event, a businessperson who always votes for the business interest will suffer the loss of credibility. Nevertheless, the perception is difficult to shake and undermines confidence in the self-regulatory regime.

But probably the most significant weakness in self-regulation is the lack of incentives to control and enforce standards. Self-regulation works in the medical and legal professions because their members are entering into a high-income homogenous group through compulsory membership.[30] Self-regulation seems to work well as long as the group of agents exerting the power is relatively small and cohesive. Various studies for different reasons come to the same conclusion: self-regulation is more difficult and less effective when it involves a large and heterogeneous group of agents.[31] This heterogeneity may be the critical reason that makes self-regulation of Internet content difficult if not impossible. The easiest solution to heterogeneity is to set the lowest standard. But that undermines the confidence of users in the self-regulatory regime. All things considered, therefore, the conditions for self-regulation of the Internet are absent.

On the Internet, where industry has come together to attempt self-regulation, the results have been unimpressive. Privacy seals and content labeling have not worked well.[32] In the area of online privacy, TRUSTe cofounder Esther Dyson said she was "disappointed in what ended up becoming of it."[33] Both types of seals have not been able to develop the critical mass needed to be useful to both consumers and site-owners.

The lack of success is not for the lack of intellectual inputs or funding. Internet self-regulation in Europe has been aided by the work of the Bertelsmann Foundation.[34] From 1999 to 2000, the foundation, funded from dividends by one of the world's largest book publishing houses, brought together a group of experts to address the issue of content filtering. The result was the creation in 2000 of the Internet Content Rating Association (ICRA) filtering platform based on PICS (Platform for Internet Content Selection).

ICRA aims to empower parents with simple tools to filter contents of web sites while addressing the concerns of censorship.[35] American civil libertarian groups such as the ACLU (American Civil Liberties Union) and the CDT (Center for Democracy and Technology) were among various bodies consulted during the development of the platform. However, more than a year after the official launch of the complete filtering tool in early 2002, fewer than 250,000 web sites had been self-labeled. The ICRA case is elaborated on later in this chapter; for now it suffices to say that one major factor for the low take-up was the decision by the Internet browsers not to incorporate more sophisticated filtering tools. While they were engaged in the browser war to outdo each other, both Microsoft and Netscape added features, among which was a filter based on a simple four-checkbox system created by the U.S.-based RSACi (Recreational Software Advisory Council for the Internet) that filtered sites for language, nudity, sex, and violence. By the time ICRA finally emerged with a more sophisticated system that had sixty boxes for the webmaster to check, the browser war was over. Neither Microsoft nor Netscape was keen to enhance its browser. That meant that the ease of use and the full power of the sophisticated ICRA system could not be realized.

Self-regulation is also failing to work in the countries that have the most extensive experience with it, so Australia and the United Kingdom eventually went for more direct legislation. In Australia, self-regulation of businesses in the area of consumer protection has been found wanting. In many instances, the rules in the self-regulatory regime have become law. The push toward self- and co-regulation should therefore be seen as an experiment. Down the road, it is likely that the regulation will be in the form of legislation.

Proposal for Intergovernmental Coordination

Internationally Agreed Offenses

What will the shape of regulation be for content? Much will depend on the type of content. Where there is international agreement it will be possible, indeed, arguably necessary, to pass legislation, which will then be backed up by law enforcement internationally. Examples of such areas of international agreement are consumer fraud and child pornography. The harm in those instances is evident. Terrorism, especially the use of violence on civilians, may be another area where governments will cooperate.

Indeed, at the time of writing in 2003, the U.S. FBI's Internet Fraud Complaint Center[36] had a link to encourage reports on terrorism.[37]

The European Union is worth monitoring for the types of online content and use of the Internet that governments agree would be offensive, if not illegal. In general, many Western European countries recognize some degree of censorship of some media content. Such censorship is often to protect the young (which many countries, including the United States, can agree on) as well as to acknowledge sensitivities such as anti-Semitic expression. The European Union position more closely resembles the situation in many parts of the world where censorship is present, varying only in degree. And it should be noted that the many European countries also protect free expression under Article 10 of the Council of Europe's Convention for the Protection of Human Rights and Fundamental Freedoms.[38] This protection is similar to but not identical to the U.S. Constitution's First Amendment. In short, Europe has checks and balances on the competing interests of censorship and free expression. This means that if the European Union can develop regulations pertaining to Internet content, because the Union would have considered its members' various competing interests, it is likely that such rules would be plausible models for other countries too.

National Offenses

In many areas, international cooperation will be difficult if not impossible to achieve. Hate speech is a good example. While practically all countries may agree that such expression is offensive, there is wide disagreement on the substance of what is offensive. The same statement uttered in two different countries can have divergent responses. However sympathetic one may be to the feelings of the nationals of one country, that same intensity of emotion is likely to be absent in other countries. The same issue applies to acts such as gambling, defamation, and political expression. What may be legal in one country may not be legal in another. This is the problem posed by the borderlessness of the Internet: it is difficult to control content coming from another country.

The problem, however, has not prevented governments from attempting to assert their sovereignty. Governments the world over want to assert their jurisdiction and sovereignty where they can do so. Few governments will simply stand idly by and allow their laws to be circumvented through a new medium. A basic rule that governments start with is that offline rules should apply online. This makes sense to the extent that the rules can be enforced. This begs the question: can governments act against a foreign-based server?

The first country to act against an Internet company with its servers located outside the country was the United States. In February 2000, the New York Supreme Court convicted online casino operator Jay Cohen under the Wire Wager Act for accepting bets on sporting events through his World of Sports online gambling operation even though his servers were located in Antigua.[39] Until then, it was thought that there was no way that governments could stop online gambling. In fact, a spokesman for the U.S. Justice Department, John Russell, was quoted in 1998 as saying that the U.S. government had neither the jurisdiction nor the resources to police international Internet gambling. Russell said: "International Internet gambling? We can't do anything about it. That's the bottom line."[40] The World of Sports case is instructive because many countries consider gambling to be the activity of a consenting adult. Going against that grain, in October 2002, the U.S. Congress proposed to outlaw online gambling as well as the processing of gambling transactions.[41] In the continuing saga, in 2003 both Antigua and Barbados complained to the World Trade Organization that the American regulations against online gambling violated international trade agreements.[42]

Enforcement of national laws on the Internet, however, will always pose a problem. In the World of Sports case, several of Cohen's friends are fugitives. This begs the question: given the difficulty in policing the Internet, to what degree is enforcement possible? A frequent comment about enforcement on the Internet is that because there are loopholes (that is, it cannot be 100 percent effective), the law should not be enforced. That is not the way of law enforcement in practice: there are unsolved murders all over the world all the time. On the Internet, the *France v Yahoo!* case is a striking example of both the capabilities and the limits of enforcement. In 2000, a French court ordered the Internet search engine Yahoo! to block French users from accessing a section of the site that auctioned Nazi memorabilia.[43] Although Yahoo! initially objected, it later banned the sale of Nazi memorabilia and other hate-related material on its site.[44] To be sure, there are peculiarities in the case. First, Yahoo! had used a French-controlled domain name—yahoo.fr. That gave the French legal system a jurisdictional toehold on Yahoo! Second, Yahoo! was targeting the French users through its advertising. On this point, the judge therefore found it unconvincing when Yahoo! said it had no way to filter out users; if French users could be directed to a French version of Yahoo!, surely Yahoo! could similarly filter its users to deny them access to Nazi memorabilia. The panel of experts formed to advise the court on the technical feasibility of blocking French users concluded that filtering would be 70 percent effective. The court accepted that and ordered Yahoo! to deny French users access to Nazi memorabilia on its site.[45] The case illustrates the point that governments are prepared to live within the practical limits of law enforcement on the Internet.

Many decry the French approach. Lawrence Lessig, apparently overlooking the 2000 World of Sports case, was quoted as saying that the *France v Yahoo!* case was the clearest example to date of a free and sovereign nation setting rules for what material can be accessed from its soil.[46] The United States differentiates between speech (of which it is more tolerant) and action (of which it is less so). And so within its own legal framework, the decision to close down an offshore online gambling site can be justified—the U.S. government is acting against commercial transactions; in contrast, the French court by going after Yahoo! was going after content, which receives much greater protection in the United States. However, not all countries agree with that divide. And so the U.S. action in the World of Sports case in fact lends legitimacy to governments that want to regulate content.

There are other reasons for countries to attempt to regulate the Internet to suit their preferences. Manufacturers of physical goods do sometimes make them available in one country but not in another. Product labels do differ between markets. If cans of drinks can be labeled differently in different languages, it is hard to see why, at least in principle, Internet content should be different. Content, after all, is most powerful when it is in the local dialect. If goods manufacturers can comply with local language requirements, it is hard to see why Internet content providers should enjoy special exemptions.

To be sure, courts are more aware of the implications of their rulings. When the Australian court in the defamation case of *Gutnick v Dow Jones* held that a publisher in the United States can be sued in Australia, the judgment was based on narrow grounds, in recognition of the peculiar nature of the Internet.[47] The judge in the case opined that the suit was to claim damages only in Melbourne, where the plaintiff resided. In other words, he was not unreasonably holding Dow Jones for damage to Gutnick's reputation around the world even though the Internet had that reach.

Legislation

Given the more limited enforcement capability on the Internet, international legislation about content has to be restricted to what is both internationally acceptable as well as technically feasible. One such area that is ripe for regulation is spam, which is really a regulation on the manner of delivery of content rather than of content itself. Many countries, especially in the more developed world, have laws against unsolicited fax advertisements. It is only logical that such laws be extended to the Internet. Spam laws, however, have the effect of driving spammers to the countries that do not prohibit spam. Therefore, once a significant group of countries begins to pass anti-spam laws, it is only to be expected that other countries will follow suit. The EU, as well as many European countries, also has laws against spam.[48] In December 2003, the U.S. CAN-SPAM Act was signed into law, and many other countries are adopting similar laws. Enforcement of these laws is possible when there is an offending party that can be traced.

Self-Regulatory Attempts in Privacy and Seals

Overall, there is likely to be less self-regulation and more legislation. There have been several high-profile failures of self-regulation in the United States. The Direct Marketing Association in the United States failed to come up with a code to protect children's privacy after several deadline extensions by the U.S. Federal Trade Commission.

TRUSTe had gross violations of privacy by two of its members—Real Audio and Microsoft—but because the violations were technically not within the terms of the agreement, no sanctions were administered.[49] When this author e-mailed TRUSTe back in 2000 to ask whether they had sanctioned anyone for violating their code, they did not reply. Since then, in June and July 2001 TRUSTe has expelled two enterprises for breaching the privacy code.[50] Until then, companies were removed from the TRUSTe program for expiry of agreements rather than for privacy violations. Nevertheless, the publicity surrounding the two offending enterprises has been low-key. And the page containing the advisories appears not to have been updated since July 2001 when this author checked in December 2003, or else there have been no other offenders since.[51] The mild reaction of TRUSTe is understandable: set too robust a level of sanction and potential members are scared off; set too low a level and consumers are not impressed and potential members lose interest in joining. To solve this dilemma, it is therefore understandable if legislators opt for a minimum standard for the industry.

Content Labeling

A discussion on self-regulation would not be complete without looking at the most earnest attempt by the Internet Content Rating Association, which was mentioned earlier in this chapter.[52] ICRA deserves attention here because it is a truly international attempt by an independent nonprofit organization to address the concerns of those who see the need to filter content while keeping in mind free speech concerns. Its mission is "to protect children from potentially harmful material and to protect free speech on the Internet."[53] ICRA can trace its origins to 1999, when the European Union started a project called INCORE (Internet Content Rating for Europe) to develop a filtering system for Europe. Of course it made no sense to be Eurocentric on the Web. So because the European Union was limited by its geographic mandate, the Bertelsmann Foundation stepped in to broaden the project to cover the rest of the globe. The Bertelsmann company publishes more than 60 percent of the paperbacks sold in the United States. It is a privately held company with a sizeable block of shares held by the foundation. Those shares earn the same dividends as ordinary shares with the difference that they do not command any votes. The foundation is therefore self-sustaining and carries out its own projects. In the best spirit of German craftsmanship, the Bertelsmann Foundation set about ensuring that the final result would be above reproach. Among its advisors and on the advisory council from its formation in 1999 was Professor Jack Balkin of Yale, who had taught a number of the lawyers working in the nonprofit organizations CDT (the Center for Democracy and Technology) and ACLU, which were lobbying against regulations that would impinge on free speech. Some of his students would later challenge him in a public forum. From Europe, the foundation had Stefaan Verhulst, then of Oxford University's Program in Comparative Media Law and Policy; Herbert Burkert, who has chaired the European Union's Legal Advisory

Board; and Ulrich Sieber, a law professor who had been on the team that defended Felix Somm, manager of CompuServe, when the company was accused of carrying pornography.

Others in the group were academics and researchers in content regulation (including this author from 1999 to 2004); also invited to the meetings were some civil liberty groups such as the CDT and the ACLU with then President Nadine Strossen attending some of the meetings. In 1999, the sophisticated content labeling system was presented at a conference in Munich and the nonprofit ICRA was formed.

ICRA uses PICS (Platform for Internet Content Selection), a protocol developed by the World Wide Web Consortium (W3C), to label the contents of a web page and the labels are then read by a filtering program. More accurately, ICRA should be called a "labeling" organization. The word "rating" comes from the idea of a movie rating, which is, also more accurately, a movie label. The ICRA system is an improvement over the system from the defunct RSACi (Recreational Software Advisory Council for the Internet). ICRA was more detailed with multiple descriptions about content instead of the limited four categories under RSACi; it was context-sensitive, with exemptions for news, for example. It also aimed to address civil liberty concerns, although after Munich, the CDT issued a report that was critical of the system. Perhaps most importantly, it was an international effort. After Munich, the ICRA board and council further developed the system by creating a downloadable filter as well as a complaints and sanctions mechanism.[54] At the ideal deployment stage, users select through a maximum of two clicks the standard of filtering that they want applied to their computer to block content deemed undesirable for their children. To move this system into widespread use, ICRA had to overcome two hurdles: getting the filtering program incorporated into the browser and getting web sites to self-label.

Of the two hurdles, the unexpectedly difficult one was to get the filtering program installed on the browser. While they were running neck-and-neck in the browser war, both Netscape and Microsoft packed their programs with filtering features. But soon after Microsoft emerged as the dominant browser, and with the dot-com downturn, both companies scaled back the addition of features. The result is that Netscape's Navigator and Window's Internet Explorer can read the old labeling codes but not the new and improved codes that ICRA had implemented. ICRA therefore had to develop a plug-in filtering software that, unfortunately, was not bug-free. Developing the plug-in also cost it precious funds and slowed down its development momentum.

As for getting web sites to label, it is an interesting case study on the impact of civil liberty groups like the ACLU and CDT. In the first instance of the *ACLU v Reno* case, in which a part of the Communication Decency Act was struck down, civil liberty groups took the position that filters were the less restrictive method of screening content. But after that section was struck down, the ACLU came up with their report "Fahrenheit 451.2: Is Cyberspace Burning?" that denounced filters, as well as labeling, as intrusions on free speech. The point of the ACLU report is best summarized in its single footnote: "While PICS could be put to legitimate use with adequate free speech safeguards, there is a very real fear that governments, especially authoritarian governments, *will* use the technology to impose severe content controls (emphasis added)."[55] According to the executive director of ICRA, after that report was presented to the press, the coverage turned negative towards the idea of labeling and filtering. Until then, the publicity had generally been positive.[56] The ACLU's concern about authoritarian governments using labels to censor the Internet has proven unfounded. Authoritarian governments distrust their own citizenry—that is why they are authoritarian. To allow web sites to self-label runs against that fundamental attitude; self-labeling, as opposed to compulsory labeling, would be trusting and empowering web site owners. Further, the ICRA system is aimed at labeling content considered objectionable to children. As it stands, it does not work on the kind of objectionable content authoritarian governments are most worried about—political discussions. The South Korean government did for a time consider compelling all Korean web sites to label using the ICRA system. If it had insisted on doing so, it would have had to create its own labeling scheme: the terms of use of ICRA prohibit compulsory labeling.

A concern raised by many is the possibility of mislabeling. This concern has turned out to be unfounded. Perhaps because the system has been well thought through, there have been few complaints about mislabeling as far as the author knows. Instead, the biggest source of complaint has been from parents who call to point out that an ICRA-labeled site is pornographic. This is a misunderstanding of the ICRA system. The site only labels. It is up to the individual parent to determine what should be filtered. Toward the end of 2003, ICRA embarked on a fundraising project for an education campaign to spur user demand and thereby spur webmasters to label.

One year after the official launch of the ICRA filter software in March 2002, the number of sites rated hovered around 250,000. Among these were high-traffic sites such as Yahoo!, MSN, and the German ISP (Internet service provider) T-Online.[57] Pornographic and online-gambling sites, wary of government regulation, have turned to ICRA. Nevertheless, the number of sites labeled is too small for a critical mass. ICRA may linger in that awkward limbo between being too prominent to fail and too small to succeed.

However, even if ICRA as an Internet-content-labeling venture fails, there have been positive spinoffs. One of them is that Internet service provider associations in Europe have formed a network, the Association of Internet Hotline Providers in Europe (INHOPE), to alert each other to illegal content on web sites in each other's countries. INHOPE can be a model for how content that is illegal, as opposed to merely objectionable, can be policed across borders. Perhaps the most significant achievement of ICRA is that it demonstrated the possibility of developing a system to label content that would address most of the concerns of civil libertarians. It is, however, an expensive venture. Further, users need to be informed and educated about the way ICRA works. The main dampener for ICRA is the lack of sufficient demand by web site owners and users for such a labeling system.

Self-Help

To round off the discussion on self-regulation, it should be noted that self-help as a remedy has been used for such areas as defamation and copyright. Self-help is not self-regulation because the principle here is that of an entity enforcing its own rights. The most significant example of this is the International Olympic Committee (IOC), which hired a company to patrol the Internet for webcasts during the Sydney Olympics of 2000 and Winter Olympics of 2002 because such webcasting would violate its exclusive license with the contracted television broadcaster. The patrols showed how it is possible to trawl the Internet through a combination of technical as well as manual tools. The IOC looks set to continue the practice, at least until 2008 when the agreement expires.[58]

Policy Models

Types of Content

The European Union probably offers the best model of policy formation for the Internet for reasons discussed earlier in this chapter. It is a grouping of countries that are sufficiently proximate in geography with cultural overlaps. From admittedly modest experience working with EU officials and observing both the process and the results, at least when it comes to Internet issues, the author has observed that the EU discusses matters in a reasonably open and transparent manner with inputs from stakeholders, NGOs, and academics. Politics of the petty and not-so-petty kind of course do intrude; this is only to be expected in any international process. And it should be noted that the EU changes its presidency every six months. This means that agreements hammered out over a few years would have to be robust. A regional grouping of culturally overlapping countries makes a lot of sense because it is easier to reach agreements and to enforce them. With clear champions, it may be possible for several regional agreements to agglomerate to arrive at some international agreements.

Probably the one clear area of international agreement concerns child pornography. Practically all countries outlaw child pornography, with Japan as the last significant holdout, criminalizing child pornography in May 1999.[59] In this arena there are champions. The nongovernmental organization ECPAT (End Child Prostitution, Child Pornography and Trafficking of Children for Sexual Purposes) has done much work to combat child pornography and child prostitution, and has held conferences, conducted studies, mounted educational campaigns, and developed programs to share best practices.

At the national level, content created in a country will have to comply with the laws of the land. It is only too easy for law enforcement to crack down in such a case. However, there is a tendency to use the ISPs as the control point. The United Kingdom has compelled ISPs to retain data for up to two years under the Regulation of Investigatory Powers Act, which was passed in 2000 to empower the police to intercept communications over the Internet. The European Parliament has passed a directive requiring ISPs to retain data for a period of between six months and two years.[60] Australia and New Zealand have passed similar laws that allow the police to intercept telecommunications, including Internet traffic.[61,62] The United States is planning to require ISPs to set up a centralized system to monitor the Internet.[63] ISPs are understandably unwilling to have the burden imposed on them. UK ISPs estimate that the cost of such data retention will amount to more than £100 million (U.S. $165 million) a year. They have instead recommended the United States approach toward data preservation, whereby ISPs keep the data of those named by government agencies.[64] Whichever approach is used, ISPs have implicitly accepted that they are checkpoints in the information highway.

Recommendations

Like several of the recommendations in the other chapters of this volume, those presented here are given to no single entity but to the Internet community at large, to national governments, and perhaps especially, to well-endowed foundations that might be able to act on some of the recommendations.

Accept that the Internet Will Become More Regulated

As the Internet moves from being an innovation to becoming more and more a part of modern life, it will inevitably attract more regulation.[65] This is a natural response to the greater importance of the Internet in the social, political, and economic life of modern society. A study by Deborah Spar of how rules eventually form to tame disruptive technologies is also instructive. She argues persuasively that rules may disappear during a period of disruptive technology but that eventually, it is in the interests of all the players to have rules.[66] In short, there is utility in rules. The mantra that "the Internet should not be regulated" has been misleading and arguably even damaging to less developed countries, some of whom have misunderstood the phrase to mean that the Internet should be free of all regulations. In fact, the United States, the country where the Internet was invented, has from early on had rules governing various aspects of the Internet except that such rules were not all called Internet regulations. Often they were extensions from the telecommunications arena, such as the Wire Wager Act previously mentioned. At the very minimum, developing countries need some "enabling" Internet regulations that would, for example, allow courts to admit electronic evidence in order to enforce contracts agreed upon through e-mail. Developing countries have also been forced to consider regulations that would punish abusers of computer and Internet systems, such as those who maliciously hack into computer systems or release damaging viruses. The Philippines, for example, could not punish the youth who released the "Love Bug" virus because it had no law against it.[67]

Educate Judges and Legislators to Ensure Regulations Are "Sensible" and "Rational"

There is of course the danger that the rules promulgated would be the kind that control information rather than empower users. Such an approach towards control would be detrimental first to users within the country and to a lesser extent to the wider Internet community. The antidote to such ill-advised moves is good advice. Attempts to hack or to offer bypass solutions[68] are counterproductive in the long run. It is far better to get governments to open up on their own volition. Along these lines, to ensure that the regulations promulgated are sensible and well-adapted to the Internet, many of the legislators and judges will need to be educated about the Internet. The Asia-Pacific Development Information Programme, under the United Nations Development Programme, regularly conducts seminars for senior government officials from the developing countries in Asia on the technical as well as legal and policy aspects of the Internet. Many of those attending, however, are so senior in government that they will most likely set broad directions rather than be involved in the nuts and bolts of law and policy formation and implementation. More officials, especially those directly involved in rule making, should be educated. Judges should also be included in the education. The German *CompuServe* case shows that what may be possible—blocking nationals from accessing certain material—can have an impact beyond the national borders.

Civil libertarians and others who are concerned about regulations per se should not be unduly alarmed. Because of the international nature of the Internet, regulations that are promulgated must be "rational" (a relative term) and workable when the Internet crosses borders. To the extent that the functions and impact of the Internet can be confined to a country's borders, it may be possible for a country to enforce rules that the larger Internet community would consider nonrational. But outside of its borders, a government that tries to regulate the Internet in a manner inconsistent with the values and technical workings of the larger community will not be using the Internet to its optimum potential and instead will find itself cut off. A study by the Carnegie Endowment for International Peace on how the Internet affects authoritarian countries such as China and Cuba shows that while the Internet is not "inherently democratic," it has elements that make for a more open government. The study points out, for example, that China is putting government services online.[69] While that in itself is not necessarily politically liberating, it does make for a more open government. In the longer term, therefore, governments are likely to make rules that are more rational in this sense.

At the International Level, There Should Be Agreements on Areas Where There Is Agreed Social Harm

There is widespread agreement that consumer fraud, child pornography, the "grooming" of children in chatrooms, and spam should be outlawed. In the developed countries, there exists legislation or pending legislation against consumer fraud and child pornography. Legislation against spam is being passed in other countries. Once a sizeable bloc of countries have these laws, they will drive the offending users to countries that do not have such legislation. Typically, these are also countries that are less capable of addressing those concerns. In effect, the strong will push their dregs out into the lands of the weak. International agreements are necessary to ensure that such ill effects do not result. Further, because the nature of the Internet requires international cooperation when it comes to enforcement, it would make the most sense to have some international agreements.

At the National Level, Tolerate Cultural Differences

The Internet community should tolerate, if not respect, the necessity for cultural differences in content regulation. Content on the Internet will invariably be framed in a cultural context: from language to tone to idioms, culture plays a significant part. Within its context, each culture's frame of reference and values make sense. As a corollary, each country also has its own national blind spots. The example previously discussed of the United States restricting online gambling, which though arguably not restricting content legitimizes the same principles that can be used to justify regulating content. The French should be allowed to block Nazi memorabilia and the Germans anti-Semitic sites *provided* they do not deny access by users in other countries. American attempts to build anti-censor programs are misguided; they will boomerang, such as by allowing Americans to access gambling sites.

The EU Offers the Best Models for International Agreement and Cooperation

In developing policies, rules, and codes, the best model to look to is Europe and in particular the European Union. There are several reasons. First, in general, the creation of policies, rules, and codes should involve a transparent and consultative process. The Internet is still developing and so inputs from a wide range of stakeholders are essential. The EU policy- and rule-making process is transparent and consultative. It accepts inputs from many sources, including NGOs. For example, the Council of Europe, an entity larger than the European Union, withdrew and amended its twenty-second draft Convention on Cybercrime when there was an outcry by the Internet community, just when the council thought it was ready to be released.[70] The inclusion of NGOs is important because they can have a significant impact, as the free speech groups in the United States have demonstrated. Second, the EU process by its very nature must allow for cultural differences. To be sure, the EU countries as a regional bloc have more cultural affinities to each other. Nevertheless, there are differences and working out these differences strengthens the final result. Third, Europe has had some success. The European tipline, INHOPE, seems to have taken off well. The tipline is run by industry, supported by the European ISP Association, which seems not to mind too much the additional cost of such an operation. And almost as a confirmation of the views described, the European Union announced the launch of a co-regulation network at the 2003 World Summit on the Information Society (WSIS).[71]

In contrast, Asia is too disparate. There is no similar rule-making body like the European Union, which leaves matters in the hands of industry. When it comes to policy matters, Asia tends to leave things in the hands of government. There are no strong industry associations. Many of the existing associations are chapters of U.S. organizations. Perhaps the closest thing to an effective association is APRICOT (Asian Pacific Regional Internet Conference on Operational Technologies),[72] which runs technical workshops to update ISPs and other industry players on the technology of the Internet. A few attempts to inject policy discussions into the conference, in some of which the author has been involved, have not gone very well.

Conclusion

The Internet is growing up. It is gaining in importance and stature but along with that maturity comes the need to be less freewheeling. The issue is not regulation versus no regulation but rather the quality of regulation. For the developed countries, there are existent laws that could be used to regulate the Internet and its content. The rest of the world will need new laws to catch up.

Is there a danger that the laws passed will benefit only the authorities and governing bodies? Yes, but there is an inherent check on such "bad laws": they show up faster and more clearly on the Internet. By its very nature of affording access to material all over the world, the Internet will make content and information more readily available compared with traditional media such as print newspapers and broadcast television or radio. The Internet will provide users even in authoritarian countries with a window to see the world. In the long run, that can only be good for these countries and for humanity.

On July 18, 2005, the Working Group on Internet Governance, a body appointed by the United Nations' secretary-general, presented a report that called for a new international forum to discuss issues concerning Internet governance. Launched in 2006, the Internet Governance Forum (IGF) may advance some of the recommendations offered here. The IGF incorporates inputs from government, the private sector, and civil society groups. Issues concerning content, such as spam, rank high among the concerns of users, industry, and government. It is the best hope that sensible and rational regulation will be discussed in order to establish international coordination and cooperation needed to minimize the harm that can occur on the Internet.

Notes

1. For an elaboration of this approach, see Wu, "Internet v. Application."

2. *ACLU v. Reno*, 929 F. Supp. 824 (E.D. Pa., 1996), affirmed, 521 U.S. 844, 117 S. Ct. 2329, 1997 U.S. LEXIS 4037 (1997).

3. Ang, "How Countries Are Regulating Internet Content."

4. Ang and Nadarajan, "Censorship and the Internet."

5. Ang, "How Countries Are Regulating Internet Content."

6. Brekke, "South Korea Blocks Geocities."

7. Ang, "How Countries Are Regulating Internet Content."

8. *Attorney General of the United States v. American Civil Liberties Union.*

9. Shapiro, *The Control Revolution*, 70.

10. *Ottawa Citizen*, "Canada Eyes Internet Regulation."

11. Johnson and Post, "Law and Borders."

12. President Clinton, Presidential Directive on Electronic Commerce.

13. Barlow, "A Cyberspace Independence Declaration."

14. The last significant holdout was Japan, which criminalized child porn in May 1999 to be in line with international norms; until then, the Japanese drew no distinction between porn involving children and adults. (*The Economist*, "Japan: The Darker Side of Cuteness.")

15. Akdeniz, "The Regulation of Pornography and Child Pornography on the Internet."

16. McCartney, "Computer Crime Issues."

17. *CNN Worldwide*, "How Police Smashed Child Porn Club."

18. McCullagh, "Politech: Transcript of DOJ 'Operation Avalanche' Child Porn Announcement."

19. Bright and Harris, "Child Porn Swoop Nets 90 Police."

20. Australian Competition and Consumer Commission, "International Internet Sweep Days."

21. Australian Competition and Consumer Commission, "ACCC Leads International Internet Sweep for Travel Offers."

22. New Zealand Commerce Commission, "Commission Joins Forces for International Sweep Day."

23. This section draws on Ang, "The Role of Self-Regulation."

24. For the collection of online materials about the case, refer to the German *Compuserve* case at http://www.qlinks.net/comdocs/somm.htm. The site contains an unofficial English translation at

http://www.cyber-rights.org/isps/somm-dec.htm. For a news source concerning the final outcome, please see Andrews, "German Court Overturns Pornography Ruling against Compuserve."

25. Fletcher, "Theories of Self-Regulation," 1993.

26. Ang, "The Role of Self-Regulation."

27. The explanation comes from Alan Davidson of the Center for Democracy and Technology in June 30, 2000, at a conference, "Self-Regulation of Internet Content," organized by the Bertelsmann Foundation, Hanover Expo 2000.

28. Federal Trade Commission, "New Rule Will Protect Privacy of Children Online."

29. Australian Broadcasting Authority, "Online Services Content Regulation."

30. van Den Bergh, "Self-Regulation of the Medical and Legal Professions."

31. Scarpa, "The Theory of Quality Regulation and Self-Regulation," 254.

32. Ang, "The Role of Self-Regulation."

33. Boutin, "Just How Trusty Is Truste?"

34. Waltermann and Machill, *Protecting Our Children on the Internet.*

35. Civil liberty groups embrace the notion of parental empowerment, although apparently with a different intent. In his book *Cyber Rights: Defending Free Speech in the Digital Age* (1998, at 233), Mike Godwin, legal counsel for the Electronic Frontier Foundation, candidly admits that the term parental empowerment was intended to be a "positive frame" to counter forces pushing for legislation against pornography.

36. Please see http://www1.ifccfbi.gov/index.asp (February 3, 2003).

37. FBI, "FBI Tips and Public Leads."

38. Please see http://conventions.coe.int/Treaty/en/Treaties/Html/005.htm.

39. U.S. Department of Justice, "Jay Cohen Convicted." According to the web page, ten other defendants had pleaded guilty previously. Cohen was subsequently sentenced to twenty-one months in jail. Scheherazade Daneshkhu, Matthew Garrahan, and Sathnam Sanghera, "Online Betting Operator Jailed," *Financial Times*, August 11, 2000, 8.

40. Matlick, "Congress Betting It Can Control the Internet."

41. Glaser, "Wanna Bet?"

42. *Washington Post*, "WTO to Investigate U.S. Antigambling Measures."

43. *The Wall Street Journal*, "Yahoo! Ordered to Bar the French From Nazi Items."

44. *The Wall Street Journal*, "Yahoo! Will Ban Hate Material and Charge Fees on Auction Sites."

45. See note 43.

46. AP News, "French Judge Orders Yahoo!'s Nazi Auction 'zoned.'"

47. *Dow Jones v. Gutnick.*

48. See the European Union section of www.spamlaws.com.

49. Boutin, "Just How Trusty Is Truste?"

50. TRUSTe, "TRUSTe Orders Aveo to Remove Privacy Seal," "TRUSTe Terminates Relationship with eHow."

51. On December 10, 2003, TRUSTe's web site carried as its most current watchdog advisory a July 2003 status report—more than four months out of date. See http://www.truste.org/news/padvisories/ (December 10, 2003).

52. The author was a member of the ICRA's International advisory council from 2000 to 2004 and a member of its board from 2002 to 2004.

53. http://www.icra.org.

54. Because of the author's familiarity with consumer complaints and his research on self-regulation, one key area of contribution concerned the complaints and sanctions mechanism.

55. American Civil Liberties Union, "Fahrenheit 451.2." Curiously enough, the footnote is missing in the online version, http://www.aclu.org/privacy/speech/15145pub20020317.html.

56. Personal conversation, Stephen Balkam, chief executive officer, ICRA, March 8, 2002, Hong Kong.

57. Yahoo! and MSN are partially labeled. See http://www.icra.org/cgi-bin/labelTester.cgi?lang=EN&url=http%3A%2F%2Fwww.yahoo.com and http://www.icra.org/cgi-bin/labelTester.cgi?lang=EN&url=http%3A%2F%2Fwww.msn.com. They have a label for their respective primary domains but have not labeled content from other domains.

58. Mayfield, "Regulating the Olympic Rings."

59. Strom, "Legislators Tighten the Ban."

60. BBC News, "EU Approves Data Retention Rules."

61. Metherell, "Parliament Rubberstamps Howard's Eavesdropping Laws."

62. Mold, "'Duty' of Telecoms to Assist Snooping."

63. Markoff and Schwartz, "Bush Administration to Propose System for Monitoring Internet."

64. Leyden, "Net Snooping to Cost UK Taxpayers £100m+ a Year."

65. Since this chapter was first written, the United Nations' secretary-general has convened the Working Group on Internet Governance, of which the author was one of the forty members.

66. Debora Spar, *Pirates, Prophets and Pioneers.*

67. Ang, "Policing Asia's Internet."

68. Boyd, "Bypassing China's Net Firewall."

69. Kalathil and Boas, *Open Networks, Closed Regimes.*

70. Reuters, "Cybercrime Treaty Draft."

71. EurActiv.com, "European Internet Co-Regulation Network Launched," cited in *TKR Communications News*, http://www.ktr-newsletter.de/.

72. APRICOT's web site is www.apricot.net.

References

Akdeniz, Y. 1997. "The Regulation of Pornography and Child Pornography on the Internet." *The Journal of Information, Law and Technology* 1, http://elj.warwick.ac.uk/jilt/internet/97_1akdz/.

American Civil Liberties Union. 2002. "Fahrenheit 451.2: Is Cyberspace Burning?" (March 17), http://www.aclu.org/privacy/speech/15145pub20020317.html.

Andrews, Edmund. 1999. "German Court Overturns Pornography Ruling against Compuserve." *New York Times*, November 18, http://www.nytimes.com/library/tech/99/11/biztech/articles/18compuserve-germany.html.

Ang, Peng Hwa. 1997. "How Countries Are Regulating Internet Content." Internet Society Annual Conference. Kuala Lumpur, Malaysia. June, http://www.isoc.org/isoc/whatis/conferences/inet/97/proceedings/B1/B1_3.HTM1997.

Ang, Peng Hwa. 2000. "Policing Asia's Internet." *Asian Wall Street Journal*, September 7, p. 8.

Ang, Peng Hwa. 2001. "The Role of Self-Regulation of Privacy and the Internet." *Journal of Interactive Advertising* 1, no. 2 (Spring), http://www.jiad.org/vol1/no2/ang/index.html.

Ang, Peng Hwa, and Berlinda Nadarajan. 1996. "Censorship and the Internet: a Singapore Perspective." *CACM* Communications of the ACM 39, no. 6 (June): 72–78.

AP News. 2000. "French Judge Orders Yahoo!'s Nazi Auction 'Zoned' so French Can't Bid," November 20, http://www.freedomforum.org/templates/document.asp?documentID=3457.

Attorney General of the United States of America v. American Civil Liberties Union, et al. 117 S. Ct. 2329 138. L. Ed. 2d 874 (1997), http://www.aclu.org/court/renovacludec.html.

Australian Broadcasting Authority. Undated. "Online Services Content Regulation," http://www.aba.gov.au/what/online/register_codes.htm.

Australian Competition and Consumer Commission. Undated. "ACCC Leads International Internet Sweep for Travel Offers." Media release, http://203.6.251.7/accc.internet/digest/view_media.cfm?RecordID=946.

Australian Competition and Consumer Commission. Undated. "International Internet Sweep Days." http://www.accc.gov.au/ecom2/inter%5Fnet%5Fsweep.htm.

Barlow, John Perry. 1996. "A Cyberspace Independence Declaration" (Electronic Frontier Foundation), http://www.eff.org/Publications/John_Perry_Barlow/barlow_0296.declaration.

BBC News. 2005. "EU approves Data Retention Rules." December 14, http://news.bbc.co.uk/2/hi/europe/4527840.stm.

Boutin, Paul. 2002. "Just How Trusty Is Truste?" *Wired News* (April 9), http://www.wired.com/news/exec/0,51624-0.html.

Boyd, Clark. 2004. "Bypassing China's Net Firewall," *BBC News*. March 10, http://news.bbc.co.uk/1/hi/technology/3548035.stm.

Brekke, Dan. 1997. "South Korea Blocks Geocities." *Wired News* (October 22), http://www.wired.com/news/email/7896.html.

Bright, Martin, and Paul Harris. 2002. "Child Porn Swoop Nets 90 Police." *The Observer* (October 20), http://politics.guardian.co.uk/homeaffairs/story/0,11026,815573,00.html.

CNN Worldwide. 2001. "How Police Smashed Child Porn Club," February 13, http://www3.cnn.com/2001/WORLD/europe/UK/02/13/paedophile.police.

European Union section. Undated. http://www.spamlaws.com/eu.html.

Dow Jones v. Gutnick. 2002. HCA 56. December 10, http://www.austlii.edu.au/au/cases/cth/high_ct/2002/56.html.

The Economist. 1998. "Japan: The Darker Side Of Cuteness," May 8, 32.

EurActiv.com. 2003. "European Internet Co-regulation Network Launched at World Information Summit," December 15, http://www.euractiv.com/cgi-bin/cgint.exe/457777-93?204&OIDN=1506856&-tt=me.

European Communities. 2001. "Communications Data Protection Directive," http://europa.eu.int/smartapi/cgi/sga_doc?smartapi!celexplus!prod!CELEXnumdoc&lg=en&numdoc=52000PC0385.

Federal Bureau of Investigation. Undated. "FBI Tips and Public Leads," https://tips.fbi.gov/.

Federal Trade Commission. 1999. "New Rule Will Protect Privacy of Children Online," October 20, http://www.ftc.gov/opa/1999/9910/childfinal.htm.

Financial Times. 2000. "Online Betting Operator Jailed," August 11, p. 8.

Fletcher, Amelia. 1993. "Theories of Self-Regulation." PhD dissertation. England: Oxford University.

Glaser, Joanna. 2002. "Wanna Bet? Feds Say Not So Fast." *Wired News*, http://www.wired.com/news/business/0,1367,55510,00.html. (February 3, 2003).

Godwin, Mike. 1998. *Cyber Rights: Defending Free Speech in the Digital Age*. New York: Times Book.

Hayes, Simon. 2003. "ISPs As Chokepoints: US Tightens Net Copyright." *Australia IT*, http://australianit.news.com.au/articles/0,7204,5896759%5e16123%5e%5enbv%5e,00.html.

Johnson, David R., and David Post. 1996. "Law and Borders: The Rise of Law in Cyberspace." *Stanford Law Review* 48: 1367–1378.

Kalathil, Shanthi, and Taylor C. Boas. 2003. *Open Networks, Closed Regimes: The Impact of the Internet on Authoritarian Rule*. Washington, DC: Carnegie Endowment for International Peace.

Leyden, John. 2003. "Net Snooping to Cost UK Taxpayers £100m+ a Year." *The Register*, January 29, http://www.theregister.co.uk/content/6/29097.html.

Markoff, John, and John Schwartz. 2002. "Bush Administration to Propose System for Monitoring Internet." *The New York Times*, December 20, http://www.nytimes.com/2002/12/20/technology/20MONI.html.

Matlick, Justin. 1998. "Congress Betting It Can Control the Internet." *Las Vegas Review Journal* (Sunday, June 7), http://www.reviewjournal.com/lvrj_home/1998/Jun-07-Sun-1998/opinion/7632214.html.

Mayfield, Kendra. 2002. "Regulating the Olympic Rings." *Wired News* (February 8), http://www.wired.com/news/technology/0,1282,50275,00.html.

McCartney, Michael. Undated. "Computer Crime Issues." (New York State Attorney General's Office), http://www.oft.state.ny.us/security/electronic%20presentations/ag%20new.pdf.

McCullagh, Declan. 2001. "Politech: Transcript of DOJ 'Operation Avalanche' Child Porn Announcement (August 8), http://lists.insecure.org/lists/politech/2001/Aug/0032.html.

Metherell, Mark. 2005. "Parliament Rubberstamps Howard's Eavesdropping Laws." *Sydney Morning Herald*, March 31, http://www.smh.com.au/news/national/parliament-rubberstamps-howards-eavesdropping-laws/2006/03/30/1143441281193.html.

Mold, Francesca. 2002. "'Duty' of Telecoms to Assist Snooping." *New Zealand Herald*, November 13, http://www.nzherald.co.nz/category/story.cfm?c_id=55&objectid=3004028.

New Zealand Commerce Commission. 2002. "Commission Joins Forces for International Sweep Day: Target Sector Is Health," http://www.comcom.govt.nz/publications/display_mr.cfm?mr_id=929.

Ottawa Citizen. 1996. "Canada Eyes Internet Regulation," November 15, sec. D15.

President William Clinton. 1997. "Presidential Directive on Electronic Commerce," http://www.technology.gov/digeconomy/presiden.htm.

Reuters. 2000. "Cybercrime Treaty Draft: Take 23," 7 am, November 13, http://www.wired.com/news/politics/0,1283,40134,00.html.

Scarpa, Carlo. 1999. "The Theory of Quality Regulation and Self-Regulation." In *Organized Interests and Self-Regulation: An Economic Approach*, ed. Bernaerdo Bortolotti and Gianluca Fiorentini, 236–260. Oxford: Oxford University Press.

Shapiro, Andrew L. 1999. *The Control Revolution: How the Internet Is Putting Individuals in Charge and Changing the World*. New York: Public Affairs.

Deborah Spar. 2001. Pirates, Prophets and Pioneers: Business and Politics along the Technological Frontier. London: Random House.

Strom, Stephanie. 1999. "Legislators Tighten the Ban on Under-Age Sex." *New York Times*, May 19, 1999, 6.

TKR Communications News. 2003. (December 15), subscription available from http://www.tkr-newsletter.de/.

TRUSTe. 2001a. "TRUSTe Orders Aveo to Remove Privacy Seal" (June 5), http://www.truste.org/news/padvisories/users_aveo_seal.html.

TRUSTe. 2001b. "TRUSTe Terminates Relationship with eHow." (July 23), http://www.truste.org/news/padvisories/users_ehow.htm.

U.S. Department of Justice. 2000. "Jay Cohen Convicted of Operating an Off-shore Sports Betting Business that Accepted Bets from Americans over the Internet" (28 February), http://www.cybercrime.gov/cohen.htm.

van Den Bergh, Roger. 1999. "Self-Regulation of the Medical and Legal Professions." In *Organized Interests and Self-Regulation: An Economic Approach*, ed. Bernaerdo Bortolotti and Gianluca Fiorentini, 89–130. Oxford: Oxford University Press.

Waltermann, Jens, and Marcel Machill, eds. 2000. *Protecting Our Children on the Internet: Towards a New Culture of Responsibility*. Gutersloh: Bertelsmann Foundation Publishers.

Wu, Timothy. 1999. "Application-Centered Internet Analysis." *Virginia Law Review* 85, no. 6 (September): 1163–1204.

The Wall Street Journal. 2000. "Yahoo! Ordered to Bar the French from Nazi Items," November 21.

The Wall Street Journal. 2001. "Yahoo! Will Ban Hate Material and Charge Fees on Auction Sites," January 3.

Washington Post. 2003. "WTO to Investigate U.S. Antigambling Measures," July 21, http://www.washingtonpost.com/wp-dyn/articles/A24490-2003Jul21.html.

9 Creating Conventions: Technology Policy and International Cooperation in Criminal Matters

Ian Hosein

A common belief is that since activities may now be global due to increased mobility, transnational trade, and transnational data flows enabled by information and communications technology, then regulating conduct is an international problem and needs international solutions. In this chapter I look at technology policy solutions to international criminal matters, which are in turn spurred by technology diffusion. The solutions are sometimes more hazardous than the original problems.

It is nontrivial to establish new policy in this domain. In fact, the interaction of actors as they negotiate, deliberate, consult, object, question, implement, or ignore proposed policies is complex. In recent history we have seen a number of rich and sometimes highly controversial national policy discourses on cryptography, free speech, and privacy, among many others. Time and again many have noted the challenges of establishing regulation in a changing technological environment, an economic environment that has potential for growth, and an environment that involves transborder data flows, among other characteristics. These are the positions of politicians, officials, business representative, lobbyists, and advocates.

In recent years we have seen international solutions arise to resolve these challenges. These solutions, however, only exacerbate jurisdictional problems while they circumvent national democratic discourse. Many national actors fail both to take notice of and to engage with these international decision-making forums. Meanwhile, international forums have all but ignored national actors. The proposed policies also fail to acknowledge the difficulties in policy making in a technological environment.

International policy dynamics may lead to pernicious solutions that mask complexity, prevent consultation, and ignore technology. The rich and sometimes controversy-filled national discourse may be replaced by the silence of adherence to international obligations, *faits accomplis*, and calls for "harmonization." When legislating in areas involving government powers, we must be cautious. If we are blind to the technology, we miss the point of it all. This is not a case of merely "updating" old laws to new circumstances. Technology significantly alters the landscape for national policy. International policy making conceals this fact. So the question I want to address in this chapter is: how can we reinvigorate discourse within this international and technological environment?

This chapter begins by presenting the challenges of establishing national policies on high-tech crime. First, I show in detail the work of the Group of 7/8 (G8) and the Council of Europe (CoE), which developed agreements ensuring that the abilities of law enforcement agencies are not hampered by borders, technologies, and legal conflicts. If the reader is in a rush I recommend moving quickly to the subsequent section where I present a synthesis of the lawful access powers proposed by these arrangements through the development of a topology of lawful access arrangements. Once we understand the nature of these demands and requirements we can better comprehend the concerns of the variety of actors involved in these policy discourses. The solutions presented by the G8 and the CoE may affect our understanding of the interaction between technology and law, and national discourse and international policy making. Strategies for reinserting deliberative democracy into these processes are recommended in the concluding sections of this chapter.

Challenges of Jurisdiction and National Technology Policy

National governments are usually entitled to enact and enforce laws within their jurisdiction; it is, after all, their sovereign right to do so. Transborder activity challenges this right. Enforceability becomes immediately questionable: activity may occur beyond the jurisdiction of national law, thus regulating national activities is fruitless and hazardous economically.

The Logic of Conventions

Data flows within transnational digital networks are a case in point. Action may occur from a distance, where the overflow of activity can occur without an individual having to physically enter the jurisdiction. Whether it is the penetration of computers or the downloading of pornography, this conduct can occur across borders, preventing law enforcement agencies, with their traditionally bordered jurisdictions, from conducting investigations and generating evidence.

From the technological perspective, the challenges are already numerous. Identifying and tracing an individual behind conduct on the Internet is nontrivial, as is identifying the alleged perpetrator's geographic location. Identifying the location of evidentiary information is also difficult: identifying which Internet service providers (ISPs) hold customer data, e-mails, and transactional data regarding an incident, and ensuring that the information is available to law enforcement officials upon request. This is presuming, of course, that the law enforcement agencies conducting this investigation have the appropriate investigatory capabilities with appropriate authorization and oversight procedures, enshrined in law and made possible by technology.

Resolving national policy challenges is no easier. Criminalizing conduct, through substantive law, such as hacking or downloading child porn may seem trivial. Yet the process of drafting bills and testing laws through court decisions has proven to be difficult for many countries. Reasons for these difficulties include vagaries in definitions, contentious updates due to new technologies and techniques, or legal challenges on constitutional values. Such a state of national flux presumes the existence of constitutional protections, and an active nongovernmental sector including industry lobbyists and human rights organizations. Without such a political culture, passing new laws would be far simpler.

Establishing state powers of investigation is also nontrivial; once it is decided, technologically, that investigative information may exist in the form of computer data, allowing for lawful access to this data usually requires new laws of interception, search and seizure, and production and preservation by third parties. This all presumes that law enforcement agencies have procedures and knowledge to deal with these situations.

Finally, when information exists within other jurisdictions, countries may be required to preserve and access the information. Much of this, arguably, needs to occur quickly in order to begin real-time access, or access to stored data before the data is deleted. This all presumes that such regimes for international cooperation exist.

This is the logic behind international agreements on high-tech crime: creating harmonized crimes, standardized surveillance powers, and regimes for governments to share this data. By doing so we avoid all the trials and trivialities of domestic law and politics.

International Regimes and Discourses and Technology

Establishing national policies amid transnational data networks and flows is a pressing technology policy issue. It is not new, nor is it as infeasible as often presumed or argued legally. Such presumptions and arguments have been responded to by legal theory and in practice. Theoretically, it is now argued that the infeasibility arguments exaggerated the problems and promises of data flows. In so doing, the arguments failed to recognize that there have long been multijurisdictional regulatory problems involving transnational transactions in other fields of law.[1]

Indeed, in recent years countries have regulated data flows, sometimes successfully. The European Union finalized a harmonizing directive on data protection in 1995 that included two articles regulating transborder data flows to defend privacy. Countries as diverse as Australia, China, and Saudi Arabia implemented censorship regulations to control what kind of information is posted, received, or both;[2] despite claims of infeasibilities and inaccuracies.[3] In 2001, a French court decided that Yahoo! was obliged to prevent French users from accessing components of Yahoo!'s sites that were illegal according to French law despite that the servers and services were in the

United States.[4] Regulating global data flows indeed appears to be no different than regulating other transnational activities.

At the same time, international cooperation in legal matters has also been on the rise. For example, between 1970 and 1997 the number of international treaties tripled; and from 1985 to 1999 the number of intergovernmental organizations rose by two-thirds.[5] International treaty making dealing specifically with criminal matters has also been on the rise.

It is thus tempting to assume that creating national laws on criminal matters involving transnational data flows is possible, and can be supported easily by international agreements. The technology policy dimensions may not be so easily dismissed, however. The introduction of the lawful access regimes implicate and have an effect on a number of actors. Regulatory burdens, conflicts of laws, implications for civil liberties, and varying demands for criminalization of activities have arisen within discourses surrounding both national and international regimes, particularly because of the involvement of technology.

Technology is a significant factor in all of these affairs. First, new challenges exist no matter how hard we try to pretend that the new technologies are similar to previous technologies. Surveillance in the age of the telegraph is different than it is in the age of digitally distributed communications. Second, the new technologies may warrant new legal techniques, which may in turn create new conflicts to our legal and social norms. Gaining access to my banking transfers is different from finding out where I have been on the Internet. The technological environment alters somewhat the powers of the state—in some cases magnifying the powers of the state, in others requiring the state to magnify its own powers to cater to technological developments.

By looking at the discourses surrounding two of the most active international forums in the arena of criminal matters and transnational data flows, the G8 and the CoE, we see a plethora of actors, a myriad of interests, and additional challenges introduced by the technology. The activities surrounding these forums involved responses and consultations, both solicited through formal processes and reacted upon through self-organization, from a number of actors including industry organizations, nongovernmental organizations (NGOs), academia, and other sectors of civil society.

Governance Tenders: G8 and CoE

Establishing even national policy in the realm of transnational data flows is not easy. When the United Kingdom tried to create an electronic commerce policy that contained law enforcement components, particularly dealing with encryption and interception of communications, sometimes fervent opposition arose. The opposing actors articulated viewpoints with varying concerns regarding the perceived imperatives of the development of networks and associated technologies and services, costs to industry for operating under a regulatory regime, impacts upon civil liberties of the considered actions, threats of jurisdictional arbitrage if burdens are too high for the regulated institutions or the regulated activity, and concerns regarding legislating within a changing technological environment.[6] The work of the G8 and the CoE can be seen as attempts to avoid some of these same challenges.

Solutions were sought particularly to deal with law enforcement interests, to maintain traditional powers to the changing information and technological infrastructures, to update and enhance traditional powers to meet the new challenges, and in some cases to introduce new powers. After years of seeking these powers at the national level, in the mid- to late-1990s a number of governments turned to international agreements such as the Wassenaar Arrangement for export controls of privacy enhancing technologies, and the Cryptography Policy Guidelines of the Organization for Economic Cooperation and Development (OECD) to meet domestic interests. The core activity on criminal matters, however, arose from the G8 and the CoE.

The G8

The work at the G8 began within efforts to combat organized crime. At the Halifax summit in 1995, the then-G7 countries established a working group to look at the issue of transnational organized crime. While the G7 concentrated traditionally on economic cooperation, summits mentioned only occasionally issues regarding national security such as terrorism and narcotics. This new "expert group" took the unprecedented step in 1996 of moving into technological issues and national law enforcement. In Paris, prior to the Lyon summit of 1996, the group of experts agreed to address the use of high technology, and the abuse of such technology by criminals, stating:

> States should review their laws in order to ensure that abuses of modern technology that are deserving of criminal sanctions are criminalized and that problems with respect to jurisdiction, enforcement powers, investigation, training, crime prevention and international cooperation in respect of such abuses are effectively addressed.... States should promote study in this area and negotiate arrangements and agreements to address the problem of technological crime and investigation.
>
> We emphasize the relevance and effectiveness of techniques such as electronic surveillance, undercover operations and controlled deliveries. We call upon States to review domestic arrangements for those techniques and to facilitate international cooperation in these fields, taking full account of human rights implications.

The G7 later endorsed the work of the expert group, the Lyon Group as it became later known, along with a forty-point statement that recommended increased mutual legal assistance and the establishment and enforcement of international treaties and conventions.

A month later at a ministerial meeting in France, G7 foreign ministers met and recommended twenty-five measures to combat terrorism, including increased international assistance, coherence, and efficiency. In 1997 at the Denver summit, the momentum increased as quickly as the mandate expanded. The foreign ministers submitted an evolved report of their previous work, stating:

> The significant growth in computer and telecommunications technologies brings with it new challenges: global networks require new legal and technical mechanisms that allow for a timely and effective international law enforcement response to computer-related crimes. To that end, we will work together to enhance abilities to locate, identify, and prosecute criminals; cooperate with and assist one another in the collection of evidence; and continue to develop training for law enforcement personnel to fight high-technology and computer-related crime.

The 1997 summit attendees agreed to the Lyon work product, and in their final group communiqué argued:

> We must intensify our efforts to implement the Lyon recommendations. In the coming year we will focus on two areas of critical concern: First, the investigation, prosecution, and punishment of high-tech criminals, such as those tampering with computer and telecommunications technology, across national borders; Second, a system to provide all governments the technical and legal capabilities to respond to high-tech crimes, regardless of where the criminals may be located.

These conclusions carried over into the 1997 Statement of Principles established by interior and justice ministers, in Washington, DC (in preparation for the summit later that year in Birmingham), where among many decisions, they agreed that

- there must be no safe havens for those who abuse information technologies;
- investigation and prosecution of international high-tech crimes must be coordinated among all concerned states, regardless of where harm has occurred;
- legal systems should permit the preservation of and quick access to electronic data, which are often critical to the successful investigation of crime;
- mutual assistance regimes must ensure the timely gathering and exchange of evidence in cases involving international high-tech crime; and
- to the extent that it is practicable, information and telecommunications systems should be designed to help prevent and detect network abuse, and should also facilitate the tracing of criminals and the collection of evidence.

These principles are supported by the following action plan to

- review our legal systems to ensure that they appropriately criminalize abuses of telecommunications and computer systems and promote the investigation of high-tech crimes;
- consider issues raised by high-tech crimes, where relevant, when negotiating mutual assistance agreements or arrangements;

- continue to examine and develop workable solutions regarding the preservation of evidence prior to the execution of a request for mutual assistance, transborder searches, and computer searches of data where the location of that data is unknown;
- develop expedited procedures for obtaining traffic data from all communications carriers in the chain of a communication and to study ways to expedite the passing of this data internationally; and
- work jointly with industry to ensure that new technologies facilitate our effort to combat high-tech crime through preserving and collecting critical evidence.

This was endorsed at the Birmingham summit in 1998. Further work was recommended for the ministerial summit to be held in Moscow in 1999. Speaking specifically with regards to the Internet, the ministers stated their concern regarding child pornography, fraud, financial crimes, traceability, and access to stored data across borders. They acknowledged that further work was required on determining access to traffic data, and this required consultation with industry. These principles set the stage for the G8-Industry discussions that were arranged for 2000 and 2001 in Paris, Berlin, and Tokyo.

G8-Industry Meetings

The G8 established a dialogue with industry representatives to discuss these issues further. The conclusions of the first meeting in Paris, May 2000, were uncertain, as industry and governments diverged in their interests and statements. The final communiqué articulated some of these concerns, including raising civil liberties and privacy on the agenda, maintaining the powers of law enforcement agencies, defining a clear and transparent regime to combat "cyber-criminality," and ensuring free and equitable market development to ensure good conditions for industry, while evaluating the effectiveness and consequences of the policies. With these conflicting values and technological options, further dialogue was deemed required, and the Lyon Group looked to the Berlin meeting.

In October 2000, government and industry representatives met in Berlin, this time with a more detailed action plan. Rather than repeating the plenary presentations of perspectives from Paris, theme-specific workshops were established. Workshop 1a and 1b concentrated on the theme of "Locating and Identifying High-Tech Criminals," with the first sub-workshop focusing on "Data Retention and User Authentication," while the second focused on "Data Preservation, Real-Time Tracing, and Machine Authentication." The theme of workshop 2 was "Threat Assessment and Prevention," and workshop 3 concentrated on "Co-operation between Industry and Law Enforcement."

The Berlin meeting ended with some further clarifications, but again representatives agreed that further work was still required. Particular concerns were articulated regarding the quantification of costs of proposed measures, the implications of technology development, continuing law enforcement concerns about the availability of data, evolving definitions of the different types of service and service providers, process and procedures in mutual legal assistance, the integrity of the data, and privacy and civil liberties. The conclusions of Berlin set the stage for Tokyo, in May 2001, as further research and consideration were required.

Along similar lines, the Tokyo meetings were in workshops: Workshop 1 on "Data Retention," workshop 2 on "Data Preservation," workshop 3 on "Threat and Prevention," and workshop 4 on "Protection of E-Commerce and User Authentication," as well as further work on "Training." The results of workshops 1 and 2 are most relevant to the focus of this chapter.

G8 on Preservation

Data preservation applies specifically to individual cases based on existing, not future, data. The discussion paper for this workshop was written by the U.S. Department of Justice; and defines *traffic data* as "non-content information recorded by network equipment concerning a specific communication or set of communications. Traffic data includes the origin of a communication, the duration, the nature of the communication activity (not including content) and its destination. In the case of Internet communications, traffic data will almost always include an IP address and port number. Traffic data does not include the content of a communication."

Depending on the type of service used, however, the traffic data may differ; as does the divide between what constitutes content and what is traffic. Preserving this traffic data is deemed an essential technique for investigations. It is tightly defined by the G8: "Upon lawful request by government or law enforcement, based on the facts of a specific case, specified historical data can be preserved to prevent its deletion pending issuance of a lawful request by government or law enforcement to disclose the data. Not future collection of data, not creating new obligations on providers to generate data not already created."

This was one of the earliest attempts to separate the term traffic data from data retention; previously the two terms were ambiguously replaceable. *Retention* was therefore applied to all future data that was to be accumulated, and applying to a nonspecific scale, for example, applying to all users of an ISP rather than simply one suspect.

Identifying the flow of traffic is also difficult. If suspicious activity was detected against a military network in Virginia the authorities would attempt to track down the unique IP address of the attacking computer. If that address pointed to a service provider in the UK the police would want to know which user that was; however, if the true user was merely using that IP address assigned by a UK service provider but doing so from a computer on another network in Germany, the law enforcement officials would be stymied unless they could quickly identify the German's user details from the UK service provider. For this reason, an option that developed at the Berlin meeting (and already existed in the CoE convention drafts) is the *preservation and par-*

tial disclosure order: "Such an order might authorize the preservation of critical traffic data by a service provider and a review of that portion of the data by the government issuing the order which would reveal whether the communication originated in the requested state, or whether instead it originated in still another state. The order would apply to a specific communication end-to-end and not to only one particular service provider in the chain."

This option has some interesting implications. First, it would require ISPs to disclose data to law enforcement authorities without receiving a specific order for their service; that is, the order applies to the entire chain of ISPs where the data packets flowed. A second implication is that these orders are proposed to be applied within a jurisdiction even if a crime has not occurred within that country; even if the perpetrator is not in that country. This is a precursor to the general issue of *dual criminality*.

The discussion paper notes, very briefly, that there are arising problems to such a scheme: "Careful thought must be given to just how this new kind of legal process would be implemented, and to the impediments to it being effective. Likewise, burdens and concerns of the private sector must be considered, as well as the effect on privacy, along with preferable alternatives."

New processes and procedures would need to be established in order to implement such orders, and a great deal of concern regarding costs and data protection arose naturally from Berlin and was reported in Tokyo. These were itemized as

- use of provider personnel to comply with preservation orders;
- use of provider personnel to provide testimony as custodian of records or technical expert;
- use of technologies associated with preservation;
- use of data storage medium and space;
- added costs if extra data is retained in complying with the request;
- liabilities under data protection laws;
- liability without "immunity" for cooperation with the government; and
- standardized fees to be paid for specific types of requests.

These issues remained problematic and unresolved, even after Tokyo.

G8 on Retention

The Tokyo workshop on data retention opened with a discussion paper written by the Canadian Department of Justice, in consultation with some industry representatives.[7] The paper reviewed the work completed thus far, and the challenges set for consideration of data retention. Particularly this document approached the issue of data retention by first considering the rationale, then the privacy challenges and conflicts, then the cost requirements, and some consideration of business models and service profiles. Finally, the document reviewed how the varying business models could perform retention and what factors would need to be considered in deciding what is retained based upon what is collected, and the arising costs and concerns.

The final report from the Tokyo workshop on retention reported that progress had not been made in resolving the strains identified in the discussion paper. Among the key challenges is the identification and definition of what type of data is being retained: *subscriber data* and *traffic data* both need to be identified, defined, and set within a regulatory framework (different laws may address treatment of different data types). Traffic data differs depending on the business model and the type of services provided. In the plain old telephone system, traffic data generally constituted the phone numbers called and the date and time of the calls, namely, generally the information found in billing records. The Canadian paper acknowledged that for e-mail, traffic data could be derived from the SMTP (simple-mail transfer protocol) and POP (post-office protocol) headers to include: "Date and time of connection of client to server, IP address of sending computer, Message ID (msg_id), sender (login@domain), receiver (login@domain), status indicator, IP address of client connected to server, Userid, and in some cases identifying information of email retrieved."

But for web access (on hypertext transfer protocol), log data can include "date and time of connection of client to server, IP source address, operation (i.e., GET command), path of the operation (to retrieve html page or image file), "last visited page," response codes."

As a result, the end-of-workshop report acknowledges:

> Data retention is a very complex and sensitive issue. The group discussed current practices and issues related to data retention in order to develop a balanced set of options for data retention. More specifically, the group discussed the impact of data retention on business models of different internet service providers, privacy and data protection implications, technical feasibility, law enforcement and consumer interests.

The workshop concluded that "[g]iven the complexity of the above noted issues, blanket solutions to data retention will likely not be feasible."

As a result of these challenges, the Tokyo workshop recommended that government consider the lawful access provisions at the national level, and particularly to perform analysis on data protection and civil liberties constraints to these powers, and also considering costs and feasibility. That is, while the report from the Tokyo workshop on data retention indicated that some progress was made in developing model preservation requests, and in consideration of model legal procedures, the conflicts of laws continued to be a controversial problem. This was particularly true in the light of data protection laws:

> Some participants commented that a lawful notice or request issued by a competent authority could override data protection provisions that otherwise prohibit retention of certain data. . . . Individual countries with data protection regimes can have different authorized uses justi-

fying the collection of data: for example, for billing purposes only; for law enforcement requests in support of investigations; and for provider protection of property or provider protection against fraud and network abuse. Disparities of this nature among countries can create uncertainties for providers who operate across borders.

This is further exacerbated by differing laws on reporting to individuals whose data has been preserved and accessed. Finally, the issue of dual criminality was raised, also under the light of domestic criminality, where some countries require not only that the crime being investigated is a crime under that country's laws, but also that the crime needed to be committed within that territory.

The workshop also concluded that further work was still required to address dispute resolution, outreach to non-G8 countries, technological change, costs to industry, the applicability of preservation requests to *ephemeral* data, and partial disclosures and real-time tracing.

G8 Conclusions

The Tokyo summit did not end on a high note; in fact the continued inability to agree left the future of such industry-government dialogue in question. In a summary report of the Tokyo meeting by the Canadian Department of Foreign Affairs and International Trade, the complexity of the issues was considered an obstacle to gaining any progress through the meetings: "While deemed to be a success, the Tokyo Conference highlights certain deficiencies in the G8 government/industry dialogue process that will have to be addressed before future meetings are convened."

The earlier perception that further negotiations with industry were needed was corrected by the events of September 11, 2001. Among articulations of "greater urgency to this work," a draft G8 response to the terrorist attacks on the United States, released in November 2001, included calls for data retention and alterations to data protection regimes to address "public safety and other social values." The G8 response also called for permitting domestic law enforcement agencies to serve foreign data preservation and real-time access instructions to domestic service providers after expedited approval, to ensure expedited preservation and real-time access to traffic data and expedited mutual legal assistance even if there is no violation of the domestic law of the requested state, and to encourage "user-level authentication" for appropriate uses. Within the G8 response to September 11 (in a footnote to the document, actually) it is noted that the procedures outlined for international cooperation are not intended to limit the right of states to require dual criminality grounds, or "other essential interests," for assistance. Otherwise, the obstacles uncovered in the G8-Industry meetings were no longer; the input to those meetings was kept, and the output, being the lack of agreement, was ignored.

These recommendations were included as official documents at the Mont-Tremblant summit of justice and home affairs ministers in May 2002. They included a call for governments to decide which information is useful for public safety purposes, drawing directly from the G8-industry workshop documents detailing the types of traffic data to be considered; a checklist for data preservation requests, procedures, and legal frameworks; and an official statement of how data protection regimes "seriously hamper public safety"; and calling for the limited retention of data. The summit leaders also addressed international treaties, including the CoE convention.

The Council of Europe

The CoE is a political organization consisting of forty-five member states from throughout Europe, and additional observer states including Canada, Japan, Mexico, and the United States. Originally established as a human rights, standardization, and cultural organization, the CoE transformed in 1993 to include *democratic security*, later expanded further in 1997 to include social cohesion and *the security of citizens and democratic values*. The CoE has previously announced recommendations regarding high-tech crime, and regarding mutual assistance, but only in 1997 did it embark on a process of creating the Convention on Cybercrime. The convention ultimately would be opened to signature by member states in November 2001 and would enter into force in January 2004.

The CoE's European Committee on Crime Problems created a new subcommittee, the Committee of Experts on Crime in Cyberspace (PC-CY) in order to draft the Convention. The mandate for the PC-CY was to create a legally binding convention that would incorporate (1) Harmonized substantive law with regard to cybercrime; (2) Consistent domestic procedural powers; and (3) A coherent regime for mutual legal assistance relating to those procedural powers. The first component is relatively uncontroversial in that it criminalizes specific activities like hacking and fraud. The establishment of new powers for the police and new measures to conduct investigations across borders gave rise to much more controversy, and I'll cover them in greater detail in this chapter.

Although the PC-CY's drafting process began in February 1997, a draft (version 19) was released only in April 2000 for public review, and some consultation followed. This draft contained many missing parts, including the powers of lawful access to real-time traffic data and communications content. In October 2000, shortly after the initial consultation period closed, PC-CY released version 22, which finally incorporated these missing sections; and version 24-2 followed in November; then version 25 in December 2000. Version 27 was released early in 2001 with some modifications, and later became the finalized text in June 2001 that was opened for signatures in September 2001.[8]

After the terrorist attacks on the United States in that same month, the convention was positioned as a means of combating terrorism.[9] At the November 2001 ceremony in Budapest thirty countries lined up to sign on.

CoE on Procedural Law

The procedural law section of the "cybercrime" convention defines a minimum standard of powers to be granted to law enforcement agencies within ratifying states. These *lawful access procedures* are the crux of much of the debate surrounding the convention.

The first lawful access procedure, in article 14 of the convention, is that all the prescribed powers must be applicable toward all "cybercrimes" (established within the convention's substantive law section). So access to traffic data, search and seizure, and the other powers in this section of the convention must be ready for use to investigate hacking, fraud, and intellectual property crimes. But these powers must also be applicable for "other offences committed by means of a computer system," and in cases where it is possible for "the collection of evidence in electronic form of a criminal offence." As just about all transactions we perform today leave an electronic data trail, the CoE convention ensures that the lawful access regime may be applied to practically any investigation. From this point onward the convention is no longer about cybercrime.

As a result, all these powers must be applied equally to all investigations of all types, except the real-time surveillance of communications and transactions. The first exception to this scope-principle is the interception of communications: as it is considered an invasive act, governments may limit its use only to serious crimes. The second exception is that governments may reserve the right to apply real-time traffic data gathering only to those same conditions applied to the interception of communications. This latter reservation is a partial recognition (however unstated within the convention itself) of the invasive nature of the procedure and the sensitivity of collected transactional information. The CoE dissuades signatory states from adopting this reservation, however, within its explanatory memorandum: "As the real-time collection of traffic data can be very important in tracing the source or destination of computer communications (thus, assisting in identifying criminals), the Convention invites Parties that exercise the right of reservation to limit their reservation so as to enable the broadest application of the powers and procedures provided to collect, in real-time, traffic data." The convention time and again allows for liberal interpretation of more controversial powers but then requires that such interpretations be minimized. It does not act similarly in interpreting police powers: rather, it always pushes for the maximizing of such powers.

Safeguards were included so as to prevent misuse of the lawful access regime, as covered in article 15. But these were included at a late stage in the convention drafting process.[10] The article ensures that the powers within the convention are subject to safeguards *under domestic law* to "provide for the adequate protection of human rights and liberties... and which shall incorporate the principle of proportionality" (art. 15.1). As considered appropriate, the safeguards may include "judicial or other independent supervision, grounds justifying application, and limitation on the scope and duration of such power or procedure" (art. 15.2). There is also some room for negotiation with industry within a clause that calls for consideration of the impact of these powers upon "the rights, responsibilities and legitimate interests of third parties" (art. 15.3), in a manner that is consistent with the public interest, that is, "public safety, public health, and other interest." There is no discussion of costs or risks to service providers, however.

The specific powers prescribed to the law enforcement agencies follow in this chapter.

Expedited Preservation of Stored Computer Data

Articles 16 and 17 require "Parties," that is, governments, to adopt legislative and other measures to order or obtain the expeditious preservation of specified computer data, including traffic data that has been stored by means of a computer system. In most countries this is new legal power. In most circumstances the recipients of these orders will be companies. This preservation shall be maintained "for a period of time as long as necessary, up to a maximum of 90 days to enable the authorities to seek disclosure" (art. 16.2). Parties must also be able to oblige the preserver of the data to keep this preservation confidential, but only for the period of time provided for by its domestic law.

When the preserved data is communications traffic data, article 17 requires the adoption of legislative and other measures to ensure that preservation is available regardless of whether one or more service providers were involved in the communication transmission (art. 17.1). This is in order to ensure the expeditious disclosure "of a sufficient amount of traffic data to enable the Party to identify the service providers and the path" of the communication transmission (art. 17.2). If a malicious hacker is routing traffic through a number of ISPs, all of these ISPs must be prepared to preserve this data.

This power of preservation is unrelated to the power of compelling data retention. If the data is not within the computer system, then it may not be preserved. The CoE explanatory report, although nonbinding, is clear that these articles of the convention do not constitute a policy of data retention.

Production Orders

In situations where specified computer data is in the possession of an individual or service provider, article 18 allows authorities to order the submission to law enforcement agents of specified computer data in that person's possession or control. In the particular case of a service provider, policy may order ISPs to submit subscriber information in that service provider's possession or control. According to the CoE, this measure, as with preservation, will apply only to the existing data, and only to the extent that the person or service provider maintains such data. There are no requirements to gather subscriber data that does not otherwise exist, nor is there an obligation to verify the identity of subscribers or to resist the use of pseudonyms, so that, for example, pay-as-you-go subscribers and open-Wi-Fi networks do not have to generate subscriber lists. The CoE is conscious to curtail the breadth of this power as it is intended for specific criminal investigations: "The provision does not authorise Parties to issue a legal order to disclose indiscriminate amounts of the service provider's subscriber information about groups of subscribers e.g. for the purpose of data-mining."

Search and Seizure of Stored Computer Data

The search and seizure of data has long been a contentious national policy issue in a number of countries yet is dealt with succinctly in article 19. A further extension to the traditional power is required (in art. 19.2) so that if the data being sought is found to exist elsewhere and is accessible from or available to the initial system, the authorities "shall be able to expeditiously extend the search or similar accessing to the other system." Notification of accessed or seized data is not discussed within the body of the convention but is raised as a possibility in the explanatory report. The PC-CY drafters considered this matter too challenging for the body of the convention, as it could create a discrepancy with existing national laws.

Another extension to this power of search and seizure, and to the power of compelling production is that the authorities must be empowered "to order any person who has knowledge about the functioning of the computer system or measures applied to protect the computer data therein to provide, as is reasonable, the necessary information" (art. 19.4). Although the language and its breadth of application are unclear, this procedure implies that if law enforcement officials want access to data that is secured, that is, encrypted, then data-system owners must render that data intelligible to law enforcement authorities. This may include the compelled disclosure of plaintext and decryption keys, a power that has given rise to significant national debate when proposed in the United Kingdom but was hardly noticed when included in the convention. Concerns regarding data security and self-incrimination naturally arise when individuals are compelled to disclose confidential data, particularly when those individuals are the suspects themselves. The PC-CY explanatory report addresses this procedure ambiguously. It uses as an example a situation where the "person" subject to a request is the system administrator, rather than the individual under investigation. That is, the language of the convention appears to apply this power to individuals and suspects, and yet the language of the explanatory report attempts to describe this power with the example of a third-party system administrator, thus avoiding any discussion of legal and constitutional conflicts.

Real-Time Collection of Computer Data

Two articles within the convention are dedicated to ensuring real-time access to two specific types of data: traffic data and content. Both discuss the adoption of legislative and other measures to empower authorities to gain such access.

The key difference between the traffic data and the content data articles (20 and 21, respectively) is that an additional safeguard is included for content interception: the adoption of legislative and other measures must be "in relation to a range of serious offences to be determined by domestic law." That is, interception is considered to be an invasive measure and may only be applied to *serious offenses*. These "serious" offenses are not defined within the convention itself and are left to national legislatures. There is already a high level of divergence on this among various national laws. As examples, in the UK a serious offense includes investigations to protect national security interests and the economic well-being of the UK, while in the Netherlands any crime that is potentially punishable by four years in prison is a serious offense worthy of interception. In Italy, the police perform up to 100,000 interceptions a year, by far the highest rate in Europe.[11]

The collection of traffic data is also invasive. As mentioned earlier (in the discussion of the convention's scope, as declared in art. 14.3), governments may reserve that powers to gain access to traffic data are restricted *at least* to the same conditions under which police may intercept communications. Although the CoE recognizes that real-time surveillance of traffic data is invasive, this course of action is dissuaded, however: "the Party shall consider restricting such a reservation so as to enable the broadest range of application of the measure of collection of traffic data." The CoE goes further to dissuade governments from applying the serious crime consideration even for interception by stating, "Parties should consider applying the two measures [of access to traffic data and content] to the offences established by the Convention in Section 1 [substantive law], in order to provide an effective means for the investigation of these computer offences and computer-related offences." In effect, the CoE is asking that copyright crimes, fraud, and the other cyberoffenses be considered serious enough to warrant the invasive procedures such as interception, and real-time collection of traffic data.

Unanswered questions remain with regard to authorization and oversight of this form of surveillance. In some countries, authorization to intercept is granted by judicial authorization; in other countries, such as the UK, warrants are signed by politicians. In the realm of traffic data divergences are even greater: in some countries, real-time access requires more rigorous oversight than others. Police access to traffic data is poorly understood and even more poorly regulated. Various levels of privacy invasion arise when traffic data is accessed, particularly due to the multiple protocols and the nature of the various communications infrastructure under consideration. The CoE acknowledges that privacy may be threatened by the granularity of the data that may include web sites visited: "the collection of this data may, in some situations, permit the compilation of a profile of a person's interests, associates and social context. Accordingly Parties should bear such considerations in mind when establishing the appropriate safeguards and legal prerequisites for undertaking such measures, pursuant to Articles 14 and 15." While this is acknowledged, however, definite action is not stated nor required.

The CoE, within the explanatory report, also notes that necessity, subsidiarity, and proportionality may be considered, as well as a limitation on the duration of interception, right of redress, and other such safeguards reflected in the European Convention on Human Rights and associated jurisprudence. However, these measures are not required explicitly or necessarily within the convention, and future signatory states outside of the CoE, as well as the observer states, are not necessarily bound by the European Convention on Human Rights (ECHR). This gives rise to concerns with regard to mutual legal assistance as these powers of surveillance are applied beyond borders.

CoE on International Cooperation

One of the primary reasons for establishing the CoE convention was to enable governments to cooperate in investigations to ensure that international cybercrimes are fought through international means. But as with the procedural powers already outlined, the mutual legal assistance components of the convention aren't limited to cybercrime investigations. Assistance between "Parties" must be afforded "to the widest extent possible for the purpose of investigations or proceedings concerning criminal offences related to computer systems and data, or for the collection of evidence in electronic form of a criminal offence" (art. 25). The only exception to this rule is extradition: the components of the convention that enable extradition only apply to cybercrimes where the maximum imposed "deprivation of liberty" is at least one year.

Governments are compelled to cooperate with one another and there are very few opportunities for refusal. Requests for assistance from one government may be refused by another if dual criminality is required by national law (art. 25.5), though this course of action is discouraged. Governments may also refuse to assist in investigations if the suspected crime is deemed to be a *political offense*, prejudices the sovereignty of the state or its security, or affects the *ordre public* or other essential interests (27.4). The existence of national safeguards must otherwise be set aside in favor of cooperating. For instance, refusal may only occur on grounds of privacy law and data protection "only in exceptional cases." According to the CoE:

> A broad, categorical, or systematic application of data protection principles to refuse cooperation is therefore precluded. Thus, the fact the Parties concerned have different systems of protecting the privacy of data (such as that the requesting Party does not have the equivalent of a specialised data protection authority) or have different means of protecting personal data (such as that the requesting Party uses means other than the process of deletion to protect the privacy or the accuracy of the personal data received by law enforcement authorities), do not as such constitute grounds for refusal. Before invoking "essential interests" as a basis for refusing cooperation, the requested Party should instead attempt to place conditions which would allow the transfer of the data.

The conditions for refusal are considered modifiable, and are to be used infrequently. The CoE goes on to state:

> The requested Party may instead provide assistance subject to conditions. If the conditions are not agreeable to the requesting Party, the requested Party may modify them, or it may exercise its right to refuse or postpone assistance. Since the requested Party has an obligation to provide the widest possible measure of assistance, it was agreed that both grounds for refusal and conditions should be exercised with restraint.

Limitations such as data protection and other national laws and conditions are thus considered impediments to the higher cause of providing assistance.

Although governments are permitted to require dual criminality as a precondition to assistance, the convention does not support it. In fact, the convention obliges governments to abandon or disregard dual criminality in the case of preservation. The CoE argues that dual criminality is "counterproductive" on a number of grounds. First, it claims that the "modern trend" is to eliminate dual criminality for all but the most intrusive powers, and the CoE does not consider orders on ISPs to preserve data as being intrusive as there is no disclosure involved. That is, the preservation order can be made at the request of another country without checking for dual criminality because the order only requires the preserving of the data, while disclosure is negotiated at a later date. Second, establishing dual criminality may take too long, and in that time data may be deleted, removed, or altered.

For real-time surveillance across borders, however, little is said regarding grounds for refusal, except that assistance is governed by domestic law or existing treaties. In the specific case of traffic data, the convention acknowledges that domestic law may minimize cooperation because of the difficulty in providing access to this form of data. A requirement is established that "each party shall provide such assistance at least with respect to criminal offences for which real-time collection of traffic data would be available in a similar domestic case" (art. 33.2). The use of "at least," according to the CoE, is designed to encourage permission for as broad assistance as possible, "even in the absence of dual criminality."

CoE Conclusions

The convention on cybercrime is among the largest conventions on substantive, procedural, and mutual legal assistance law ever created. It is important to repeat that the convention deals with cybercrime only in its first section; the procedural law and mutual assistance regime apply generally to all investigations involving data in an electronic form. This sets the CoE convention apart from other international treaties on legal assistance: they usually focus investigative methods on criminal acts defined within the treaty (e.g., on trafficking, terrorism, etc.). Because this convention deals with computer-related issues, the procedural powers are defined around the existence of computer-data, not necessarily the occurrence of computer crime. This convention is not about cybercrime.

Each of the powers within the convention is defined loosely, allowing for maximum interpretative flexibility. One can interpret this flexibility as allowing for a comfortable fit (often minimization of powers) with existing domestic laws and protections. Yet the reservations included to minimize the powers are dissuaded as often as possible, and the failure to establish minimum protections and require dual criminality raises doubts and concerns among other actors.

A Topology of Lawful Access

The G8 and the CoE are essentially pursuing a set of lawful access powers, with some overlap. These may be grouped together under the categories of Preservation and Retention of Data, Disclosure and Search and Seizure of Data, and Real-time Access to Data and Communications.

Each of these powers provides a set of technology policy challenges, however. These will be synthesized as follows, describing the form of the power, the actors who are affected and regulated, the conditions of use and authorization issues, and some implications.

Preservation and Retention of Data

The CoE and the G8 both establish preservation schemes for data, particularly transactional data such as traffic data, and other volatile forms of data. Of the two international arrangements, only the G8 proposes retention.

The differences between *preservation* and *retention* are mostly of scope and timing of application. Preservation is often *invoked* or applied, while retention is a general practice or legislatively required.

Retention is a practice in which all data (or a policy-specified subset) is retained for all individuals, transactions, or communications. The idea is that this data may be used for unspecified future purposes. Retention involves past, present, and future data for it is a constant practice of retaining data as it is generated, for some extended period of time. As a result, it involves the accumulation of a massive amount of data, likely to be costly to store, and even more costly to search through for identifiable data upon request by governments.

Preservation is a practice applied only to existing data that is already collected by a service provider for the purpose of providing a service. The order for preservation of this data is made while conducting a specific criminal investigation, possibly dealing with a specific individual or communication. It does not deal with future or past data as it is based on a specific request rather than a normal operating practice. The purpose of the preservation request, as we have seen previously, may not always be well defined, if defined at all, particularly if the order is arising from a request for international cooperation and assistance.

The data being retained or preserved is commonly considered traffic data, subscriber data, communications-related data, or transactional data. For each data type, differing levels of protection and oversight may be required, but are often poorly thought out. The mere existence of these data stores creates a security risk, as well as a data protection liability. Moreover, as some of this data may be legally protected under legal privilege, for example, it may leave the service provider vulnerable to legal attacks.

In some cases, preservation may be conducted within the breadth of application of retention. For example, after the terrorist attacks on September 11, 2001, and July 7, 2005, the UK National High-Tech Crime Unit requested (for voluntary adherence) that UK communication service providers preserve all existing data currently within their data stores on all users in case that data would be helpful at some point in the subsequent investigations. That is, as the perpetrators were not yet identified it was considered prudent to store the data on all individuals until the investigations progressed. This level of data preservation and retention was considered to be a proportionate response because of the gravity of these heinous acts.

Since then the European Union has taken the lead on retention. As part of its responses to attacks in the United States, Madrid, and London, the EU implemented in December 2005 a directive requiring member states to implement data retention laws. All communications service providers are to retain the data on communications transactions made by their clients for between six and twenty-four months. This data can then be accessed by officials in all member states for investigations as they see fit.

Disclosure, Search/Access, and Seizure/Securing of Data

Separate from ensuring data is available is the set of powers to ensure that authorities may gain access to the data upon request. Some of these access powers are traditional, such as access to stored and subscriber data. Others involve new powers, such as partial disclosure powers and compelled disclosure of protected data. In each case, however, these powers have modern implications that may require enhancement of powers, and conflict with existing domestic legal protections.

Oversight and authorization for each of these powers may vary based both on the sensitivity of the data, and the level of coercive action required. Access to subscriber data (e.g., who owns the IP address in question) often involves minimal oversight, while access to transactional data (e.g., which web sites were visited) and protected data (i.e., decryption keys and decrypted text) may involve additional oversight/authorization.

Access to stored data and subscriber data may be subject to expedited requests and may involve requests from foreign officials. The scope of the requests can be for specific data or generalized access. Both may involve access by authorities or disclosure to authorities directly upon request while others may require judicial intervention.

Stored data may be subject to modification or deletion and thus may require search/access and seizure/securing provisions. It may also relate to data that has been preserved/retained. The power may apply to third parties (i.e., service providers and other data collectors) as well as to individuals (i.e., as a coercive measure against suspects). The type of data stored may be personal files and communications; and/or specific or general transactional data whose form and nature is dependent on the type of service provided, as well as the business model of the service provider.

The new power of *expedited partial disclosure* of traffic data raises new and challenging issues. In particular it entails the need to identify all of the service providers involved in a transmission. In such a situation, according to the CoE, the implementation of such partial disclosures involves three possible processes (to be decided by national law). The first is for the authorities to approach each service provider in the chain to identify the next provider in the chain, each with separate orders. This is considered as "unduly time consuming." The second (and preferred by the CoE) is to obtain a single preservation order that would be applicable to all service providers that are subsequently identified as being within the chain of a specific communication. Such a *roving order* has received a hostile reception when proposed in the United States and the UK in the past. The third process involves giving an order to a service provider that includes an additional order for that provider to notify the next service provider in the chain of the existence and terms of the preservation order; the second service provider would pass on the notice similarly.

Partial disclosure is intended to occur on an expedited basis to identify a characteristic of this data that may assist in identifying further data for preservation and/or accessing. The purpose of the investigation leading to the partial disclosure is not always identified, however; and international cooperation may not be in accordance with domestic law of the requested state.

Access to protected data can involve approaching third parties for data held on their systems relating to a specific individual, or applying coercive measures to individual suspects. In the case of third parties, permitting law enforcement authorities access to protected data can involve disclosing a password or granting access to files that only privileged users may access, such as system administrators. It may also involve access to encrypted data that involves decryption by these third parties; or access to decryption keys belonging to third parties, such as service providers, that protect data other than the specified data being sought. If the order is applied to individuals (which is ambiguous within the CoE, optional under the OECD guidelines on cryptography policy, and intentionally avoided by the G8), then this is a coercive measure to require individuals to provide plaintext of encrypted data, or the means to decrypt that data, such as through providing decryption keys. However, these keys may protect data other than the specified data being sought, and may harm general trust and security. Additionally, this may conflict with the rights against self-incrimination enshrined in some constitutions, charters, and conventions.[12]

Real-Time Access to Data and Communications

The process of adapting old surveillance practices to new technologies has been fraught with complexities. Real-time surveillance of telephone transactions was relatively simple: communications only existed for a brief moment across telephone lines so real-time surveillance was required. Surveillance of "computer communications" involves a wider variety of monitoring mechanisms. For instance, the surveillance of emails, chat sessions, and web-browsing all involve different procedures, some in real-time.

First there is the grey area of access by law enforcement agencies to stored communications. Communications may be accessed in real-time, as with the telephone system, or they can be accessed when stored, as with the postal system. Applying the same rules to access to stored communications as to stored data is likely to be problematic. Access to communications that are stored on mail servers or locally often requires greater protection than access to other stored data. This is because communications are often protected within laws and jurisprudence to a higher degree than other stored data because of its deemed "sensitivity."

There are also varying regimes of protection depending on where the communications are found (e.g., e-mail server v. local computer) and the state in which they are discovered (e.g., messages that are unread by the user v. messages that have been read and are stored). In some situations stored communications are treated as stored data and thus susceptible to search and seizure. For example, Canada's Department of Justice considered treating stored, but read, communication as stored data. According to their documents: "This stage is similar to the situation where a person, having read a letter, files it into a filing cabinet rather than throwing it into the garbage. Obtaining an e-mail at this stage is more analogous to a seizure than it is to an interception." The practice on stored communications varies from country to country, however. Australia, for example, tried to pass a law in 2002 to access stored messages in transit without a court order; but the bill was rejected by the Senate.

Real-time interception of communications is an invasive and often intensive procedure that grants authorities access either through a third party or directly to the content of communications as they are communicated. The exact definition of "communications" or "content data" is interpretively flexible, however. Interception is often specified for limited use, such as in the undefined concept of *serious crime* (as in Europe), or a list of offenses (as in the United States), or a list of punishments (as in Australia and the Netherlands), or a list of interests (as in the UK). Authorization procedures vary often, sometimes greatly. Moreover, it is technologically challenging to identify a specific data stream for interception; and this gave rise to much of the controversy surrounding the FBI's Carnivore system, installed at many ISPs to sort through all traffic in order to identify specific communications of a specific individual.

"Communication" may also mean the set of transactions undertaken while online, for example during a web-surfing session. Real-time access to traffic data is also an intensive and invasive procedure that grants authorities access either through a third party or directly to the transactional data involved in a given service. The type of data collected differs based on the type of service (http, ftp, mail, web mail, mobile phone location data, etc.). As a result, the definition of traffic data is interpretively flexible as well. The terminology surrounding this power also differs; sometimes it is referred to as transactional data, lifestyle data, pen-registers, or trap-and-trace, to name a few; and this is emblematic of the varying sensitivity of traffic data, its varying treatment and local sensitivities. Like interception, accessing data in real-time is technologically challenging for both the government and the service providers.

The Discourse

Policy discourse at a national level is valuable to the development of laws, as mentioned earlier. Through these discourses, interests are debated and procedures are negotiated to create a settlement. At the international level, however, discourse is limited. For both the G8 and CoE arrangements, access to the negotiations was limited, and sometimes hazardously so.

Industry representatives were formally invited into the consultation process at the G8 with the Paris, Berlin, and Tokyo summit meetings. After some pressure, civil society organizations were invited by the U.S. delegation in Berlin and Tokyo: while a consumer organization interested in fraud-related issues was invited to the Paris meeting, the Center for Democracy and Technology was invited to the Berlin meeting and the American Civil Liberties Union attended the Tokyo meeting. No other delegation included civil society. As a result, non-industry actors knew little of the events at the G8, let alone their outcomes.

The Council of Europe did not actively solicit comments, except for the publication of an email address for interested parties to submit comments on one of the earlier drafts and a one-day session in Brussels in April 2001. The CoE believed that consultation should occur at the national level. When the CoE was harangued for its lack of openness, its secretary general responded that the CoE is not an organization that customarily performs consultations. The CoE also responded that the release of a draft convention was actually a novel move for this treaty-making organization. At the national level, open meetings were conducted by the U.S. Department of Justice, upon request; while the Canadian Department of Justice and UK Home Office held closed meetings with industry representatives only, as civil society was not invited formally.[13]

After the Berlin G8-Industry summit of October 2000, an ad hoc coalition of primarily U.S. industry and civil society officials and representatives formed to discuss both international arrangements. This coalition included large and small telephone companies, Internet service providers, software companies, law firms, and nongovernmental organizations. Eventually the coalition reached out to industry organizations and NGOs in other countries. Most important, however, the coalition reached out to the U.S. Department of Justice (DoJ), and the DoJ began to reach out in return.

The discussions with the DoJ proved enlightening. The DoJ educated the ad hoc coalition on international cooperation; and in return was informed on industry concerns. These concerns included some of the definitions of crimes, but focused mostly on the procedural law and international cooperation. As the CoE finalized the text, the U.S. negotiators were able to convince the other members of the CoE drafting committee to adopt amendments based on the concerns of the coalition, including article 15 on safeguards.

Despite some concessions by the CoE, the coalition remained concerned, however, and the DoJ generally tried to listen. When pushed by the ad hoc coalition to call on the CoE for further amendments, the DoJ representatives returned from the next drafting meeting to say that they had effectively expended any remaining goodwill of the other PC-CY drafting committee members. That is, DoJ reported back to the ad hoc coalition that there was a growing resistance within the CoE to the changes being recommended by the U.S. delegation.

In response to the resistance, the DoJ representatives said that they would take to the next drafting committee meeting any submissions, letters, and comments that the ad hoc coalition wished to send along. The coalition responded with an open letter to the U.S. government, to be taken to the CoE, to show its support for the DoJ in the hope of giving additional support and voice to further changes. The letter stated:

> Once the process was opened up in 2000, our coalition and others have undertaken to consult broadly with the Department of Justice and those it has convened, and have offered substantial critique of the draft treaty, and suggested changes which are essential for gaining support from the coalition's participants. We did have serious concerns about the original process which was undertaken in a closed manner, as was typical with the CoE. However, once the efforts of our DoJ were successful in opening up the process, we welcomed and continue to support the efforts which the DoJ continues to engage in, in consulting and seeking to work toward changes in the treaty. We want to again express our appreciation for the commitment and extensive efforts of the DoJ and others from government, to achieve the changes which have been obtained.
>
> However, the treaty, as presently drafted, is not yet supportable, in spite of those changes which DoJ and others have worked to obtain. We fully appreciate that this is an international environment, and that the U.S. alone cannot champion international private sector concerns.

The ad hoc coalition also drafted a "fact sheet," warning the negotiators from other CoE countries that if the recommendations of the DoJ and the ad hoc coalition were not attended to, then the convention as a whole would suffer as the United States would be unable or unlikely to ratify the convention. The fact sheet stated:

This fact sheet describes several issues of concern to the European negotiating states if the CoE Cyber-crime Convention is ratified by the principal European negotiators, but rejected by the U.S. either through withdrawal from the process, or through failure to ratify.

- Today, much of international Internet traffic including traffic to or from Europe and the Asia-Pacific is routed through the U.S. Internet backbone. Moreover, a very significant percentage of Internet content resides on computer servers in the United States. This makes the U.S. a critical player in cybercrime investigations.
- If the U.S. is not a participant in the regime of the Convention, European law enforcement interests involved in preparing this Convention will not receive the benefit of the significant procedural cooperation vehicles provided by the Convention with regard to data preservation, traffic data, seizure of evidence, and interception.
- Should the U.S. refuse to ratify the Convention the voluntary cooperation extended today to European law enforcement in preliminary investigations, for example, may become slower and more difficult, as different regimes of cooperation develop among countries that are and are not parties to the Convention.
- Cooperation may also be hampered because U.S. and international companies will be less certain of the legal implications of European requests for assistance.

Specific industry organizations also submitted their own comments and voiced individual concerns. According to NetCoalition.com, a group representing a number of technology companies: "The Convention contains no provision limiting European jurisdiction over Internet sites located solely in the U.S. The drafters of the Convention have made clear their intention to extend the Convention's harsh liability structure to hate speech, even though it is protected by the First Amendment in the U.S."

Another industry group, Americans for Computer Privacy, argued that although there were benefits to the convention they remained concerned: "Throughout the negotiations we have expressed concern that the treaty must not either by its terms or as a pretext for other countries' actions . . . violate the legitimate privacy rights of Americans at home or work in today's electronic world."

U.S. industry organizations were not alone. The coalition eventually included some European telecommunications companies and Internet service providers,[14] and international organizations. These actors also became active in voicing concerns. The World ISPA forum, for instance, had concerns about international cooperation among the various countries that would be parties to the convention, and the effects on civil liberties. According to one statement: "[The convention] should include a requirement on each country to offer an oversight mechanism to ensure human rights are being protected within their territory. . . . In addition, the draft Convention does not permit countries ratifying the Convention to take any reservations on any of these points."

The European Telecommunications Networks Operators (ETNO) association was also concerned with international cooperation. In their call for changes to the convention they stated:

Therefore, ETNO considers it essential that any proposed measures meet the following requirements:...

- Application of a dual criminality test in cases of cross-border cooperation.
- Cross-border co-operation should not lead to operators or ISPs receiving directly orders from a foreign law enforcement body.

The majority of these industry concerns did not lead to substantial changes in the final version.

Civil society was also active. A number of organizations were integral parts of the ad hoc coalition, including the American Civil Liberties Union, the Center for Democracy and Technology, the Electronic Frontier Foundation, the Electronic Privacy Information Center, and Privacy International. These NGOs also cooperated under the aegis of the Global Internet Liberty Campaign (GILC) to write a number of letters to the CoE chairman, each letter signed by at least twenty other organizations worldwide. The October 2000 letter argued:

We believe that the draft treaty is contrary to well established norms for the protection of the individual, that it improperly extends the police authority of national governments, that it will undermine the development of network security techniques, and that it will reduce government accountability in future law enforcement conduct.... Different countries have different procedures, admittedly, but now is the opportunity to harmonize them, on the condition that we assure a high level of consistency on individual rights protections.

When the CoE released draft version 24.2, GILC responded with another letter,[15] this time stating, "To our dismay and alarm, the convention continues to be a document that threatens the rights of the individual while extending the powers of police authorities, creates a low-barrier protection of rights uniformly across borders, and ignores highly-regarded data protection principles." In a later letter responding to the call for submissions by the DoJ, the ACLU, Electronic Privacy Information Center, and Privacy International[16] argued that the amendments that were made to date, even the inclusion of safeguards (article 15), were insufficient:

We recognize that the legal protections have been modestly improved in Article 15 by the reference to various other international instruments, but we still believe that the protections it affords are not adequate to address the significant demands and requirements for privacy-invasive techniques in the rest of the Convention.... A vague reference to proportionality will not be adequate to ensure that civil liberties are protected. We recognize that countries have varying methods for protection of civil liberties, but as a Council of Europe Convention drafted in consultation with other democratic nations, this document missed an important opportunity to ensure that minimum standards consistent with the European Convention on Human Rights and other international human rights accords were actually implemented.

The draft convention was finalized shortly after these letters were received, and as a result no subsequent amendments were made.

Other nontraditional actors also contributed ideas and voiced concerns, including epistemic communities and nonexecutive governmental bodies. Early in the G8-Industry consultations some computer experts and academics were invited to speak, particularly at the Paris summit of 2000. Similarly, early in the CoE discourse a letter was drafted, signed, and submitted by a group of leading researchers and experts on their concern "that some portions of the proposed treaty may inadvertently result in criminalizing techniques and software commonly used to make computer systems resistant to attack."[17] The contentious parts of the convention were clarified in a later draft, the group of experts welcomed those changes generally, and the organizer of the letter recommended that people still interested in the convention look to the GILC's comments and letters, as "it is worth reading as it addresses a broader range of concerns." From that point onward, the epistemic community remained relatively silent.

Meanwhile, the leading nonexecutive governmental body in both discourses has been the European Commission's Article 29 Working Party on Data Protection. This collection of EU privacy regulators commented on some of the G8 proposals, particularly on preservation powers and on G8 calls for "balancing" data protection regulations. The Article 29 Working Party responded with a reminder that any proposals must be consistent with the European Convention on Human Rights: "The legal basis must precisely define the limits and the means of applying the measure: the purposes for which the data may be processed, the length of time they may be kept (if at all) and access to them must be strictly limited. Large-scale exploratory or general surveillance must be forbidden. It follows that public authorities may be granted access to traffic data only on a case-by-case basis and never proactively and as a general rule."

The working party was also concerned about the divergent laws even within the European Union regarding the collection of traffic data, and recommended some form of standardization. Regarding the CoE, it commented on version 25 of the convention by releasing an "opinion." The working party was concerned with the weak article 15 safeguards that were being insufficiently harmonized, and the failure to require effective safeguards. The opinion articulated concerns that non-CoE countries may sign on and would not have ratified the ECHR. Similar concerns applied to the failure to require dual criminality in international cooperation, and other forms of limitations that meet the tests of necessity, appropriateness, and proportionality. After this opinion, though, the Article 29 Working Party was relatively mute.

The media reception to the CoE convention and the G8 work was not always warm. A number of news articles appeared speaking of the "globalization of law enforcement," "the snoopers' charter," and "the long arm of government." Growing concerns with privacy and due process caused the CoE to speak out publicly regarding these concerns. In a press release on the day of the signing of the convention, the CoE stated,

"The text covers only specific criminal investigations, and certainly does not lend itself to the setting up of an Orwellian-style general electronic surveillance system.... [The Convention had been drawn up with care so as to strike] a precious balance between the requirements of criminal investigations and respect for individual rights."

Over twenty countries signed on to the convention on that very day. There are now few mentions of the controversy in the media, and the convention appears as a stand-alone authoritative document.

The ad hoc coalition has also been active since then, particularly with respect to the additional protocol on hate speech; but the level of activity is drastically lower since the main text of the convention was closed in the summer and fall of 2001. Civil society organizations continue to monitor the developments, awaiting the convention's introduction into national parliaments for ratification, or the G8's policies into national law.

Since the events of September 11, there has been a deluge of legislation with components of both the CoE and the G8 recommendations embedded within anti-terrorism policies. As examples, the USA PATRIOT Act, commonly known as the Patriot Act, created legal procedures for lawful access to Internet traffic data, and legal standing to the controversial interception device Carnivore; the United Kingdom passed the Anti-terrorist, Crime, and Security Act 2001 that allowed for data retention, later followed by a number of European countries and eventually the European Union as a whole; and the Canadian Anti-Terrorism Act 2001 allows for spontaneous data sharing between countries and exempted police databanks from data protection regulations. The wait continues, however, for formal ratification procedures to begin in all of these countries.

Not about Cybercrime: Discourses and International Regimes

The CoE and the G8 policies were developed to respond to, among other things, the jurisdictional curse to state sovereignty due to transborder activity. As transborder activity necessarily reaches beyond the jurisdiction of national law, this would have a chilling effect on the very creation of new laws. When the United Kingdom tried to regulate electronic commerce in the interests of law enforcement, articulations and objections arose naturally on the basis of infeasibilities, regulatory burdens, and competition, among others. When the United States tried to regulate the development of cryptography, off-shore developers and foreign competitors seized on the opportunities presented by regulatory arbitrage. The solution, offered by the G8 and the CoE, is harmonization and international agreements: if all countries pass similar laws, arbitrage risks would be abated.

If there is anything to learn from the events surrounding the United States withdrawal from the Kyoto protocol, it is that the ratification by the United States is key;

and this relies on a politically equitable and just distribution of burdens. Whether "just" and "equitable" mean much in this era of wars on terrorism remains to be known, particularly as the discourse surrounding the G8 and the CoE has dissipated significantly since 2001. The issues raised by the discourse are unlikely to disappear or become irrelevant, however, and this is due to the lawful access regime, and the arising technology policy challenges.

On Discourse and Access

The two discourses surrounding each international forum and set of agreements varied in form, although the constitution and results were similar.

The G8 pursued closed summit-based meetings until the G8-Industry meetings of 2000 and 2001. The membership of delegations to the G8-Industry meetings was decided by national governments, sometimes in consultation with industry members. With the exception of the U.S. delegation, however, all nongovernment attendees were from the private sector; the United States invited a consumer protection NGO to Paris, the Center for Democracy and Technology to Berlin, and the ACLU to the last meeting in Tokyo. There was also some media presence and coverage following these events. After the Tokyo summit, however, further consultation was not pursued, while after September 11, consultation ceased and deliberation and decision making returned to the level of ministerial and heads-of-state summits.

The CoE held a consultation session where comments on one of the early drafts could be submitted to an e-mail address, while some CoE representatives spoke at conferences and to the press. Generally there was no active consultation by the CoE, with the exception of a one-day meeting in April 2001. Activity from interested organizations was much more organic, with coalitions formed in the United States, through the assistance of the U.S. Department of Justice. After the convention draft was finalized in Spring 2001, discussion generally quieted down as signing ceremonies were deemed relatively formal, and actors looked to ratification processes as the next forum for discussion. The media, who had been more active in covering the CoE than the G8, also generally quieted down.

With the exception of some changes, such as the inclusion of article 15 on safeguards, as a direct result of the work of the Center of Democracy and Technology, few changes resulted from the consultation process of the CoE on the basis of civil liberties. Industry cost and scope concerns were attended to slightly more by the CoE. Meanwhile, most substantive changes to the G8 agenda by the consultation with industry were ignored in the post-September 11 environment.

It would be fair to say that in both processes many of the same actors voiced concerns, with the exception of nongovernmental organizations. NGOs did not substantively comment on the G8 process, with the exception of those invited by the U.S. delegation. In the CoE discourse, NGOs played a much larger role, at the national level in the United States particularly with some response from the DoJ, and directly to the CoE through the GILC; but with very little response.

While the U.S. Department of Justice was remarkable in comparison to the other ministries of justice and home affairs through inviting NGOs to the G8 and holding open meetings with coalition members, and even speaking at public conferences, there was still some concern. As the CoE drafting meetings were closed, it was not possible to be sure that the DoJ was actually representing interests other than its own. According to a leading expert in the United States, James X. Dempsey from the Center for Democracy and Technology, "no one from outside the room can know what the Justice Department is saying and whose interests they are representing." Even as the United States reported back to the ad hoc coalitions, there was no way through which their work could be verified, and thus it relied too much on the good faith of the coalition members. After all, the DoJ holds many interests of its own;[18] expecting it to clearly represent the interests of others may be optimistic, or naïve.

The DoJ is not alone, however, as all the actors and constituencies held a rich set of interests. Not all NGOs opposed content regulation. Not all industry organizations opposed the substantive law definitions (particularly copyright protection). Similarly, it can be said that not all justice and home affairs representatives supported international cooperation; or other ministries and departments may have opposed some of the requirements for fear of harming commerce. This is likely to be endemic of any political discourse in a democratic system.

What is disappointing is that neither the CoE or the G8 cultivated discourse; and now one has to ask what is the value of national discourse once a treaty is sent for full ratification, for two reasons. First, because of the size and breadth of the CoE convention and G8 proposals, most actors can find something that meets their interests within the agreements, such as procedural law to investigate hacking crimes, and substantive law regulating fraud. Secondly, international "obligations" can be seen as *faits accomplis* and full ratification is usually required; countries may not really change the substance of a treaty at the national level once the treaty is agreed upon. National discourses at this stage appear relatively useless, or disingenuous.

On Technology Policy and Lawful Access

The problem with the breadth of the agreements and the appearance of a *fait accompli* is that there are many subtleties involved in the lawful access procedures. The procedural powers discussed here are either new powers, or due to technological details, involve a departure from traditional powers. Access to protected data is new, as is data retention. Access to traffic data, through preservation or real-time, represents a more invasive act because the innocuous information held by telephone companies contrasts sharply with the information that ISPs may collect on a user's web-browsing habits.

As an indication of the importance of national discourse for technology policy, consider the United Kingdom where all of these procedural powers exist on statute already. The Regulation of Investigatory Powers Act 2000 created powers of interception of communications, access to traffic data, and forced disclosure of protected data and decryption keys; but this legislation was developed through a rich discourse that resulted in a large number of amendments. Interception costs were discussed in detail, resulting in an altered regime to minimize risks. Traffic data was accepted as being sensitive in some circumstances, and as a result highly sensitive information such as web site addresses were treated as content instead of the lower authorization bar of traffic data. Self-incrimination, burden of proof, and security risks introduced originally under forced disclosure and production rules were changed to minimize harm to trust, security, and fundamental human rights. Once the technological and legal details were explained to parliamentarians (with a dose of lobbying of course), a less unreasonable law was developed. In contrast, the data retention powers enshrined in the Anti-Terrorism, Crime and Security Act of December 2001 lacked sufficient discourse (as was common among many laws enacted in that post-September 11 environment), and the conflicts with human rights, technological feasibilities and challenges, and regulatory and cost burdens remain grave concerns for many of the actors.

As it stands in the G8 and CoE arrangements, due to vague formal policy language, the lawful access provisions regard traffic data on the Internet to be as sensitive as traffic data over telephone systems; but then the technical language within the procedural documentation shows how quite specific this data may be. Even the CoE acknowledges that traffic data may be more sensitive depending on the communications medium, but the convention does not reflect this realization (nor does the U.S. Patriot Act).

The same applies for real-time access to communications and content; the type of data that may be collected is quite different from the data collected from interception of the postal system, or interception of telegraphs, or interception of conversations. As the technology changes, the nature of the power transforms; as does the structure of its collection. Radio interception can occur without any interruption; originally, telephone interception required access to telephone lines and switches; interception of modern telephone communications requires built-in interception capabilities;[19] interception of communications over the Internet requires a number of mechanisms, sometimes located in a number of countries—for example, a suspected hate promoter in France who uses Hotmail based in the United States.

A good indication of how the technology changes the legal environment is the introduction of the power of expedited preservation and partial disclosure of traffic data. This new form of warrantry, however, needs to be applied to a number of service providers, perhaps in a number of jurisdictions. The disclosure must occur immediately, without regard to proportionality, the possibility of contemplating whether the disclosure is appropriate, or whether the crime being investigated is indeed a crime. As a result, France could require a U.S. ISP to expeditiously preserve and disclose information regarding an American citizen without having to disclose the nature of the crime being investigated; and similarly in return.

The ambiguities in the text of the convention and within the G8 fails to represent any of the lessons learned from the UK process and other detailed national discourses, and may lead to countries adopting practices and language blindly, under the veil of international obligations. Even the best-case scenario is that each set of actors within each country will now have to re-invent the lobbying/advocacy wheel that was formed in these other national discourses, such as in the United Kingdom surrounding the Regulation of Investigatory Powers Act 2000; this is a lot to expect from many G8 and CoE countries where civil society and industry actors are not as well resourced or attentive to such issues.

On Mutual Assistance and Cross-Border Policy Making

There are also subtleties to international cooperation that may not be noticed within national discourses. While mutual legal assistance is increasing, it is also an opaque area of transnational activity. The treaty-development forums are relatively closed, as the G8 and the CoE proved to be; and the details of international law are not always well known to nationally oriented NGOs. What the fair-trade movement has achieved for the WTO and other forums has not been seen in criminal matters. Instead, more worrying trends are appearing.

The CoE's mutual legal assistance regime is the product of many years of work. According to Fijnaut (2001), in the 1980s the CoE attempted to create a mutual legal assistance treaty allowing for cross-border pursuit, surveillance, and confiscation of goods, and cross-border interception of communications; but no agreement could be reached because of its unwillingness to "modernize." "The result was that the CoE, in effect, put itself out of contention in the reorganization of international mutual assistance in Europe.... The German State Secretary Schomburg openly declared at an international congress in 1990 that it was high time for international mutual assistance to be modernized and that, at the level of the member states involved, there was no longer any place for the principle of speciality, the principle of double punishability or grounds for refusal."

The Council of Europe managed to create such a "modern" convention in 2001 under the aegis of cybercrime. This was possible because, unlike other conventions and treaties that agree on harmonized substantive law, that is, common criminalization of activities, it was also defining a legal regime for investigating a "new" infrastructure involving transnational data flows. The CoE felt that it was therefore natural, if not imperative, not only to ensure that the powers of investigation were created to be used in cooperation with other countries, but also to allow these powers to be used for all forms of investigations. As a result, the convention and its mutual legal assistance regime, as a national policy device, remain controversial. The CoE convention is poorly entitled: it is not really about cybercrime.

Again, this international arrangement is appealing to national governments because it appears to deal with disastrous and publicized issues of cybersecurity and trust, and therefore ratification is necessary in order to combat hacking and child pornography. It is also appealing because it requires that the powers of investigation be applicable for any crime; thus saving the time of national parliaments by killing a number of birds with one stone. Finally, both the G8 and CoE arrangements provide model language for national policy implementation; instead of having to come up with localized language, the intentionally ambiguous and flexible language of the CoE and G8 may be used. Local appropriation may involve interpreting the text to create fewer protections and safeguards, further fracturing legal consistency and harmonization.

Finally, the increase of international cooperation may have worrying effects on national due process protections, perhaps leading to a situation where governments may use regulatory arbitrage to their advantage. If the procedures for national law enforcement agencies appear too onerous, they too may seek the assistance of other jurisdictions to circumvent their own national laws. An example of this, within a controversial context, is the case of Zacarias Moussaoui, the suspected twentieth hijacker on September 11. Having arrested him in August 2001, the Federal Bureau of Investigation wished to gain access to his computer; however they felt that they could not justify this search adequately under U.S. law (which turned out to be an erroneous assumption). Rather, the FBI concocted a plan to send Moussaoui to France where French officials could more easily gain access to his computer to send the data back to the U.S. officials.[20] According to reports, the extradition was set for September 17, 2001.

This "modernization" and simplification of international cooperation is a sign of things to come. In January 2003, a draft bill developed by the U.S. Department of Justice combined the international war on terrorism with criminal matters and asked for the authority to request search warrants in response to international requests, particularly traffic data when there is no treaty. The bill stated, "The United States therefore may find itself in a situation where it cannot assist a foreign government in one of its criminal investigations, which is hardly an effective way of encouraging foreign allies to assist our own counterterrorism investigations" (sec. 321).

The draft bill also addressed the problem of mutual legal assistance treaties drafted around specific topics and framed for specific eras: "Many of the United States' older extradition treaties contain 'lists' or 'schedules' of extraditable offenses that reflect only those serious crimes in existence at the time the treaties were negotiated. As a result, these older treaties often fail to include more modern offenses, such as money laundering, computer crimes, and certain crimes against children. While some old treaties are supplemented by newer multilateral terrorism treaties, extradition is possible under these newer treaties only if the other country is also a party to the multinational treaty, leaving gaps in coverage" (sec. 322).

The draft bill calls for the ability to extradite without the need for a treaty. Otherwise, the Department of Justice warns "that the US can become a 'safe haven' for some foreign criminals, and that we cannot take advantage of some countries' willingness to surrender fugitives to us in the absence of an extradition treaty; these nations usually require at least the possibility of reciprocity" (sec. 322). The United States ratified the convention on cybercrime in April 2006. The DoJ is interested in ensuring its capabilities adhere to the spirit of the convention, and even going further.[21]

The fear of creating "safe havens" for criminality must be countered with the legal realities of our time: no amount of harmonization will reflect the diversity in law that is endemic to our international system. After all, each country, despite numerous similarities in their laws, has different procedures and different constitutional requirements and interpretations. The UK, for example, only requires ministerial authorization for interception of communications; in the United States a judicial warrant is required. The U.S. constitutional protection against self-incrimination may prevent lawful access to decryption keys. According to Goldstone and Shave (1999),[22] "Such compulsion may implicate the Fifth Amendment's privilege against self-incrimination. Therefore, compulsion of an unrecorded password may not be a reliable process for U.S. law enforcement to access plaintext of encrypted data, while other countries may safely presume the legality of such compulsion when designing their procedures."

Situations such as these become even more complex if we consider the varying legal systems of the forty-five member states of the Council of Europe, plus the legal systems of the additional observer states, all with different procedural laws; the more countries, the more complexity in the legal issues.[23] According to Goldstone and Shave, this is not uncommon necessarily:

> The variation among procedural laws can be exacerbated by direct conflicts among the procedural laws of different countries. This problem is best exemplified by the scenario presented in [*United States v. Bank of Nova Scotia*], where a Canadian bank was held in civil contempt for failing to comply with an order enforcing a grand jury subpoena duces tecum notwithstanding the fact that compliance with the subpoena would have required the bank to violate a Bahamian bank secrecy rule. As more companies take advantage of computer networks to operate internationally, those companies increasingly become subject to the laws of multiple nations. As more investigations of crime committed over those networks are conducted—and as the laws regulating privacy of electronic data evolve—more conflicts are sure to arise.

Interestingly, these differences in legal systems were acknowledged by the G8 and the CoE, but instead of arguing for greater constraint in the prescribed powers, the differences were used to highlight the problems in requiring dual criminality for mutual legal assistance.

On Concluding Balance

With multiple actors and the plural interests operating within this maturing area of international law and challenging domain of technology policy, it is alarming that, within the discourse, rhetoric regarding "balance between privacy and security" continues to permeate. Both institutions stated that their solutions were ideal balances of individual rights and societal protection.

While there are many interesting properties of the discourses that have arisen, one of the most remarkable is the complexity of interests and issues. In the days of state-owned telephone operators and simpler markets based on monopolies and the according technological innovation, discussion of balance may have been appropriate. In our economic, legal, and technological environment there are many additional actors. Laws passed regarding surveillance of communications now affect small service providers and large providers in vastly different ways. With mutual legal assistance, we may be placing companies at the mercy of laws in other parts of the world. These lawful access regimes create worries among industry regarding costs and technological feasibility at every step of the way.

As a result, the options and varying factors within a policy discourse are plentiful. Reducing the terms surrounding the two supposed opposite poles of privacy and security reflects two simplistic issues in comparison to the harsher economic and relatively uncompromising technological conditions of our times. A richer discourse might have brought the myriad of interests forward for a more enlightened discussion; rather, at the international level, the discourse was minimized and simplistic statements on balance were thus left uninterrogated, to the benefit of few.

Recommendations on Policy Discourse

The interesting effect of the CoE and the G8 work in this area is that some in civil society and industry have become more aware of issues relating to mutual legal assistance and international cooperation, issues that previously were often beyond their scope of attention. Meanwhile, these international governmental organizations are accustomed to operating under closed circumstances. Due to the issue of the regulation's effects on civil liberties and on the operating costs of responding to foreign requests and on implementing surveillance techniques, and because of the technological infrastructure required, the general public is becoming involved in a usually arcane body of law and literature. Accordingly, these international organizations must be more open to promote discourse. This discourse must include nonstate actors, and when a diversity of ideas is added to the discourse we may abandon much of the old rhetoric.

The task ahead for those interested in participating in the upcoming national discourse is challenging. Policies negotiated at the international forums are now coming to national legislatures in the form of sometimes specific, but often broad statements and mandates for new laws and powers. There are means for reinvigorating discourse, even within these closed forums; and opportunities for these nongovernmental actors to still effect change.

Practice and Propose Liberal Interpretations of Instruments

Some components of the conventions and policy proposals are, due to legal differences among countries, ambiguous and vague. While this ambiguity and vagueness is likely to be used to the advantage of the policy setters, civil society and other actors may also interpret the texts proactively to meet their own interests. In a sense, the purpose is to generate *creative compliance* in the interests of civil liberties and due process rights. This requires some expertise, however.

Guidance and Policy Libraries

As more countries introduce similar policies due to these international pressures and obligations, and as some countries may conduct more thorough national dialogues, some changes may be achieved in these improved circumstances. Reports on the quality of the national discourses, the changes in proposed policies, and reports of the input by civil society and industry can be used as educational material to be replicated in other countries. Academics and policy researchers may play a role in providing such guidance and policy libraries for reference.

Internationalize Policy Attention

Due to various constraints, many within civil society and industry organizations have been focusing on national events and not on international developments. As these policies are being negotiated at international forums, greater attention is required beyond immediate borders. This often reaches beyond the mandate of most NGOs, however.

International Institutions to Place Obligations for Dialogue on States

Increasing transparency of international institutions is a common goal nowadays. A different approach would be to place pressure on international institutions to require, either through practice, policy, or treaty texts, that member states must have national dialogues with interested parties. This in turn may increase the transparency of the international institutions. The U.S. Department of Justice should be credited for its earlier work with industry and civil society during the later stages of the CoE convention drafting.

Require Transparency in Mutual Legal Assistance

Attention to criminal activity and investigations is also generally limited to within the jurisdiction of a nation-state. Reporting practices of governments on the use of surveillance techniques is generally regarded as interesting only in domestic situations. The area of mutual legal assistance needs to be opened up for research and analysis at the practical level, to see the procedures being followed, the interests pursued in negotiations, and barriers to cooperation.

Include as Many Voices as Possible

Assuming that all state interests are the same is a mistake. Often the trade and industry arms of governments, and even the human rights arms of justice and home affairs departments may have objections or different opinions regarding international policies and their negotiations. In the discourse surrounding the CoE, the alignment of industry and civil society in the United States was powerful, although not necessarily significant, as it has been in other policy discourses. NGOs and industry organizations may wish to seek out other communities, such as bar associations (which played interesting roles in the negotiations of a number of anti-terrorist laws in 2001), and other such epistemic communities. Most important, NGOs must, if they have not already, broaden out to reach nontechnologically focused civil society.

To conclude, the final recommendation is a rather conceptual one: we must abandon exclusive concerns with technology as though it was something particularly separate from the world. With the waning interest in the new economy and the intellectual criticisms of "cyberspace as a place," we must accept that technology is no longer separate but rather an integral part of our daily lives. Considering the interest in the CoE convention on cybercrime as meeting the daily concerns regarding security of computers while allowing for surveillance in all cases of criminal investigations suggests that it is misleading to focus only on "cyber" issues, since governments are no longer doing so as well. Privacy and civil liberties are not the exclusive concern of technology policy discourses; nor should these policies be the exclusive concern of technology policy experts in civil society.

This is not at all a recommendation to abandon concerns regarding technology. In fact it is interesting to note that the powers of law enforcement agencies are in many cases increased by the changing technological infrastructures. This leads to an interesting dialectic in the debate: to focus or not to focus on technology. The interests of wider society may not be captivated if the discourse is focused on technology; and the true implications of the policies will not be understood. Yet at the same time, these policies are affected and even extended by the technological conditions, constitutions, and changes. So while the G8 and CoE policies are not only about high-tech or cybercrime, the hazardous effects of these policies are only exacerbated by the technology.

Therefore a reasonable and appropriate discourse would acknowledge the intent of the policies as being about the wider law enforcement and national security interests, while also acknowledging the role that technology plays in changing regulatory environments. Laws may need to be updated to allow for lawful access to data within new technological infrastructures; but laws also need to be updated to allow for adequate protection of other interests such as civil liberties, regulatory burdens, and technological innovation. We have focused much on lawful access and poorly on the others.

Notes

Components of this work were supported by resources made available by the American Civil Liberties Union, Columbia University, the Electronic Privacy Information Center, Industry Canada, the Markle Foundation, the London School of Economics, and the Social Science Research Council. The author is grateful for the assistance of a number of individuals and organizations who were involved in this research and writing process. Most notable are my colleagues in civil society, especially David Banisar, Barry Steinhardt, and David Sobel, with whom I have collaborated on a number of occasions, and shared information and ideas. In industry, Stephanie Perrin and Jacques St-Amant, members of the Canadian Delegation to the G8, and members of the ad hoc coalition were helpful. In government, most notable were Betty Shave (U.S. DoJ) and John Fennel (UK Home Office), as well as Jim Ladouceur (Industry Canada). Within academic circles, I would like to thank Hayward Alker and Martin Dodge for their advice, and the Information Technology and International Cooperation fellows and the other researchers at Columbia University's Institute for Social and Economic Research and Policy, who suggested a number of research approaches.

1. For an excellent exposition of the techno-optimist critique, see Goldsmith 1998a,b.

2. See Australian Broadcasting Authority 1999; BBC 2002; and Lee 2001.

3. See Electronic Privacy Information Center 1997; Clarke 1999. Dogcow, "Evading the Broadcasting Services Amendment (Online Services) Act 1999."

4. See Akdeniz 2001.

5. According to Patrick 2002.

6. Hosein and Whitley 2002 has an exposition of the articulated arguments in the United Kingdom's debates surrounding encryption policy.

7. The author of this chapter contributed to this document.

8. An associated document was also created, the Explanatory Report, but this is a nonbinding statement of the CoE's intentions and some elaboration. It too changed over time.

9. The response to the attacks from the Committee of Ministers called for greater international cooperation and for ratification of the convention. Committee of Ministers, "Communiqué on International Action against Terrorism," 2001.

10. In version 25, from December 2000.

11. See Privacy and Human Rights 2005 from the Electronic Privacy Information Center and Privacy International, available at www.privacyinternational.org. Alternatively see the European Digital Rights (EDRi) resource on wiretapping in Europe at http://www.edri.org/issues/privacy/wiretapping.

12. Beatson and Eicke 1999 is a legal opinion on this issue in the United Kingdom, based on a draft bill that later developed into the Regulation of Investigatory Powers Act 2001.

13. The author had interactions with the UK Home Office representative on an individual basis, but not as part of a larger consultation program.

14. It is also worth mentioning that some of the U.S. companies also operate in Europe and articulated their concerns accordingly.

15. The author of this chapter wrote this letter on behalf of GILC.

16. Along with the October 2000 letter, the author of this chapter contributed text to this intervention.

17. See Spafford 2000.

18. For an interesting exposition of its confused mandate, see Burnham 1996.

19. In the United States this was enabled under the U.S. Communications Assistance for Law Enforcement Act 1996.

20. The Congressional testimony of Eleanor Hill recounts this story in detail; see Hill 2002.

21. This was made clear in a DoJ press release regarding the leaked document.

22. Betty Shave works at the DoJ, and was the lead contact for the government-coalition discussions.

23. This point has been made previously by Anderson (1989), among others.

References

Ad Hoc Coalition. 2001a. "Examples of Reasons CoE States Should Be Concerned about U.S. Ability to Support CoE Ratification: Delivered to the Department of State." Washington, DC, March 29.

Ad Hoc Coalition. 2001b. "Letter to John Sopko, Acting Assistant Secretary for Communications and Information, U.S. Department of Commerce." March 6.

Akdeniz, Yaman. 2001. "Case Analysis of League against Racism and Antisemitism (Licra), French Union of Jewish Students, v. Yahoo! Inc. (USA), Yahoo France, Tribunale De Grande Instance De Paris, Interim Court Order, 20 November 2000." *Electronic Business Law Reports* 1, no. 3 (2001): 110–220.

Americans for Computer Privacy. 2001. "Letter to the Honorable John Ashcroft." Washington, DC, June 22.

Anderson, Malcolm. 1989. *Policing the World: Interpol and the Politics of International Police Co-Operation*. Oxford: Clarendon Press.

Article 29 Working Party. 1999. "Recommendation 3/99 on the Preservation of Traffic Data by Internet Service Providers for Law Enforcement Purposes." Brussels: European Commission.

Article 29 Working Party. 1997. "Working Document: Working Party on the Protection of Individuals with Regard to the Processing of Personal Data Notification." Brussels: European Commission.

Australian Broadcasting Authority. 1999. "Broadcasting, Co-Regulation and the Public Good." October 29.

BBC. 2002. "China Internet Firms 'Self-Censoring.'" *BBC News Online*, July 5, 2002, 16:14 GMT 17:14 UK.

Beatson, J., and Tim Eicke. 1999. "In the Matter of the Draft Electronic Communications Bill and in the Matter of a Human Rights Audit for Justice and FIPR." Available at www.fipr.org/rip/ripaudP3.html.

Burnham, David. 1996. *Above the Law: Secret Deals, Political Fixes and Other Misadventures of the U.S. Department of Justice*. New York: Scribner.

Canadian Delegation. 2001. "Discussion Paper for Workshop 1: Potential Consequences for Data Retention of Various Business Models Characterizing Internet Service Providers." Paper presented at Tokyo. G8 Government-Industry Workshop on Safety and Security in Cyberspace, May.

Clarke, Roger. 1999. "Subject: ABA Demonstrates Its Ignorance to the World." Forwarded to the Politech Mailing List, message titled FC: More on Australian official demanding Net-regulation—demonstrating ignorance to the world, November 3.

Committee of Ministers. 2001. "Communiqué on International Action against Terrorism." November 8. Strasbourg: Council of Europe.

Council of Europe. 2001. "Convention on Cybercrime, CETS No. 185." Available at www.conventions.coe.int/treaty/en/treaties/html/185.htm.

Council of Europe. 2001. "Convention on Cybercrime Explanatory Report." Strasbourg. Available at www.conventions.coe.int/treaty/en/reports/html/185.htm.

Council of Europe Press Service. 2001. "The Convention on Cybercrime, a Unique Instrument for International Co-Operation." Budapest, November 23.

Council of Europe Press Unit. 2001. "First International Treaty to Combat Crime in Cyberspace Approved by Ministers' Deputies." Strasbourg: Council of Europe, September 19.

Department of Justice, Industry Canada, and Solicitor General Canada. 2002. "Lawful Access—Consultation Document." Ottawa: Government of Canada, August 25.

Dogcow. "Evading the Broadcasting Services Amendment (Online Services) Act 1999." 2600 Australia, 1999.

Electronic Privacy Information Center. 1997. "Faulty Filters: How Content Filters Block Access to Kid-Friendly Information on the Internet." Available at www.epic.org/reports/filter_report.html.

Escudero-Pascual, Alberto, and Ian Hosein. 2004. "The Hazards of Technology-Neutral Policy: Questioning Lawful Access to Traffic Data." *Communications of the ACM* 47 no. 3: 77–82.

European Telecommunications Network Operators Association. 2001. "ETNO Reflection Document on Data Protection and Privacy Aspects of the Draft Council of Europe Convention on Cyber-Crime." European Telecommunications Network Operators. Available at www.etno.bc.

European Union. 1995. "Directive 95/46/EC of the European Parliament and the Council of 24 October 1995 on the Protection of Individuals with Regard to the Processing of Personal Data and on the Free Movement of Such Data." Official Journal L 281, November 23, pp. 31–50.

Fijnaut, Cyrille. 2001. "Transnational Organized Crime and Institutional Reform in the European Union: The Case of Judicial Cooperation." In *Combating Transnational Crime*, ed. Phil Williams and Dimitri Vlassis, 276–302. Portland, OR: Frank Cass Publisher.

G7. 1988. "Political Declaration." Toronto: G7 Summit, June 19–21.

G7. 1989. "Declaration on Terrorism." Paris: G7 Summit, July 14–16.

G7. 1996. "Declaration on Terrorism." Lyon: G7 Summit, June 27.

G7. 1993. "Tokyo Summit Political Declaration: Striving for a More Secure and Humane World." Tokyo: G7 Summit, July 6–9.

G8. 1997. "Communiqué." Denver: G8 Summit, June 22.

G8. 1999. "Ministerial Conference of the G-8 Countries on Combating Transnational Organized Crime." Moscow, October 19–20.

G8 Foreign Ministers. 1997. "Foreign Ministers' Progress Report." Denver: G7 Summit, June 21.

G8 Government-Industry Workshop. 2001a. "Report from Data Preservation Workshop." Tokyo: Group of 8 Conference on High-Tech Crime, May 22–24.

G8 Government-Industry Workshop. 2001b. "Workshop 1 Report: Potential Consequences for Data Retention of Various Business Models Characterizing Internet Service Providers." Tokyo: G8 Government-Industry Workshop on Safety and Security in Cyberspace, May.

G8 Justice and Interior Ministers. 1997. "Meeting of Justice and Interior Ministers." Washington, DC: G8, December 10.

G8 Justice and Interior Ministers. 2002a. "Data Preservation Checklists." Mont-Tremblant: G8 Summit, May 13–14.

G8 Justice and Interior Ministers. 2002b. "G8 Statement on Data Protection Regimes." Mont-Tremblant: G8 Summit, May 13–14.

G8 Justice and Interior Ministers. 2002c. "Principles on the Availability of Data Essential to Protecting Public Safety." Mont-Tremblant: G8 Summit, May 13–14.

G8 Lyon Group. 2001. "Recommendations for Tracing Networked Communications across National Borders in Terrorist and Criminal Investigations (Draft)." G8, November 19.

G8 Lyon Group. 2000. "Un Dialogue Entre Les Pourvoirs Publics Et Le Secteur Privé Sur La Sécurité Et La Confiance Dans Le Cyberespace." Communiqué Du G8 (Groupe De Lyon)." Paris: G8, May 17.

Global Internet Liberty Campaign. 2000. "Member Letter on Council of Europe Convention on Cybercrime." October 16.

Global Internet Liberty Campaign. 2002. "Member Letter on Council of Europe Convention on Cyber-Crime Second Protocol." February 27.

Global Internet Liberty Campaign. 2000. "Member Letter on Council of Europe Convention on Cyber-Crime Version 24.2." December 12.

Goldsmith, Jack L. 1998a. "Against Cyberanarchy." *University of Chicago Law Review* 65: 1199–1250.

Goldsmith, Jack L. 1998b. "Symposium on the Internet and Legal Theory: Regulation of the Internet: Three Persistent Fallacies." *Chicago-Kent Law Review* 73: 1119–1131.

Goldstone, David, and Betty-Ellen Shave. 1999. "Essay: International Dimensions of Crimes in Cyberspace." *Fordham International Law Journal* 22 (June): 1924–1971.

Haas, Peter M. 1992. "Introduction: Epistemic Communities and International Policy Coordination." *International Organization* 46, no. 1: 1–35.

Hill, Eleanor. 2002. "The FBI's Handling of the Phoenix Electronic Communication and Investigation of Zacarias Moussaoui Prior to September 11, 2001." Testimony to the Joint Congressional Committee on 9/11 Inquiry, Washington, DC, September 24.

Hosein, Ian, and Edgar Whitley. 2002. "Developing National Strategies for Electronic Commerce: Learning from the UK's RIP Act." *Journal of Strategic Information Systems* 11, no. 1: 31–58.

Lee, Jennifer S. 2001. "Companies Compete to Provide Saudi Internet Veil." *New York Times*, November 19.

Lemos, Robert. 2000. "Cybercrime Treaty Still Doesn't Cut It." *ZDNET*, December 13. Available at www.news.zdnet.com/2100-9595_22-526382.html.

NetCoalition.com. 2001. "The Cybercrime Convention Will Harm U.S. Internet Users and Businesses." January.

OECD. 1997. "Cryptography Policy: The Guidelines and the Issues." Organisation for Economic Cooperation and Development.

P8. 1996. "Senior Experts on Transnational Organized Crime Group Recommendations." Paris: P8 Summit, April 12. Available at www.g7.utoronto.ca/crime/40pts.htm.

Patrick, Stewart. 2002. "Multilateralism and Its Discontents: The Causes and Consequences of U.S. Ambivalence." In *Multilateralism and U.S. Foreign Policy: Ambivalent Engagement*, ed. Stewart Patrick and Shepard Forman, 1–45. Boulder, CO: Lynne Rienner Publishers, Inc.

Privacy International, American Civil Liberties Union, and Electronic Privacy Information Center. 2001. "Letter of Concerns Regarding Version 27 of CoE Convention on Cybercrime." June.

Purdy, Dan. 2001. "Report of the G8 Government/Private Sector High Level Meeting on High-Tech Crime." Tokyo, Japan: Department of Foreign Affairs and International Trade, May 24.

Schulzki-Haddouti, Christiane. 1999. "'We Also Want to Make a Guide for Other Countries.' Interview with Scott Charney, Chairman of the G-8 Work Group 'High-Tech Crime.'" *Telepolis*, June 11. Available at www.jya.com/g8-charney.htm.

Secretariat of ISPA. 2001. "URGENT–ISPA–Data Preservation request–US atrocities." A message to ISPA news mailing list. September 13.

Spafford, Eugene. 2000. "Awareness Program for the Draft Convention on Cyber-Crime." Purdue University (cited February 11, 2003). Available at http://www.cerias.purdue.edu/homes/spaf/coe/index.html.

Spafford, Eugene, et al. 2000. "Proposed International Cyber Treaty: Statement of Concerns." Mimeo.

U.S. Department of Justice. 2003a. "Domestic Security Enhancement Act of 2003." Draft, section-by-section analysis, 87, January 9. Washington, DC.

U.S. Department of Justice. 2003b. "Statement of Barbara Comstock, Director of Public Affairs." Washington, DC: Press release from U.S. Department of Justice, February 7.

U.S. Delegation. 2001. "Discussion Paper for Data Preservation Workshop." Tokyo: Group of 8 Conference on High-Tech Crime, May 22–24.

World ISPA Forum. 2001. "Open Letter." World ISPA Forum Secretariat, April 18. Mimeo.

10 Privacy in the Digital Age: States, Private Actors, and Hybrid Arrangements

Henry Farrell

Privacy has emerged as a key regulatory issue in the wake of the information and communications revolution. New technologies have brought new problems; they have made it more difficult for individuals to maintain their privacy (or for other actors to protect it on their behalf), while also giving rise to complex issues of global regulation.

The right to privacy, however it is defined, rests on the individual's ability to control information about himself or herself, and how that information is disseminated and used. Advances in information and communications technology have had profound consequences for individuals' ability to exercise that right. New technologies, including, but not limited to, the World Wide Web (WWW) make it far easier for third parties to gather information about behavior, and potentially to link this information to specific individuals. Data mining and information sifting techniques, together with access to computing power, make it easier to analyze that information and make it useful. These new technologies make it far more difficult for individuals who value their privacy to maintain it. The sheer volume of information, and of uses to which information can be put, also make it more difficult for specialized agencies (such as data protection commissioners) to protect individual privacy.

Not only has communication technology made it more difficult to protect privacy in and of itself, but the vast increase in cross-border data flows has generated new problems of international governance. Different countries have adopted very different approaches to privacy protection. For example, the member states of the European Union (EU) have adopted increasingly stringent legal measures to protect privacy,[1] most notably the EU-level Data Protection Directive (a discussion follows). Many other states, such as the United States, have either opted for an approach that privileges voluntary forms of self-regulation, or have not adopted any substantial measures at all. In states within the developing world, privacy legislation is typically a low priority, compared to more pressing, material needs. These radical differences of approach are increasingly the source of international disagreement, as communications technologies such as the Internet lead to increased interdependence among countries.[2] States may reasonably be concerned that their particular approach to privacy protection may be undermined if firms or individuals export data outside their jurisdiction and process it there. Thus, for example, the European Union has introduced measures in its Data Protection Directive that threaten to block data flows to countries that do not provide "adequate" privacy protections. However, countries that have different or no means for privacy protection may for their part feel threatened by the efforts of other countries to create a high international threshold for privacy.

What explains this varying pattern of privacy regulation? In this chapter, I argue that privacy regulation has always had a strong international component. To adapt Peter Gourevitch's famous analysis, domestic privacy regulation is best captured through the "second image reversed"; that is, through examining how international factors may translate into domestic outcomes. However, the causal impact of international factors is likely to depend on two key intervening variables. First of these is existing national institutional traditions, which affect whether individual states do, or do not, take up specific modes of protecting privacy. Second is state bargaining power; stronger states may be able to force weaker states to reform their domestic policy, regardless of these weaker states' underlying preferences. While I do not propose an explicit model with hypotheses about the circumstances under which the one or the other factor will be most important, I do show how these factors in combination provide a very considerable degree of insight into the development of institutions governing privacy at different instances over the last thirty years. I also show the relevance of these factors for policy, and for the viability of different policy recommendations in the current international context.

How have the institutions protecting privacy developed over the last thirty years? In a first phase of development, an "epistemic community" of policy experts developed a set of fair information practices that served as a template for comprehensive domestic laws in many domestic contexts. However, one key actor—the United States—was unwilling to accept a comprehensive privacy law based on fair information principles. Because of U.S. bargaining strength, the United States was able to go it alone, without acceding to external pressures.

In a second phase, states sought to create international instruments to protect privacy on the basis of their existing domestic institutions and preferences. However, irresolvable divergences of interest among powerful states led to stalemate, and the creation of two international instruments, one of which is nonbinding, but which commanded assent among advanced industrialized democracies, and the other which involved strong binding commitments, but was not acceded to by the United States and other non-European states.

In the current phase of development increased pressures from interdependence are again leading to changes in the privacy agenda. On the one hand, European Union strong-arming is leading many countries to converge on an EU model of data protection law. EU pressure has further induced the United States to enter into an international arrangement that allows U.S. firms to comply with privacy standards that have been set in negotiations with the European Commission. On the other, new pressures (especially in the wake of September 11) are leading to a downgrading of privacy standards in many countries, including EU member states that had previously pushed for strong privacy standards.

Evolution of the Privacy Debate and International Institutions

Current controversies over privacy have their roots in debates on privacy that began in the late 1960s and early 1970s, and in states' differential response to these debates. These began in worries about how state administrations might use mainframe computers.[3] The demand of state bureaucracies for technical means to collate and analyze individual-level information seemed insatiable, and there were few legal safeguards to protect privacy.[4]

Concerns over privacy were especially strong in continental Europe, which had recent memories of how personal information had been abused during the era of National Socialism. They were also present in the United States, with its strongly individualistic political tradition. In due course, they were manifested in formal legislation across a large number of industrialized countries, intended to provide formal protections for individual privacy.[5]

Experts in the field soon began to reach a rough consensus about how privacy might best be protected through "fair information principles." This cross-national "epistemic community" played an important role in realigning domestic debates within various countries, and in encouraging states to adopt laws that instantiated appropriate principles of privacy protection.[6] The fair information principles that these experts promulgated laid down basic guidelines about how information ought to be treated, in order best to protect individual privacy. Although different countries legislated for privacy at different times, and formulated their legislation through quite different processes, there was a remarkable degree of convergence across states at the level of principle.[7] However, there were notable differences in states' enthusiasm for privacy protection. In some states, such as the United Kingdom, privacy legislation was as much motivated by the fear of external difficulties as by enthusiastic debate of the ideas put forward by privacy experts.[8] In other contexts, most notably the United States, fair information principles exerted a weaker influence on policy debates than elsewhere.

These differences began to manifest themselves in clear policy divergences, as the debate over privacy shifted to include commercial as well as governmental uses of personal information. As firms began to develop their own computerized databases, and to use them for commercial purposes, it became clear that state administrations were not the only potential invaders of privacy. Even where there was rough consensus about appropriate principles of privacy protection, there were stark disagreements among states about how and where these principles should be implemented. Many mainland European countries began to develop comprehensive data protection laws that sought to provide broad legal protection to privacy across a variety of social and economic arenas. In contrast, the United States (and, at a later juncture, Japan and North Korea) introduced legislation that applied privacy standards to the federal government, but failed to extend comprehensive legal protections to the private sphere, instead relying on a patchwork of self-regulation and narrowly based laws. In the U.S. case at least, this failure to adapt a comprehensive approach can be traced back to a different constellation of state-private actor relations, and to the hostility of influential business actors to new legislation that would curtail their ability to use personal information. Instead, business actors suggested self-regulation as a viable means to privacy protection, and in some cases went so far as to introduce self-regulatory schemes that were rather more notable for rhetoric than for substantial consumer protections.

Thus, in this first phase, it is clear that international factors—the creation of a transnational epistemic community of privacy experts—did have an important influence on domestic outcomes. As Colin Bennett demonstrates, these experts succeeded in creating a rough consensus around "fair information principles" that then played an important part in guiding the creation of national legislation in many advanced industrialized democracies. However, the influence of this epistemic community was itself limited by domestic factors. In some national contexts—most notably many of the countries of mainland Europe—it proved possible to adopt comprehensive laws applying fair information principles. In other contexts, such as the UK, governmental actors were less enthusiastic to adopt comprehensive new laws but recognized that it was probably in the interests of the UK to introduce legislation, given the increasingly close economic and political connections between the UK and mainland Europe. In the United States, in contrast, both existing institutional frameworks and the opposition of powerful domestic actors meant that the U.S. government was disinclined to introduce comprehensive laws. Furthermore, in contrast to the UK, the United States was not deeply embedded in a dense web of relations with countries that were introducing comprehensive privacy protections, and was furthermore in a strong bargaining position vis-à-vis other industrialized democracies. U.S. governmental actors thus had neither strong internal nor external motivations to adopt comprehensive privacy legislation.

These disparities led to considerable disagreements among advanced industrialized democracies in a second stage of debate—when states sought to build upon their preexisting national legislation to create international institutions in the sphere of privacy. The international debate on privacy came to the fore as the result of increasing interdependence; as firms began increasingly to move data over national borders, international commercial transfers of data became the subject of controversy. Over the late 1970s and early 1980s, many states began to worry that transborder data flows (TDF) threatened their ability to exercise their sovereign authority successfully.[9] More specifically, even though the level of cross-border data flows was still relatively limited at this point,[10] some states were concerned that their domestic privacy legislation would be undermined if firms were able to transfer personal data to a laxer jurisdiction. This led to disagreement between the United States, which had weak to nonexistent legal protections for privacy in the commercial arena, and those European states that had enacted comprehensive data protection laws. The latter pushed for strong international instruments to protect privacy, while the United States sought to water down these proposals, which some U.S. commentators perceived as barely masked protectionism.[11]

These disagreements led to arguments over whether binding international standards on privacy were appropriate. Debates were conducted in two main forums; the Council of Europe (CoE), and the Organization for Economic Cooperation and Development (OECD). The Council of Europe, a broadly based organization, has a membership consisting of European states with the Vatican, the United States, Canada, Japan, and Mexico as observers. Member states of the CoE adopted the 1981 Convention for the Protection of Individuals with regard to Automatic Processing of Personal Data. This convention laid out rules for information privacy that reflected fair information principles. It also forbade parties to the convention from stopping data flows to other such parties for the sole purpose of protecting privacy, except when the other party did not protect certain kinds of data that were protected in the originating country. The convention did, however, allow adhering countries to block data flows to third-party jurisdictions that had not signed up to it.

The OECD also reached agreement in 1980 on the so-called OECD Guidelines Governing the Protection of Privacy and Transborder Data Flows of Personal Data. In contrast to the convention, the guidelines were not substantively binding on the states that had agreed to them. The guidelines may best be interpreted as an effort to woo the United States by stages into signing up to an international arrangement on privacy.[12] However, the OECD guidelines had few teeth, precisely because of the desire of other states to persuade the United States to agree to them, and because of U.S. bargaining power (which was rather stronger in the OECD than in the Council of Europe). The result was predictable. While the United States was a signatory to the guidelines, its efforts to implement them were limited to (short-lived and largely ineffective) exhortations to U.S. firms to abide by them.[13]

By the mid-1980s, the international policy debate on TDF was, for all intents and purposes, over. The United States and large multinational corporations had successfully clamped down on debate on multilateral measures to control data flows.[14] However, these data flows themselves continued to increase in volume, especially between advanced industrialized economies. The continued differences in approach to privacy protection led to important policy dilemmas, which nonbinding statements of intent, like the OECD guidelines, did little to address. Countries that placed a high premium on formal laws to protect privacy, such as Germany and France, legitimately feared that cross-border data transfers could undermine their laws, by allowing actors to transfer personal data to jurisdictions with weak or nonexistent privacy protection, and process the data there. However, any efforts by these countries to control data flows would in turn have consequences for third-party countries with different approaches to privacy.[15]

Thus, at the end of this second phase, disagreements between advanced industrialized democracies had led to the creation of two international instruments in privacy protection. One of these instruments, the Council of Europe convention, provided for comprehensive—and binding—rules protecting privacy. The countries adhering to this convention were thus willing to accede to strong international rules. The second international instrument, the OECD guidelines, was nonbinding, and thus of uncertain consequence. This allowed states such as the United States to sign up to the guidelines, which they perceived as aspirational, if not indeed a complete dead letter. Again, because of the United States' disproportionate bargaining power, American negotiators were able to avoid signing up to real commitments that would have obliged domestic change. Rather than resolving disputes among industrialized democracies over privacy protection, the OECD guidelines papered over them.

The third—and current—phase of privacy regulation began in the early 1990s. It is a product of the wave toward comprehensive privacy protection in many industrialized democracies in the first phase, and the failure of these democracies to create a binding agreement that would include countries such as the United States in the second. In order to cement, harmonize, and rationalize national data protection laws, the European Union conducted discussions throughout the early 1990s on a comprehensive EU-level framework. This culminated in the EU's Directive on the Protection of Individuals With Regard to the Processing of Personal Data and on the Free Movement of Such Data, more tersely and conveniently dubbed the Data Protection Directive, which took effect in late 1998. This directive had its origins in European worries that the EU's creation of a single internal market was being hampered by differences in data protection regulations; some EU member states were using data protection law to block data transfers to other member states that had weaker privacy protection. The directive was drafted to ensure that all member states had broadly similar privacy standards in their domestic legislation, and thus to remove any justification for blockages of data flows within the EU.[16] However, the directive is more than just an intra-European housecleaning exercise; it has also had important consequences for actors outside the European Union. The directive mandated the European Commission to decide, under the strictures of a comitology procedure,[17] whether third countries had "adequate" protection for personal data or not. In cases where the Commission found that a country did not have adequate protection, member states were enjoined to prevent data transfers to that country, except under highly specific circumstances laid out in the directive.

This led immediately to controversies with countries such as the United States, which did not have data protection laws, or which had data protection laws that were likely to be judged inadequate by the Commission. This controversy was especially acrid because of the potential knock-on effects of the directive for the regulation of e-commerce. The prevailing wisdom in the United States, and, to a lesser extent, in other advanced industrialized economies, was that e-commerce should only be regulated by government insofar as was absolutely necessary.[18] The Data Protection Directive, even though it had been conceived and drafted before the e-commerce revolution, was perceived by U.S. business and policy makers as an EU effort to reimpose government control over e-commerce, and thus as setting a potentially dangerous precedent. While U.S. firms and others successfully lobbied to weaken the directive's requirements (originally, the directive had required that third-party jurisdictions have "equivalent" rather than merely "adequate" protections),[19] it was still clear that the directive could have quite substantial consequences for third-party jurisdictions.

Between 1998 and today, the European Union has negotiated with various non-EU states over the circumstances under which the Commission would be prepared to find their systems of data protection "adequate" and thus obviate the threat of data flow blockages. Negotiations with the United States received the most attention; the U.S. administration initially argued strongly against the directive, and sought to encourage an alternative approach based on self-regulation and so-called "privacy seal" organizations, such as TRUSTe and BBBOnLine.[20] The United States also expressed itself willing to use its bargaining strength to prevent the EU from imposing a solution upon it, and EU negotiators recognized that they were unlikely to bring through major domestic reforms within the United States. However, the EU still refused to accept pure self-regulation as sufficient. EU-U.S. negotiations culminated in the so-called "Safe Harbor" arrangement, in which the EU withheld a general adequacy judgment from the United States, but announced that specific U.S. firms, which voluntarily signed up to an agreed set of privacy principles, would be considered to have satisfied the requirements for adequacy.[21] Enforcement of these principles involved a mixture of public and private actors; self-regulatory organizations such as TRUSTe and BBBOnLine could provide an initial line of defense, while the U.S. Federal Trade Commission and EU's data protection commissioners also played an important role.[22]

The EU has been much less inclined to make concessions in its discussions with weaker trading partners.[23] The consequence has been that countries in the developed world, and increasingly within the developing world, are seeking to implement legislation that reflects EU priorities, in order to be declared adequate.

However, even while the EU has been pushing other countries to implement stronger data protection laws, it has been weakening its own protections, as have other countries, in the wake of the events of September 11, 2001. Policy makers' perceptions of the relationship between security and privacy have shifted so that measures that would previously have been unthinkable, because of their negative consequences for privacy, have been implemented with little debate. Many of these measures have little to do with the prevention of terrorism. State security and policing services are succeeding in getting privacy-intrusive measures passed that they have advocated for years, regardless of whether or not these measures address terrorism directly.[24] Even those states that have had comprehensive privacy legislation in the past are substantially weakening their protections. In particular, new battles are beginning to develop over traffic data retention, concerning government requirements that telecommunications companies and Internet service providers (ISPs) retain information on their clients' communications traffic, and make that information available to the state for law enforcement and anti-terrorist purposes.

As of yet, these new pressures have not led to the creation of extensive multilateral arrangements, with the partial exception of the Council of Europe's Convention on Cybercrime (which was, however, drafted before September 11; see further, Hosein, chapter 9). However, more worryingly, they have led to the creation of transgovernmental regulatory networks,[25] which increasingly seem to be driving privacy policy—and not in a privacy-friendly direction. In short, if an epistemic community of privacy experts helped drive the international convergence on data protection principles at an earlier juncture, officials in justice, home affairs, and security ministries and agencies are now playing a similar—but much less privacy-friendly—role in driving many pertinent areas of policy. Furthermore, one may predict that these officials will likely play a more important policy role than privacy experts over the medium term, precisely insofar as they play a more direct and important role in policy setting. More generally, privacy advocates face a very substantial challenge if they wish to develop new instruments to hold state actors accountable for privacy violations, especially in the current political climate, where privacy-intrusive security measures are perceived by many as legitimate.[26]

Current Debates on Privacy

As discussed earlier, privacy questions are being debated at a variety of levels, both domestically and internationally. While these levels intersect, they do so in somewhat confusing ways. Accordingly, in this section, I set out to describe the two key arenas in which the privacy debate is playing out at the moment: international relations among states (which may further be subdivided into relations among advanced industrialized democracies, and relations between advanced industrialized democracies and countries in the former Eastern bloc or developing world); and domestic relations within advanced industrial democracies between the state and commercial actors and citizens.

Relations among States

International relations among states in the sphere of privacy involves two main subsets of relations, each of which has a quite different logic. Relations *among* advanced industrialized states receive the most attention in the literature, and generate the most public disputes, but relations *between* these states and states in the developing world and elsewhere have the potential to generate substantial policy problems in the future.

As described in the previous section, advanced industrial democracies have frequently disagreed over privacy regulation. Important differences in how these states regulate privacy have led increasingly to international disagreement. In some cases, there are substantial divergences; for example, differences between countries that have comprehensive privacy laws based on fair information principles and associated formal mechanisms of protection (data protection commissioners), and countries that do not. In other cases, differences are less pronounced (specific differences, for example, in how fair information principles are applied in the legislation of different countries). In both instances there are pressures for convergence.

Currently, the main motor force for convergence is rather different than it was at previous junctures. The key factor is not a community of policy experts, or agreed multilateral instruments, but the European Union's Data Protection Directive, and its external consequences.[27] Since the directive has come into force, third-party jurisdictions have increasingly found themselves forced to adapt privacy standards along European lines.[28] In the eyes of some, this is leading to the creation of a new international privacy regime.[29]

Even given the notorious elasticity of the concept of regime,[30] this may be overstating the case. The end result is an array of bilateral negotiations; the EU's requirement of "adequacy" still permits a considerable degree of variation across different systems.[31] However, the EU's demands upon its trading partners are still resulting in an upward ratcheting of privacy standards across the developed world, even if this is not leading to general convergence upon a regulatory endpoint.

The precise consequences of the Data Protection Directive for third countries vary according to (a) the fit between the country's existing privacy protections and EU demands, and (b) the country's bargaining strength vis-à-vis the EU. The Safe Harbor arrangement, discussed in the previous section, reflects the unique bargaining strength of the United States, as well as that country's profound unwillingness to introduce comprehensive privacy legislation along EU lines. Thus, the Safe Harbor serves as a kind of "interface solution,"[32] minimizing conflict between the EU's emphasis on formal enforcement, and the U.S. self-regulatory approach. It remains to be seen whether

Safe Harbor will provide a long term solution to EU-U.S. disagreements over privacy; to date, relatively few U.S. firms have signed up to the arrangement. However, it is significant as an exemplar of a new trend toward mixtures of state and private enforcement as a solution to international policy problems.

Countries with less bargaining leverage than the United States have had difficulty in reaching compromises that allow them to maintain their existing approach to privacy. This has led to some disgruntlement, especially where the EU asks countries to make substantial changes to their existing systems. Australia, for example, has a federal privacy commissioner, and has introduced legislation enhancing privacy protections, which in part seeks to respond to European demands. Australia's Internet Industry Association has furthermore sought to implement a code of privacy practice based on the Safe Harbor principles. However, European data commissioners[33] and European Commission officials still find that Australian protections fall considerably short of adequacy, and clearly believe that they can wring further concessions from the Australian government. Australian officials have responded that they are being held to a higher standard than the United States,[34] with the implication that they have grounds for a WTO action.[35] Other countries have had less difficulty in conforming to EU requirements. Canada and Switzerland have received positive adequacy judgments from the Commission, while New Zealand's privacy laws are close enough to EU requirements to require only minor legislative changes. There still remain some states where EU pressures do not seem likely to result in legislative changes in the near future. It is possible but by no means certain that there may be an upward ratchet effect that may eventually lead to changes in such states; multinational firms that have to obey EU rules in any event may acquiesce, or even actively press for changes in these countries over the longer term.[36] In the shorter term, efforts to use EU standards to provide internal leverage within these systems, as well as efforts to use existing programs (such as Safe Harbor in the United States) may have at least a palliative effect.

In some respects, relationships between industrialized states and middle-income states within the developing world resemble relations among industrialized states themselves. The same is true of the new democracies of Central and Eastern Europe. However, there are clear differences. Most obviously, there are disparities of power. Even where developing countries have relatively substantial markets (as, for example, India), they still tend to have less clout in bilateral or multilateral negotiations than their size would suggest. But there are also differences as well in the extent to which they are concerned with privacy as a policy issue. Colin Bennett's judgment in 1992, that privacy is primarily a concern for industrialized countries, still has some force.[37] Most countries in the developing world have little interest in formal protections for privacy as such. Authoritarian regimes have little more interest in promoting privacy than in promoting other political rights that might have destabilizing implications for their rule, while poorer democracies usually have more pressing material needs to address, and limited resources with which to address them. There are some notable exceptions to this generalization. The Republic of South Africa, for example, does enshrine the right to privacy in its constitution, although it has yet to enact a data protection law; there is vigorous political debate over the extent to which the state should be able to monitor private communications. Hong Kong, insofar as it can be considered to be part of the developing world, has strong and comprehensive legislation protecting privacy, as well as a privacy commissioner. However, these are among the few exceptions to a more general pattern.

Even if developing and middle-income states historically have had little interest in privacy policy, they increasingly are finding that such an interest is forced upon them, for external reasons. Again, the European Union's Data Protection Directive is a key factor. Countries that wish to maintain good trade relations with the European Union, and encourage inward investment, are finding themselves obliged to conform to Europe's external requirements. Argentina, for example, has introduced comprehensive legislation along European lines, while Peru has introduced specific sectoral legislation following the European model, and has established a commission to draft more comprehensive reforms.[38] Other countries, such as India, face increasing pressure to enact legislation from their own firms, which are fearful of losing important markets.[39]

Nowhere was EU influence more marked than among the former states of the Eastern bloc, many of which have recently become EU members. In order to prepare for membership, candidate countries were required to enact comprehensive privacy legislation along the lines laid out in the directive as part of the *acquis communautaire*, the set of formal obligations associated with EU membership. Thus, they not only had to enact laws that reach the less onerous standard of adequacy, but also laws that closely approximate the laws of existing EU members. Those countries still outside the EU, but with the desire to join and a reasonable prospect of membership in the short term (Romania, Bulgaria), face the same requirements.

In summary, the EU's directive is having a substantial external effect on middle-income countries within the developing world, and is substantially determining the privacy laws of candidate countries within the former Eastern bloc. Where its effects are more ambiguous are in the poorest developing countries, as well as some middle-income countries (such as Russia) where the rule of law is at best imperfectly established. Some of these countries are seeking to respond to the EU's perceived demands. However, there is little substantial likelihood that they will receive adequacy judgments, which depend not only on the formal protections offered to privacy, but also on the degree to which these protections are likely to be implemented in practice. Countries in which legal institutions are underfunded, or ineffective, will find it difficult to conform to EU requirements. On the one hand, they cannot credibly guarantee that comprehensive privacy laws will be enforced if they are enacted. On the other hand, they are ineligible for Safe Harbor style solutions, even if the EU adopted a more liberal approach to negotiating such solutions. Commission officials have made it clear that such solutions are only appropriate in well-functioning legal systems.[40]

If one leaves the problems of weak states to one side, the preceding might suggest that the long-term international outlook for privacy is positive; a ratcheting upwards to Europe-set standards. However, an important set of complicating factors has recently begun to emerge, due to state initiatives to combat crime and terrorism. The Council of Europe's Convention on Cybercrime, which was ratified in 2004, requires participating states to legislate for new surveillance capabilities on the Internet and to use these capabilities where necessary to cooperate in criminal investigations.[41] This instrument was formulated in a poorly publicized series of negotiations in the period leading up to September 2001. It is rapidly becoming a sort of international "lowest common denominator" agreement structuring the relationship between privacy and law enforcement, with implications that go well beyond Council of Europe members. For example, justice ministers of the Organization of American States have recently recommended that OAS member states consider the advisability of acceding to the Convention on Cybercrime.

Since the events of September 11, pressures to curtail privacy in the fight against terrorism have increased dramatically. The U.S. administration has demanded—and received—changes in European policy that have important consequences for the existing EU privacy regime. In a letter dated October 16, 2001, the U.S. administration requested that the European Union "consider data protection issues in the context of law enforcement and counterterrorism imperatives," and that the EU institute a series of policy changes, including the modification of draft legislation so as to allow the retention of traffic data.[42] The EU has acquiesced to this request. EU-level legislation protecting the privacy of traffic data has been dramatically weakened, while European member states have agreed to a wide-ranging exchange of police and security information with the United States through Europol. Finally, the United States successfully demanded that the EU weaken the Data Protection Directive by allowing airlines flying to the United States to share passenger name record data with U.S. authorities. In reaching a compromise with the United States over this issue, the European Commission has very likely overstepped its legal authority, according to an initial opinion from the European Court of Justice's advocate-general.

It is likely that over the next few years, the fight against terrorism will serve as a reason (or excuse) for new international, multilateral, and bilateral initiatives that will substantially weaken privacy protections in "high" privacy countries such as those of Western Europe. Policy makers speak of a necessary tradeoff between privacy and security; while this arguably mischaracterizes the complex relationship between the two, it is likely to shape public debate and relevant public policy over the coming decade. Thus, there are new international pressures for the weakening of privacy protections in many jurisdictions; I return to this point in the next section.

Privacy Debates within Advanced Industrialized Democracies

The international disagreements over privacy described earlier intersect with disagreements at the domestic level in many Western democracies.[43] The specific constellations of these disagreements vary according to political setting and issue area. Two sets of disputes are especially important. First are continuing arguments over the proper balance between law and private enforcement. Second are new disagreements over governments' access to information on people's behavior on the Internet and other communication networks.

First, in domestic contexts where there are weak or nonexistent laws governing business's use of personal information, privacy advocates have sought comprehensive legislation to prevent privacy abuses by firms. Firms, in contrast, have typically lobbied against such legislation, claiming (with more fervor than credibility) that market forces and self-regulation suffice to protect consumer interests. Here, privacy advocates seek to defend the interests of citizens against abuses by firms, which in turn have tried to use their clout with the government to block change. A new set of battles is being fought over the appropriate mix between government and private enforcement, as industry groups and web seal organizations expand the range of self-regulatory privacy regimes.

Here again, the United States is the key test case of a prominent state without effective and comprehensive privacy laws. In the late 1990s, U.S.-based privacy advocates had hoped to use external pressures, especially the EU's Data Protection Directive, to press for domestic privacy laws that would meet international standards. However, they faced formidable obstacles. The U.S. political system makes it notoriously difficult to enact major reforms, because of its many veto points,[44] and it furthermore privileges business over consumer interests to an extraordinary degree. The U.S. administration, far from acceding to European demands, instead proposed an alternative vision of privacy protection on the WWW, which would rely on self-regulation.[45] In the short to medium term, the administration proposed reliance on self-regulatory web seal organizations, which would award seals to web sites that adhered to certain privacy standards. U.S. administration officials argued that this would provide consumer protection—consumers could choose only to do business with web sites that had signed up to these seal programs. Ira Magaziner, the architect of the U.S. white paper, argued that self-regulatory organizations would eventually be superceded by technological tools, which would give individuals control over precisely how they shared their information.[46] The administration's position reflected the views of key figures in the U.S. information technology industry, who were vigorous proponents of a hands-off approach to the regulation of e-commerce.

U.S. administration pressures, together with the desire of some businesses to garner favorable publicity, led firms to sign up with two privacy seal organizations, TRUSTe (originally called Etrust), and BBBOnLine.[47] Industry associations such as the Direct

Marketing Association (DMA) later set up their own schemes to make it easier for members to comply with Safe Harbor. However, privacy advocates viewed these programs as an entirely unsatisfactory substitute for comprehensive laws; TRUSTe in particular came under heavy fire for perceived gaps in enforcement.[48] Advocates continued to press for federal legislation protecting consumer privacy in online (and, if possible, offline) business transactions. In mid-2001, they appeared to be making some progress; privacy issues occupied a prominent position on Congress's legislative agenda. Many observers predicted that comprehensive legislation would be passed, albeit with weaker protections than advocates would like, if only to preempt the possibility of legislation at the state level. However, in the wake of the events of September 11, much of this momentum was lost; while privacy legislation in the medium term is still possible, it is by no means certain. Recent scandals involving the loss of consumer data to criminals by large firms have reinvigorated debate, but legislation faces substantial obstacles and opposition from business.

Both BBBOnLine and TRUSTe have marketed themselves outside the United States. The logic of this is clear; if the WWW is indeed a global phenomenon involving transnational commercial transactions, then the market for web site privacy certification is also global. However, many countries are uncomfortable with the idea that foreign private entities should be guarantors of privacy standards, leading to renewed international discussions about the appropriate mix between public and private enforcement in privacy protection. This debate is complicated by differing notions of the relationship between public and private, and between government regulation and self-regulation.

Many countries with strong traditions of privacy law devolve certain aspects of privacy protection to nonstate actors. However, this practice is better described as "co-regulation" or "private interest government"[49] than self-regulation; it involves government specifically delegating certain public interest tasks (with accompanying procedures of oversight and responsibility) to private sector associations. Typically, governments exercise strong forms of oversight.[50] This differs markedly from the Anglo-American concept of self-regulation, in which business self-regulation is seen not as a means of implementing government regulation, but as a substitute for it.[51] Arrangements such as that prevailing in Australia, in which the federal privacy commissioner may grant approval to industry codes that appear to embody appropriate principles, stand somewhere between co-regulation and self-regulation.

Differences between these models of public-private interaction have led to wide variation in officials' attitudes to international web seal organizations. Some national-level privacy commissioners have cautiously welcomed web seal organizations as a possible means to protect individual privacy in international transactions, where it is difficult for national officials effectively to exercise their powers.[52] Others have vigorously disagreed. Many officials from non-Anglo-American political systems have expressed considerable doubts about schemes such as TRUSTe, which they see as embodying weak principles and providing inadequate enforcement with little opportunity for external oversight.

These disputes are likely to continue, and leave web seal organizations in an unenviable position between states and firms. On the one hand, these organizations must persuade state authorities that they present an effective means to protect privacy in order to protect and extend their limited realm of private authority. On the other, they face pressure not to administer principles too harshly, from the very firms on whom they rely on revenue. It appears that, at least in some instances, web seal organizations are finding themselves increasingly drawn into closer relationships with governments, as they become enmeshed in "hybrid arrangements," which mix state oversight with private enforcement.

The EU-U.S. Safe Harbor agreement is the best existing example of such an arrangement; it rests on principles that have been negotiated between the EU and United States, but are in part enforced by web seal organizations (or other providers of dispute resolution services).[53] Such arrangements offer states some advantages. They may make it easier to resolve regulatory clashes between states,[54] while increasing states' leverage over private international commercial relations. However, as privacy advocates have been swift to point out, they may also lead to new problems of transparency and accountability. Lines of responsibility are typically blurred; it is hard to hold either private enforcers or government overseers liable for their actions. Furthermore, these arrangements are highly nontransparent with convoluted decision-making procedures. Because they lie between politics and markets, they are only weakly subject to the democratic restraints of oversight associated with the one, and the pressures of market choice associated with the other.[55] Finally, it is far from clear that self-regulatory organizations along the lines of TRUSTe and BBBOnLine are sustainable in the long term as a front line of enforcement, unless there is a credible threat of government regulation that might induce firms to regulate themselves.

Second, there are emerging conflicts over the circumstances under which governments can access and use information on their citizens. These result less from worries about government use of its centralized databases (although such worries persist) than from new concerns about how governments may require private actors such as ISPs and telcos (telecommunications companies) to accede to surveillance technologies, to retain information on their customers, and to provide it to the government on request. These battles have received most debate in the United States, where there is a highly active privacy community that has publicly excoriated initiatives such as the FBI's "Carnivore surveillance system," the U.S. administration's failed Total Information Awareness project, and the National Security Agency's widespread monitoring of U.S. citizens. However, the most important fights are taking place in other parts of the world (although they are at least to some degree the result of U.S. government pressures in the "war on terrorism"). State security and law enforcement agencies across the developed world have used the events of September 11, and associated pressures from the United States for increased intelligence, to press for new laws permitting them massively to expand their information-gathering capacities. In particular, these agencies are pressing, often successfully, for the removal of restrictions on their ability to gather data about individuals' behavior on the Internet. Further, they are building on international instruments, such as the Council of Europe Convention on Cybercrime, as a means to forestall domestic opposition.

Thus, for example, in Canada, the government used the Convention on Cybercrime as a rationale to propose new rules on "lawful access" to communications data.[56] Current legislative developments in the European Union pose even more serious problems for individual privacy; complicated transnational procedures of law making make it difficult to hold governments accountable for proposals to increase access to personal data. In the wake of September 11, the European Council (the body directly representing member state interests within the European Union) has pushed successfully for the elimination of certain kinds of personal data, persuading both the European Commission and European Parliament to abandon their earlier opposition to these changes.[57] Rules that previously mandated the destruction of traffic data held by European ISPs and telcos after three to seven days, have been eliminated. This has made it possible under European law for EU member state governments to oblige ISPs and telcos to retain traffic data, and to make it available to law enforcement authorities and security services. The European Commission is likely to introduce a mandatory regime of data retention for all EU member states, which seeks in large part to respond to pressures from the justice and home affairs ministers of the member states. Member states are furthermore pressing for this legislation to be negotiated between the European Council and Parliament through an "early agreement" provision that would short-circuit open debate in favor of direct negotiations between a rapporteur appointed by the European Parliament and the member state holding the presidency. More generally, the area of justice and home affairs is likely to see a substantial expansion in cooperation over the coming decade. Transgovernmental networks that deal with substantive political issues within the EU have negative consequences for democratic legitimacy; this is all the more so when these networks come to include external actors from interested parties in the United States and elsewhere.

The impetus toward data traffic retention, in the EU and elsewhere, is one manifestation of a wider transformation in the relationship between states and ISPs. The latter are increasingly expected to act as agents on behalf of the state, in contexts such as copyright protection,[58] content regulation,[59] and the detection and prevention of crime. Such cooperation poses serious risks to individual privacy; in Ian Kerr's provocative description, ISP is coming to stand less for Internet Service Provider than Internet Secret Police.[60] Information on these relationships between states and ISPs and telcos, and what they involve, is difficult to come by. What safeguards there are appear to be minimal. Again, it can be seen that hybrid arrangements involving both states and private actors pose new policy problems. Unlike Safe Harbor, these arrangements are designed to share information rather than to protect it, but they involve similar issues of accountability, transparency, and legitimacy.

Recommendations

In the preceding discussion, I have sought to describe the complex relationship between technology and privacy as it has evolved over the last thirty years. I have examined how both power relations among states and existing institutional trajectories within them have influenced domestic and international outcomes. In this section, I turn to policy recommendations; given the realities of international politics, what means may we best use to protect privacy?

Extending Privacy Protections Internationally

As long as some states have lax or nonexistent privacy standards, there exists the risk of a lowest common denominator effect. The EU Data Protection Directive—despite its flaws—has had a significant positive effect in raising the international bar of best practice. Safe Harbor is inadequate when judged against the standards of the directive itself, but it is a significant improvement on the domestic *status quo ante* that participating U.S. firms observed, and it may help build pressures for change over the medium term. Thus, solutions of the Safe Harbor type should be encouraged for jurisdictions such as the United States, South Korea, and Japan that are unwilling or unable to introduce strong, comprehensive formal legislation.

A second problem is less directly pressing, but may have significant long-term consequences. Privacy issues may lead to nontariff barriers, which hamper the exchange of services with less-developed countries that have weak privacy laws or poor enforcement. This is increasingly relevant as more data processing services are contracted out to countries in the developing world. Two possible arrangements might prevent this if they are accepted as a valid means to ensure privacy—Safe Harbor-style arrangements or contracts. Here, I provisionally recommend a reliance on contracts[61]—Safe Harbor-style arrangements require an impartially functioning legal system that may not be present in some developing countries. Contracts, which may be enforced in other jurisdictions than the third country in question, are more flexible, and provide firms in these countries with a means of complying with higher requirements, without making demands for institutional change that are unrealistic in the short term.

Strengthening of International Mechanisms of Privacy Protection

There is a clear and urgent need for mechanisms to protect privacy at the international level. As should be clear from the earlier discussion, the national and international arenas intersect in ways that make it increasingly difficult for national-level actors to protect privacy without reference to international politics.

Privacy advocates, and the EU-U.S. consumer interest umbrella group, the Trans-Atlantic Consumer Dialogue (TACD), have advocated an international privacy convention that would embody strong and enforceable privacy standards.[62] While this remains a laudable long-term objective, it is impossible to achieve under current configurations of power and interest. A more practical response might be the strengthening of democratic oversight and consumer representatives' voice at the international level; this expansion must be accompanied by expanded oversight, to ensure that international meetings do not short-circuit democratic accountability. Such oversight would ideally involve a mixture of strengthening of accountability at the national level through national parliamentary committees.

More broadly, the international role of consumer representative organizations should be greatly strengthened. The TACD serves as an excellent and successful example of how consumer organizations may articulate a common position on international issues of e-commerce regulation, but its membership is limited to EU and U.S. organizations that wish to address the EU-U.S. relationship. Public Voice is underfunded, only partially representative, and lacks the formal "status" of the TACD. A wider partnership would address consumer and privacy issues across a variety of multilateral settings, serving both as an "official" voice for consumer interests, and as a counterpart to the Global Business Dialogue on Electronic Commerce (GBDe) and similar organizations. It would also provide an opportunity for consumer interests from developing countries to have their voice heard in international forums. As discussed already, the developing world is likely to have a different set of priorities regarding privacy, at least in the short term; these priorities are currently given short shrift.

Increased Accountability Requirements for Public-Private Actor Relationships (Hybrid Arrangements)

As discussed earlier, many of the new policy issues on the privacy agenda involve emerging, hybrid relationships between public and private actors. Such relationships have diffuse lines of responsibility; they weaken democratic accountability by devolving important functions to private actors.

The Safe Harbor arrangement, in its first two years of operation, illustrates this problem. There is remarkably little information available on what procedures are followed in evaluating and adjudicating complaints under Safe Harbor, and some evidence to suggest that the European Commission (the relevant European administrative body) has not made serious efforts to monitor day-to-day enforcement of the arrangement.[63] Problems of transparency and accountability are likely to be worse still in situations involving security or police investigations.

Thus, this chapter makes the following recommendations. First, there should be strong reporting requirements when international public-private partnerships are used to achieve important policy goals. There should be regular reports detailing both general procedures and specific actions. These reports should be published, and the relevant actors should be explicitly accountable to democratic assemblies (such as, in the case of Safe Harbor, the European Parliament).

Second, public-private cooperation in the sphere of security and policing should only be mandated where absolutely necessary. These public functions of the state are, quite simply, too important to be delegated to private actors, except where there is a burning and immediate need. In instances where they are necessary, they should be limited, targeted, and proportional to the need at hand. They should not involve generalized and diffuse increases in the policy aegis of security services or the responsibilities of private actors or both. Further, they should be subject, insofar as is compatible with any legitimate need for secrecy and confidentiality, with the reporting requirements outlined above. In any event, regular reports should be published, indicating the general patterns of enforcement. Furthermore, there should be strong oversight mechanisms, involving independent third parties, to ensure that public-private cooperation does not invade privacy more than is absolutely necessary for specific and legitimate purposes.

Conclusion

The regulation of privacy involves both international and national dimensions. In this chapter, I have argued that the relationship between the two in different conjunctures is best understood by concentrating on two key factors—existing institutional traditions and power relations among states. Existing institutional traditions in various countries help explain their initial preferences, and their relative willingness to adopt comprehensive privacy laws that were proposed by a cross-national policy community of privacy experts. Countries in mainland Europe, which had previously existing traditions of extensive business regulation, and which had recent experience of Nazi abuses of privacy, were relatively willing to regulate. Countries such as the United States, which had weak traditions of business regulation, and a different constellation of state-private actor relations, were not.

Power relations—and the relative bargaining strength of states—help explain the circumstances under which states have accepted, or failed to accept, external privacy rules that do not accord with their previously existing traditions. Powerful states in strong bargaining positions have little incentive to accept such rules, which would be domestically costly and difficult to implement.[64] Weaker states may have little choice but to accept externally imposed constraints, even when they have little desire to do so, when they believe that more powerful states will retaliate against them if they do not.

This explains the persistent unwillingness of the United States to accept externally imposed rules on privacy. Given the U.S. relationship of dominance with most of its trading partners, it has few incentives to make domestic concessions, especially when such concessions would involve major institutional changes. While the EU has been successful in pressing the United States to accept a "hybrid" solution, this solution does not formally involve any change to existing U.S. institutions. Other countries, which are weaker vis-à-vis the EU than the United States, have had little success in winning Safe Harbor–style concessions. Countries in weak bargaining positions—most especially EU applicant countries—have had little choice but to accept the EU's basic position on data protection.

These twin forces not only help explain the circumstances under which international pressures result in domestic change, they also help explain the circumstances under which domestic preferences are (or are not) instantiated in substantial, binding political agreements. The failure of an epistemic community to produce complete convergence at a previous juncture meant that there was substantial divergence in goals between European states on the one hand, and the United States and a few other countries on the other, over whether or not there should be binding international rules covering data transfer. The United States, which had both different preferences from most other countries and an impregnable bargaining position, refused to participate in any binding international agreements, resulting in two international arrangements, one which was inclusive and nonbinding, the other binding but only involving a more limited club of states.

This analysis furthermore suggests clear limits to the expert-driven processes of policy convergence around strong privacy standards that Bennett observed at an earlier stage.[65] Epistemic communities are only likely to succeed in persuading countries to adopt comprehensive privacy legislation under relatively limited conditions—that is, where such policy solutions instantiate previously existing social goals, and are compatible with broad national institutional frameworks governing, for example, state-private actor interaction. Where such conditions do not apply, as, for example in the United States, the advice of nonpolitical experts will at best have limited political effect. New international forums for discussion, such as the Internet Governance Forum that emerged from the UN World Summit on the Information Society, appear to provide a more privacy-friendly arena for discussion than more traditional international institutions,[66] and may perhaps affect international debates over privacy in the future. However, in the absence of fundamental changes in the perceived interests of the United States and other key actors they are unlikely to have substantial short-term consequences.

The current-day politics of privacy are driven by two countervailing forces. On the one hand, the European Union has enjoyed some success in bringing countries that are in some way dependent on it to introduce comprehensive privacy laws. This suggests a gradual upward ratcheting in privacy regulation around the world. On the other hand, the events of September 11 have very considerably strengthened the hand of security elites in the various advanced industrialized democracies. These elites have successfully pressed for the introduction of new laws that limit personal privacy, both in their particular domestic contexts and through various transgovernmental policy networks. In many cases, emerging domestic and international security frameworks involve close—and perhaps sometimes incestuous—relations between states and private actors that have strongly negative implications for privacy. To use less academic language, it is arguable that these changes are eating out the heart of the emerging privacy framework, even as this framework is apparently being extended to new parts of the world. Unlike privacy experts at an earlier juncture, security elites are at the very heart of the state apparatus; they thus seem to be enjoying more general success in altering domestic and international practices.

Notes

1. Although, as this chapter discusses later, some of these protections are being quietly watered down and abandoned.

2. Farrell 2003.

3. See also Bennett 1992.

4. See also Flaherty 1979.

5. Bennett 1992.

6. Ibid.

7. Bennett 1992; Mayer-Schönberger 1997.

8. Bennett 1992.

9. For a jaundiced but convincing account of the burgeoning and collapse of the TDF debate, see Drake 1993.

10. Bennett 1992.

11. See also Drake 1993; Bennett 1992.

12. Drake 1993.

13. Ibid.

14. Ibid.

15. See also Farrell 2003.

16. Regan 1999.

17. For a discussion of the comitology system, see Bergström, Farrell, and Héritier 2007.

18. Farrell 2003.

19. Regan 1999.

20. I discuss privacy seal or web seal organizations in greater detail below.

21. See Farrell 2002, 2003; Heisenberg and Fandel 2002; Kobrin unpublished; Long and Quek 2002; and Shaffer 2000, for discussion of the Safe Harbor negotiations.

22. For a more detailed discussion of Safe Harbor, see Farrell 2003.

23. See the discussion later in this chapter.

24. I owe this point to Maria Farrell.

25. Slaughter 2003.

26. See also chapter 9.

27. The Data Protection Directive is, of course, itself an international instrument, but the EU is typically considered by international relations scholars as a single actor in its external aspects, rather than a multilateral forum.

28. Bennett (1992) argues that Britain introduced data protection legislation largely in consequence of perceived external pressures, but argues that it was the exception rather than the rule at this phase of development.

29. Heisenberg and Fandel 2002.

30. Haggard and Simmons 1987.

31. Indeed, there is still some divergence of implementation within the European Union; EU directives, unlike regulations, tend to set general principles, which may be applied in different ways.

32. Scharpf 1994; Farrell 2002, 2003.

33. See Data Protection Working Party, Opinion 3/2001 on the Level of Protection of the Australian Privacy Amendment (Private Sector) Act 2000, Available at http://www.europa.eu.int/comm/internal_market/en/dataprot/wpdocs/wp40en.htm.

34. See Peter Ford, "Implementing the Data Protection Directive: An Outside Perspective," available at http://www.europa.eu.int/comm/internal_market/en/dataprot/lawreport/speeches/ford_en.pdf.

35. The General Agreement on Trade in Services (GATS) has an exception for measures taken to further data protection, but requires that these measures not be applied in a discriminatory fashion. See also Shaffer 2000; Swire and Litan 1998. Thus, if the EU were to block data flows to Australia, the Australian government would have a prima facie case for WTO action.

36. Shaffer 2000.

37. Bennett 1992.

38. Andrews and Privacy International 2002.

39. See ibid., for discussion of the nascent debate in India.

40. Author's interviews with European Commission officials.

41. See chapter 9.

42. Letter available at http://www.statewatch.org/news/2002/feb/useu.pdf.

43. See also Farrell 2002.

44. Tsebelis 2000.

45. See "White House Framework for Global Electronic Commerce," available at http://www.ta.doc.gov/digeconomy/framewrk.htm.

46. Interview with Ira Magaziner, conducted September 21, 2000.

47. See also Farrell 2003.

48. TRUSTe's credibility was badly damaged by its failure to punish member firms such as Microsoft and Real Networks for violations of their customers' privacy, which fell outside TRUSTe's formal ambit of enforcement.

49. Streeck and Schmitter 1985.

50. However, note that private-interest government is strongly associated with corporatist forms of interest intermediation and policy making, which some democratic theorists find to be problematic.

51. See also Bach and Newman 2004.

52. Interview with Malcolm Crompton, federal privacy commissioner for Australia, September 8, 2000. For further discussion, see Cavoukian and Crompton 2000.

53. The EU's interest in Safe Harbor was in part spurred by its belief that the arrangement would allow it to influence the codes of web seal organizations. See Farrell 2003.

54. See also Scharpf 1994.

55. I further note that market forces provide only poor protection for many kinds of individual and collective rights, even (and especially) when they work in an unconstrained manner.

56. See http://www.canada.justice.gc.ca/en/cons/la_al/.

57. Increases in the power of the European Parliament (the extension and refinement of the so-called "codecision" procedure), were supposed to result in greater transparency and democratic accountability. In many cases, however, they have not had this effect, instead increasing the power of rapporteurs and power-brokers within the two main parties to reach agreement with member states behind closed doors. See also Farrell and Héritier, 2004.

58. This is a key issue in ongoing legal battles between the U.S. music industry and ISPs over whether the United States Digital Millenium Copyright Act obliges ISPs to hand over the names of customers who are suspected of downloading illegal music. ISPs argue that this would effectively oblige them to police the behavior of their customers, and breach customer privacy on a massive scale.

59. See Frydman and Rorive 2002.

60. See http://www.cacr.math.uwaterloo.ca/conferences/2002/isw-eleventh/kerr.ppt.

61. Model contracts have been approved by the European Commission for various forms of data exchange.

62. See http://www.tacd.org/cgi-bin/db.cgi?page=view&config=admin/docs.cfg&id=97.

63. This is representative of a more general set of problems that the Commission faces; while it is responsible for enforcement across a wide variety of EU policy areas, it must delegate many aspects of implementation to third parties, and has scanty resources to monitor how these third parties behave.

64. I do not take account here of more complex dynamics between arenas, which may allow underprivileged actors in one arena to use leverage in another. See also Farrell 2002.

65. Bennett 1992.

66. Drake and Jørgensen 2006.

References

Andrews, Sarah, and Privacy International. 2002. *Privacy and Human Rights 2002: An International Survey of Privacy Laws and Developments*. London: Privacy International.

Bach, David, and Abraham Newman. 2004. "Self-Regulatory Trajectories in the Shadow of Public Power: Resolving Digital Dilemmas in Europe and the United States." *Governance* 17: 387–413.

Bennett, Colin. 1992. *Regulating Privacy: Data Protection and Public Policy in Europe and the United States*. Ithaca, NY: Cornell University Press.

Bergström, Karl-Fredrik, Henry Farrell, and Adrienne Héritier. 2007. "Legislate or Delegate?: Bargaining over Implementation and Legislative Authority in the European Union." *West European Politics* 38, no. 2: 338–366.

Cavoukian, Ann, and Malcolm Crompton. 2000. *Web Seals: A Review of Online Privacy Programs*. Canberra/Toronto: Office of the Federal Privacy Commissioner/Office of the Privacy Commissioner.

Drake, William J. 1993. "Territoriality and Intangibility: Transborder Data Flows and National Sovereignty." In *Beyond National Sovereignty: International Communications in the 1990s*, ed. Kaarle Nordenstreng and Herbert I. Schiller, 259–313. Norwood: Ablex.

Drake, William J., and Rikke Frank Jørgensen. 2006. "Introduction." In *Human Rights in the Global Information Society*, ed. R. F. Jørgensen, 1–49. Cambridge, MA: The MIT Press.

Farrell, Henry. 2002. "Negotiating Privacy across Arenas—The EU-US 'Safe Harbor' Discussions." In *Common Goods: Reinventing European and International Governance*, ed. Adrienne Héritier, 105–126. Lanham, MD: Rowman and Littlefield.

Farrell, Henry. 2003. "Constructing the International Foundations of E-Commerce: The EU-US Safe Harbor Arrangement." *International Organization* 57, no. 2: 277–306.

Farrell, Henry, and Adrienne Héritier. 2004. "Interorganizational Cooperation and Intraorganizational Power: Early Agreements under Codecision and Their Impact on the Parliament and the Council." *Comparative Political Studies* 27 (10): 1184–1212.

Flaherty, David. 1979. *Privacy and Government Data Banks: An International Perspective*. London: Mansell.

Frydman, Benoît, and Isabelle Rorive. 2002. "Regulating Internet Content Through Intermediaries in Europe and in the United States." *Zeitschrift für Rechtssoziologie* 23, no. 1: 41–59.

Haggard, Stephan, and Beth Simmons. 1987. "Theories of International Regimes." *International Organization* 41, no. 3: 491–517.

Heisenberg, Dorothee, and Marie-Helene Fandel. 2002. "Projecting EU Regimes Abroad: The EU Data Protection Directive as Global Standard." Paper prepared for delivery at the 2002 annual meeting of the American Political Science Association, Boston, August 29–September 1.

Long, William J., and Marc Pang Quek. 2002. "Personal Data Privacy Protection in an Age of Globalization: The US-EU Safe Harbor Compromise." *Journal of European Public Policy* 9, no. 3: 325–344.

Kobrin, Stephen J. 2002. "The Trans-Atlantic Data Privacy Dispute: Territorial Jurisdiction and Global Governance," working paper.

Mayer-Schönberger, Viktor. 1997. "Generational Development of Data Protection in Europe." In *Technology and Privacy: The New Landscape*, ed. Philip E. Agre and Marc Rotenberg, 219–241. Cambridge, MA: The MIT Press.

Regan, Priscilla. 1999. "American Business and the European Data Protection Directive: Lobbying Strategies and Tactics." In *Visions of Privacy*, ed. Colin Bennett and Rebecca Grant, 199–216. Toronto: University of Toronto Press.

Scharpf, Fritz W. 1994. "Community and Autonomy. Multilevel Policy-making in the European Union." *Journal of European Public Policy* 1, no. 2: 219–242.

Shaffer, Greg. 2000. "Globalization and Social Protection: The Impact of Foreign and International Rules in the Ratcheting Up of U.S. Privacy Standards." *Yale Journal of International Law* 25: 1–88.

Slaughter, Anne-Marie. 2003. "Global Government Networks, Global Information Agencies, and Disaggregated Democracy." Unpublished paper.

Streeck, Wolfgang, and Philippe C. Schmitter. 1985. "Market, State, Community and Associations? The Prospective Contribution of Interest Governance to Social Order." *European Sociological Review* 1: 119–138.

Swire, Peter P., and Robert E. Litan. 1998. *None of Your Business: World Data Flows, Electronic Commerce, and the European Privacy Directive.* Washington, DC: The Brookings Institution.

Tsebelis, George. 2000. "Veto Players and Institutional Analysis." *Governance* 13, no. 4: 441–474.

11 Intellectual Property Rights, Capacity Building, and "Informational Development" in Developing Countries

Christopher May

We are frequently told that the world has entered not only a new millennium but also a new age: the information age. The production of material goods is no longer most important for continued economic development. Now it is the application and use of knowledge-based informational resources that drives forward growth and productivity gains. The impact of this "information age," however, varies across the global system to a great degree, not least because the potential global scope of the Internet is far from being fulfilled. Most obviously intellectual property rights (IPRs) have become a concern for developing countries because of the AIDS pandemic in sub-Saharan Africa and the wealth effects on the distribution (and use) of medicines caused by IPRs. In this case the conflict of interest between the owners of pharmaceutical patents and poverty-stricken groups of AIDS sufferers is acute, if politically difficult to resolve.[1] However, IPRs also impact on the governance of global electronic networks (through, for instance, disputes over digitized content, the control of industry standards, and domain name allocation). Developing countries' governments need to be aware of how IPRs will impact on their ability to capture the benefits of the information age. Intellectual property is the key legal instrument underpinning power and wealth in the new millennium. If developing countries fail to master the politics of IPRs, their ability to take advantage of the burgeoning global information society will be compromised.

Since 1995 the Trade-Related Aspects of Intellectual Property Rights (TRIPS) agreement overseen by the World Trade Organization (WTO) has been the locus of the global governance of intellectual property. Broadly, this agreement represents a cooperative solution to the "problem" of international piracy and theft of valuable IPRs, building on, and adding to, the previous regime overseen by the World Intellectual Property Organization (WIPO). This new system has been designed and is controlled by the dominant rich countries. These powerful countries have constructed a "one-size-fits-all" solution to *their* problems. This system functions well for those countries that are well developed, have considerable IPR-related resources, and are major players in the field of information and communications technologies (ICTs). However, for poorer countries lacking sophisticated systems of innovation or well-developed means of providing access to informational resources, the system is far from satisfactory, nor self-evidently supportive of their developmental goals.

Although this agreement does not determine national legislation, to be TRIPS-compliant, WTO members' domestic law must support the protections and rights laid out in the TRIPS agreement's seventy-three articles. The agreement is a commitment to uphold certain standards of protection of IPRs and to provide legal mechanisms for their enforcement. Furthermore, the WTO's robust dispute settlement mechanism encompasses disputes about IPRs. Prior to 1995 there were long-standing multilateral treaties in place regarding the international recognition and protection of IPRs overseen by WIPO. These were regarded as essentially toothless by developed countries' governments (especially the United States) in the face of widespread "piracy" and disregard for the protection of non-nationals' intellectual property. Although some members of the WTO remain in the TRIPS agreement's mandated transitional period (recently extended to 2016 for pharmaceuticals), it establishes for the first time a potentially global settlement for the protection of IPRs.

Policy makers and negotiators from the developed countries who shaped the TRIPS system claim that the global governance of IPRs is a technical issue best left to the experts. This is not the case: the issues at the heart of the international recognition and enforcement of IPRs are profoundly political. The governance of IPRs affects the distribution and exercise of power in the global system, influencing and shaping the capacities of countries to achieve their developmental goals. The current global political settlement, brokered during the Uruguay Round of trade talks (and still subject to continuing negotiations, both at the WTO and WIPO), is not a "done deal" with only its implementation outstanding; rather, it is the site of continuing and significant political contestation and disagreement. In this chapter I briefly introduce IPRs for those unfamiliar with them. I then lay out the mechanisms for their global governance, and discuss the extensive assistance that is available to help new- and existing-developing-country members of the WTO reach legal compliance with the TRIPS agreement. I conclude by assessing how the politics of IPRs may play out in the realm of globalized electronic networks.

As the political mobilization around the issue of patents on AIDS medicines has demonstrated, there is a need to reestablish that countries at different levels of economic development may require very different levels of protection for IPRs. Indeed, prior to the TRIPS agreement this was the de facto position, with the levels and scope of protection of IPR left largely to domestic policy processes. At WTO ministerial meetings in Doha, Cancun, and more recently Hong Kong, developing countries have started to (re)assert their national interests in the freer flow of knowledge and informational resources (encapsulated in products and services). This has returned the discussion of IPRs to the political realm, where it belongs.

However, it is not merely in negotiations and multilateral forums that the ideology of private rights to property in knowledge and information has been promoted. Quietly, but persistently, "capacity building," aiding WTO members to establish compliance with the TRIPS agreement, also privileges this model of knowledge ownership. These programs are intended to ensure developing countries construct and manage the high levels of protection mandated by the TRIPS agreement, but more accurately represent a form of political education for policy makers, legislators, and law enforcers. While this advice may help developing countries construct the national IPR regimes the richer members of the WTO would like to see throughout the world, it is far from clear they represent the best models for poor countries at their stage of development.

What Is Intellectual Property?

Intellectual Property's General Character

When knowledge or information or both become subject to ownership, IPRs express ownership's legal benefits: the ability to charge rent for use; to receive compensation for loss; and demand payment for transfer. Intellectual property rights are subdivided into a number of groups, of which three generate most discussion: industrial intellectual property (patents); literary or artistic intellectual property (copyrights); and trademarks. Conventionally, the difference between patents and copyrights is presented as between a patent's protection of an idea and copyright's protection of the expression of an idea. Patents cover such things as technological innovations and novel industrial products, while copyright protects literary and artistic creations. Trademarks are the exclusive identifying names or signs that companies use to distinguish their products or services. Other forms of IPRs include geographical indicators and computer chip layout designs.[2]

Intellectual property laws support the rights of individuals over their creative endeavors, but this is balanced with a recognition that extensive social and economic benefits flow from their circulation. There is always a danger that the public good of access to, and use of, knowledge or information will be circumscribed by the price demanded by the "owners" of IPRs. Therefore, balancing private rewards and the public interest is at the heart of all intellectual property legislation, and is often expressed through time limits on IPRs. Unlike property rights in material things, IPRs are formally temporary: once their term has expired they enter the public realm (or in the case of trademarks they have to be renewed, demonstrating continued use). Where this is deemed not a sufficient support for the public good, governments have historically reserved the right to compulsorily license patents, or have allowed "fair use" of copyrighted material.

Compulsory licences allow a government to transfer a patent to a non-owning manufacturer, who may produce the product without paying a license fee (or only a much lower one). Recently, this has been at the center of debates about generic drug manufacturers in Africa and Asia "pirating" AIDS drugs patented by U.S. and European pharmaceutical companies. While compulsorily licenses are allowed under article 31 of the TRIPS agreement, members of the WTO differ over the interpretation of their scope and circumstances that prompt their legitimate use. In copyright, across the world common practice and legislation have recognized that the use of copyrighted goods may sometimes "infringe" owners' claimed rights. However, for instance, the use of excerpts for education and criticism, or the recording of TV programs for private use, have been supported by the law despite frequent complaints by the content industries. In both cases, policy makers and legislators in the past have argued that the public good of easing the restrictions on use instituted by IPRs in specific instances outweighs the social utility gained through the continued protection of private rights, although in recent U.S. and European legislation this position has been progressively eroded.

The most important aspect of IPRs is their formal construction of scarcity (as related to possible use) where none necessarily exists. Knowledge and information, unlike material things, are not necessarily rivalrous, and thus coincident usage does not detract from utility. Take the example of a hammer (as material property); if I own a hammer and we would both like to use it, our utility is compromised by sharing use. I cannot use it while you are, you cannot while I am, our intended use is rival. For you to also use my hammer, either you have to accept a compromised utility (relying on my goodwill to allow you to use it when I am not) or you must also buy a hammer. The hammer is scarce. However, the idea of building something with hammer and nails is not scarce. If I instruct you in the art of simple construction, once that knowledge has been imparted, your use has no effect on my own ability to use the knowledge at the same time. We may fight over whose turn it is to use the hammer, but we have no need to argue over whose turn it is to use the ideas of cabinet construction, for these are nonrival. Ideas, knowledge, and information are generally nonrivalrous.

In cases where knowledge may produce advantage for the holder (information asymmetries), by enabling a better price to be extracted, or by allowing a market advantage to be gained, information and knowledge *are* rivalrous. If there was perfect information (universal access), then the knowledge holder's benefits would evaporate. However, rivalrous knowledge is not necessarily a social benefit: information asymmetry produces market choices that are not fully informed and can be inefficient, or even harmful. Thus, when information is "naturally" rivalrous, the social good may be best served by ensuring that it is shared, not hoarded. For instance, if used-car dealers were required to reveal *all* they knew about cars they were selling, this would likely reduce the price they could obtain for much of their stock, but to the customers' benefit. However, generally knowledge and information are nonrival and hence it is difficult to extract a price for their use. This is why a legal form of scarcity is introduced, to ensure a price can be obtained for the use of knowledge.

Justifying Property in Knowledge

Significant time and effort is spent telling stories about intellectual property that are meant to justify its existence as a set of legal rights and support this imposition of scarcity.[3] These narratives involve three claims for the usefulness of making knowledge and information property. The first argument is that effort deserves reward. This draws on a long line of political theory suggesting that where man has improved nature he deserves to have property in the fruits of this effort. This started as John Locke's argument about property rights in previously common land being awarded to the diligent cultivator, and has now become a more general argument that effort requires reward. For IPRs this justification is expressed both as a reward and an incentive. Only by allowing innovators and creators ownership rights over their creations can we reward their efforts (and by doing so encourage further effort). Thus the construction of scarcity serves the social need to encourage effort and innovation. The second story suggests IPRs reflect the rights of individuals to own the products of their own mind, as part of their self-identity. Individuals must own the intellectual property in the products of their mental activity, because it is *their* mental work that has produced it. This argument is frequently alluded to in arguments about piracy and "theft" of content.

The third narrative of intellectual property reflects the capitalist character of modern society. Here the argument is concerned with the benefits of introducing markets into any particular area of social existence. Markets, we are told, promote efficiency of use and therefore if we want to ensure that ideas and knowledge are used efficiently, for the maximum benefit of society, we must introduce markets for knowledge and information. This ensures that those who value knowledge and information most highly will pay most for it (rewarding the innovators) and will be also forced by a competitive market to enhance their efficiency in using these resources. The imposition of scarcity promotes efficient use, because knowledge can be costly to produce, and the drive to enhance efficiency itself produces further surplus to spend on more knowledge creation. Thus, it is asserted that without IPRs there would be little stimulus for innovation. Why would anyone work toward a new invention, a new solution to a problem, if they were unable to profit from its social deployment? Not only does intellectual property reward intellectual effort, it actually stimulates activities that have a social value, and therefore supports the social good of progress. By encouraging and rewarding the individual creator or inventor, societies will continue to develop important and socially valuable innovations.

These stories appear in various combinations and in various ways. Whenever IPRs are contested, disputed, or merely discussed, they are (re)told and have become part of the "common-sense" view of treating knowledge as property. Indeed, these narratives of the rights of owners and the social benefit of recognizing these rights have had and continue to have a major impact on the globalized governance regime for IPRs.

From International to Global Governance of Intellectual Property

The history of intellectual property itself stretches back to fifteenth-century Venice, and for the first 350 years was almost entirely a national issue (May and Sell 2005). In the middle of the nineteenth century, however, the possibility of international protection of intellectual property (and specifically patents) became the subject of a forthright political debate between supporters and abolitionists. The abolitionists focused on three main (still familiar) arguments. First, they noted that claims that there was a "natural" right to own intellectual property obscured the very necessary legal *construction* of IPRs (their scarcity as property was hardly natural!). Second, although arguments about just rewards had some weight, the abolitionists noted that seldom were these rewards distributed fairly, and seldom did the innovators receive them. Third, despite the claimed "incentive to invent," patents were actually a disincentive to rival inventors once first inventions were patented. Furthermore, the abolitionists noted humankind had seemingly been quite innovative throughout history without recourse to intellectual property (Machlup and Penrose 1950).

In the end, supporters of IPRs won, mobilizing an agenda of justification similar to the one relied on today. The abolitionists had seen the nonavailability of patents in some jurisdictions as giving those countries an unfair advantage, and concluded that patents should be completely abolished to reestablish free trade. Conversely, the internationalist position sought to widen the scope of patents to halt, in the words of John Stuart Mill, "attempts which, if practically successful, would enthrone free stealing under the prostituted name of free trade" (Mill 1871, II, 552). The move to internationalize IPRs was therefore a direct response to a debate regarding patents' shortcomings, and the growing recognition of the injustice of the "theft" of knowledge. Diplomacy between the major trading nations went on to establish two intellectual property agreements that broadened the governance of IPRs beyond national borders.

The Rise of Internationalism in Intellectual Property Governance

The two sets of conferences focusing on the international coordination of protection for IPRs resulted in the Paris Convention (covering patents), which was completed by an interpretative protocol in Madrid in 1891, and the Berne Convention for the Protection of Literary and Artistic Works (1886). These agreements were rooted in the round of international commercial exhibitions that followed the Great Exhibition in London in 1851. As manufacturers and inventors increasingly exhibited their wares abroad, they perceived the advantages to having their patents recognized outside their own country. Similarly, international meetings of authors during the latter decades of the nineteenth century revealed to many the losses they incurred through piracy and unauthorized editions of their work. Hence, both groups had brought political pressure to bear on their governments regarding the international recognition of IPRs, and prompted the diplomatic efforts that produced these new international treaties.

While these conventions grew and developed over the subsequent century, as early as 1893 the common issues across both had led to the establishment of a combined secretariat. The secretariat functioned under various names until the establishment of WIPO at the end of the 1960s. As an agency of the United Nations since 1974, WIPO also administers other international treaties covering intellectual property (including trademarks, geographic indicators, and industrial designs) and is responsible for promoting technology transfer by supporting the recognition of IPRs in developing countries.[4]

These conventions aimed to ensure that the rights of owners could be easily exercised in foreign jurisdictions (national treatment), utilizing common processes and levels of protection. However, not only did the conventions themselves (and thus WIPO's secretariat) have no explicit or binding rules on enforcement, there also was no settled and robust mechanism for the settlement of disputes between members (such as those regarding the protection offered non-nationals) (Matthews 2002, 11). Members enjoyed enormous discretion over how they legislated to protect IPRs. Many potential signatories of the various conventions who were IPR-importers did not perceive accession to be in their immediate national interest. Given most countries' uneven enforcement of WIPO's various treaties, few governments regarded it sensible to litigate international disputes, as this would merely open their own practices to scrutiny (Braithwaite and Drahos 2002, 61).

The differences between the various members' perceptions of their national interests undermined attempts in the 1970s and 1980s to establish a workable dispute settlement procedure. Dispute settlement was possible through the International Court of Justice, but this route was seldom used as it was neither speedy nor effective in the resolution of disputes. Growing concerns over piracy among important industrial sectors (especially the content industries) in the richer, developed countries continued to be frustrated, even as IPRs were moving steadily to the center of the commercial concerns of a number of such sectors. This international phase of governance, with states signing some of WIPO's agreements and not others based on their perception of their national interest, came to an end in the mid-1990s. During the Uruguay Round of multilateral trade negotiations a new *global* phase of governance was finally initiated by the developed countries.

The Globalization of Intellectual Property's Governance

The main political pressure from the developed countries to include intellectual property in the Uruguay Round was a response by the content industries to a series of technological innovations. These were centered on new information and communications technologies (ICTs), which enhanced both the possibilities of an international (commodity) trade in information- and knowledge-related goods, but also enlarged the possibilities of "theft" and "piracy" (May 2000, 81–85). Trade negotiators from the developed countries also argued that the complex of twenty-four multilateral treaties administered by WIPO produced too much rule diversity. This did little to stimulate developing countries' interest in including IPRs in multilateral trade negotiations.

To "encourage" a change of heart, the U.S. trade representative threatened (and imposed) bilateral trade sanctions (under the Special 301 section of the Omnibus Trade and Tariff Act, 1988). This bilateral pressure largely overcame the considerable resistance to the TRIPS agreement (led by Brazil and India). The power to severely disrupt developing countries' limited export opportunities by closing off U.S. domestic markets using Special 301-linked sanctions proved decisive. However, this very large stick was combined with the carrot of a promise to open up agricultural markets and an offer to abolish the Multi-Fibre Arrangement, which constrained developing countries' textile exports (May 2000, 88). The developing countries generally lacked the expertise and resources to fully resist this sort of pressure. Thus, when the developing countries joined the new WTO they had to accede (with some transitional arrangements to be sure) to the TRIPS agreement as well.[5]

The understandable shortcomings of developing countries' representation in the pre-TRIPS negotiating forums, alongside the then less extensive reach of the new Internet-related ICTs (which really only accelerated in the mid-1990s), resulted in computer-related issues being dominated by the representatives of the United States, European Union, and Japan (Stewart 1993, 2290–2291). The few developing countries that had signed the Treaty on Intellectual Property in Respect of Integrated Circuits ("The Washington Treaty"), negotiated under the auspices of WIPO in 1989, *did* argue that there should be no additional protection for integrated circuit designs. However, in the TRIPS agreement the term of protection was extended to ten years (from seven) and enforcement of rights against importation of computer chips with illegally produced topography were introduced (ibid., 2307).[6] The limited efforts developing countries *did* make in the negotiations that led to the TRIPS agreement focused on pharmaceuticals and bio-piracy, rather than electronic networks and their supporting technologies. Developing countries' governments were certainly aware of issues around ICTs and knowledge flows (witness the diplomatic efforts that had focused on demands for a new information and communications order in the previous decade), but it seems possible that their negotiators failed to fully appreciate the impact electronically networked communications would have in the following ten years.

The keystone of the TRIPS agreement itself is the adoption in the realm of IPRs of principles that are central to the WTO (like the GATT before it): national treatment (art. 3); most-favored-nation treatment (MFN) (art. 4); and reciprocity. Reciprocity had been the main diplomatic principle behind the WIPO-managed system, but the inclusion of national treatment and MFN introduced new elements to the international governance of IPRs. Under the auspices of WIPO, the various treaties and conventions had widely varying lists of signatories; with MFN all such specialized agreements apply to all members on accession to the WTO and the TRIPS agreement. Furthermore, national treatment ensures that favoritism accorded to domestic inventors or prospective owners of IPRs relative to non-nationals is forbidden. This is an important shift, as many existing IPR systems had favored domestic "owners" either through legal or procedural means. Indeed, in the past, many countries (such as the United States in the nineteenth century) allowed piracy of non-national intellectual property by awarding protection to nationals who were known not to be the original innovators.

The TRIPS agreement, like the arrangements overseen by WIPO (and domestic law in the developed countries), is built on the clear understanding that there is considerable social benefit to be gained from the treatment of knowledge and information as property. This underlying norm of commodification suggests there is a direct metaphorical relationship between things and ideas when it comes to using them in economic relations. This requires the rendering of knowledge and information into property, and as the preamble to the TRIPS agreement notes, "intellectual property rights are private rights" (GATT 1994, A1C:2). To maintain this normative position, within discussions of IPRs and the resulting legislative settlements, considerable weight is laid on the three narratives of intellectual property already briefly discussed.

The norms encapsulated in these narratives are then used to support a set of bargains between private rights to reward and the recognized public or social value of the dissemination of knowledge and information. Thus article 7 of the TRIPS agreement establishes that: "The protection and enforcement of intellectual property rights should contribute to the promotion of technological innovation and to the transfer and dissemination of technology to the mutual advantage of producers and users of technological knowledge and in a manner conducive to social and economic welfare, and to a balance of rights and obligations" (GATT 1994, A1C:5). The problem is that while these normative objectives are well rooted in industrialized and developed countries (building on enlightenment notions of individuality and the social value of individualized endeavor), this is far from the case in other areas of the global system.[7]

Nevertheless, the TRIPS agreement is not a model piece of legislation that is to be incorporated directly into national law. Rather, it sets the minimum standards for the national legislation of all WTO members. It does not preclude members managing protection for IPRs in their own manner except where this violates the agreement's articles in some way. While the character and scope of intellectual property is modified to some extent by the agreement (especially in the realm of computer programs), the main area of discontinuity with prior practice is in the national protection of IPRs. By bringing intellectual property under the purview of the WTO (although WIPO retains considerable diplomatic importance), the TRIPS agreement stipulates that differential protection of IPRs (or more often their nonprotection) should not be used to disrupt trade flows (art. 3.2) (GATT 1994, A1C:4). Overall this globalized extension of the protection of IPRs represented a major triumph for the U.S. pharmaceutical, entertainment, and informatics industries that was largely responsible for getting intellectual property on the agenda of the Uruguay Round of trade negotiations.[8]

The contemporary global regime for the management of IPRs, through the TRIPS agreement, foregrounds trade issues at the expense of developmental, or other social good–related, issues. Any disputes with regard to the legal protection of IPRs across borders are now subject to the dispute settlement mechanism of the WTO. If it is established that national legislation is not TRIPS-compliant, then the country that has initiated the action is able to impose trade sanctions (across *any* trade sector) to secure compliance. Complaints can only be made by signatories (i.e., states) and in the judicial process; currently (although this may be subject to change), only states can make representations concerning the dispute. Given companies' enhanced access to trade negotiators, this results in the privileging of commercial interests. Although there is considerable flexibility within the TRIPS agreement, reflected in the preamble's recognition of the "underlying public policy objectives of national systems for the protection of intellectual property, including developmental and technological objectives" (GATT 1994, A1C:1), this has been circumscribed by the actions of the most powerful members of the WTO.

Most important, the statutory authority on which the U.S. Special 301 provisions are based notes that it was "amended in the Uruguay Round Agreements Act to clarify that a country can be found to deny adequate and effective intellectual property protection *even if it is in compliance* with its obligations under the TRIPS agreement" (USTR 2003, 9; emphasis added). The likelihood of enjoying any flexibilities in the agreement that differ from U.S. strategic demands as regards IPRs is very limited. Dispute settlement has certainly been enhanced through issue linkage at the WTO (allowing powerful trading countries to use access to their markets as a lever for legislative harmonization). However, WIPO also continues to act as a forum for the advancement of more focused treaties.

After the TRIPS agreement had become international law, WIPO's efforts at continued harmonization of IPR laws led in 1996 to the WIPO Copyright Treaty and the WIPO Performances and Phonograms Treaty (both came into force early in 2002 with over thirty formal ratifications).[9] These treaties have considerable impact on ICTs and the Internet as both include measures related to the legal protection of anticircumvention software, which is meant to preclude piracy (and unauthorized use) of digital files. Both also seek to widen the scope of IPRs and the scope of control of distribution of content that IPR owners can expect from domestic legislation. However, while these require signatories to bring their domestic legalization into compliance with the treaties, developing countries have been slower to sign them. Thus, the treaties while having some effect in the developed countries, currently have less impact elsewhere. However, they are likely to be increasingly important in the future as more developing countries reach higher levels of Internet connectivity. Apart from these treaties, WIPO continues to provide, as it has done for decades, considerable support to developing countries to bring their legislation, judiciary, and enforcement procedures into line with the developed countries. Under the provisions for technical cooperation in the TRIPS agreement (article 67), "capacity-building" programs have taken on a renewed importance.[10]

Incorporating Developing Countries into the TRIPS "Regime"

Although there may be evidence that some developing countries have adopted a WTO-based political economic agenda regarding the trade in material goods and commodities (Ford 2002), this is much less evident in the realm of IPRs. The crucial notion of "owning" knowledge and information is quite alien to many non-Western cultures. A number of agencies have been particularly active in offering transitional assistance to help developing countries address this "problem" by constructing the legal capacity to fulfill their TRIPS-related obligations. The development of administrative practices and enforcement mechanisms is important for the protection of IPRs, but the foundation of any capacity building is the legislation itself. Without compliant legislation the best administration and enforcement will do little to ensure developing countries fulfill their formal undertakings on IPRs. Therefore it is little surprise that extensive aid resources have been directed at the drafting of developing countries' domestic IPR-related legislation.[11]

The Support Offered to Developing Countries

One of the most important programs is the Co-operation for Development Programme through which WIPO provides support and training for countries developing the legal structures that the TRIPS agreement's undertakings require. This program has two distinct elements, an assistance program and the maintenance of a documentation collection. The Collection of Laws section at WIPO has centralized the archiving of legislative texts. These are available electronically to all members, to aid in the drafting of their own legislation. The assistance program is conducted by express agreement with WTO and is explicitly aimed at transitional countries to help them draft TRIPS-compliant legislation.

Assistance may take a number of forms: the submission of a draft law on any aspect of industrial property or on copyright and related rights; comments or studies on draft laws or on existing laws as regards their compatibility with relevant international treaties; or general legal advice on intellectual property law (WIPO, undated). Draft laws and other legal instruments often circulate between a member's government and

WIPO itself a number of times before a final version is settled on. Assistance can include visits to the country by officials or invitations for key legislators or civil servants or both to Geneva for consultations. After the law has been enacted, WIPO offers national workshops on the adopted legislation, judicial symposia, and training for enforcement officers.

To give an appreciation of the scale of WIPO's capacity building and technical support operations, between January 1998 and June 2001, WIPO provided the following technical assistance:

- 2,087 intellectual property officials received training in awareness building and human resources development (1,451 from Africa, 383 from Asia-Pacific, 225 from Arabic-speaking countries, and 28 from Haiti);
- Thirty-four countries received assistance in building up or upgrading their intellectual property offices with adequate institutional infrastructure and resources, qualified staff, modern management techniques, and access to information technology support systems; and
- Thirty-two countries were beneficiaries of WIPO assistance on legislation in the areas of intellectual property, copyright, and neighboring rights and geographical indications (WIPO, undated).

More recently, during 2003, for example, WIPO reported that more than 17,000 representatives from 98 developing countries participated in 228 meetings, seminars, and other training sessions, while WIPO staff undertook approximately 300 missions to developing countries to offer support and assistance in implementing various aspects of IPRs (WIPO 2003, 5). The extent of this work reflects the agency's view that a "clear and balanced view of the Agreement enables [governments] to assess the conformity of their existing national legislation vis-à-vis the provisions of the TRIPS Agreement." While this dwarfs most other provisions, other agencies are also providing important sources of support.

European Patent Office (EPO) programs center on training, advice, and assistance, as well as the provision of patent (and other IPR-related) documentation. Although the EPO limits most of its work to "awareness raising" and focuses on direct trading partners, it also has its own academy, where in 1999, for instance, 422 trainees from 80 countries attended 23 courses. The EPO has a strong outreach program arranging seminars for patent attorneys, judges, and administrators in Eastern Europe and across Asia. Perhaps most important, given the recent Chinese accession to the WTO (and therefore the TRIPS agreement), the EPO has worked with the Chinese State Intellectual Property Office to improve its in-house training (EPO 2000, chapter 5). One of the key elements of the EPO's activities has been the provision of European national legislation and collections of patent applications as learning materials. These collections are intended to help legislators draw up laws that reflect the procedures and protections found in European law. As with WIPO's activities, training involves the composition of model laws and the importation of specific elements of European law.

Unlike WIPO and the EPO, the World Bank is a little more cautious in its support for the universalization of the TRIPS standards of IPR protection. In the 2002 *Global Economic Prospects* report, the chapter on IPRs concludes: "While promising some eventual benefits, the new [TRIPS] regime is asymmetric in its likely effects across countries. Low-income economies may expect to incur net costs for some time, suggesting that patience and assistance are needed, along with programmes to limit potentially negative effects in areas such as new medicines" (World Bank 2002, 148). Nevertheless, it has included IPRs in its own wider legal training program, and continues to help countries develop the legal capacity to establish TRIPS-compliant legislation and practices, with recent programs in Brazil, Indonesia, and Mexico (alongside legal programs sponsored by the WTO). The World Bank also works at the more general level of developing the judicial and legislative capacity for the "rule of law," now seen as a central element of "good governance."

In the realm of bilateral aid, the U.S. Agency for International Development (USAID) now spends around a quarter of its annual budget on legal and regulatory training including a major program focusing on trade policy/regulatory activities, and capacity building related to the accession to the WTO. Between 1999 and 2001, USAID provided $7.1 million in aid related to compliance with the TRIPS agreement. This took the form of technical assistance from the U.S. Patent and Trademark Office (USPTO) to help countries bring their domestic legislation into compliance with the TRIPS agreement. Assistance ranged from assessments of draft laws to recommendations regarding existing laws and the manner in which they can be brought into compliance. The USPTO also runs a visiting scholars program that includes hands-on training in the administration of intellectual property law. It has assisted a number of countries with seminars and training programs for officials and legislators (including Kenya, Ghana, Mozambique, India, Brazil, Poland, Mexico, Russia, Georgia, Lithuania, Macedonia, Malaysia, Sri Lanka, Thailand, Uzbekistan, Oman, the Dominican Republic, Lebanon, and Cyprus) (USAID 2002, chapter 2; DeLisle 1999).

The consolidation of global IPR regulation centered on the WTO and WIPO has severely marginalized the UN Conference on Trade and Development (UNCTAD). This reflects the critical line UNCTAD had taken previously to the IPR-related demands of the developed countries (which were finally codified in the TRIPS agreement). As Braithwaite and Drahos note, therefore, UNCTAD, "the one UN organ with high levels of analytical expertise on trade and intellectual property has largely become irrelevant in affecting intellectual property standard-making" (2000, 68). However, the agency continues to provide analysis and advice to developing countries, while NGOs, like Quaker United Nations Office, Oxfam, and Action Aid, also offer policy advice (although this is not necessarily uncritically supportive of TRIPS compliance).

Problems with Legal "Assistance"

The TRIPS model of IPR governance may not self-evidently serve developing countries' immediate best interests, even though this is the model that capacity building aims to reproduce. Paradoxically, in an attempt to ensure their clients are not caught up in costly IPR-related trade disputes with developed-country members of the WTO, the staff of WIPO have often encouraged developing countries to adopt legislation that goes beyond the formal requirements of the TRIPS agreement (Drahos 2002, 777). This has been prompted by the position of the USTR that compliance with the TRIPS agreement is *not* sufficient to avoid U.S. trade sanctions. Additionally, bilateral trade agreements with the United States have required a number of developing countries to adopt "TRIPS-plus" legislation. By helping countries bring their legislation into line with U.S. demands, assistance programs have undermined the possibilities of diplomatic (and democratic) engagement with the TRIPS agreement itself.

The provision of ready-made legislative models and extensive "support" does little to help develop legal capacity in client countries that may have a more flexible (or critical) view of the WIPO/TRIPS model. By training and supporting a certain trajectory of legal development, WIPO and other agencies are seeking to ensure that criticism is muted or rendered outside the realm of accepted (governmental, policy-oriented) opinion. However, the demands of the TRIPS agreement seldom fit with developing countries' previous political and legislative traditions. There has been a critical political response in many developing countries, with recent debates ranging from perceptions of bio-piracy, to concerns about pharmaceutical products, from costs of useful software, to the "theft" of traditional knowledge. Thus, political pressure is often directly opposed to the models that assistance programs are promulgating, creating continued political tensions between developing countries' governments and domestic interests. Furthermore, these debates, and their effects on the legitimacy of IPRs in developing countries, impact directly on the governance of the international electronic trade in IPRs. The political, economic, and ethical (social cost-related) arguments raised by these questions are also relevant to the flow of digitized informational products.

Global Electronic Networks and Intellectual Property Rights

Many multilateral institutions now stress that improved communications (utilizing new ICTs, and specifically the Internet) are a key aspect of development. The World Bank's 1998–1999 *Knowledge for Development* report lays great emphasis on the role of the Internet and linked digital technologies (World Bank 1999, chapter 4), as do the UNCTAD report *Knowledge Societies* (Mansell and Wehn 1998) and the G8's Digital Opportunities Task Force report (DOT Force 2001). In one sense, all development is informationalized development: development is the application of (new) knowledge (and information) to existing or historical social, political, and economic problems.

However, one of the central arguments of the 1998–1999 World Bank report is that new ICTs "hold great potential for broadly disseminating knowledge at low cost, and for reducing knowledge gaps both within countries and between industrial and developing countries" (World Bank 1999, 57).

Indeed, there has been considerable discussion of the Internet as a public good supporting the dissemination of valuable knowledge and information for development (see, e.g., Adamson 2002; Lessig 2001; Spar 1999). As Jerome Reichman stresses, the "increased use of electronic publication via the Internet already allows even the latest-comers to access the most advanced thinking and methods in certain fields [note deleted]. These information networks thus become critical tools for breaking through the neo-mercantilistic fences that increasingly surround innovative products and processes in the technology-exporting countries" (1997, 85). Although there may be technological problems (linked to infrastructural development), here I want to emphasize that these knowledge flows are also compromised by the continuing commoditization of knowledge and the widening scope of IPRs.

TRIPS and "Informational Development"

When the TRIPS agreement was being negotiated, many of the issues central to the "informational development" agenda were relatively under-recognized, especially by developing countries' trade negotiators. Despite calls for a new world information and communication order in the 1980s, during the Uruguay Round the full potential of the Internet had not yet been fully appreciated. Thus, for many developing countries' negotiating teams this aspect of the TRIPS agreement was seen more as an item for cross-sectoral bargaining rather than anything that would have an immediate impact on their developmental potential. However, IPRs have a considerable (and growing) salience for developing countries' interactions with global electronic networks. Broadly speaking, there are two linked issues where IPRs may have a serious impact: on access to valuable knowledge and information for non-ICT-related development goals (new agricultural- or health-related information, for instance); and on the ability to utilize and further develop ICTs and related technologies themselves.

Developing countries may have handicapped themselves by acceding to the TRIPS agreement. This is especially the case regarding the costs of obtaining electronic tools (software), which need to be compatible with the software used elsewhere in the network. Certainly, there are legitimate arguments about the possibility of technical transfer being facilitated by protecting the owners rights, so that they are willing to license new and innovative technologies to developing countries' users. Conversely, the character of the Internet allows owners of operating systems to enjoy monopoly rents when these technologies become the standard interfaces for connectivity and knowledge dissemination. Although the Internet may encourage the proliferation of content, as the domination by Microsoft demonstrates, the considerable network effects of communications infrastructure have allowed a virtual monopoly based on the software underpinning the system to be established. Thus, much, if not all, software used to access the Internet is based on proprietary technologies.

Open source (nonproprietary) software is increasingly available but as yet represents only a tiny minority of deployed software. It has yet to win the full confidence of many developing countries' users who need to carefully choose how to use their limited resources (and need to maximize the potential interactivity from such investment). However the lower cost base and, perhaps more importantly, the ability to amend programs for specific needs (without infringing IPRs) suggests that developing countries' users would be well served by a move to open-source technologies.[12] Already, in the developed countries the LINUX open-source operating system has captured a significant and growing proportion of company servers for this reason. Additionally, in the wake of the increasing incidence of viruses aimed at machines running Microsoft programs, open-source alternatives may also become a more secure, as well as cheaper, option for users with limited resources to deploy on connectivity.

In the past the cost problem was often side-stepped by pirated software, which is widely available in the urban centers of many developing countries. This might be regarded as a necessary, if illegal form of technology transfer when confronted with relatively high prices of legitimate software. Yet this pragmatic response is dwarfed by the vast financial transfers from the poorer countries to the rich due to the current IPR system. Even *The Economist* has noted that the evidence "suggests that inflows of foreign direct investment may rise when intellectual property rights are strengthened. In the meantime, however, governments of poor countries are being asked to co-operate in a redistribution of global income that will cost them hundreds of millions of dollars" (*The Economist* 2001). Although the gains and benefits that developing countries can expect from the global information society are in the future, the costs of protecting IPRs are all too immediate.

The trade in ICT technologies in developing countries can be easily characterized as rent taking by companies that have already fully recovered their costs of development and made significant profits in developed-country markets. Under previous national legislation, high social costs might have prompted the legitimate recourse to some form of compulsory licensing, but under the TRIPS agreement such strategies have been severely constrained. Furthermore where the source code of software is protected (which while "encouraged" bilaterally is not actually mandated by the TRIPS agreement), this inhibits reverse engineering of specific programs. This further restrains development as reverse engineering has been the traditional method of technology transfer, and in the past allowed local innovators to improve off-the-shelf technologies to reflect local conditions.

Therefore, not only are the very tools that are central to "informational development" expensive, the previous (albeit illegal) methods for taking advantage of them also are being withdrawn under immense political pressure from the United States and the EU. However, there is no global mechanism to deal with the high social costs of this protection, and no global agreement on the social value of ICT-related technology transfer over and above their market provision by private owners.

Content Management and the Developing Countries

The likely undertakings for the enforcement of content management technologies that will be required from developing countries in the near future may be particularly onerous, especially in light of the increasing use of digital rights management (DRM) software. The deployment of DRM is being pushed by the content industries (and others) as a way of limiting the impact on their business of faultless digital copies. Digitalization has widened the scope for unauthorized use and (perfect) copying of content. Hard DRM aims to make such practices impossible, while soft DRM technologies merely enhance surveillance of users to facilitate legal actions against infringers. The 1996 WIPO Copyright Treaty, which many developing countries are under pressure to adopt, includes under articles 11 and 12 significant measures as regards DRM and their enforcement as IPR-related technological solutions to piracy.

The U.S. Digital Millennium Copyright Act (DMCA) was the first piece of national legislation to explicitly support DRM, most importantly with regard to anticircumvention measures. Under the DMCA any attempt to bypass limitations on use in a hard DRM technology is illegal. This represents a major challenge for "fair use" as even circumvention of technological limitations to allow a legal usage is rendered illegal by the act (Gross 2002). This was followed by similar legislative provisions in the European Union Copyright Directive. The propagation of legal protection for DRM through bilateral TRIPS-plus agreements suggests that the technology may consolidate (or even worsen) the wealth-linked limitations on the distribution of information and knowledge in developing countries.

The Commission on Intellectual Property Rights report (commissioned by the British Department for International Development, and published in 2002) concluded that DRM technologies "rescind traditional 'fair use' rights to browse, share or make private copies of copyrighted works in digital formats, since works may not be accessible without payment, even for legitimate uses. For developing countries, where Internet connectivity is limited and subscriptions to on-line resources unaffordable, it may exclude access to these materials altogether and impose a heavy burden that will delay the participation of those countries in the global knowledge-based society" (CIPR 2002, 106).

The extra controls that subscription online services and copy-protected products allow content owners have dissipated the hard-won compromises for users that have been encapsulated by previous understandings of fair use. Fair use recognizes that sometimes private rights are bought at too great a social cost. The monopoly rights enjoyed by the "owner" are therefore constrained by allowing some nonremunerated fair use. However, DRM will have (and is already having) the effect of shifting this previously widely accepted private/public bargain firmly in the direction of greater protection of so-called owners rights, with a considerable diminution of social rights of public access. This has also prompted considerable criticism from European consumers' rights groups as well as a number of campaigns around "digital freedom" in North America.

However, it is not completely evident that DRM represents a comprehensive technical fix to the problem of piracy of content. A group of Microsoft employees (although, explicitly noting that the conclusions may not be those of their employer) famously suggested that the spread of what they refer to as the "darknet" of illegal content transfers and interactions "will continue to exist and provide low cost high-quality service to a large group of consumers. This means that in many markets, the darknet will be a competitor to legal commerce" (Biddle et al. 2002, paragraph 5.2). This would be a two-sector world of content distribution: one where those wishing to remain legal interact and purchase content, and one where intellectual property norms are essentially absent. While WIPO and the WTO clearly are aiming to keep the developing countries in the first group, for many something akin to the second position may be more developmentally efficient. Only when countries reach a level of economic development where the benefits of "ownership" rights of IPRs outweigh the developmental costs of recognizing them may they then wish to instigate full protection of IPRs.

Although the potential method of delivery of content has changed with the arrival of the Internet, historical ironies abound. As Ruth Gana points out, Europeans and Americans only began to protect foreign content relatively recently, because governments recognized that the "availability of literature at affordable rates is crucial for meeting educational objectives that are a vital part of the developmental process" (Gana 1996, 327).[13] To support their developmental objectives these countries explicitly limited the benefits of copyright protection to domestic authors, which effectively rewarded the piracy of foreign authors' works. The United States only entered the international copyright system a decade or so prior to the negotiations that produced the TRIPS agreement, although by then measures had been enacted bilaterally to protect some non-U.S. authors and rights holders. Only at a certain level of development does the widespread protection of IPRs in content and technology actually aid further development. While developing countries might well need (and perhaps should) enter the global system mandated by the TRIPS agreement, that time has not yet come.

There are also difficulties in the interaction of trademarks and domain names. Here disputes have been handled by WIPO in conjunction with ICANN through the Uniform Domain Name Dispute Resolution Policy (see Mueller and Woo, chapter 14). Conversely with geographical indicators and in the protection of traditional knowledge, forms of IPRs may be able to serve the interest of the developing countries (CIPR

2002, chapter 4). For traditional knowledge, at least, the robust protections available under the TRIPS agreement may help developing countries profit from the use and commoditization of their cultural assets.

Where developing countries have significant domestic innovative capacity, then these industries may well benefit from TRIPS-compliant legislation. If, like India, they have growing software and new ICT-related sectors, then the protections that cause problems for poor users may at least support innovators and service companies in the global market, but equally they may choose the increasingly attractive open-source alternative (May 2006b). However, it is far from clear that this is a significant issue for most developing countries at present, most of which tend to be importers of IPR-related good and services. This is not to suggest developing countries are deserts of creativity or innovation, only that given the network effects of knowledge and information from developed countries, imported knowledge resources are often of great developmental utility, alongside existing and developing local knowledge and expertise.

Recommendations

In the realm of the global governance of IPRs generally, while IPRs impact on a number of sectors and issues, these should be dealt with together rather than only as special and separate cases. Thus, the lessons learned by developing countries' policy makers, NGOs, and other interested parties regarding specific IPR-related issues need to be generalized and applied across the TRIPS agreement. Up until now, while capacity building has been conducted at the general level of legislation, political engagement has really only been evident in separate issue areas, most obviously around the AIDS pandemic (linked to pharmaceutical patents) and bio-piracy (linked to the activities of global agricultural business). Developing countries stand to gain if the current settlement for IPRs is more generally examined and opened up for renegotiation. This will require a similar level of political and diplomatic engagement that has been mobilized around the AIDS pandemic, but the very success of this effort should encourage its replication and widening. While this may seem a tall order, if developing countries are to advance in a global society that increasingly values information and knowledge assets (and resources), then they have little alternative unless they are willing to accept the continued reproduction of global (informational) inequality.

Three further recommendations stem from this more general conclusion. Developing countries' trade negotiators and legislators need to fully exploit the existing flexibility of the TRIPS agreement; use future international trade negotiations to establish differential treatment regarding protection of IPRs; and emphasize the need for a robust recognition of fair use.

Exploitation of Flexibility

The provision of training assistance and other avenues of capacity building needs to be carefully monitored by developing countries' policy makers. There is a clear need to develop legal expertise to allow legislators, policy makers, and civil servants to be fully conversant with the regulatory model the TRIPS agreement implies. However, legislators need to ensure that the possibilities of legislative flexibility are recognized and exploited. As previously noted, elements of the agreement's preamble and specific articles (such as article 31 on the use without authorization of the rights holder) contain significant possibilities, as the Doha Ministerial Declaration demonstrated with regard to health emergencies. As the agreement is becoming more familiar to activists, developing country negotiators, and policy makers, so too its legal structure (the result of considerable negotiation in the Uruguay Round) is being revealed to be more flexible than the developed countries claim. Developing countries must not merely accept the model of compliance that is furthered (implicitly and explicitly) through capacity-building activities.

Certainly there is no legislative model that will be equally useful for all developing countries; different governments in discussion with national (and international) stakeholders may want to vary the scope and length of protection in light of agreed national developmental priorities. Furthermore, despite their good intentions the TRIPS-plus model being "supported" by WIPO's program needs to be firmly resisted, as do the efforts of the USTR to include TRIPS-plus measures in bilateral trade negotiations. Here trade negotiators need to recognize the important developmental impact of IPR-related provisions and not merely treat them as a bargaining counter that can be sacrificed for other (albeit important) gains.

Differential Treatment

The notion of applying different levels of compliance depending on local problems and wealth levels has already been the subject of considerable speculation and proposals during the post-Doha debates in the TRIPS council regarding health emergencies. However, given the emphasis on knowledge for development, developing countries may find that accessing information (not least of all that related to the production and use of "crucial" ICTs) is subject to wealth effects. Given the international history of IPR protection, in the realm of international diplomacy there is already a significant interest in developing a campaign to establish differential treatment. (At the very least, this should aim to replicate the extension of the transitional period for pharmaceutical patents to the entire scope of the TRIPS agreement.)

Certainly, differential protection would require a clear and transparent mechanism for adjudging the point at which any country crossed the differential treatment threshold, and such a mechanism would need to be both equitable and accountable to the members of WTO and other stakeholders. Differential protection is a radical suggestion in the current climate but is already being discussed quite widely by interested NGOs and by developed country academics and policy advisors. As this was essentially the position prior to the Uruguay Round, it cannot be said to be unprecedented.

Fair Use

Especially as regards digital content, there is a need to build on the (now threatened) fair use aspects of Western law in this area. Many of the arguments regarding the use and dissemination of information and knowledge find precursors in debates about IPR law in the developed countries, in both the nineteenth and twentieth centuries. The principle of fair use and the notion of the public domain need to be rescued from the category of residual rights to which they have increasingly been relegated by neoliberal visions of individualized knowledge creation.

Specifically, as regards the governance of global electronic networks, there is already a considerable movement in software and information dissemination toward open source and away from the proprietary model (May 2006b). There is pressure from the owners of proprietary software and technologies not to adopt open standards, but there are also suppliers who do support this technology. Developing countries' policy makers need to investigate this movement and assess the potential to side-step the IPR regime by utilizing Linux and other open-source technologies. Developing countries' governments might also usefully consider extending capacity-building requests to open-source ICT engineering.

The social model of shared innovation and non-ownership can be supported by the legal infrastructure of the TRIPS agreement through "copyleft" licenses that preclude the claim of ownership of any version of open-source software. This may also dovetail better with developing countries' own cultural history and aspirations. Indeed, given copyleft's dependence on copyright law to maintain non-ownership, it may represent a key point where the TRIPS agreement dovetails with developing countries' traditional legal culture. In this regard Phillipe Aigrain has concluded that "copylefting serves the public good especially for any software that plays or may later play a critical role in the activities of an information society" (Aigrain 2002). The use of copyleft or GPLs (general public licenses) is one way to ensure that developmentally crucial software is not re-propertized or enclosed by private commercial interests. Certainly developing countries that are in the position of developing software specific to their own needs should carefully consider balancing this sort of open access "protection" with the dominant forms of IPRs that privilege private rights.

Conclusion

In the only copyright-related dispute (over music royalties) to reach a full WTO dispute panel in its first eight years, the United States decided to pay the EU compensation (just over $1 million) rather than comply with the ruling to change its domestic law. This has led Jane Ginsburg to conclude that the "TRIPS will be a supranational code only for those too principled or too poor to opt out by paying compensation" (Ginsburg 2002, 119). Those developing countries subject to bilateral and multilateral pressures may find they are held to more exacting standards on IPRs than the richer developed members of the WTO. Furthermore, the current (TRIPS-mandated) social bargain between private rewards and pubic benefits at the heart of the legal construction of IPRs is inappropriate to many developing countries' developmental aims. On one side many developing countries' elites and governments are keen to join the international trading community and see the need to adopt the increasingly universalized rules of the system as part of this process. This is especially the case for electronically delivered services, where India and the Caribbean states are developing significant sectoral capacity. But, conversely, there are vocal constituencies that have prompted a political response less supportive of an unqualified adoption of TRIPS-related standards, such as that in the global pharmaceutical sector.

The agenda of possibilities embedded within the TRIPS agreement has been set by the developed countries in light of their current practice and the interests of their leading industrial sectors. This neither recognizes that needs of the developing countries at various "stages" of development may be very different, nor that the developed countries themselves, when they were developing, were not held (or willing to be held) to such stringent standards of protection. As a negotiating bloc, developing countries reluctantly allowed the TRIPS agreement to be included in the general bargaining in the Uruguay Round, hoping that any costs to their economies would be far outweighed by the movement on the international trade in agricultural products and textiles promised by the developed states in return. The opening of textile markets continues to be partial, while the deal of agricultural products has proved as illusive as ever.

Developing countries' policy makers and trade negotiators have now realized that the costs and constraints that compliance with the TRIPS agreement brings are heavier than they may have envisioned. They have started the very necessary process of reexamining their commitments under the TRIPS agreement and as indicated by the Doha declaration are looking at ways in which the flexibility in the agreement can be exploited. These countries have been able to utilize significant advise from NGOs in the developed countries, which have provided considerable legal assistance. This has led developing country members of the WTO to focus on those aspects of the TRIPS agreement that build on the previous arrangements for the governance of IPRs. Negotiators and governments are now focusing on the preamble to the agreement, which stresses flexibility for developing countries and public policy objectives. These aspects have been downplayed by the richer members of the WTO, but are increasingly seen, alongside article 31's recognition of special circumstances of "national emergency" as routes to a more development-friendly interpretation of the agreement. Certainly, the TRIPS agreement still privileges private rights, but like IPRs more generally, within the agreement, the recognition of a public good is also present. Now it is for developing countries to emphasize this aspect of the agreement in the face of continued pressure from commercial interests. This may not be easy, but it is by no means impossible, or unprecedented!

Notes

I would like to thank the participants in the Governing Global Electronic Networks workshop hosted by the Open Society Institute in Budapest, November 2002, and especially the editors of this volume, for their comments and suggestions on previous drafts of this chapter.

1. See May 2002 for a fuller discussion of this issue.

2. Space precludes a full discussion of the character of the various forms of intellectual property, but see May and Sell 2005, 5–11.

3. See May 2000, 22–29.

4. I have discussed the WIPO at length elsewhere; see May, 2006a.

5. Extended discussions of the negotiations that led to TRIPS can be found in Matthews 2002, chap. 2, and Stewart 1993, 2245–2333.

6. Peter Drahos sees this as a clear case of forum shifting. The original success by developing countries to limit the scope of the Washington agreement at the WIPO was undermined by the shift to the pre-TRIPS/WTO negotiations where the U.S. and EU negotiators had another chance to make the protection more robust (Drahos 2002, 780).

7. See, for instance, the discussion in Burkitt 2001 regarding the cultural specificity of copyright laws and the considerable variation of norms underlying legislation even between developed countries themselves.

8. For discussions of the role of the private sector in this process see Drahos and Braithwaite 2002; Matthews 2002; and Sell 2003, chapter 5.

9. A full and detailed, though perhaps overly positive, discussion of the treaties can be found in Reinbothe and von Lewinski 2002; my discussion of the political resurgence of the WIPO is set out in May 2006a, chapter 6.

10. The following treatment of capacity building draws on May 2004, 2006a.

11. Interestingly, de Lisle (1999) discusses the various U.S. programs (although not those supported by multilateral institutions, and with a focus on aid to post-Communist regimes) and concludes that although there is a wide diversity in experience and success, programs related to IPRs have actually been among the less successful attempts to export legislation and practice *directly* related to U.S. models.

12. I have discussed the potential of free and open-source software to serve the developmental needs of developing countries at some length in May 2006b.

13. See also May and Sell 2005, chaps. 5 and 6.

References

Adamson, Greg. 2002. "Internet Futures: A Public Good or Profit Centre?" *Science as Culture* 11, no. 2: 257–275.

Aigrain, Philippe. 2002. "A Framework for Understanding the Impact of GPL Copylefting vs. Non-copylefting Licenses." *Technology, Innovation and Intellectual Property*, no. 4. Available at http://www.researchoninnovation.org/tiip/currentissue/aigrain.htm.

Biddle, Peter, Paul England, Marcus Peinado, and Bryan Willman. 2002. "The Darknet and the Future of Content Distribution." Available at http://www.briefhistory.com/footnotes.

Braithwaite, John, and Peter Drahos. 2000. *Global Business Regulation*. Cambridge: Cambridge University Press.

Burkitt, Daniel. 2001. "Copyrighting Culture—The History and Cultural Specificity of the Western Model of Copyright." *Intellectual Property Quarterly* no. 2: 146–186.

Commission on Intellectual Property Rights (CIPR). 2002. *Integrating Intellectual Property Rights and Development Policy*. London: CIPR/DfID.

de Lisle, Jacques. 1999. "Lex Americana?: United States Legal Assistance, American Legal Models and Legal Change in the Post-Communist World and Beyond." *University of Pennsylvania Journal of International Economic Law* 20, no. 2 (summer): 179–308.

Digital Opportunities Task Force (DOT Force). 2001. *Digital Opportunities for All: Meeting the Challenge*. Available at http://www.dotforce.org/reports/DOT_Force_reportV5.oh.doc.

Drahos, Peter. 2002. "Developing Countries and International Intellectual Property Standard Setting." *Journal of World Intellectual Property* 5, no. 5: 765–789.

Drahos, Peter, and John Braithwaite. 2002. *Information Feudalism. Who Owns the Knowledge Economy*? London: Earthscan Publications.

The Economist. 2001. "Markets for Ideas." April 14, 96.

European Patent Office [EPO]. 2000. *EPO Annual Report*. Munich: EPO. Available at http://www.european-patent-office.org/an_rep/2000/html.

Ford, Jane. 2002. "A Social Theory of Trade Regime Change: GATT to WTO." *International Studies Review* 4, no. 3 (fall): 115–138.

Gana, Ruth L. 1996. "The Myth of Development, The Progress of Rights: Human Rights to Intellectual Property and Development." *Law and Policy* 18, nos. 3/4 (July/October): 315–354.

General Agreement on Tariffs and Trade [GATT]. 1994. *Final Act Embodying the Results of the Uruguay Round of Multilateral Trade Negotiations*. Geneva: GATT Publication Services.

Ginsburg, Jane C. 2002. "Berne Without Borders: Geographic Indiscretion and Digital Communications." *Intellectual Property Quarterly* no. 2: 111–122.

Gross, Robin D. 2002. "Copyright Zealotry in a Digital World: Can Freedom of Speech Survive? In *Copyfights: The Future of Intellectual Property in the Information Age*, ed. Adam Thierer and Clyde Wayne Crews. Washington, DC: Cato Institute, 189–196.

Lessig, Lawrence. 2001. *The Future of Ideas. The Fate of the Commons in a Connected World*. New York: Random House.

Machlup, Fritz, and Edith T. Penrose. 1950. "The Patent Controversy in the Nineteenth Century." *Journal of Economic History* 10, no. 1 (May): 1–29.

Mansell, Robin, and Uta Wehn. 1998. *Knowledge Societies. Information Technology for Sustainable Development*. Oxford: Oxford University Press/UNCTAD.

Matthews, Duncan. 2002. *Globalising Intellectual Property Rights: The TRIPS Agreement*. London: Routledge.

May, Christopher. 2000. *A Global Political Economy of Intellectual Property Rights: The New Enclosures?* London: Routledge.

May, Christopher. 2002. "Unacceptable Costs: The Consequences of Making Knowledge Property in a Global Society." *Global Society* 16, no. 2 (April): 123–144.

May, Christopher. 2004. Capacity Building and the (Re)Production of Intellectual Property Rights." *Third World Quarterly* 25, no. 5: 821–837.

May, Christopher. 2006a. *The World Intellectual Property Organisation: Resurgence and the Development Agenda*. London: Routledge.

May, Christopher. 2006b. "Escaping TRIPS' Trap: The Political Economy of Free and Open Source Software in Europe." *Political Studies* 54, no. 1 (March): 123–146.

May, Christopher, and Sell, Susan. 2005. *Intellectual Property Rights: A Critical History*. Boulder, CO: Lynne Rienner Publishers.

Mill, John Stuart. 1871. *Principles of Political Economy*, 7th ed. London: Longmans, Green, Reader and Dyer.

Reichman, Jerome H. 1997. "From Free Riders to Fair Followers: Global Competition under the TRIPS Agreement." *New York University Journal of International Law and Politics* 29, no. 1: 11–93.

Reinbothe, Jorg, and Silke van Lewinski. 2002. "The WIPO Treaties 1996: Ready to Come into Force." *European Intellectual Property Review* 24, no. 4 (April): 199–208.

Sell, Susan. 2003. *Private Power, Public Law: The Globalisation of Intellectual Property Rights*. Cambridge: Cambridge University Press.

Spar, Debora L. 1999. "The Public Face of Cyberspace." In *Global Public Goods. International Co-operation in the Twenty-first Century*, ed. Inge Kaul, Isabelle Grunberg, and Marc A. Stern. New York: United Nations Development Programme/Oxford University Press, 344–362.

Stewart, Terence P. 1993. *The GATT Uruguay Round. A Negotiating History (1986–1992)*. Deventer: Kluwer Law and Taxation Publishers.

United States Agency for International Development (USAID). 2002. *Trade Capacity Survey Report.* Available at http://www.usaid.gov/economic_growth/tradereport.

[Office of the] United States Trade Representative (USTR). 2003. *2003 Special 301 Report.* Washington, DC: USTR. Available at http://www.ustr.gov/sectors/intellectual.shtml.

World Bank. 1999. *World Development Report 1998/99—Knowledge for Development.* New York: Oxford University Press.

World Bank. 2002. *Global Economic Prospects and the Developing Countries.* Washington, DC: International Bank for Reconstruction and Development.

World Intellectual Property Organization (WIPO). Undated. "Assistance in the Field of Intellectual Property Legislation." Available at http://www.wipo.int/cfdiplaw/en/assistance_ip.htm.

World Intellectual Property Organization (WIPO). 2003. *Annual Report.* Geneva: WIPO.

III The Participation of Nondominant Stakeholders in Network Global Governance

12 *Louder Voices* and the International Debate on Developing Country Participation in ICT Decision Making

David Souter

The international policy environments for information and communication technologies (ICTs) and for international development have historically been distinct. Until the late 1990s, decisions made by international institutions governing ICTs—particularly telecommunications—were primarily technical and this remains true today of the majority of work within the International Telecommunication Union (ITU) and more recent Internet governance bodies such as the Internet Corporation For Assigned Names and Numbers (ICANN) and the Internet Engineering Task Force (IETF). International development institutions meanwhile focused either on sectoral issues or on broad principles (whether macroeconomic growth or poverty reduction) to which ICTs were, until recently, considered tangential. International decisions concerning ICTs were largely made by ICT governance bodies by technical specialists from the industrial countries that had extensively deployed them—people who were largely absent from the discourse in development agencies, which rarely included the ICT sector.

The late 1990s and opening years of the twenty-first century, however, have seen increasing concern expressed at the relative lack of participation by developing country stakeholders in international ICT decision making. This concern results from two separate developments: an increased belief in the significance of ICTs for social and economic development; and an increased perception in the international community generally that the impact of international decisions of any kind on developing countries will depend significantly on the extent to which developing country stakeholders—from government and other socioeconomic groups—participate in decision making. Two events during the first years of the new century highlighted this increased concern.

The first was the establishment, in the year 2000, of the G8's Digital Opportunity Task Force (DOT Force), purposely created as a (relatively) informal multistakeholder group from both industrial and developing countries that would be capable of addressing international ICT questions from diverse points on the development spectrum.[1] A report written for the DOT Force during 2002, *Louder Voices*, investigated the reasons for the weakness of developing-country participation in international ICT decision making and made recommendations—to the DOT Force itself, the subsequent UN ICT Task Force, and other agencies—aimed at improving this situation.[2] The findings of this report are discussed in detail later in this chapter.

The second "event," the World Summit on the Information Society (WSIS) that took place in two phases (December 2003 and November 2005), illustrated the extent to which ICTs had entered into mainstream development thinking by mid-decade—but also how far there was still to go before international decision-making processes fully comprehended the possibilities and (at least as importantly) the limitations of these new developmental tools.[3] A crucial problem evident at WSIS, as in many other international forums concerned with ICTs and development, was the fundamental paradigm gap between international institutions (and, for that matter, government ministries) primarily concerned with ICT decision making and those primarily concerned with development issues. The communities populating these different institutions (and ministries) do not yet share common definitions and objectives in this area of policy; and it is a major challenge for the international institutions concerned with it to move beyond these constituent communities' current and diverse expectations of ICT4D (as Information and Communications for Development is now commonly abbreviated) toward an understanding that is more informed, more realistic, and therefore more achievable. Such an understanding might be better abbreviated as ICD (Information and Communication for Development), shifting the focus from technology to development outcomes.

This paradigm gap should not be surprising. Technical specialists and development professionals have different perspectives and different mindsets in many policy areas, from power generation to food standards. Most obviously, ICT professionals view the value of ICTs primarily in terms of their technical potential, while development specialists are more concerned with the achievable impact they might have on their own particular priorities, whether these lie in health, education, agriculture, social welfare, or any other area of development policy.

Interest in ICT4D is also very recent. As late as 1997, information and communication technologies were considered by many development specialists to be luxuries, incapable of transforming lives and livelihoods or delivering mainstream development goals in lower-income countries. Most development agencies regarded them as at best marginal to the interests of the poor—desirable contributors to national infrastructure, perhaps, but best left to market forces and the private sector. Some even saw them as inimical distractions to poverty-focused development, summed up by the antiglobalization slogan, "you can't eat a laptop."[4]

Attitudes changed around the time of the G8's Okinawa Charter on the Information Society and the United Nations Millennium Declaration.[5] The Okinawa Charter, and the subsequent G8 DOT Force, declared access to knowledge and information through

ICTs to be a prerequisite for effective development, while the establishment of the UN ICT Task Force in 2001 attached them to the achievement of the mainstream developmental Millennium Development Goals (MDGs) set out in the UN declaration.[6] Bilateral donors, too, began to develop ICT strategies, built initially around pilot projects, subsequently around the objective of "mainstreaming" ICTs in more traditional areas of development activity—that is, focusing on the value of ICTs as instruments for delivering objectives in health, education, and other poverty-related areas rather than on ICTs as a sector in themselves. The first phase of the World Summit on the Information Society represented, perhaps, the apogee of this new enthusiasm for ICT4D—though the overhyping of ICTs' potential by some enthusiasts during the WSIS process may in time prove to have been the starting point for a more realistic reappraisal of their potential and limitations in development activity.[7]

The international institutional drivers of this new "ICT and development" orthodoxy—the DOT Force and the UN ICT Task Force—were distinctive in two ways (in comparison with established international forums on ICTs and other international governance bodies). One is that they were purposely multistakeholder in character, including private sector and civil society participants, from both industrial and developing countries, as well as governments. This established a precedent that proved controversial during the preparatory phase of the first (Geneva) phase of WSIS, because it was not accepted there by a number of governments with more restrictive views of the different roles of government, the private sector, and civil society. The second important distinctive feature of these new forums is that they brought together representatives from both sides of the paradigm gap, that is, from institutions responsible for development and for telecoms/ICT policy.

Whether any of the main institutions involved in international ICT policy making has yet successfully achieved a genuine and informed balance of participation between industrial and developing countries is questionable. Certainly the outcome texts of relevant meetings, WSIS in particular, emphasize the importance of such balance, and WSIS itself probably came closer to achieving it than other forums. However, there were times during the WSIS process when the gap in thinking between developing and industrial countries seemed very substantial indeed—notably during a dispute over the desirability of establishing an ICT-specific Digital Solidarity Fund during the first phase of the summit. The engagement of different developing countries in WSIS was also highly variable, with a small number of larger countries much more prominent—especially during the second-phase discussion of Internet governance—than any Least Developed Countries (LDCs).

Enhancing developing country participation is, of course, a much more complex matter than simply increasing the number of delegates from developing countries in major decision-making meetings. Decisions are made over periods of time in many different forums, through informal as well as formal interactions, and on the basis of complex national and corporate policy-making processes involving many kinds of expertise. Effective participation therefore requires far more than mere presence when decisions are made—it requires detailed understanding of the issues involved on the part of delegates and the organizations or governments they represent; substantial analytical capabilities; extensive networking; and committed engagement with decision-making processes over a significant period of time.

These issues were assessed in depth in a study commissioned for the DOT Force and the UN ICT Task Force by the UK Department for International Development (DFID) in 2002. Called *Louder Voices*, this study was undertaken by Panos London (the UK partner in the international Panos Institute) and the CTO (Commonwealth Telecommunications Organisation), with the support of a substantial group of international experts, mostly from developing countries.[8] It had four main objectives:

- to map the institutions and processes responsible for international decision making on ICTs and their role in development policy;
- to assess the effectiveness of developing country participation in them;
- to identify obstacles facing developing countries at the international, regional and national levels;
- and to recommend actions that could be taken—by all relevant parties—to strengthen developing country participation in international decision making in the future.[9]

The study was built around detailed case studies of three international institutions—the ITU, the World Trade Organization (WTO), and ICANN—and of six developing countries—Brazil, India, Nepal, South Africa, Tanzania, and Zambia.[10]

This chapter summarizes and comments on the findings and recommendations of the *Louder Voices* study. It then reviews their relevance in the light of the World Summit on the Information Society and makes recommendations that build on the *Louder Voices* findings for the post-WSIS environment.

Mapping the International ICT Decision-Making Universe

At the heart of the *Louder Voices* analysis, and of this commentary, lies the nature of the international ICT decision-making universe. Like the universe of astronomy, this is vast, diverse, complex, and continually changing as new technologies, services, and markets appear, evolve, and disappear. It has been expanding rapidly since the "big bang" of ICT development in the 1990s—in which computer technology, the Internet, and mobile telephony transformed the potentialities of the existing information technology sector. It contains a large number of diverse centers of gravity (or influence)—slowly evolving giants like the United Nations institutions, including the ITU, and dynamic but unstable newcomers like the Internet's diverse forums for international

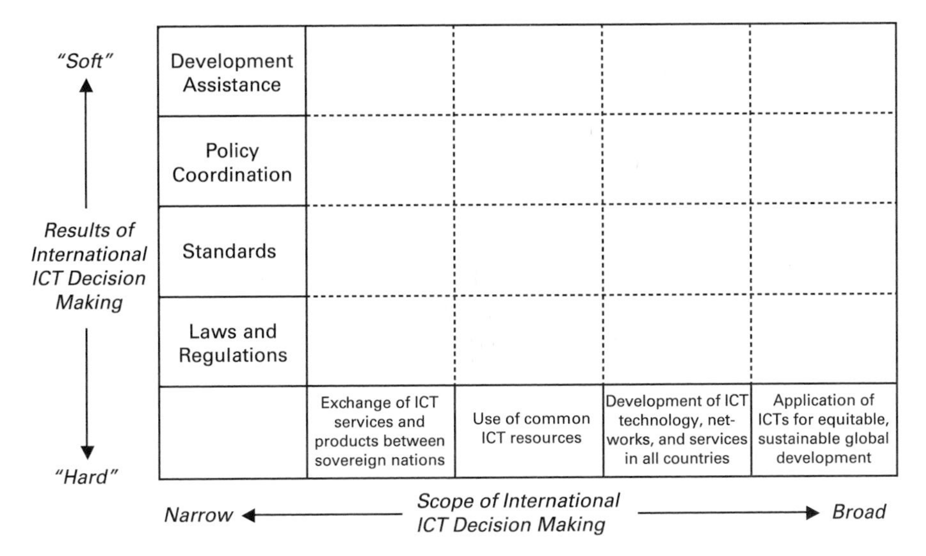

Figure 12.1
Issues matrix: International ICT decision making

governance (or at least coordination)—many of which are light years apart in thinking, style, and ways of working. There are even "black holes" into which technologies, corporations, and policy issues occasionally disappear.

Mapping this universe is anything but easy. To help conceptualize and visualize it, the *Louder Voices* study constructed a matrix that juxtaposed the scope of decision making—from the narrow exchange of services and products, through the use of common resources and development of international networks, to the broad application of ICTs in social and economic development—against its results—from "hard" or precise laws and regulations, through international standards and policy coordination, to the imprecise or "soft" outcomes of development assistance. This matrix is shown in figure 12.1.[11]

Different agencies have different responsibilities and powers within this matrix/map. Those of the three core organizations studied in depth by the *Louder Voices* team are shown, for example, in figure 12.2. Of these, the ITU has broad responsibilities across almost the whole surface area of the map—though its influence and impact vary in intensity;[12] the role of the WTO is narrower in scope, but also extends across the range of outcomes from strict regulation to development assistance; while ICANN's remit is strictly limited to one segment, where it nevertheless has very considerable power.[13] These areas of decision-making competence overlap, creating scope for interinstitutional conflict as well as confusion for all but the best informed and most experienced participants and observers.

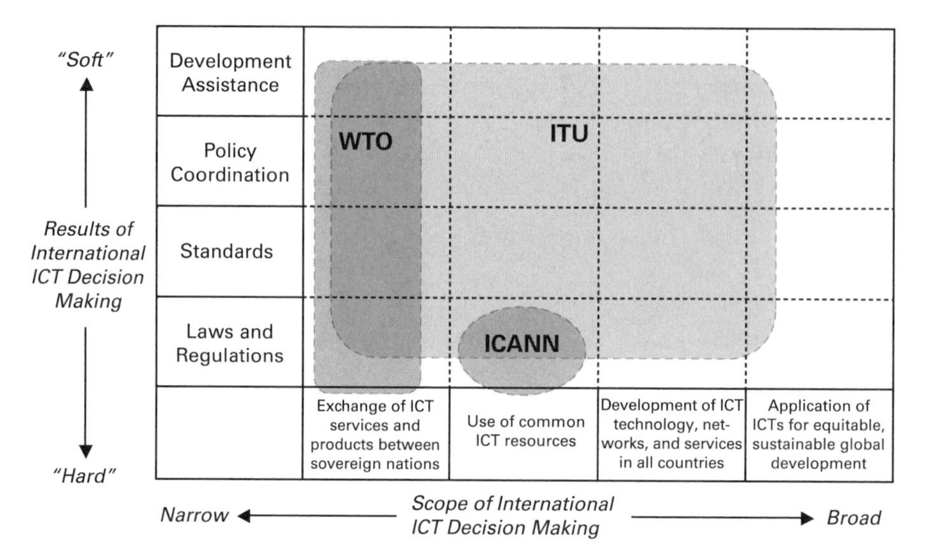

Figure 12.2
Activities of three core organizations

Figure 12.2 includes only three of the myriad of international agencies involved in ICT decision making. The role of others is illustrated in figures 12.3 and 12.4. Figure 12.3 is concerned with other key members of the United Nations family involved in ICT decision making—the United Nations Development Programme (UNDP), UNCTAD, UNESCO, and the intellectual property governance agency WIPO—and with other Internet governance bodies. Adding these adds more areas of overlap, potential confusion, and institutional rivalry, more scope for ambiguity in decision making, and more difficulty for governments and other stakeholders wishing to influence the outcomes of this expanding decision-making universe. Figure 12.4 adds yet another layer of complexity, including (for example) the private sector standardization forums that now play a very important part in determining how manufacturers turn technologies into products and services.

This matrix illustrates complexity, however, not chaos. To continue the astronomical metaphor, there may be many different centers of gravity in the ICT decision-making universe, but they do have their own orbits, interact within separate spheres of influence and responsibility (albeit these sometimes seem to be in separate galaxies), and act in most cases according to reasonably predictable rules of procedure (laws of physics?). The *Louder Voices* matrix/map is helpful in defining the core responsibilities and outer limits of the gravitational pull—the policy-making impact—of different agencies; in enabling them to focus their own priorities; and in guiding governments

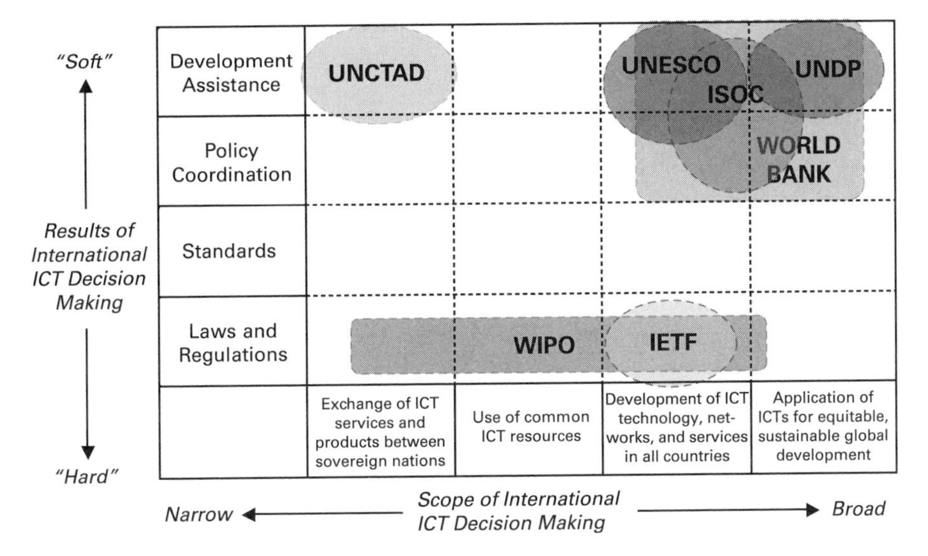

Figure 12.3
Other UN agencies and private sector organizations

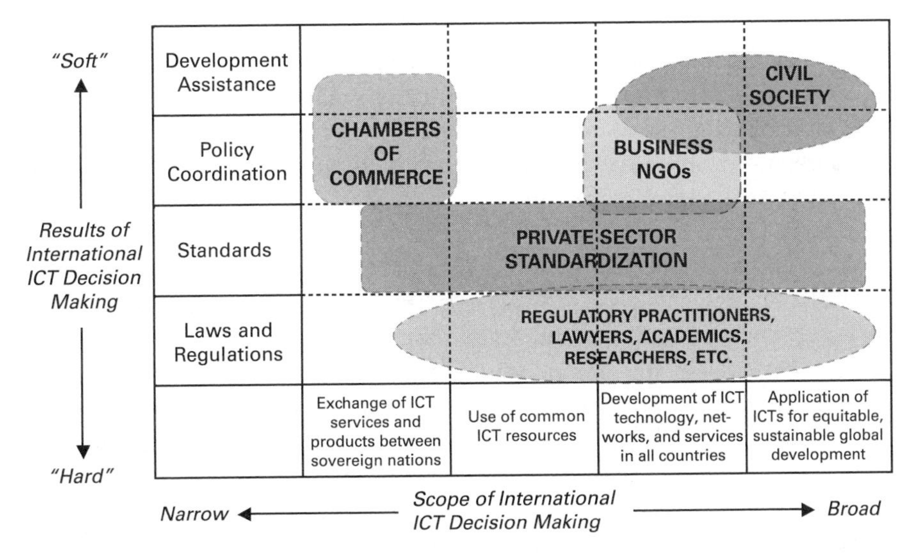

Figure 12.4
Additional nongovernmental actors

and other stakeholders seeking to influence those aspects of the international decision-making agenda that are of most importance to themselves. As such, the authors of the *Louder Voices* report hoped that it would provide a useful—and even entertaining—tool.

Two further crucial distinctions need to be drawn between different agencies in the international decision-making universe. These are (a) between intergovernmental and nongovernmental actors, and (b) between agencies primarily concerned with ICTs and those whose primary focus is development.

Traditionally, international governance has been built—not just in ICTs, but in all such sectors—around equality of sovereign states. In any intergovernmental agency, each member (country) has had one vote, regardless of demographic or economic size, and that vote has been exercised by its government. This is still the central decision-making principle of many of the agencies included (or potentially included) in the matrix/map, including United Nations agencies such as the ITU (although private sector members now have significant impact on standardization questions there).[14] The Internet, by contrast, is "governed" by largely self-appointed and self-regulating groups of individuals and organizations outside government control (or even influence). This is even true of ICANN, the Internet body that exercises powers most related to national identity and so most similar to those held by intergovernmental agencies in other contexts.[15] Private sector forums made up of manufacturers are increasingly responsible for setting standards for equipment, consumer products, and services, independently of the ITU and other intergovernmental bodies. And some of the new technologies and services involved—the Internet itself and community radio for two examples—intrinsically empower consumers vis-à-vis any authority that might wish to regulate their impact on society.

The second distinction, between ICT-focused and development-oriented agencies, is also very important. As noted earlier, and in spite of high-level rhetoric in partnerships like the DOT Force and the UN ICT Task Force, there has been little convergence between the major international institutions involved. Decision making in ICT-oriented agencies like the ITU is still dominated by telecommunications sector specialists (whether from government or the private sector); while many development agencies have little understanding of how such telecoms decision makers think. The 2002 World Telecommunication Development Conference showed this distinction clearly. In spite of efforts by the ITU's Development Bureau (and many useful outcomes), the conference was sparsely attended by development agencies, and government delegations were overwhelmingly composed of industry and communications ministry personnel.[16] This preponderance of ICT rather than mainstream development actors was repeated at the World Summit on the Information Society, where it was also complicated by the prominent role played by diplomatic drafters leading the development of summit outcome texts.[17]

These problems concerning the participation of developing countries are by no means unique to the ICT sector, but are common to all complex technical areas of international governance—such as international trade arrangements, food standards, or human and animal disease management. They are exacerbated in the case of ICTs, however, by the very rapid pace of technological change that characterizes the sector—and by the changes in available products and services, in industry structure, and in the analysis and impact of different policy options that result from it. What seems one year like a solution to major problems—the Global Mobile Personal Communications by Satellite (GMPCS) systems that were deployed in the late 1990s, for example—can easily turn into a liability the next. This adds a further dimension to the difficulty of addressing deficiencies in developing countries' participation where ICTs are concerned, but also suggests that successful approaches in the ICT sector may prove valuable exemplars elsewhere.

Assessing the Impact of Developing Countries in International Decision Making

The key issue addressed by the *Louder Voices* study was the perception, in the DOT Force and elsewhere, that developing countries—and particularly LDCs—play little part in international ICT decision making. They were, it was believed, underrepresented in international meetings, unengaged in international institutions, uninvolved in their decisions, and unable, in many cases, either to identify the challenges and opportunities international decisions had for their national environments or to respond accordingly. Four questions arose for the study from these assumptions: Were they true? Did they matter? What were the obstacles that led to underrepresentation in decision making? And what action, if any, could be taken to alleviate them?

As previously noted, the *Louder Voices* study addressed these questions through a combination of detailed case studies—of three core institutions (the ITU, WTO, and ICANN); of three specific areas of decision making (IP telephony [the ITU], ccTLD domain name disputes [ICANN], and WTO accession negotiations); of two international meetings (the 2002 World Telecommunication Development Conference and the 2002 ICANN Assembly[18]); and of six individual countries (Brazil, India, Nepal, South Africa, Tanzania, and Zambia)—as well as through questionnaires and interviews designed to elicit the views of developing country participants in decision-making forums.

Overall, interviewees from developing countries in the *Louder Voices* study were clear that the role they were able to play in international decisions affecting their countries was limited both by the structure of international decision making and by weaknesses in national policy-making expertise and practice. There is little to suggest that its findings would be different today, although some developments have taken place and are considered later in this chapter.[19]

Physical participation by developing country delegations is most evident in forums that are primarily concerned with general policy matters affected by ICTs and with development assistance—particularly in the United Nations agencies. Developing country participation is also relatively high in some of the more specialized forums such as the ITU, WTO, and WIPO that deal with legal and regulatory matters of concern to governments. Where they are present in these intergovernmental forums, however, developing country delegations are usually too small and often too inexpert to exert much influence in their debates; and those that do play a significant part, like Brazil and India, are large countries that have long had strong technical capacity and have different characteristics and aspirations from smaller or Least Developed Countries.[20] Representation is substantially weaker in nonintergovernmental forums such as those concerned with the Internet or with determining standards for new technologies. For all its "democratic" structure (historically, at least, giving equal weight to all comers) and in spite of its efforts to hold meetings in developing countries, such countries are underrepresented, for example, at meetings of ICANN. With few high-technology manufacturing companies of their own, they are almost wholly absent from the forums making decisions that affect the next generation of technology development and design.

These problems are compounded by the interface between formal and informal decision making. A developing country representative does not simply need to be present at the final decision-making meeting in order to make an impact. She also has to be involved in the preparatory meetings, informal gatherings, coffee breaks, and networking and lobbying, with a team of specialist advisors at her beck and call if she is to make a success of the drafting and informal negotiations that underpin any international decision-making process. Few developing countries can afford to invest as much in these processes as any medium-sized industrial country manufacturer.

In any event, by the time ICT issues surface at the international level, they have generally been filtered through several non-decision-making processes—through technological standardization, market research, the economics of production and supply and actuality of customer demand, and the policies and regulations of leading ICT countries and regions, as well as negotiations between major suppliers and trading blocs. As the *Louder Voices* report puts it, "International ICT decision-making is a 'bottom-up' process in which those who are not present when decisions are made in a sense may have more influence than those who are, even among developed countries."[21] This problem is illustrated in figure 12.5.

Underrepresentation is important, and not merely for reasons of international equity, or just for developing countries themselves. The ICT environments in developing countries differ substantially, and in many different ways, from the ICT environments of industrial countries. Many developing countries have different and difficult terrain: desert, mountain, archipelago. Their populations have different patterns of social and

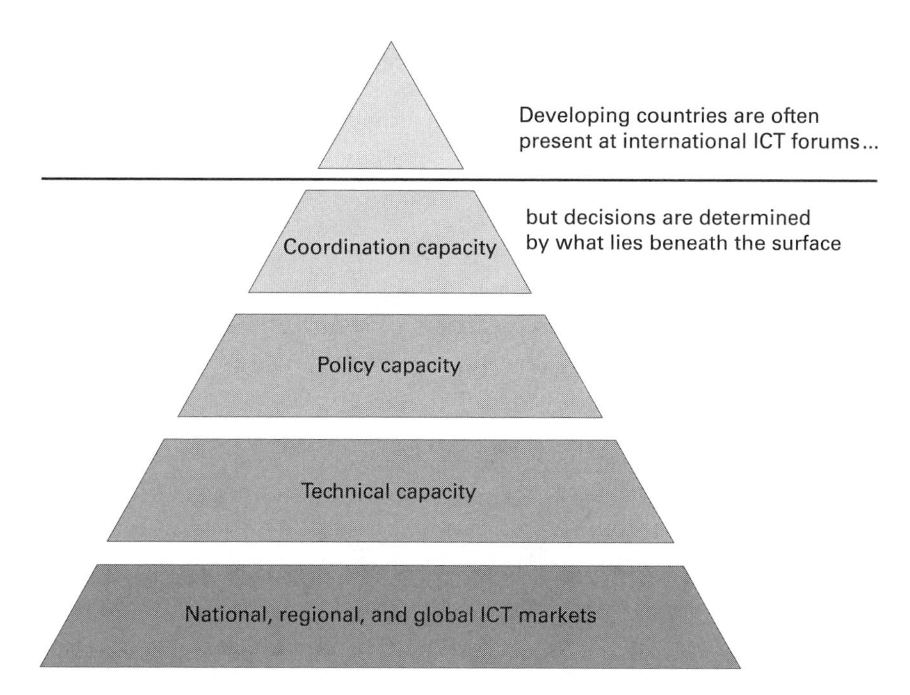

Figure 12.5
The "bottom-up" decision making process

economic behavior, based on different cultural factors or national economic characteristics. Most important of all, they have different markets for ICTs: markets that are often based primarily on shared or public access, even to facilities like mobile phones, rather than on mass domestic ownership; low-income markets in which choices to use ICTs and patterns of use differ markedly from the high-income markets found in member countries of the Organization for Economic Cooperation and Development (OECD). These developing country environments are therefore likely often to require different approaches and different policies, even different technological choices from those most appropriate to high-income mass markets for communications in OECD countries.

Developing country and low-income markets are also likely to respond differently to the products and services that are made available within them than politicians, regulators, business leaders, and development specialists used to high-income industrial country markets may predict. Business leaders and telecoms specialists from industrial countries will tend to make decisions in private sector standardization forums or in international bodies like the ITU that are based on the high-income industrial country markets with which they are familiar. Even so, their record in predicting technologies and markets is mixed: very few anticipated, for example, mass use of the Internet or mobile telephony, or the popularity of short message service (SMS) text messaging. It is self-evidently more difficult for such industrial country experts to predict the technical requirements or market behavior of low-income economies—where, for example, lower demand may make alternative network technologies more appropriate or where cultural factors such as language differences may limit the scope for technology use (e.g., for menu-driven complexity in mobile handsets). Incorporating the characteristics of developing country demand is, in fact, as much in the interest of industrial country manufacturers as it is of developing country consumers, for it is the latter—regardless of relatively low income levels—who offer the largest future markets for the phone-makers and computer companies of North America, Western Europe, and the Pacific Rim.

The problem of developing country underrepresentation is not unrecognized in international forums themselves. Although some industrial countries have not always seen increasing developing country participation as desirable—perhaps believing that developing countries lack the technical, policy, and regulatory capacity to add value to the process and anxious to avoid "unnecessary" delay in decision making—most now recognize the value of broadening participation and fostering informed input from the developing world. However, the means adopted to achieve this have not always been effective—often addressing the appearance of representation (e.g., through fellowships allowing individuals from specific countries to attend plenary conferences) rather than its substance (facilitating meaningful participation throughout a decision-making process).

Obstacles to Developing Country Participation

Developing country interviewees in the *Louder Voices* study identified a wide range of obstacles to effective involvement. Some of these relate specifically to the way in which international institutions work and structure participation within themselves. More, however, are concerned with national weaknesses in policy making and analysis, which give those representatives who are involved at the decision-making hubs too little information, too little knowledge, and too little guidance on what to do or say. However well delegates may perform during meetings, they will be unable to represent their countries' interests effectively if their understanding of these is not reinforced by serious strategic policy-making institutions and processes at home, including participation from the whole range of national stakeholders and expertise. Participation, in short, is only part of effective representation, which also requires policy analysis, policy dialogue, and liaison with other countries and delegations that share concerns and interests.

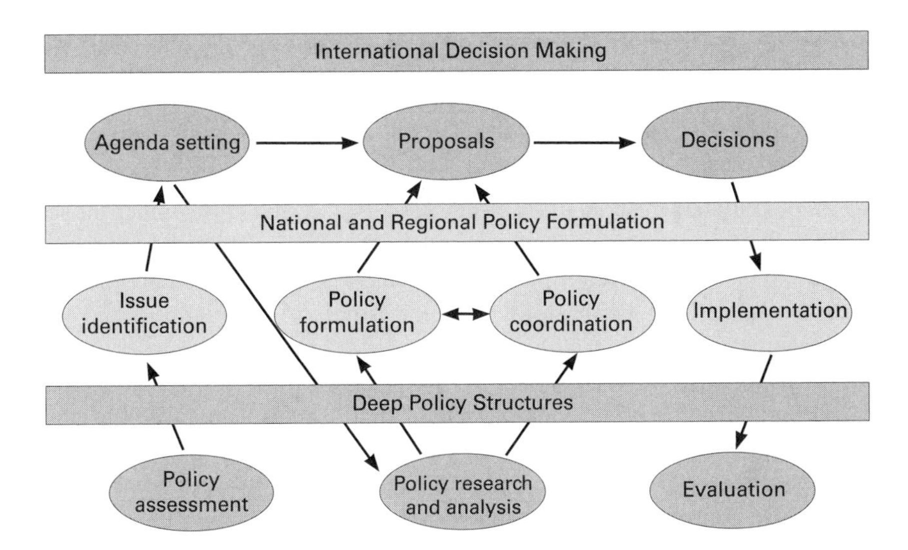

Figure 12.6
Requirements for effective participation

The nature of the problem at the national level is illustrated in figure 12.6, which shows in graphical form the complex web of resources and inputs that are necessary for effective participation (and that give industrial country delegations the edge in international forums). Effective delegations benefit from powerful and coherent national and regional policy formulation processes that identify priority issues, coordinate and synthesize the views of different stakeholders, and establish a position that can be sustained through lobbying and negotiation over the weeks and months preceding a decision as well as at the decision-making meeting itself. (An example here might be the decision-making processes undertaken within the European Union.) These policy formulation processes are in turn underpinned by deep policy structures—the analytical capabilities that allow policy makers to understand the implications of different options and choose the most effective strategies for their countries.

It will always be difficult for developing countries to secure this degree of complexity and comprehensiveness in policy making, not least because expert human resources are in short supply, but this does not mean that little can be done to improve the quality of representation. One key issue is prioritization. Major industrial countries need this high level of policy making across the board because they have deep interests in many different areas of ICT policy. Developing countries, by and large, have fewer interests. They can afford to prioritize—to identify the limited range of issues on which decisions are sufficiently significant for them to invest substantial resources, and to focus explicitly on those issues, then develop the policy capacity required to have an impact and involve the full range of stakeholders in national policymaking. They can also seek to secure the support of regional peers and wider developing country groupings that can have more impact if they act collectively in informal as well as formal decision-making gatherings, and to provide fuller support to the delegate or delegates attending all of the forums involved.

Interviewees for the *Louder Voices* study clearly understood that it was much more valuable for a developing country to have real influence in two or three areas of key significance than to have an ineffective presence in a larger number of decision-making processes—particularly if the lead role on different issues of importance could be shared among countries within a region, with each developing appropriate expertise on behalf of the regional group as a whole. They were also clear, however, that far too little prioritization along these lines took place, and that available expertise was currently spread too thinly and too indiscriminately to have the impact that their governments desired.

Within this overall context, participants in decision-making processes identified five main areas of weakness in international decision making, each with national and international dimensions.

First, interviewees cited *lack of awareness* of the importance of international ICT decisions for national policy—at all levels, from the highest reaches of government to the owners of small and medium-sized businesses and individual citizens. The basic problem, in the words of the report, was that "people ... do not see—in concrete, practical terms—how ICT can make a difference to their lives, and how it can contribute to their development as individuals, as members of families, and as citizens of a country."[22] The more arcane the subject—Internet domain names, for example—the more lacking the awareness.

There are good reasons for this. Evidence of the impact of ICTs on individual empowerment, social and economic development, and the livelihoods of poorer citizens is still sketchy and anecdotal.[23] Many decision makers grew up in an age when ICTs played little part in policy and much less of a role in society than they do today. They lack the personal experience and acculturation to ICTs that give intuitive understanding of their potential, while the link between high-policy decisions and real impact on individual lives seems at best indirect and often opaque. For example, decisions on international telecommunications gateways do not appear to have major developmental impact and are not generally taken by development specialists—but they do have real significance for Internet pricing and therefore for the accessibility of Internet-based knowledge.

Lack of awareness does not only lead to underassessment of the potential of ICTs. There is also a risk of it having the opposite effect—of senior decision makers who do not understand the limitations of ICTs overestimating their potential for achieving social and economic change. It has become routine for assessments of an ICT's value in development to acknowledge that it is "not a panacea," but under-researched over-optimism remains widespread, as was evident in many contributions to the first session of the WSIS. While ICTs do offer major opportunities in development, the extent to which many of their advantages can be achieved depends to a considerable degree on the extent to which ICTs are already diffused within society, and on complementary factors such as language capabilities, the underlying educational and skill levels within society, the openness of regulatory frameworks, and the capacity of government and business to adopt organizational approaches that maximize ICTs' potential. Many senior decision makers worldwide underestimate the importance of these complementary factors.[24]

Part of this problem can be addressed by demystifying ICTs. Defining anything as technological creates barriers of understanding for nontechnologists. So, in this case, does focusing on "new ICTs," such as those requiring digital technology. Policy makers need to understand the place of these new ICTs within a wider spectrum of established information and communication media, such as broadcast radio, which are also affected by ICT policy making, and of the patterns of use that citizens make of information and communication resources of all kinds in securing their livelihoods and seeking to improve their quality of life. Policy makers should also pay more attention to the value of ICTs' end product—information/knowledge—and how this can be maximized in an appropriate policy framework that understands, enhances, and builds on citizens' real lives and aspirations.

More specific action can also be taken in this area. More could and should be done, for example, to research and disseminate real instances of the impact ICTs can have on development objectives (positive and negative, for there is at least as much to learn from failure as from success). International decision-making agencies could provide much more independent, authoritative technical and policy research on the developmental, as well as commercial, implications of major issues requiring decision. They could also provide comprehensive nontechnical information to the broad stakeholder community, and disseminate their internal analysis and policy formulation in more open and accessible forms.

The second critical area of weakness identified by *Louder Voices* interviewees was the *lack of technical and policy capacity*. Most developing countries do not have expertise available in many of the areas under discussion, particularly new policy issues such as migration to IP-based networks or the governance of electronic commerce. This is not a problem unique to ICT policy, but it is exacerbated in the case of ICTs by rapid changes in available technologies, products, and services. There are no quick or easy solutions. Capacity building is a long-term challenge that requires investment by government, the private sector, and civil society in building expertise within a country as a whole. International agencies—and industrial country actors—can assist by recognizing the value for them of effective input into decision making from the whole range of political, economic, and geographical environments affected by decisions, and by directly supporting capacity building through training courses and other means. Some, such as the WTO, already pay some attention to this—although it must be recognized that training courses are no substitute for experience in the cut-and-thrust of real-life intergovernmental discourse, and that effective capacity building needs to focus as much on the issues requiring input as the negotiation and other processes required for making that input effective. Delegates need to know what to negotiate for as well as how to negotiate.

A further problem in this area is that some international ICT forums do not allow participation by some kinds of stakeholders in their decision-making processes or national delegations. Specialist expertise is in short supply in most developing countries, but more is often available in the private sector, for example, than in ministries or regulatory commissions. This is particularly true where the Internet is concerned. Preventing developing countries from making use of this expertise in international meetings further undermines their capacity to make an impact.

A third key area of weakness identified by interviewees for the *Louder Voices* study is the *lack of easy, affordable, and timely access to information*. The complex and diverse nature of international ICT decision making is illustrated by the mapping exercise of decision-making bodies described earlier in this chapter. It is difficult even for industrial countries to keep up-to-date with all of the issues involved, and many specialist employees, research institutes, think tanks, and consultancies benefit from commissions to support their national delegations at the ITU and elsewhere. The problem is much greater for developing countries, though they are as affected by the lack of readily digestible (and thus usable) information as they are by information overload.

Lack of easy, affordable, and timely access to information and analysis about ICT-related issues, forums, and processes was consistently identified by *Louder Voices* interviewees as an important barrier to developing country participation. To play any meaningful part in international decision making, participants from developing countries need to know the background to issues under discussion, the views of the main players within the sector, the analysis underpinning consideration of those issues, and the key points within them that are the focus of dissension or where small policy changes may have major implications later. They need this information to be easily accessible—sufficiently brief to be absorbed quickly in a busy schedule, but sufficiently comprehensive to permit meaningful engagement in debate. They also need access to analytical capacity to answer questions, assess the implications of different issues and options for developing country economies and societies, and advise on the most effective and cost-effective ways of achieving goals that are worth pursuing.

Addressing this problem was one of the key areas of focus when the *Louder Voices* authors came to develop their recommendations; and it is one in which it is crucial for international agencies to listen to the requirements of developing country partners and participants that will make use of any new information and analytical resources made available. One message emphasized by many interviewees was that what they as individuals found most lacking was not access to the full range of available information, but rather, reliable sources providing succinct précis and objective analysis of the issues under international discussion, the outcomes of international meetings, and the content of international reports. Their key requirement was for resources to provide sufficient basic knowledge for them to make an informed judgement of what was significant for their countries, to give an informed opinion on all such issues on behalf of their government or sector, and to select more detailed sources of information in those cases where an issue was of sufficient importance to merit more detailed (or proactive) attention. The scarcity of useful information, in short, should be addressed by supplementary resources related directly to developing country stakeholders' needs, not replaced by information overload.

Fourth, and most important, developing country interviewees for the *Louder Voices* study identified *weaknesses in national ICT policy processes* as critical causes of ineffectiveness in international decision making. These drew more comment than any other barriers to effective participation identified by the study. At the national level, general weaknesses reported in case studies and by interviewees included lack of political leadership, the absence of an effective national ICT strategy, ineffective coordination between different government departments and agencies with ICT responsibilities, and the absence of ICT policy processes open to participation by all relevant stakeholders or based on public discussion and debate.

More specific weaknesses were reported in the preparatory processes for international meetings, in particular:

- inadequate dissemination of information about meetings to affected government departments and other stakeholders;
- inadequate consultations with interested and affected parties, and little or no preparation of agreed national positions on the basis of such consultations;
- noninclusion on national delegations of the technical and policy experts best qualified to advise national spokespersons and to engage in effective lobbying and negotiation;
- inadequate dissemination of information about the outcomes of international meetings to interested parties and the general public; and
- the absence of any effective structure to ensure that the national challenges and opportunities resulting from international decisions were identified and grasped by national actors—whether government agencies, private businesses, or civil society organizations.

In the words of one country case study undertaken for the project, "A more participatory national policy process, greater use of capacity outside of Government, regional leverage of common interests to influence agenda-setting, swing or veto votes, and general lobbying by organisations would improve developing country participation."[25] The opportunity to leverage common interests by developing common regional positions was emphasized by many interviewees. Individual country delegations lack the scale to engage effectively in major decision-making processes alone, especially where industrial countries and major businesses are represented by numbers of specialists. Individual countries often lack authority in institutions in which they do not have a longstanding record of engagement, especially if they also have not played a major part in other international forums.

These difficulties can be substantially reduced if individual countries and delegations work with partners, within and even beyond their own geographical regions, which share common concerns and interests. Regional groupings can share expertise, develop common positions, divide responsibilities at decision-making forums, and coordinate with potential allies more effectively than individual nation-states, particularly where they also have joint voting power to deploy (though this last can have negative as well as positive effects).

These weaknesses in internal—national and regional—policy processes are more significant factors in leading to poor policy participation by developing countries than *weaknesses in international policy processes*—about which there were relatively few complaints by *Louder Voices* interviewees. Nevertheless, there are undoubted problems for developing countries—and, indeed, for any countries with small delegations or limited technical and policy capacity or both—in the way that many international organizations structure their decision making.

Developing countries generally send small delegations to international meetings, because of the high cost involved, the need to retain decision-making capacity in-country, and the shortage of relevant expertise available. The fact that different organizations often schedule overlapping meetings, or that meeting agendas schedule important sessions in parallel, imposes additional burdens on developing country participants (as well as on smaller industrial countries or private sector organizations with wide-ranging interests). This problem is exacerbated if rules of procedure exclude certain classes of stakeholder from participating in decision-making processes.

Financial barriers are also significant. If effective engagement in one particular decision requires participation in six or seven meetings at the regional or global level, plus informal lobbying at other gatherings not immediately concerned with the issue itself, it becomes difficult for governments to justify the cost. However, experience suggests that funds usually are found for meetings that are considered sufficiently important at a senior level, and that improvements in prioritizing issues through national policy-making processes will lead to a sharper focus of both financial and human resources.

There is a strong case, on grounds of equity and access, for international organizations to hold more meetings in developing countries—as ICANN, for example, has done. As ICANN's experience shows, however, this does not necessarily result in higher levels of developing country participation. Nor does holding meetings in developing regions reduce the average cost of participation to developing countries: lower costs for countries within the region in which a meeting is held are offset by higher travel costs for those from other developing regions than would be incurred by them in visiting a global center such as New York or Geneva.

It might be more useful, therefore, for international organizations to give more emphasis to regional preparatory meetings in decision-making processes. Attending these genuinely does cost less and has the added value, for developing countries, of facilitating regional networking and the sharing of resources in policy development and implementation. As noted earlier in this chapter, policy positions that are the result of regional discussion and carry the weight of regional opinion will be considerably more influential when the final global "horse-trading" takes place.

The fifth area of weakness is related to the *financial resources* allocated by some international agencies to assist participation by developing country personnel. Interviewees frequently expressed concern that these "fellowships" and other resources were not used effectively and often resulted in "the wrong people attending meetings for the wrong reasons."[26] While recognizing that it was difficult for international organizations to set eligibility rules for sponsorship programs, many experienced delegates felt strongly that governments needed to pay more attention to selecting appropriate participants, monitoring their participation, supporting them while at meetings, and disseminating meeting outcomes on their return.

The *Louder Voices* Program of Action

In summary, the *Louder Voices* study identified five main areas of weakness in decision-making processes as a result of its case studies and interviews with participants in international decision making, most of which concerned national rather than international constraints. It summarized these obstacles, and the overall approach required to address them, as follows:

> Changes at the international level alone cannot compensate for underlying weaknesses in the ICT strategy and policymaking processes of developing countries—the root causes that must be addressed to enhance international participation. In particular:
>
> ▪ Given the limited technical, human and financial resources available to developing countries, it is very important to focus on issues that have the highest impact on national development strategies. This in turn requires better information than is currently available about items on the international ICT decision-making agenda, and much stronger capacity to research, analyse and assess these items in terms of development strategies and priorities.
>
> ▪ Generic technical assistance and training cannot substitute for "the real thing"—e.g.
>
> • the development of independent, professional policy research and analysis capacities;

- multi-stakeholder processes geared to the development of policy on issues up for decisions;
- exchanges, secondments and attachments between policy and regulatory authorities on a North–South and South–South basis;
- involving qualified, relevant technical and policy experts in international decision-making meetings.
- Although the international playing field is severely tilted against developing countries, many have technical/policy capacity that is not effectively engaged because of current practices at the national and international levels, either because "the wrong people" attend meetings, or because the private sector and civil society are excluded from participation.
- There may be significant attitudinal obstacles to stronger participation that are very difficult to change unless national and international policy processes are changed—the feeling that agendas are set elsewhere, that important decisions are made by small groups behind closed doors, that the private sector cannot be trusted, that civil society is a disruptive force.
- Throwing money at the problem (e.g. through travel funds that are not contingent on qualification and performance) will not solve it, particularly if it winds up in the wrong hands. The study results suggest that, even in poor countries, money will be found if ICT is given policy priority, and that expenditure of developing country resources is likely to provide an incentive for making more effective use of existing technical and policy resources.[27]

A fundamental conclusion from this study is, therefore, that developing country participation in international ICT decision-making forums cannot be strengthened without first strengthening national ICT policy capacity. How can this be done? The *Louder Voices* report included sets of recommendations for developing country governments themselves, for international institutions, and for the global development community.

Individual governments are encouraged by the report to include international policy and regulatory issues in the e-strategies that many have been preparing, often with the support of international agencies such as the UN Economic Commission for Africa or international partnerships such as the DOT Force-initiated e-Policy Resource Network (e-PolNET). This will only be effective if e-strategies are themselves effectively integrated with national development strategies, such as the Poverty Reduction Strategy Papers prepared for the International Monetary Fund and the World Bank—which has not always been the case.

Global institutions tend to seek global solutions that encompass all aspects of any problem. However, processes are more important in this context than e-strategies: progress needs to be made quickly and can be made incrementally, without waiting for the conclusion of a grand design. In the fast-changing world of ICT technology and policy, waiting for consensus on global solutions is often counter-productive. Developing country policy makers need enhanced opportunities for policy development today more than they need aspirational statements about optimal approaches to agenda setting for tomorrow. However worthwhile, the latter will not impact on issues such as Internet deployment in the next two years.

Governments can take immediate action to improve their current policy-making processes on ICTs in seven key areas:

- improve information flows and policy coordination between different government departments and agencies with ICT responsibilities;
- promote informed public discussion and debate through both general and specialised media;
- include all relevant stakeholders in policy making on an issue-by-issue basis;
- encourage participation of experts from the private sector and civil society in national delegations to international decision-making forums;
- share information, expertise and experience on a subregional and regional basis;
- implement knowledge management techniques to ensure that information gained through participation in international ICT decision-making forums is captured, disseminated to relevant stakeholders, and made accessible to other interested parties such as the media; and
- review arrangements for the selection and management of international delegations.

International institutions could assist by improving their processes and procedures in ways that would facilitate meaningful participation by developing country delegates. By promoting awareness of the developmental potential of ICTs and by providing better information and building capacity on their policy-making processes and critical issues, international institutions could do much to help governments and other stakeholders in developing countries to identify clearly the issues that are of real importance to them. By diversifying the location and format of decision-making forums, they could also foster more effective developing country participation and regional discussion among developing countries on those issues during the decision-making process. In particular, international institutions could arrange:

- to hold more stages in decision-making processes at a regional level, in order to reduce costs of participation and facilitate regional coordination;
- to avoid simultaneous scheduling of important meetings within the ICT sector and of simultaneous sessions within their own meetings where these will compete for the attention of the few personnel attending from individual developing countries; and
- to ensure that their procedures allow all sources of developing country policy and technical capacity to participate in decision making, whether they come from government, the private sector, or nonprofit organizations.

The *Louder Voices* Recommendations

The *Louder Voices* report concludes with a series of recommendations that form a proposed program of action that could be implemented by a consortium of development agencies within the framework established by the G8 DOT Force, the UN ICT Task

Force (to whose Working Group 1 on ICT policy and governance the report was referred), and the WSIS.[28] This section of the chapter describes the *Louder Voices* program of action as it was presented in June 2002. Later in this chapter we reassess its recommendations in the post-WSIS context.

The *Louder Voices* program was not intended as a comprehensive approach to address all of the problems identified in the *Louder Voices* report, but to identify a small number of inexpensive initiatives that could be undertaken quickly and would facilitate work by developing country actors to enhance their own awareness, understanding, and policy capacity. Its six initiatives focused on the information needs and analytical capacity requirements of developing country stakeholders.

The first—and most expensive—of the *Louder Voices* recommendations was to establish a *network of small regional centers of expertise in international ICT issues, based in developing country regions*. The action program suggested these might ultimately include two in Africa, two or three in the Asia/Pacific region, one in the Middle East, one in Latin America, and one in the Caribbean—though the priority areas should encompass the most LDCs and least effective international input, with a particular initial focus on Africa.

The objective of this recommendation was not to construct major new institutions that would adopt policy positions or align themselves with the views of any particular country, international institution, or vested interest. On the contrary, it was suggested that the centers would add value by providing objective information, guidance, and analysis equally to all stakeholders (government, private sector, and civil society) in all countries of their regions. To maintain independence and integrity, each regional center would probably need to be located in an academic institution, governed by strict codes of practice, and managed by a board clearly representative of all three main stakeholder groups and diverse countries within its region. In time, these governance arrangements could be reinforced by a coherent management structure involving all participating centers or institutes.

The *Louder Voices* partnership believed that a significant start could be made in each region by a team of four—including a telecommunications analyst, an ICT/Internet analyst, and an information resources specialist. The regional centers would require initial funding from development partners but could become self-financing in time by undertaking commissioned research on behalf of regional stakeholders—provided that care was taken to ensure this did not bring their neutrality or objectivity into question.[29]

The second *Louder Voices* recommendation was for *readily accessible, Internet-based information resources*. As noted earlier in this chapter, developing country interviewees indicated strongly that their key requirement was for ready access to accurate and objective information that required little bandwidth to access and download, was easy to navigate as well as comprehensive, and pointed the way to other resources on issues they might wish or need to pursue further. A simple-to-use one-stop shop with resources on a single web site, in other words, would add more value than a complicated portal built around hyperlinks to other graphics-heavy, information-intensive sites. (Unless well managed, the latter can in practice inhibit rather than facilitate research by under-resourced policymakers in developing country governments and stakeholder groups.)

Discussions with interviewees in the *Louder Voices* study suggested that a resource of this kind should have two main components: a regular news-oriented summary giving relevant information on current meetings, issues, and reports (both comprehensive and archived); and a library of key documents in international ICT issues accessible directly from the site rather than through hyperlinks (which require constant maintenance to avoid going out of date). Content generation could, of course, in time become the responsibility of the network of centers of expertise resulting from the first recommendation.

Complementing these information resources, the report recommended a *program of small-scale, largely (but not exclusively) country-specific, research on international ICT issues and national ICT policy*. Again, the objective here was not to build a new institution, but to establish a mechanism to fund quick research projects, undertaken by developing-country research institutions, on the impact of particular ICT issues in individual countries. To ensure that resources were focused on issues of immediate relevance, the report recommended that projects should be limited to $25,000 in budget or to a maximum of sixty person-days; that findings should be made available within one month of completion; and that findings should be disseminated through the web site and the centers of expertise that might be established as a result of previous recommendations. Such a fund could be put in place very quickly indeed, with backing from development partners. Ownership by the global South could be secured and (as with Internet resources) direction of the work undertaken could be transferred to a network of centers of expertise if and when this program became established around the world.

Rigorous research work on this relatively small scale could be particularly important in enhancing awareness and realistic understanding of international ICT issues among nongovernment and non-ICT-sector stakeholders, in particular ministries responsible for development policy areas and civil society organizations. However, the *Louder Voices* recommendations also recognized that specific assistance was required by civil society bodies, particularly in smaller and Least Developed Countries, if they were to acquire the capacity to make significant contributions to national policy making. The fourth recommendation of the *Louder Voices* study therefore was for a *program to promote understanding of international ICT issues in developing country media and research institutes*, with the aim of improving the quality of public debate about ICT policy issues and establishing self-supporting networks of expertise within individual countries. Small partnerships of IT professionals have been influential in guiding ICT policy in a number of developing countries (e.g., the e-ThinkTank in Tanzania[30]). This recommendation would have the added value of broadening the professional expertise within such groups to include specialists in other areas of social and development policy.

Fifth, and bringing together these streams of increased awareness, knowledge, and understanding, the *Louder Voices* report recommended the development of *model national ICT policy-making processes* that could be adopted—or adapted for their own requirements—by policy makers in individual countries. The aim of such model processes would be to bring together the experience, positive and negative, of countries that have established internal processes for developing national input into international meetings and of countries that have initiated their own national ICT strategy development, in order to assess their outcomes and to pilot better ways of developing policy through the kind of broad stakeholder dialogue regarded as essential by so many interviewees during the course of the study.

These five recommendations focused primarily on national processes—on increasing awareness, information, analytical capacity, and dialogue within countries rather than at international meetings themselves. The sixth and final *Louder Voices* recommendation addressed the international institutions themselves.

As noted earlier in this chapter, the *Louder Voices* report called for international institutions to examine their decision-making processes and identify ways in which they could be made more open and inclusive to participants from developing countries—for example, by increasing the importance of regional pre-meetings; changing meeting schedules (and venues); facilitating participation by smaller and less experienced delegations in informal, committee, and plenary sessions; and providing direct support to developing country delegations.

This last support can take many forms, including information resources and guidance on issues at decision-making meetings themselves as well as capacity-building initiatives on individual issues, processes, and negotiating skills. One form that has proved relatively popular in the past—not least because it is easy to implement and can be simply quantified—is the provision of fellowships, that is, funding for the travel, accommodation, and subsistence of delegates from developing countries at meetings they would not otherwise attend.

Two problems, however, were identified by interviewees with this rather blunt instrument of representation. The first is that fellowships alone do not make even the most experienced and expert delegates effective. Unless institutions back up fellowships with training, advice, and other support, individuals can easily feel isolated, unable to contribute effectively in informal lobbying and working group processes, unfamiliar with procedures, and wary of making statements or asking questions. Equally important, many interviewees felt that fellowships were often ineffective because selection procedures in-country did not lead to the choice of appropriate and relevant personnel. Some concern was also expressed about the risk of institutions offering fellowships selectively, with potential implications for the outcomes of controversial votes.

Nevertheless, fellowships are regarded as valuable in building expertise and understanding of the decision-making process, and the *Louder Voices* report therefore recommended the development of a *code of practice on fellowships and other financial support to delegates attending international decision-making meetings*. Such a code would help developing countries to select appropriate experts, as well as set out the support that they would require before travelling, the arrangements for dialogue between them and their governments/countries during meetings, and the process by which they would report back to national stakeholders on the outcomes of the meetings in which they participated. As well as making it more likely that "the right people attended the right meetings," this code would help to anchor their work in the national policy dialogue needed to ensure effective input into international meetings as well as effective reporting and implementation of the outcomes of those meetings. Such a code could also reach beyond national borders to facilitate regional cooperation in international forums, with delegates from within a region acting together to support common, coherent positions developed through regional dialogue.

Such a code of practice would only be effective if it had the full support of delegates and countries benefiting from fellowships—and would therefore have to be developed by a representative group of developing countries (albeit with some involvement, in defining objectives and building infrastructural capacity, by the institutions that offer fellowships). The *Louder Voices* report suggested that this role might be performed by the developing country members of the DOT Force.

Conclusions and Recommendations

The international ICT decision-making agenda is increasingly complex and difficult for all participating governments and other stakeholders. It is particularly difficult for developing countries with limited financial and human resources available to analyze, prioritize, and participate in the widening range of institutions making decisions that affect their ICT futures. Substantial as these problems are in principle, the *Louder Voices* study showed that they are made much greater by weaknesses in international and, especially, in national decision-making processes. Some relatively straightforward actions could be taken—by national governments, international decision-making bodies, and development initiatives—to enhance the inclusiveness of international decision making. Specific recommendations made in the report, and summarized in this chapter, have been put before the international community.

It is important to recognize that, like everything else within the ICT sector, the *Louder Voices* report and recommendations were products of their time. They assessed the influence and impact of developing countries in international ICT decision-making institutions at a particular moment—the first six months of 2002, after the global significance of ICTs had become widely understood but was still fresh and new to senior policy makers. The study's recommendations were made to the concluding session of the DOT Force, the first major international forum specifically to address "ICT for development" issues from the perspectives of both the ICT sector and the development community—and one that brought together participants from all three major stakeholder groups (government, the private sector, and civil society) in both developing and industrial countries. Those recommendations were designed for implementation in the short-term aftermath of the DOT Force, as developing country stakeholders and international agencies alike began to address the shortcomings of existing participation.

Time, in the ICT for development universe, moves very quickly. Yesterday's profound insight quickly become today's orthodoxy and tomorrow's misleading assumption. Subsequent experience, including that of the WSIS, suggests that the analysis in the *Louder Voices* report remains valid—in particular, its assessment of critical factors underpinning weak participation and their impact on decision making; its focus on national policy-making processes as well as international institutions themselves; and its recommendations for action by governments and international agencies to address inadequacies of understanding, research, information, analysis, and process. The report has been widely applauded for its insight, and has been influential in many different forums. The context surrounding it has changed, however, in four important and closely related ways.

First, there have been significant changes in the internal policy-making processes of a significant number, though by no means a majority, of developing countries. Quite a few have developed new national ICT strategies or e-strategies, sometimes including a wide (or at least wider) range of stakeholders in this process. These strategies are often highly ambitious and all-embracing, and some at least are open to criticism on the grounds that they overestimate the potential of ICTs or underestimate the need for skilled personnel to bring them to fruition. There has also been a tendency for e-strategies to be developed by ICT professionals and ICD advocates rather than integrated fully with Poverty Reduction Strategies and other national development plans that have the full engagement of mainstream development professionals. Nevertheless, some countries have engaged in significant multistakeholder national policy making, providing potential models for other countries and encouraging nongovernment stakeholders to gain understanding of the issues and offer to participate themselves.

In some senses, this parallels the international community's experience following the inception of the WSIS process, which is at the heart of the second important change in context since the *Louder Voices* report was published. WSIS led to more intense activity around issues concerning ICTs and development and around international ICT decision making (particularly with regard to Internet governance), though much of the policy development for WSIS in individual countries was rushed and lacked wide stakeholder involvement, or even extensive dialogue between government departments.[31]

At the same time, the very fact that WSIS occurred has led to considerable rethinking and more sophisticated analysis of the potentialities and limitations of ICT4D in a number of multilateral and bilateral development institutions, which will help to inform future decision making. More people in more countries are now familiar with ICD issues, enabling them to participate more effectively in both national and international discourse. Some precedents were also offered for wider stakeholder participation in international ICT decision making. The first phase of WSIS led to the establishment of two independent forums to tackle questions it had found intractable, and, while the Task Force on Financing Mechanisms followed a rather conventional model for the resolution of international disputes of this kind, the Working Group on Internet Governance (WGIG) took a more open approach, engaging the private sector and civil society on relatively equal terms and opening much broader consultation spaces than are generally available. WGIG's innovations are widely regarded as having made a positive contribution, at least to the quality of its work, but it is unclear whether its multistakeholder approach will transfer readily to other areas of decision making—in particular those that are more government-controlled today than is the Internet. WSIS follow-up processes, including the Internet Governance Forum established by the summit's Tunis Agenda, will provide evidence about potential long-term impacts during the two years following the WSIS.

The third significant change concerns donor agencies' views of ICTs in development. While remaining optimistic about the potential of ICTs in general, multilateral and bilateral development agencies have increasingly emphasized the mainstreaming of ICTs in traditional development sectors as the prime focus of their interest in them. This tendency has been reinforced by the central importance in development policy of the Millennium Development Goals, which set targets for reductions in poverty and disadvantage to be achieved by or around 2015. Development agencies are increasingly focused where ICT4D is concerned on identifying how ICTs might facilitate the achievement of these MDGs.

This has, if anything, reinforced the significance of the paradigm gap between ICT agencies and development agencies—a gap that was less affected by WSIS than might have been expected, as can be seen for example from the very limited attention paid to ICTs in the UN's Millennium Review Summit or the UNDP's *Human Development Report* published to assess progress toward achieving the MDGs. Nevertheless, the prevalence of mainstreaming in development agency thinking suggests that the agencies and their personnel have begun to assimilate ICTs into their established poverty-related paradigms. With a few exceptions, such as the inclusion of suggested

MDG-related targets for ICTs in the ITU's 2003 *World Telecommunication Development Report*, there is less evidence that ICT agencies and their personnel have begun to assimilate development thinking into theirs. The importance of reducing the paradigm gap between these two sets of institutions remains substantial, but the terms of engagement between them have changed.

This is closely related to the fourth significant change that has taken place since the *Louder Voices* authors reported to the DOT Force in 2002. The DOT Force had considerable impact because it was the first serious and substantial attempt—at least since the ITU's Maitland Commission in 1984–1985[32]—to produce a global perspective and recommendations on the intersection between ICTs and development. It was also markedly informal in character. It included representatives from different stakeholder communities. More important in this context, it included only a selection of countries—the G8 members and eight developing nations—and was therefore able to reach conclusions and recommendations that bound only its own members, not the international community as a whole. Its "decisions" were therefore very different from the more binding decisions of more conventional international bodies, such as WSIS, the universal-membership agencies of the United Nations system, or organizations such as the World Trade Organization with the authority to regulate fundamental aspects of their member-countries' conduct de jure and much of the rest of the world's de facto.

Different aspects of what are now called ICTs have, of course, for many years been handled by global or near-universal-membership bodies—most notably, telecoms standards and international regulation by the ITU, and, more recently, telecoms trade by WTO. The UNDP and the World Bank had both undertaken significant work on the role of ICTs in development. The ITU has also played a role in the relationship between telecommunications and development, through its quadrennial World Telecommunication Development Conference (WTDC) and its Telecommunication Development Bureau (BDT)—although for many telecoms professionals representing their countries at WTDCs, these have more to do with the development of telecommunications than with telecommunications for development.

The success of the DOT Force, however, changed this environment by bringing ICT4D much more directly to the attention of the established global institutions and national decision makers, and this in turn has changed the character of the IC(T4)D debate. Two new, albeit temporary, institutions have played an important part in this change. The first of these, the UN ICT Task Force, was in many ways a natural successor to the DOT Force—though the two overlapped in time. The UN ICT Task Force, which ran from 2001 to 2005, was essentially created to bring IC(T4)D issues into the UN family mainstream—to engage other UN specialist agencies and the governments of all UN member states. Although it assumed some features of the DOT Force, notably multistakeholder participation, its identity as part of the universal-membership UN system gave its conclusions more weight: they naturally assumed the mantle of global consensus. For this reason, they may have been more ponderous—slower to emerge, and more cautious when they did.

The second new institution was the World Summit on the Information Society, a four-year process of international discourse focused around the two formal summit meetings held in Geneva in December 2003 and Tunis in November 2005. The world summits were, in essence, gatherings designed to establish consensus on issues of perceived global importance, and to initiate (if possible) concerted plans of action to address them. They effectively absorbed the issue with which they deal into the mainstream of UN and global decision-making processes.

This was particularly evident during the first phase of WSIS in a number of ways that have further implications for the context for developing country participation in international ICT decision making. One was the simple and now universal recognition of the importance of the issues. As they were perceived to be more important, debate around them became political as well as economic, technological, and developmental. Diplomats as well as technical specialists became concerned about the wording of international agreements. The interaction between ICTs and development was complicated in WSIS by concerns about its interaction with human rights, by calls for the reallocation of resources, and especially by demands for adjustments to the international governance regime for the Internet—the three issues that were fundamentally contested in the first phase of the WSIS.[33]

The politicization of ICT issues in WSIS has a number of potential ramifications. One is that much international decision making concerned with ICTs and ICT4D may, in the future, be more formal, taking place in global institutions and therefore bound by the structures of those institutions (not all of which, e.g., are friendly toward civil society). The scope for informal initiatives like the DOT Force, which made a virtue of its nonuniversal character, may, as a result, have been reduced.[34] On the other hand, the WSIS debate on Internet governance was invigorated by the informality and multistakeholderism of the WGIG's structure and working practices. It is possible that these characteristics may follow through into the (non-decision-making) Internet Governance Forum established by the WSIS final outcome documents, and influence other ICT decision-making forums in turn.

The second potential ramification concerns the impact that politicization of ICT decision making, if it happens, may have on the participation of developing countries. More politicized decision making is more susceptible to influence from and tradeoffs with issues outside the ICT and ICT4D contexts themselves. As during WSIS, in other words, the big issues in international ICT decision making may be determined in the future partly by attitudes and developments in other areas of international policy (e.g., in the WTO) or as a result of general international political alignments. This may increase governments' level of attention to the major decision-making forums (and in some cases make them more concerned with restricting participation in these to government officials). It is less likely to increase attention to decision-making processes that have less impact on short-term public consciousness or international politics but may have more on long-term technical and developmental outcomes—for example, the Internet's technical specialist bodies.

A third important ramification follows from this: the likelihood that, post-WSIS, much more attention will be paid to the control as well as the content of international ICT decision making. One of the most contentious issues throughout both phases of the WSIS was the role of ICANN, the informal governance body that manages Internet domains. Many governments saw this as a matter of national sovereignty whose international governance should be managed by an intergovernmental agency (perhaps, though not necessarily, the ITU). There is also significant debate about the scope for intergovernmental control of Internet content. Contests about these governance issues were not resolved by WSIS; they could well proliferate in its aftermath, continuing to divert policy makers—as they did during the second WSIS phase—from concerns about the technical and developmental impact of international ICT decisions.

The issues raised during the description and analysis of the *Louder Voices* report in this chapter nevertheless remain very important in the post-WSIS environment. Developing countries, particularly small developing countries and especially nongovernmental stakeholders, are poorly represented in international decision making and have little influence on international decisions about the future of ICTs that affect both their deployment and their impact on development. Poor participation results from poor national policy-making processes, lack of awareness of the potential role of ICTs and the importance of international ICT decisions, resource constraints (both human and financial), and inadequate regional coordination. These problems can be addressed by developing countries, and international decision making would benefit—alongside developing countries themselves—if they could enhance the effectiveness with which the concerns and interests of low-income and developing markets are represented.

The paradigm gap between ICT and development thinking also remains important. While some development institutions have mainstreamed ICTs into development policy, there is less evidence that ICT institutions or policy makers have absorbed the concepts of development into their policy processes or fully understood the complementary factors required to achieve development as an outcome of ICT deployment.

The broad conclusions and recommendations of the *Louder Voices* project remain valid in this context, but need adaptation in the light of changes in the international decision-making environment that have taken place since they were presented to the DOT Force in 2002.

The introduction of regional centers of expertise; new Internet-based information resources; research; and awareness-raising programs proposed in the report were intended for implementation as nearly as possible to the immediate aftermath of the DOT Force process. In practice, only two Centres in International ICT Policy were established following the report, both in Africa and neither with the kind of remit that had been envisioned. The case for a network of similar institutions, sharing common characteristics and active in different developing regions, remains strong, but that for creating them anew is now less so. The WSIS process, in particular, has stimulated the development of research capacity; and institutions now exist in most regions that could assume this role or build it into their existing work. Post-WSIS, it would seem more appropriate to develop a network among research institutes that have already been established or have developed ICT-related programs of their own. This could be supported, at relatively little cost, by development agencies and decision-making institutions alike.

Coordinated research into the impact of ICTs, web-based information services, and awareness-raising among opinion leaders, especially in the media, are also all still important for the development of a broad understanding of the potential of ICTs within national stakeholder communities; and injections of development funding in these areas could do much, at relatively little cost, to enhance the quality of policy debate and decision making in developing countries. As with possible centers in international ICT policy, however, it would probably be more appropriate and cost-effective four years on from the *Louder Voices* report to use such funding to reinforce the capacity of institutions that have emerged around and since WSIS rather than to create new ones from scratch.

The ICT and ICT4D debate suffers from two apparently contradictory characteristics. On the one hand, decisions are often taken in the absence of the research and analysis needed to understand their impact fully enough for those decisions to be reliable. On the other hand, practical decisions are often delayed in the interest of achieving consensus about the best way forward. Both of these factors need to be addressed by the international community. Investment is needed in research that will increase understanding of the real impact that ICTs have or can have at all social and economic levels, from macroeconomics to rural livelihoods, and of the complementary factors that can maximize or minimize that impact. Without such research, poor decisions will be made, investment will be misdirected, and the hopes and aspirations of both ICT professionals and those to whom they have promised benefits will be unrealized. But the pace of change in ICTs is so fast that decisions cannot be deferred for long, simply because the technology and other characteristics that seemed appropriate one year can be outmoded by the next. One of the lessons of the DOT Force/*Louder Voices* experience may well be that, in the fast-moving world of ICT decision making, the risk of acting quickly may be less than the risk of taking time. Decision makers in national policy making and international institutions need to find ways of making decisions in this area that are both quicker *and* better informed, as well as more inclusive. That is a particular challenge for ICTs and ICT4D, above and beyond the challenges of enhancing international decision making in other sectors, and it is one that needs to be addressed urgently if the world community is to look back in twenty years on the ICT revolution as a time of opportunity realized rather than missed, of social and economic benefit achieved rather than unattained.

Notes

1. The DOT Force report is available at http://lacnet.unicttaskforce.org/Docs/Dot%20Force/Digital%20Opportunities%20for%20All.pdf.

2. MacLean et al., *Louder Voices*.

3. WSIS documentation is available at www.itu.int/wsis.

4. The British campaign for debt relief, Jubilee 2000, symbolically burnt a laptop on the beach at Okinawa during the G8 meeting that established the DOT Force.

5. G8, Okinawa Charter on Global Information Society, available at www.dotforce.org/reports/it1.html; United Nations General Assembly, United Nations Millennium Declaration, available at www.un.org/millennium/declaration/ares552e.pdf.

6. The UN ICT Task Force's documentation is available at www.unicttaskforce.org. The Millennium Development Goals are available at www.un.org/millenniumgoals.

7. A brief summary of the first WSIS session and its outcomes can be found in Souter, "The View from the Summit."

8. The study was coauthored by Don MacLean, former director of strategy at the ITU; James Deane, then executive director of Panos London; and David Souter and Sarah Lilley, then chief executive officer and research assistant respectively at the CTO. Case studies were undertaken by African, South Asian, and South American consultants, and the project as a whole was supported by an advisory committee of international specialists, mostly from developing countries.

9. MacLean et al., *Louder Voices*, 3.

10. Three of these countries were chosen because they had played a relatively substantial part in decision making—Brazil, India, and South Africa; two because they had not—Nepal and Zambia; Tanzania because, although it had exerted little impact in the past, it was represented on both the DOT Force and the UN Task Force. In addition, the study focused on three key issues—ccTLD disputes, IP telephony, and WTO telecoms offers—and on participation at specific meetings including the ICANN Assembly and the World Telecommunication Development Conference held in 2002.

11. See also MacLean et al., *Louder Voices*, 10–16.

12. The matrix/map could, of course, be adapted to illustrate these differences in intensity.

13. See chapters 5 and 14.

14. See chapter 5.

15. See chapter 14.

16. Much the same was true of the follow up conference in 2006, in spite of its billing as the first major international meeting to address ICT/ICD issues after WSIS.

17. Participants in the World Telecommunication Development Conference are listed at www.itu.int/ITU-D/conferences/wtdc/2002/doc/listParticipantswtdc02.pdf; those for WSIS at www.itu.int/wsis/geneva/participation.html and at http://www.itu.int/wsis/documents/doc_multi.asp?lang=en&id=2294|0.

18. There was no comparable WTO meeting during the timeframe for the study.

19. See the Conclusions and Recommendations section of this chapter.

20. This became an issue during discussions on Internet governance during the second phase of WSIS, in which a number of large developing countries played a prominent part. The WSIS debate on Internet governance was, of course, highly politicized in comparison with many other international decision-making forums.

21. MacLean et al., *Louder Voices*, 17.

22. Ibid., 20.

23. The impact of ICT investment on macroeconomic growth has been particularly difficult to discern, as famously noted in Robert Solow's 1987 remark that "you see computers everywhere except in the productivity statistics" (*New York Times*, July 12, 1987). Recent research by the OECD has brought forward convincing evidence for industrial countries, but suggests that the impact of ICT investment is very dependent on complementary factors such as the scale of ICT deployment, network externalities, the regulatory context, and the availability of skilled personnel. In OECD, *ICT and Economic Growth: Evidence from OECD Countries, Industries and Firms* (Paris, 2003). For assessments of the implications of this for developing countries, see D. Souter, "ICTs and Economic Growth in Developing Countries," OECD Development Assistance Committee Network on Poverty Reduction paper, DCD/DAC/POVNET(2004)6.

24. See OECD, note 23, and Souter, note 23.

25. Country case study of South Africa conducted by Alison Gillwald, cited in MacLean et al., *Louder Voices*, 23.

26. MacLean et al., *Louder Voices*, 24.

27. Ibid., 25.

28. Ibid., 26–28. These proposals were elaborated in unpublished documentation for the UK's Department for International Development (DFID) and other funding agencies.

29. Two Centres in International ICT Policy somewhat along these lines were established as part of the Catalyzing Access to ICTs in Africa (CATIA) initiative launched by the DFID and other donors in 2003. One of these, managed by Panos West Africa, is located in Dakar, Senegal, and

covers the West and Central African regions; the other, managed by a consortium coordinated by bridges.org, is located at the Makerere University in Kampala, Uganda. While the establishment of these centers largely resulted from the recommendations of the *Louder Voices* report, they were given substantially different mandates and do not in practice deliver the range of information resources envisioned in the initial report.

30. Its work is documented at www.ethinktanktz.org.

31. An assessment by the author of developing country and multistakeholder participation in the WSIS process will be published by the Association for Progressive Communications in late 2007.

32. ITU, "The Missing Link." The Maitland Commission, initiated by the ITU, investigated the discrepancy between industrial and developing country teledensity (the pre-digital equivalent to the "digital divide") and called for action to be taken to promote telephony as an instrument of economic development.

33. For a summary, see Souter, "The View from the Summit."

34. At least if governments are to be involved. It may still be possible for business or civil society interests to convene a Maitland-style inquiry that would be less constrained by intergovernmental politics.

References

Drake, William J., ed. 2005. *Reforming Internet Governance: Perspectives from the Working Group on Internet Governance*. UN ICT Task Force: New York.

Gilhooly, Denis, ed. 2004. *Creating an Enabling Environment: Toward the Millennium Development Goals*. UN ICT Task Force: New York.

Government of Canada. 2002. "Digital Opportunities for All: Report of the G8 Digital Opportunity Task Force." Available at http://e-com.ic.gc.ca/epic/Internet/inecic-ceac.nsf/en/gv00133e.html.

International Telecommunication Union. 1984. "The Missing Link: Report of the Independent Commission for World Wide Telecommunications Development." ITU: Geneva.

International Telecommunication Union. 2005. "WSIS Outcome Documents." ITU: Geneva.

MacLean, Don, David Souter, James Deane, and Sarah Lilley. 2002. *Louder Voices: Strengthening Developing Country Participation in International ICT Decision-Making*. London: Panos. Available at www.panos.org.uk/images/books/Louder%20Voices.pdf.

Milward-Oliver, G., ed. 2005. *Maitland + 20: Fixing the Missing Link*. Bradford on Avon.

Souter, David. 2004. "The View from the Summit: A Report on the Outcomes of the World Summit on the Information Society." *Info* 6, no. 1: 6–11.

13 The Ambiguities of Participation in the Global Governance of Electronic Networks: Implications for South Africa and Lessons for Developing Countries

Tracy Cohen and Alison Gillwald

With the erosion of traditional vertical systems of government, multilateral organizations have been one of the main drivers behind the shift to more horizontal forms of governance. This is true globally of the information communication technology (ICT) sector too. Yet the structures and decision-making processes of most of the multilateral organizations themselves continue to be characterized by more traditional intergovernmental systems with regard to agenda setting and decision making in international forums. Despite some gestures toward greater participation by nonstate actors, outcomes remain determined by the official positions of sovereign states. Within these largely unreformed arrangements, developing countries, although formally equal to their developed country counterparts, and generally familiar with the vertical systems of power associated with these institutions, in practice remain marginalized from the content of these determining processes. This often results in developing countries, by omission or passive participation, tacitly endorsing decisions, which from a general public-interest point of view may be seen as contrary to their interests.

With regard to the governance of the Internet, it is not only the substance from which most developing countries are far removed, but the form too. Newer organizations that have sought alternative structures and decision-making processes have been viewed as exclusionary and unrepresentative. Unsuited to the bureaucratic and statist structures associated with the United Nations and the Bretton Woods institutions, the governance of the Internet historically has been undertaken through the organization of private sector interests. To date this has not provided an effective alternative system of governance. With strong allegiance to the United States Department of Commerce, the governance arrangements for the Internet have been plagued by similar problems of lack of participation, representation, and influence by developing countries. With the crisis in confidence in ICANN (International Corporation for Assigned Names and Numbers), appeals for greater government participation to improve the efficacy of Internet governance have not come from developing countries alone. Several countries with established Internet markets have called for government intervention as evidenced by the governance debates arising out of the World Summit on the Information Society (WSIS).

It is also evident that the pursuit of unrestrained market access policies by multilateral agencies at the global level has sometimes been hypocritical and self-serving: while multilateral agreements espouse procompetitive and fair principles in relation to domestic markets, the same principles are not equally applied at the international level. Free-market arguments tend to be followed only when they serve the dominant interests in the global economy. A case in point are the restrictions on the free flow of human capital so critical to the knowledge economy. While arguing through the agencies of global governance for open access to markets and the free flow of international investment capital required to permit global markets to efficiently allocate resources, the allocation of human resources—the core resource of the knowledge economy—is highly regulated in order to protect what are regarded by many developed nations as their national interests.[1] Under the General Agreement on Trade in Services (GATS),[2] the two main methods of restricting the free movement of human capital (the temporary mobility of skilled and professional people) are in the form of quantitative restrictions such as quotas regulating visas and nonquantitative restrictions such as licensing requirements, economic needs tests, professional qualifications, and technical standards limiting the ability of such persons from delivering their services.[3] Thus, beyond addressing lack of participation by developing countries in global governance arrangements, this chapter also highlights the need for global governance that will embark on proactive regulation to facilitate fair competition at the global level.

Using South Africa as a case study, this chapter examines why and how South Africa in particular, and developing countries in general, become bound by international arrangements without the associated benefits being unequivocal. Our intention is to assess that participation by measuring the implementation of those policies, their effectiveness, costs, and obstacles, in the South African domestic context. We also highlight the costs of not participating effectively in international reform efforts.

In doing so, we examine the institutional arrangements of two agencies that have become increasingly responsible for global electronic network governance—the International Telecommunication Union (ITU) and the World Trade Organization (WTO). We note the importance of ICANN in the debate, but due to minimal participation by developing countries to date, do not address it here. ICANN is dealt with in greater detail by Mueller and Woo in chapter 14. As the United Nations agency charged with responsibility for co-coordinating international communications, the ITU has traditionally overseen telecommunications. While the agency assumed some overall responsibilities for the setting of technical standards and spectrum usage, relations between countries tended historically to occur bilaterally, usually between the monopoly operators. With the emergence of the WTO, discussed by Guermazi (chapter 5), and the inclusion of telecommunications in the GATS, the ITU was forced to reassess its validity as the primary international agency responsible for a rapidly transforming sector. The upshot of this was several rounds of reform within the ITU, as described by McLean (chapter 2), and the emergence of an agency more aligned in nature with the newer multilateral agencies within the UN and Bretton Woods systems.

ICANN, on the other hand, emerged from an entirely different lineage. As the successor to the telegraph and the telephone, the governance of the Internet should logically have fallen to the ITU. However, the dynamic nature of the Internet, and the innovative responses of those driving the frontiers of its development to overcoming technical and regulatory barriers, jarred with the highly bureaucratic nature and statist bias of the organization, and with its industrial-age approach to regulation. Under the auspices of multiple U.S. enterprises, most notably the Department of Commerce, ICANN was established during the rise of the dot-coms as a self-regulated private sector initiative that was vigorously defended against state interference. ICANN did not provide an effective alternative mechanism for global regulation of electronic networks. Lack of representation and the same inability of less developed nations, even with emerging economies, to participate meaningfully, has wracked the credibility of the organization. Following the crash of the dot-coms and growing concern about the lack of financial public accountability of firms in the light of the centrality of the Internet to the knowledge economy, the need for a more integrated and effective system of governance is being argued for both within and outside of ICANN. While it is clear that an entirely new, flexible, innovative, and integrated approach to the regulation of the Internet is required, the need to uphold the fundamental public interest principles of regulation are increasingly being called for with regard to Internet governance. The responses to these concerns in the outcomes of the WSIS continue to be a disappointment to many.

Though quite different in their size and nature, all of these organizations tend to be dominated by a few of the most industrialized countries—particularly the United States—whose interests are reflected in the policy agendas and outcomes of decision-making processes.

A three-pronged explanatory framework follows to explain the historical, resource, and domestic structural constraints that have a negative impact on the effective participation of developing countries in these global institutions of governance.

However, developing countries specifically are not presented as unwitting victims of processes that conspire against them, or as necessarily devoid of recourse within these global institutional arrangements. We seek to provide a more nuanced understanding of the power relations, institutional levers, and contested interests that have produced the current set of outcomes at the international level, and of how these can be adjusted to provide more favorable public interest results for developing countries.

This chapter argues that a number of mechanisms exist within the relevant organizations and treaties that give effect to their mandates. These can, subject to certain conditions, be utilized to increase developing country participation and influence and optimize domestic policy outcomes. At the national level, of course, diverse interests cannot be portrayed as homogenous and the chapter further explores how smaller players can indeed utilize international instruments and commitments undertaken to challenge entrenched incumbents.

Based on research conducted in South Africa with key organizations and institutions involved in ICT policy formulation, the chapter outlines a number of recommendations for strengthening participation by South Africa, and indeed other developing countries, in international forums concerned with global network governance, and for optimizing the opportunities that can result from international coordination.

Finally, the chapter argues that the transformation of the governance of electronic networks globally, together with the identification of areas requiring regulation to create a fairer competitive global playing field, would produce associated efficiency gains at the international level. These gains, in turn, could provide a new framework for international governance that is more effective, participatory, and representative and genuinely capable of contributing to sustainable global development within the ICT sector.

South Africa

South Africa (SA) presents a compelling case study for examining the power and ability of a developing country to participate effectively in global electronic governance regimes and as such may provide lessons for other developing countries. Universal extrapolation is limited, however, given that SA, although classified as a middle-income country by the United Nations, simultaneously exhibits features of both an underdeveloped country and a fully industrialized economy. It is neither fully one nor the other and as a result, provides a granulated example of the complexities that underlie international governance arrangements.

In terms of relative power within the governance structures identified, SA manifests both dominant and nondominant characteristics. With the exception of limited circumstances within the ITU, SA may be broadly characterized as nondominant. Yet, in relation to regional and continental governance issues generally, SA, given its relative economic strength and strategic geographic location, is overtly dominant, oftentimes drawing sharp criticism for its efforts to extend domestic macroeconomic policy within the region and on the continent.[4] Thus, SA plays at multiple levels on the global stage with influence in some areas and marginalization in others, displaying tensions between its different roles domestically, continentally, and internationally. It is not, however, our objective to resolve these tensions, merely to highlight them and draw out the implications for other developing countries that a study of this nature presents.

Socioeconomic Indicators

South Africa spans the southern tip of the African continent, and is divided into nine geographic and political regions.[5] Historically known for its gold and diamond mining, SA has the strongest economy and is the most advanced industrial nation in Africa. The population was estimated at 46.8 million people[6] and the gross domestic product (GDP) per capita was approximately $12,159 (purchasing power parity, or PPP)[7] in 2005. In 2005, GDP at real constant prices was R [rand] 1,108 billion with a 4.9 percent growth rate for that year[8] and almost triple that of R431.5 billion in 1994[9] when the apartheid regime fell to the African National Congress (ANC) following the first democratic elections in the country.[10] Primarily based on mineral and energy resources, the economy is also grounded in manufacturing, agriculture, and commerce, but it remains highly dependent on foreign investment,[11] an objective to which the country's macroeconomic policies—including privatization and the promotion of international trade and exports—is targeted.[12] Low investment levels in SA have been consistently viewed as a primary cause of suboptimal economic growth:[13] at the start of 2000, President Mbeki undertook to promote economic growth and foreign investment and advance the pace of the restructuring state-owned enterprises (SOEs) in transport, electricity, and defense.[14] Known as "parastatals" in SA, these companies continue to play a critical role in the current phase of economic restructuring where the focus has shifted from macroeconomic stability to the development of a sustainable growth trajectory.[15] Parastatals or SOEs will also play an increasingly significant role in regional economic integration in the South African Development Community (SADC) by providing a synthesized platform for transport, communications, and energy in the region.

While many of the basics are in place for the economy to develop at a high, sustained rate, SA unfortunately faces enormous challenges with regard to crime, unemployment,[16] health—specifically HIV/AIDS,[17] housing, inflation,[18] and poverty.[19] The UN Development Programme (UNDP) Human Development Report ranked South Africa at 120 for 2005, noting that the areas requiring urgent attention are health, poverty reduction, and education.[20] The legacies of structural underdevelopment of portions of the population along racial lines are reflected in the gross income disparities that continue to exist, with the poorest quintile of households accounting for less than 4 percent of the total income, compared to the richest quintile accounting for 62 percent of total income.[21] With such marked differences, South Africa records one of the highest Gini Coefficients in the world, measured in 2005 on 2000 data at 57.8, exceeded only marginally by a cluster of Latin American countries, headed by Brazil at 68.

Following the review of a decade of democratic government in 2004 and the return of the ANC to power for the second time in the 2005 elections, the government announced a new economic strategy—the Accelerated and Shared Growth Strategy (ASGISA). ASGISA's main goal is to achieve an annual economic growth rate of at least 6 percent of gross domestic product by 2014, reduce unemployment to below 15 percent, and halve the poverty rate to less than one-sixth of households based on sustained and strategic economic leadership from government, and effective partnerships between government and stakeholders such as labor and business.[22]

South African Communications Policy: Domestic Goals above All Else

As this chapter focuses on international institutions such as the ITU and WTO with reference to the GATS[23] and with specific focus on the Fourth Protocol on Telecommunications to that treaty,[24] it is apposite to briefly explain the evolution and state of telecommunications policy in South Africa.

The country is a full member of the ITU and a signatory to the WTO, including the GATS instruments on telecommunications. Liberalization of the domestic market by 2003 was the major commitment under the latter. It is noteworthy that reform in telecommunications not only occurred within the bounds of the more fundamental and broader transformation to democracy but also was viewed as the vehicle with which the country would "leapfrog" developmental stages and be delivered into the information era, replete with a market attractive to international investors and operators.[25] A limited exclusivity on all basic telecommunications services and facilities for the incumbent, majority state-owned operator, in return for a universal service obligation to double the network, would further ensure that the underlying and elemental public interest mandate, particularly of affordable access, would be put into effect. This broad characterization is not trivial: the politics of reform and transition were inextricably tied to an ethos of consultation and transparency. Horwitz deftly applies Huntington's transition theory to the SA context to reflect how SA's telecoms sectoral reform process amounted to a "transplacement" in which the dynamic produced by the negotiations between the powerful interest groups accounted for the outcome.[26]

While consultative practices in policy making domestically continue, they have taken on different forms, and following the highly consultative policy development process that resulted in the white paper on telecommunications described subsequently, this increasingly has been cited as procedural only in the subsequent consultation processes for the 2001 Telecommunications Amendment Act and the 2006 Electronic Communications Act. We return to this point further on but highlight here one aspect of the structural deficiency we see as inhibiting effective participation: while there are factors and constraints implicit in multilateral approaches to governance, such as financial, technical, and informational barriers, studies have indicated a concomitant paucity of transparency and effective governance at the domestic level that aggravates this deficit. Respondents in she South African case study of the *Louder Voices*[27] report (discussed in chapter 12 of this volume) highlighted the need for a "more participatory national policy process" in order to strengthen decision making and enable better participation in international governance structures.[28] While we will address this in more detail, we highlight its importance here in noting the history of policy making in ICTs in SA and the context in which participation in domestic policy has evolved. While certainly more acute in many other African countries, this problem is not limited to developing countries alone.[29]

Affected by the deregulation wave sweeping through international telecommunications markets from the late 1980s, a phased-in approach to the introduction of competition was forecast in a visionary government policy statement on telecommunications in 1995, which was later crystallized into legislation in 1996.[30] This policy document articulated a commitment to the ideal that telecommunications is not simply an aspect of development, but rather a precondition for its success.[31] Thus, with the oversight of an independent regulator,[32] competition would gradually be phased in while allowing a limited exclusivity period for the incumbent courier Telkom SA Ltd. to concentrate on the rollout of service to previously disadvantaged areas.[33] Other aspects of the policy suggested a cautious approach to gradual liberalization, which included the introduction of competition on the fringes of the network initially and later at the core. This notion saw articulation in the opening of the enhanced service and private telecommunications network markets, subject to various restrictions on the acquisition of facilities from the incumbent and an overarching prohibition on "bypass" of the basic telecoms network. More recent expression of the vision was facilitated by an amendment to the principal legislation, which contemplated, initially, one license for a public switched network operator and the issuance of new license classes for small regional network operators in underserviced areas. It also introduced a multimedia license for the incumbent signal distributor, which was in addition given a carrier-of-carriers transmission license to operate an international gateway.[34] Underpinning this is a shift, in emphasis in government policy from universal service objectives to underscoring "the maximization of state assets," as part of a broader plan to restructure state-owned enterprises.[35]

Importantly, the policy process from 1995 provided for a legislative end, at least in theory, to the historical delivery of telecommunications services on a monopoly basis, initially through a government department and later through a commercialized entity, Telkom SA Ltd., which until 1997 was owned entirely by the state.[36] Partial privatization first occurred through the conclusion of a strategic equity partnership with a U.S. and Malaysian-based consortium that would supply the access capital required to fund expansion and the management capacity and technical expertise to ensure it met the development challenges ahead.[37] Further marginal divestiture was effected through a sale of equity to a black empowerment group and, more significantly, through an initial public offering, which saw approximately 28 percent of the company listed on the SA and New York stock exchanges in 2003.[38] Beyond structural changes affecting financial operations and corporate governance, it should be noted that the locus of power and the mechanics of the monopoly remain—until a competitor is licensed—completely unchanged.

SA Internet

The information technology industry in South Africa is around the twentieth largest in the world. The Internet industry when judged by number of nodes is ranked internationally between the thirtieth to thirty-fourth largest.[39] While Internet users per 100 people in SA is measured at almost 14 percent, this figure is severely skewed toward the white minority and is still largely restricted to major urban cities. The figure is expected to improve to one in ten people with Internet access in SA by 2006.[40] Comparable data reflects averages of one in every two people having Internet access in developed countries like Canada, Hong Kong, Singapore, South Korea, the United States, and those of the European Union.[41] However, outside of the island nations, and excluding the corporate market, SA still has the highest Internet usage penetration on the African continent of 7.89 percent, which reflects an average of Internet access for one in 160 people or five million users.[42]

This notwithstanding, the SA Internet industry is a dynamic and highly competitive one, facing the same trends toward consolidation experienced internationally in recent years. There is one shared private sector Internet exchange (INX) in Johannesburg as well as a Telkom INX, for the incumbent currently does not peer with the rest of the private sector.[43] Only a few large companies dominate the local access market, with five first-tier or top-level operators (companies that fully manage their own national networks and at least part of their own international bandwidth) and a higher number of second-tier ISPs dependent on the local and international bandwidth of the top-level operators; and a number of third-tier operators (virtual ISPs that do not operate networks, but merely market and sell Internet services). While ISPs required a license to operate,[44] one of the larger problems facing the industry is its overwhelming statutory reliance on the incumbent for subscriber lines and facilities for both its residential and corporate clients. Dial-up access still dominates the SA residential market with a total subscriber base of just over one million,[45] although the rollout of dedicated leased lines (ISDN) increased by 20 percent in 2003—mainly in the corporate market—surpassing dial-up growth rates.[46] The rollout of both wireless and fixed broadband[47] is still in its infancy, and while still prohibitively expensive for residential users has experienced high growth since its introduction in 2002 (see table 13.1). Telkom is currently the only provider of asymmetric digital subscriber line (ADSL) service on the market and there is a three-month waiting list. In comparison, mobile operators offer wireless broadband within three hours, leading to higher growth rates for wireless broadband compared to ADSL. However, Telkom still dominates the broadband market with its ADSL offering.[48]

Table 13.1
Broadband penetration in South Africa

Year	2004	2005
Telkom	47,624	95,290
MyWireless	3,500	16,000
iBurst	N/A	24,000
Vodacom	N/A	44,000
MTN	N/A	12,000
TOTAL	51,124	191,290

Source: Gillwald, Esselaar, and Stork 2006.

Another means of Internet access in the country, suggesting some overlap with dial-up service, is via corporate use and through access provided by educational institutions.[49]

With the first ANC-appointed director general in the Department of Communications active internationally and attuned to global technological development, the Electronic Commerce and Transactions (ECT) Act was passed in 2002—some would argue prematurely and without the necessary capacity to see through its implementation. Until its introduction, all domain and namespace management in SA was done through the Internet community and via the cooperation of a namespace administrator, with two entities: Uniforum SA and Namespace ZA.[50] The new act proposed wide-ranging changes to this system and established a statutory body, called the ZA Domain Name Authority, with a board of directors appointed by the minister of communication in consultation with the industry that has assumed responsibility for all matters associated with the ZA domain name space, including national interaction with ICANN.[51]

The ECT Act incorporates traditional e-commerce legislation (following the UN Commission on International Trade Law, or UNCITRAL, model law on electronic commerce as well as legislation in numerous other jurisdictions), electronic signature law (a somewhat confusing amalgam of the UNCITRAL model law on electronic signature and the EC Electronic Signature Directives), computer evidence, computer crime, consumer protection, voluntary data privacy provisions, limitation of liability of service providers, domain name issues, and various others. The omnibus nature of the ECT Act is due to trying to address all identified obstacles to electronic commerce and electronic government in South Africa. Its very broad nature is its greatest strength as well as its greatest weakness, as numerous provisions of the ECT Act have still not come into effect, almost four years after the act was passed.

As a member of the domain.name board, Mike Silber, point outs, part of the reality of South African government electronic initiatives is that they are uncoordinated and disparate. Overall responsibility in terms of the ECT Act rests with the minister of communications. However, responsibility at a cabinet level for electronic government seems to be shared also with the Departments of Public Service and Administration (DPSA) and Home Affairs. In addition, numerous other national, provincial, and local government departments have launched individual projects, without any indication of planning or central coordination.[52]

This act should not be confused with the Electronic Communications Act which came into law in July 2006, following several rounds of telecommunications and broadcasting reform over the last decade. This law seeks to deal with the challenges of the converging broadcast and telecommunications trends, and the provision of new electronic communications services. Its history lies in a hurried consultative process in which a convergence bill was placed before Parliament that had been drafted by several committees, largely from industry, though with some participation by the regulator and trade unions. The resulting inherent contradictions in the draft legislation meant the first bill was returned to the Department of Communications for revision. With the continued licensing delays of the second public switched network operator and the underserviced area licenses enabled by the 2001 act, the bill included a proposed restructuring of the regulatory arrangements for the sector. This was meant to respond to criticism that the licensing and regulatory delays that had plagued the sector were a result of the structural conflict of interests that underpinned the position of the minister as responsible both for the policy of the sector, with powers to veto legislation, and for the fixed-line incumbent, the most dominant player in the sector. The regulator's budget being at the discretion of the department was also dealt with by proposing that it raise its own funds from licensing fees that historically had gone directly to the treasury. This positive attention to the institutional arrangements for the sector, which had been widely identified as being the cause of its poor policy outcomes, was short-lived. In the next round of the convergence bill, all references to the regulator had been expunged and a new bill, the ICASA Amendment Bill, was introduced, allowing for the appointment of the council directly by the minister of communications and not by the president on the advice of Parliament as had been the case. After considerable political wrangling on this issue, this is how the act was passed. Because of the constitutional protection of broadcasting and the requirement that its regulation be independent, the president, however, did not assent to this legislation when he did to the Electronic Communications Act in 2006, and he sent it back to Parliament for consideration. The resultant comprise is a messy arrangement, with the appointment being the responsibility of both the Parliament and the Ministry, as is the performance management of councillors of the authority controversially included in the act.

This continued fettering of the regulator may impact negatively on the wide-ranging endeavors of the new act to open up the sector. The primary object of this act is to provide for the regulation of communications in the republic in the public interest, but specifically to promote and facilitate the convergence of telecommunications and broadcasting signal distribution. It continues to highlight universal access but with a focus on connectivity to enable access to enhanced services in the context of greater investment, competition, and innovation in the communications sector. It continues to focus on issues of redress with regard to racially inequitable ownership within the sector. The objective of providing a clear allocation of roles and assignment of tasks between policy formulation and regulation within the communications sector is addressed through the removal of the minister's veto powers on ICASA regulations and the handing over of all licensing powers except for individual network operator licenses. New licensing categories include class licenses that should enable license registration on the basis of a set of threshold requirements determined by the regulator. There are also complete license exemptions for services to be determined by the regulator.

International Governance Mechanisms

With the preceding background sketched, this section of the chapter examines the institutional arrangements of the WTO and the ITU with respect to telecommunications, and where relevant, the Internet. The details of these structures' modes of operation and processes have been dealt with in the chapters by Guermazi (5), MacLean (2), and Mueller and Woo (14), respectively. Hence, our treatment of these mechanisms is restricted to what is necessary for the current examination.

At the outset, it is important to clarify that there is no single governance mechanism for global electronic networks. Similarly, there is no single global electronic network. ICTs and their various applications that encompass the notion of global electronic networks are manifold, precluding their treatment in a single overarching framework agreement. As such, our study is concerned less with one or another structure or institution of governance, but rather on the *participation of developing countries* in a number of those key forums that collectively suggest broad governance of several and, indeed, diverse global electronic networks that to a greater or lesser degree are interconnected both from a technical and policy point of view.

In restricting our focus in this chapter, we exclude a number of other international institutions that, it should be noted, to varying degrees, have an impact on the ICT sector—namely, the World Intellectual Property Organization (WIPO); the UN Conference on Trade and Development (UNCTAD) and other international initiatives such as the World Economic Forum (WEF); the UN ICT Task Force (UNICT-TF); and the G8 Digital Opportunity Task Force (DOT Force). Nonetheless, concerns regarding effective participation apply with respect to key global players and alliances, as well as formal governance structures, because of the potential impact the former can have on the policy outcomes of the latter.

The WTO and Its Telecommunications Instruments

The WTO's General Agreement of Trade in Services (GATS)[53] and specifically, the Fourth Protocol on Basic Telecommunications (often referred to as the Basic Telecommunications Agreement, or BTA),[54] seeks as its primary objective the liberalization of domestic telecommunications markets in basic services.[55] While allowing concessions to developing countries and permitting limitations and exemptions to the nondiscriminatory principles of national treatment and most-favored nation (MFN), the Fourth Protocol requires all national government signatories to commit to a time frame for introducing competition into domestic markets. This agreement follows from the 1994 Annex on Telecommunications,[56] which saw liberalization of enhanced and value-added services, but few commitments to the competitive supply of public switched services. The uneven bargaining power among member states was particularly evident in the passage of this agreement in which the United States, a main driver of the process, stalled negotiations, citing as the reason the "lack of critical mass" willing to similarly open their telecommunications markets. Similar pressure is evident in the passage of the GATS Financial Services Agreement.[57] In that case, the United States threatened to take an unlimited MFN exemption and pursue regional and bilateral agreements in financial services, because of what it perceived as an equally acute version of the "free rider problem" alluded to earlier. Countries with closed markets could maintain their systems while having access to other members' concessions and open markets, while simultaneously excluding reciprocal treatment of the latter. While the basic telecommunications negotiations were later jumpstarted with a negotiating guideline in the form of a reference agreement on regulatory principles—the so-called regulatory reference paper—most developing countries maintained limited monopolies in fixed-line services and scheduled full exemptions with regard to accounting rates and satellite services.[58] Table 13.2 summarizes the content of the reference paper.

Subsequent service rounds commenced in 2000 but were stalled initially by the collapse of the WTO Ministerial Conference in Seattle. The Doha Ministerial Conference in 2001 reaffirmed the value of granting concessions for least-developed and developing countries and also confirmed the right to special and differential treatment outlined in the concerns of least-developed countries in the Zanzibar declaration that same year. New market access offers and requests originally were slated for completion by April 2003, with a deadline for finalization of the process of January 1, 2005. However, following the suspension of the negotiations in July 2006, indefinite delays to this timeframe are expected.

SA's GATS Commitments

Having sketched the contextual history of SA's telecommunications reform process, the country's participation in international governance structures, particularly the WTO, should now be seen in light of its domestic policy dictates. Hence, it should be

Table 13.2
Reference paper on regulatory principles

Competitive safeguards	Prevention of anticompetitive practices, such as anticompetitive cross-subsidization, using information obtained from competitors anticompetitively; withholding technical support from service suppliers.
Interconnection	Mandatory upon request at technically feasible points on a nondiscriminatory and timely basis; procedures and interconnection agreements to be publicly available; recourse to dispute settlement.
Universal service	Transparent, nondiscriminatory and competitively neutral and not more burdensome than necessary.
Public availability of licensing criteria	Including terms and conditions; timeframes for decisions and reasons upon request.
Independent regulators	Separate from and not accountable to suppliers of basic services.
Allocation and use of scarce resources	For frequency, numbering, and rights of way. Procedures to be objective, timely, transparent, and nondiscriminatory.

noted that in terms of SA's participation in the GATS accession and related telecommunications treaties, both its 1994 GATS undertakings with respect to enhanced services, and its 1998 undertakings with respect to basic telecommunications, the country never assumed commitments that went beyond those developments envisioned in its national telecommunications reform process. A summary of the country's GATS commitments for both the Annex on Telecommunications and the Fourth Protocol is presented in table 13.3. While it would be useful to examine SA's conditional offerings for the Doha Round of further services liberalization, the failure of this round to conclude suggests that the new legislative amendments previously outlined and the implications those may have for conditional offerings cannot be appropriately explored at this time.

In 1994, in response to the adoption of the Annex on Telecommunications, SA undertook to allow market access, subject to certain limitations, in value-added/enhanced services.[59] The Schedule of Specific Commitments retained the prohibition on bypass of the fixed-line incumbent's facilities for the routing of both domestic and international traffic; and made explicit two important factors—one, that no formal policy on enhanced services was as yet in place, and two, that Telkom remained the de facto regulator under the legislation preceding the 1996 reforms.[60]

Following accession to the 1996 Fourth Protocol, the schedule once again resembled a precis of domestic telecommunications policy, giving grist to the mill for those who criticized the "standstill" value of such commitments[61] and the resulting "regulatory freeze."[62] Perhaps lending support to this view, it is also fair to say that almost until the end of the five-year period contemplated by the limitations in the Schedule of

Table 13.3
SA GATS commitments 1994–1998

1994: GATS and the Annex on Telecommunications	1998: Fourth Protocol and the Regulatory Reference Paper
▪ Service sector to be opened—Value-added/enhanced services (VANS). ▪ No bypass of SA facilities for routing both domestic and international traffic. ▪ No formal policy in place with respect to VANS; Telkom is the de facto regulator. ▪ International VANS services only to be supplied with the consent of Telkom.	▪ Service sectors to be opened—facilities-based and PSTN; mobile cellular and satellite services. ▪ No bypass of SA facilities for routing both domestic and international traffic. ▪ Telkom monopoly on all facilities and basic voice services until December 31, 2003, thereafter duopoly. ▪ 30 percent cap on foreign investment in suppliers. ▪ Resale to be liberalized during 2002–2003. ▪ Cellular service duopoly until 2000. ▪ Feasibility of additional satellite-based service provider to be examined by December 21, 2003. ▪ Full adoption of the regulatory Reference Paper.

Specific Commitments, Telkom's exclusive right to provide telecommunications services remained untouched. While more properly the subject of another argument, this however does not detract from the utility of the agreement and the undertakings in providing a fixed timetable for implementation.[63]

Possibly as a result of this flexibility, or of a generous interpretation of meaningful commitment to the principles espoused by the Fourth Protocol, SA on the whole met the commitments contained therein. As part of the reform process the country established a national regulatory authority in 1996, prior to signing the Fourth Protocol. It has also established a cost-based interconnection regime in compliance with its commitments, although this has not been satisfactorily enforced at the domestic level. The country has also licensed a third mobile entrant;[64] is in the process of establishing a policy for satellite-based systems, and is on the cusp, albeit later than planned, of licensing a fixed-line competitor to Telkom.

The SA government was, however, twice cited in 2001 and 2002 in the United States trade representative's (USTR) annual Section 1377 Review of restrictive trade practices in telecommunications, although the USTR never requested a dispute settlement body (DSB) panel to hear its concerns.[65] These pertained to the incumbent refusing to supply AT&T Global Networks (a U.S.-based company) with facilities for their value-added network services (VANS) operations on the basis that AT&T was allegedly violating the Telkom exclusivity on voice. The essence of the claim made by the USTR against SA was threefold: first, Telkom's refusal to provide facilities was contrary to

SA's WTO obligations to provide market access and national treatment for VANS.[66] It should be noted that the discrimination AT&T alleged was no different from that being experienced by local VANS. Second, in terms of article VIII:2 of GATS, SA was obligated to prevent Telkom from abusing its monopoly position when competing in the supply of a service outside the scope of its monopoly rights. Finally, under the provisions of the annex, SA was required to ensure that U.S. VANS suppliers receive "access to and use of public telecommunications transport networks and services on reasonable and non-discriminatory terms and conditions."[67] Following a number of court hearings, the matter was resolved through closed-door negotiations.[68]

Notwithstanding the citations by the USTR, we assert that SA has largely complied with the overall commitment undertaken in its GATS accession, which may lend support to Chayes and Chayes' contention that "compliance [with an international treaty] is the normal organizational presumption."[69] It may also serve to reflect the value of the critique, which suggests that the Fourth Protocol is merely comprises of policy statements that reflect intended changes at the domestic level regardless of GATS accession.

Nonetheless, clearer evidence of compliance is manifest with regard to most of the six regulatory principles, particularly universal service; allocation and use of scarce resources; and the requirement to ensure that where services are licensed the criteria and license terms and conditions are publicly available. Where concerns have emerged over competitive safeguards, interconnection policy, and independence, these have their roots in difficulties experienced with domestic implementation of competition principles common to many countries, including managing the market effects of a vertically integrated dominant incumbent operator that competes downstream with suppliers that are required to obtain their facilities from the incumbent. This is complicated, at least from an independence point of view, by the fact that the minister of communications—the person responsible for policy in the sector, remains the majority shareholder in the incumbent. Further developments and possible dilution of the independence of the regulator introduced by an amendment to the ICASA Act in 2006 have not, due to the wording of the regulatory reference paper, materially affected compliance with the requirement for an independent regulator.

Insofar as the undertaking to ensure market access by licensing another facilities-based PSTN operator, delays in this regard caused the greatest concern for compliance and did in essence, if not in law, compromise the country meeting its December 2003 deadline. A second network operator was finally licensed in December 2005. Full enterprise operations are expected in 2007, although some whole services were on offer from late 2006. Far greater implications of this potential "violation" are the effect of the perpetual postponements and setbacks in the licensing of a competitor on the future of fixed-line competition; and the adverse effects on users and service providers dependent on wholesale services and facilities from the incumbent, and consumers who continue to experience high prices for fixed-line service and rental.

There are doubts as to whether SA has met the commitment to liberalize resale services, but it might be argued that the vague wording of the Schedule of Specific Commitments, coupled with various domestic efforts to liberalize aspects of resale, through facilities sharing and enhanced services, may vitiate any claim to overall noncompliance. Certainly a set of ministerial directives announced in September 2004, to compensate for the stalled liberalization of the market anticipated by the then convergence bill being returned by the legislature to the Department of Communications, sought at least to placate those concerned about the failure of the country to comply with its commitments, in spirit if not in the letter of the law.

Ultimately however, an assessment of SA's implementation of its GATS commitments with respect to telecoms suggests that the country has followed its own policy trajectory, girded by the timetable negotiated in the undertakings at the multilateral level, and has not adhered to some external, imposed stricture for telecom policy. It has, however, remained mindful of that discipline and the GATS telecoms instruments have been used repeatedly in domestic quarters to suggest that policy reform comply with and meet its requirements. This function is not an insignificant one.

Beyond an implementation checklist, two issues require further comment: first, despite the fact that GATS requires various measures to be enacted "beyond the borders" inside a member state, there are many aspects of telecoms policy that quite properly cannot be dealt with at the multilateral level. This suggests that attention must turn to matters of capacity building and skills development at the national level, which we attend to in the following pages.

The International Telecommunication Union

While participation in the activities of the WTO is limited to government representatives and accession is binding, the ITU has less formal binding relations with its members. Outcomes take the form of resolutions and recommendations that provide guidelines to member states on how to conduct their telecommunications activities, although the imperatives to comply, particularly with the strictly technical rulings around standards, may be compelling, if only to ensure interoperability, interconnectivity, or avoidance of spectrum interference. Yet, even with regard to what may be seen as relatively nonpolitical issues relating to spectrum and equipment, standards adopted in one of the three zones into which the globe is divided for purposes of governance, tend to be set in the North and adopted as default standards by nation-states of the South, where they may not operate optimally under the very different geographical and weather conditions. In addition to the broader historical and structural constraints, to be identified in the following pages, such practical considerations are overridden by a range of factors including the politics and economics of the equipment supply industry as well as corruption and ineptitude.

South Africa's participation in the three arms of the ITU—the radio sector (ITU-R), the technical standards sector (ITU-T), and the development sector (ITU-D), like that of many developing countries, is uneven. The costs of having permanent representatives in Geneva are high, though South Africa appointed a minister plenipotentiary in Paris, responsible for covering the ITU. However, the focus of this participation was on the ITU-D where developing country issues and influence tend to be stronger. But even at these meetings, mediated through study groups focusing on particular issues, which make recommendations to the bureaus' decision-making bodies, participation by developing countries is patchy. The ITU has introduced fellowships to address this, but without the necessary technical or political expertise to work the institutional system, these on their own do not overcome the problem.

In terms of regular participation by delegations from South Africa, like other countries, meetings for the radio sector of the ITU, which deals with issues relating to increasingly valuable spectrum, are the most widely attended. With more resources than many other developing countries, SA is able to field a relatively large delegation for the World Radio Conference. Yet even this is insufficient, as critical decisions are taken in parallel sessions that are held sometimes over twenty hours a day over several weeks. Few African countries or even countries with South Africa's relatively flush resources are able to spread themselves sufficiently to participate in all decisions. This is, however, one of the few areas where the Southern African Development Community (SADC) countries have successfully sought to coordinate their strategies at the regional level in an attempt to be more effective.[70]

While SA shares the limitations of effective participation that hinder many developing countries, it has, as in other international institutions, often exerted greater influence out of proportion to its size, frequently through providing leadership on specific issues. Until recently, South Africa's plenipotentiary minister in Paris, responsible for the ITU, Lyndall Shope-Mafole, served as the first woman head of the ITU Council, its highest executive body. (She is now director general of the Department of Communications). In that capacity she was able to highlight developing country issues on the ITU agenda. But such gains are limited in that they appear tied to personalities rather than ensured through institutional guarantees. Shope-Mafole's replacement, a far more junior official, has had none of the same impact.

The ITU itself has had to adjust to broader reforms in international telecommunications, the most significant being in response to the global shift from monopoly service provision and the entry for the first time, outside of the U.S. market, of nonstate network operators in the 1980s and 1990s. This compelled the ITU to turn its focus from primarily technical issues relating to state operators, to supporting the liberalization process in member states and in particular, the establishment of independent regulators.

Traditionally, as a United Nations body, membership was built around sovereign nation-states, but during this reform period, private sector membership was permitted albeit without equal voting rights. The unintended consequence of this is that it has tended to strengthen the participation by developed economies, such as the United States, which have by and large more active private participation in the industry and in some cases the policy-making process. With telecommunications firmly on the WTO agenda, the ITU was also prompted to reexamine its governance role and two years ago, the ITU Council established a special task group, headed by Shope-Mafole, to explore the transformation of the ITU to meet its new challenges including the participation of developing countries in decision making.

Preferences, Power, and Policy Making

Outcomes of participation by developing countries in multilateral agency processes have tended to be caricatured by those opposed to participation as the product of cooption and the abdication of national sovereignty and interests to international markets and their agents. Although it is fair to say that concerns regarding the diminution of sovereignty are not limited to developing countries, the precise focus of concerns from developed and developing countries tend to reflect quite different anxieties. For example, one manifest concern is that participation in global governance regimes like the WTO might limit a country's overall influence within or control over an area of trade or restrict the ability of domestic groups to influence national policy-making processes.[71] Other concerns suggest that less dominant countries might not have their voices and preferences heard sufficiently, or at all, and are merely part of the process to lend legitimacy to the appearance of inclusiveness and transparency, while their domestic agendas are not benefited, or worse, are negatively impacted on by that participation.

Yet those in favor of participation tend to perceive the pressure brought by the "beyond-the-border" or "deep integration measures"[72] required by international agencies as much-needed relief to besieged operators seeking to meet the pent-up demand for services in developing markets, which can be frustrated by governments protecting either public or private monopolies.

While such concerns are legitimate, these polarized positions fail to recognize the complexity and nuances associated with these developments and therefore fail to leverage this understanding strategically in global, regional, or national forums. Developing countries have largely been absent or invisible in the governance of global networks. Although many African countries are members of the international agencies responsible for global governance, their impact, both on agenda setting and decision making, has been negligible and their participation through membership might serve to bind them to decisions and outcomes contrary to their interests.

The reasons why countries may find themselves in this situation are complex and result from a different set of circumstances in each case. Indeed, aggregating all developing countries, including emerging economies and least developed countries and the range of interests vested in them, is in itself problematic. However, some uniting factors do explain why more developed countries tend to be insiders and developing countries tend by and large to be outsiders.

While there may be broad similarities in preferences in increased trade and investment within the WTO, member states may prioritize anticipated outcomes from multilateral trade agreements differently. For example, increased access to foreign markets may not be a priority for developing countries in telecommunications, though its correlate, foreign direct investment in domestic markets, is likely to be high. While higher trade-related revenue might be more generally applicable, full exploitation of comparative advantage in service industries is unlikely to be an issue for countries that demonstrate little comparative advantage in the ICT sector. Even among those states in agreement over the merits of increased cross-border trade in services, member states remain highly polarized on the speed, nature, and modality of opening domestic services markets, and the competing tension produced by the pursuit of another domestic policy objective, such as the goal of universal service to address racially skewed rollout of telephony, or the promotion of empowerment equity in licenses to address apartheid inequities.[73] As such, the question needs to be located within a broader contextual analysis of other factors that create barriers to effective participation. Three interrelated explanations to this query involve historical, resource, and domestic structural constraints, as follows.

Historical Constraints

The reasons for how and why developing countries become bound by international agreements without necessarily benefiting from doing so are far more complex than can be fully explored here. This section suggests, however, three main pillars to develop a framework for understanding this result, which is ultimately encompassed in the historically unequal power relations that exist between developing countries and developed nations. As regime shifts have occurred, particularly over the last decade, this situation has been exacerbated.

While a range of noninstitutional (economic, cultural, and geographical) reasons and a number of institutional theories offer explanations for this,[74] we argue that these uneven power relations tell a broader story detailing a history of systemic imbalances in development between countries. These flow from multiple historical, political, and economic factors, from colonialism to more recent manifestations through, for example, the structural adjustment policies and conditional aid packages of the International Monetary Fund and the World Bank.

South Africa's own evolution of macroeconomic policy is illustrative: initial sentiment toward economic transformation was originally expressed in the ANC's Reconstruction and Development Programme (RDP), a socio-economic framework that sought to place the priority of meeting citizens' basic needs and the final eradication of apartheid as its priorities. Its vision was based on six principles constituting the political and economic axioms that girded its philosophy: sustainability, human capital as a primary resource, peace and security, democratization, nation building, and the critical linking of reconstruction and development as an integrated process.[75] Problematic with respect to economic orthodoxy, the redistributive focus of RDP was replaced in 1996 by the Growth Employment and Redistribution policy (GEAR), which promotes clear neoliberal disciplines far more in tune with the World Bank's reform toolkits for development, including fiscal austerity through an emphasis on deficit reduction via expenditure restraint, coupled with tight monetary policy and rapid liberalization. Critics, however, identify GEAR as a significant contributor to escalating unemployment and a negative long-term economic outlook.[76] How precisely this occurred is contested terrain. One critic summarizes the shift to a neoliberal orientation as a result of "flavored" training and capacity development programs effected through aid seminars, workshops, and fact-finding trips financed by business and foreign development agencies.[77] Others suggest that the ANC leadership abandoned the RDP because SA's economic woes, precarious international position, and the concerns frequently articulated by domestic investors required a more orthodox approach.[78] The matter is more complicated than space here allows for; however, this shift undoubtedly animates at least to some extent the domestic effects of compliance with conditional aid, loan agreements, or creating investment conditions that do not take into account the complex domestic social and political framework in which such policies are effected. This can give rise to what Dolowitz and Marsh term "coercive" rather than "voluntary" policy transfers, in which the role of international consultants, usually insisted upon by donor countries, most clearly manifests the problem.[79] Postapartheid South Africa has demonstrated some ability to balance the conflicting domestic and international pressures, often buckled under to by other countries due to tied aid or international loans. By the end of the first decade of democracy and half decade of GEAR, the country was showing steady economic growth, though this continued to be jobless and the evidence of its equity compelled the ANC, returned to power in 2003, to devise a more equitable economic policy, the Accelerated and Shared Growth Strategy for SA, within the constraints of the international orthodoxy.

In the case of telecommunications, though, the balancing act has not been as successful. The international reform model devised by multilateral agencies and sold to developing countries in the 1990s consisted of three elements—all reflecting a move consistent with the establishment of a "regulatory state"—the privatization of the state-owned public switched network; the liberalization of other market segments whether mobile, VANS, or ISPs; and the establishment of an independent regulator, with some acknowledgement of universal access strategies in less mature markets. In practice, however, in many developing countries as in South Africa, privatization was focused on at the expense of the other two components, with negative impact on the growth of the sector and the economy as a whole, as the benefits of competition were not realized. Privatization, generally with a period of exclusivity to attract investment, offered protection to incumbent revenues, provided the state coffers with a much-needed injection, and allowed investors highly privileged access to the market on erroneous conventional wisdom that this would contribute to affordable universal access. An increasing body of literature has identified the significance of having a well-established regulatory framework prior to privatization to meet affordable-access objectives and that the introduction of competition is likely to deliver on this objective, as competitive mobile markets in particular in developing countries have shown.[80] The error of this dominant reform paradigm has even been acknowledged by former bank stalwarts. Reflecting on this issue, John Stiglitz points out: "From today's vantage point, the advocates of privatization may have overestimated the benefits of privatization and underestimated the costs, particularly the political costs of the process itself and the impediments it has posed to further reform. Taking that same gamble today, with the benefits of so many more years of experience, would be much less justified."[81]

Resource Constraints

At the heart of any explanation of suboptimal outcomes for developing nations lies the general lack of capacity, including both human and financial resources, to ensure meaningful participation within global governance structures. There are at least two ways to consider this: first, that the emergence of a global economy and global network industries have created a need for the performance of far more complex public or governance functions at the global level than existed under the bilateral sovereign-state international relations that preceded it. This shift to a global, "supranational" style of regulation in turn necessitates new institutional arrangements that may—as in all cases of institutional design—give rise to a number of problems. The structure of the institution, its decision-making procedures, dispute resolution mechanisms, and the potential for lack of cooperation by all participating members, will at varying times present different obstacles to effective participation.

Historical reasons suggest a dominance of industrialized and specifically G8 countries in multilateral agencies, whose governments and corporations already exercise a disproportionate influence over global economic affairs. In such circumstances, we can assume disparities in power and influence and the absence of a level playing field for all members to such governance arrangements. Here, capacity problems are more acute: even if the historical, economic, and political factors were variable, such that the playing field was substantively more level, the capacity problems faced by developing countries are not eradicated by addressing institutional design deficiencies alone. Various resource impediments, as articulated in the *Louder Voices* study detailed in chapter 12 of this volume, will remain in place. For example, countries that more recently joined existing global governance organizations will not necessarily have the institutional experience readily available that they require to be effective. They are also unlikely to have the technical knowledge, and if they do it tends to reside within the incumbents whose sole interests are then represented at international negotiations that impact critically on domestic market structure and sector regulation. While the United States has large delegations of negotiators to attend ITU or WTO meetings, including a number seconded from industry, many developing countries will attend with one or two government officials, if they are represented at all. Details of deliberations and proceedings particularly at the WTO remain unclear, prohibiting effective participation in discussions or negotiations. While official WTO policy shuns suggestions of infamous "green room" negotiations, such practices are still reported from country representatives.[82] Moreover, in many cases, timely and continual access to information is also scant.

Second, and more overtly, developing countries lack financial resources to meet the costs of attendance at global forums, including travel and accommodation costs in a foreign (and usually significantly stronger) currency; to acquire and process information; and to obtain the necessary skills and capacity to contribute and ensure an impact on agendas and decision making. Also critical is the need to address the costs of an "empty desk" within already underresourced governmental departments at home, which is often the case when human resources and capital in specialized areas are scarce.

For example, it has been noted that the WTO is a "member driven organization," and since it has no permanent executive board, "delegates from member countries are actively involved in its day-to-day activities. If they are not, their interests are ignored."[83] To participate in "the continuous dialogue" that makes up the WTO processes requires at least forty-five meetings per week in Geneva, which patently places significant pressure on resource-strapped smaller and developing countries.[84] South Africa's Department of Trade and Industry (DTI) has a permanent representative with a staff of four in Geneva to prepare for negotiation rounds and monitor operational developments, while many of the smaller developing countries have no representation at all. This lack of presence is an important inhibitor of meaningful participation as lobbying and the positions of developing countries are often fashioned through country ambassadors in Geneva rather than through regional meetings or organizations.

Cooperating and caucusing at the regional level or among other developing countries has been identified as a major tool for effective participation in international negotiations, as evidenced at the 2003 Cancun ministerial meeting where developing countries were able to confront the hypocrisy of the United States, the European Union, and Japan on agricultural subsidies. Further evidence of this ability has been witnessed in the breakdown and suspension of the negotiations in 2006. However, this level of organization is the exception rather than the rule, as the human and financial capital required for ongoing and effective regional cooperation is frequently absent and developing countries have rarely utilized this avenue of coordination.

Taken together, the facts and consequences of lack of capacity are central to the broader governance debates, both germane to addressing the lack of meaningful participation by developing countries, and, to a large extent, curative, notwithstanding financial limitations.

Domestic Structural Constraints

The third pillar of constraint to effective participation by developing countries originates from context-sensitive domestic pressures and factors endogenous to a number of developing countries. Where present, these may frustrate the effective garnering and application of those resources where they do exist for developing countries to more effectively participate in global rule making and governance. Not least of which is the existence of effective *national* governance structures and processes that foster meaningful participation at the global level. This factor has a number of elements to it, and to some degree, animates Lodge and Stirton's contention that problems of institutional capacity domestically are less about the forces of globalization than the manner in which domestic institutions are embedded within a network of social relations within a country—that is, how institutions are embedded within sectoral structures and how the linkages between them operate to increase or reduce institutional capacity.[85]

At a practical level, this genre of constraint manipulates the selection procedures and structures for determining who should attend meetings, where nepotism or corruption is evident. This might result in the wrong people attending meetings and failing to represent a country's interests satisfactorily because of a lack of qualification or experience.[86]

Another aspect is the failure of national governments to utilize the skill and talent that exist within the ICT policy domain but may reside within industry, academia, or nongovernmental organizations (NGOs). We have already noted this point, regarding respondents to the *Louder Voices* study in SA who cited the failure of government to draw on the wider pool of resources in the sector. Indeed, SA studies suggest that substantial capacity exists domestically in the private sector, academia, and NGOs, yet is not harnessed by government in the preparation for and participation in international meetings. The predominantly state-owned incumbent on the other hand is drawn on extensively for international meetings, for which it is able to mobilize considerable resources to influence the agenda and outcomes in its more narrow interests.

In this case, government is less likely to draw on industry experience and skill, as the U.S. trade representative (USTR) does within the United States, because the commitments made and the positions entrenched at the WTO in respect to telecommunications operate as a form of regulatory "subsidy" and were designed to ensure that the incumbent received the maximum protection possible within the bounds of the agreement.[87] Likewise, the information downstream is perceived to have blockages and many respondents said that they received more information on international meetings from fraternal associations in Europe or other parts of the world than from their own government departments or agencies. The lack of opportunity for more inclusive private sector participation in the structures and processes of the global governance institutions themselves further complicates this. By and large delegations to ITU meetings from developing countries do not include private sector representation or NGOs, and few other than the very biggest private operators are able to pay the fees to participate in their own right. With the rise of the social movements in response to the negative impact of globalization, there has also been increasing pressure on the ITU to include NGO membership, but to date this has not occurred unless NGOs are willing to pay the very high "small m" private membership fee.[88] With private sector participation on the ascendancy and the exclusion of social agencies and weak government participation, concern about the lack of public interest representation in the system of governance is not unwarranted.

There appear also to be other developed and developing countries' orthodoxies that impact on international behavior and practice. As Helleiner and Oyejide have found, developed or industrialized countries historically place far more responsibility and trust for economic "governance" activities in international regimes working voluntarily in their collective interest than is typically the case with developing countries. Similarly, developed countries appear to be more responsive generally to NGO and private sector representations than are the governments of developing countries. They note that "in the emerging discussions of global governance arrangements, [developed country] NGOs are likely to make their voices heard whereas the interests of the people of the developing countries will be represented only by their governments (or, in some instances, by Northern NGOs purporting to speak for the disempowered in the developing countries)."[89] This was evident, for example, in the DOT Force meetings where NGO and donor agencies were well represented in the delegations of a number of the developed countries, and particularly influential among some, such as France, while the delegations of the developing countries, India, Egypt, and South Africa appear to have had no NGO representation at all.

A further problem in terms of domestic structural constraints is that policy formulation at the local level may often be misallocated within government and that departmental capacity at the international level is uneven between and within departments. Various SA respondents highlighted this fact as a concern with respect to e-commerce policy and resulting legislation. Under current arrangements, this policy arena falls to the SA Department of Communications. Many have voiced concern that the trade and commercial issues contained in such legislation would be better served by responsibility for this policy and its implementation being carried out by the Department of Trade and Industry (DTI). Others, however, did note that the record of DTI in the area of ICTs had been poor and the interventions so ineffectual that at least under the Department of Communications policy and legislation were more likely to eventuate. Similarly, until fairly recently, responsibility for SA's communication commitments under GATS fell to one or two individuals within DTI and the Department of Communications. With the fundamental change in government in 1994 and the transfer and retrenchment of various personnel as part of initiatives at transforming the civil service in the 1990s, much of the institutional learning acquired by those individuals has since left these departments.

While optimum allocation of policy responsibility where potential for overlap exists may be a matter of conjecture, the lack of policy coordination in formulation and especially at the level of implementation does suggest that more cooperation at the national level needs to occur to ensure a unified and cohesive policy is put forward at the global level. By necessity, this should involve more consultation with and utilization of a broader skills base in the ICT sector.

Winners and Losers

We have asserted that a few developed nations dominate the agendas and outcomes of global governance organizations in the sphere of global electronic networks. However, the precise manner in which this occurs will vary between governance institutions and also according to the nature of the issue at hand. In general, however, the interests of these countries tend to be reflected in the agendas and outcomes of these organizations, even those with one-nation, one-vote systems. Like many highly bureaucratic organizations, formal processes within these institutions seldom determine outcomes and results. Rather, many of these countries have through various international treaties, conditional aid, or loan agreements imposed arrangements on developing countries based on 'Washington Consensus' ideals without consideration for local enterprise, development gains, or the complex domestic social and political framework within which ICT policy must be effected.[90]

This is not to suggest that prior to these global interventions many African countries had operated flourishing public communication networks. In most developing countries telecommunications has represented a scarce resource, bestowed not as a right but as a privilege on the economically or politically favored.[91] After seventy-five years of public provisioning of telecommunications, the teledensity of the African continent for fixed lines, the main form of public provision over this period, has put a phone in the hands of fewer than two out of every hundred people. Further, the rationale of foreign coercion should not suggest that developing countries are always unwitting victims of these processes. There have been several instances where developing countries have used their voting powers to secure sometimes questionable outcomes. Within the ITU, African countries are regarded as being responsible for constraining recommendations that would have supported the earlier introduction of innovative, if costly, Global Mobile Personal Communications via Satellite (GMPCS) or recently the more affordable Voice over Internet Protocol (VoIP) technologies, in order to protect their often highly inefficient incumbents. Syria and Togo were responsible for holding up the approval process in ITU Study Group 3 that sought to curb pricing abuses by mobile operators, with regard to access and call origination on public mobile networks; voice call termination on individual mobile networks; and the wholesale market for international mobile roaming on public mobile networks.[92] Nonetheless, as mentioned in the introductory comments, the pursuit of unrestrained market-orientated policies at the global level has often been hypocritical and self-serving for a number of developed countries. While multilateral agreements like the Fourth Protocol espouse pro-competitive and fair principles in relation to domestic markets, the same principles often are not equally applied at the international level. The drive to expand local presence in foreign, and often highly concentrated telecommunications markets is an apt example: where large multinational telecommunications companies are dominant, the tolerance for monopolistic conditions to exist is higher, yet where trying to enter new markets dominated by other monopolistic players, the same tolerance is substantially reduced, prompting countries trying to secure entry or facilities in foreign markets, to resort to international forums where their interests are securely represented. Hence, the USTR uses its section 1377 annual review to galvanize industry concerns into threatened formal complaints at the WTO, where its influence is more marked than within a foreign country's domestic courts. In recent years, among those highlighted in the report were SA and Mexico, the latter escalating to a dispute to be determined by the WTO's dispute settlement body (DSB).[93] This suggests that if there is to be an organization with dispute settlement powers in the area of ICTs, there must be fair competition at all levels, possibly also prompting a review of the accessibility and utility of the DSB for developing country use. A basic example of the problem suggests that with financial resource constraints, the costs of hiring private representation are prohibitive, and while some developed countries offset these costs through the affected industry, in most developing countries many of the affected industries are either financially weak or in the small- and medium-scale sector, negating that option with which to recoup these costs.[94] India has recognized the problem of legal representation in the context of the DSB and proposed various remedial measures for reforming DSB practice in favor of developing countries. These would include increased time periods for developing countries to make submissions to the panels and action by the WTO to establish a de minimus level of alleged trade impairment for a developed country before it can initiate any dispute proceeding against a developing country.[95]

Again, this need to make the playing field more level is particularly important because although reciprocity theoretically exists in multilateral agreements, the *actual achievement* of reciprocal market access by developing countries to developed markets is sharply limited by their uneven levels of development.

Yet it is not always a case of winner takes all. While the current status quo in the global governance of electronic networks might not ordinarily favor participation or policy outcomes for developing countries, there are mechanisms within the relevant organizations and the treaties that give effect to their mandates, that can be utilized to increase developing countries' participation and influence. For example, within the GATS itself, the historical, selective application of the national treatment principle and within the Fourth Protocol, the ability to selectively list limitations on market access, has allowed a country like South Africa to list exemptions and limit market entry in order to protect its incumbent while it was tasked with an important public service expansion program. However, the success in optimizing these avenues of benefit is entirely dependent on resolving where possible, and balancing where not, the demands of the three-pronged constraint model previously outlined.

If so little stands to be gained for developing countries by accession in the case of the WTO, one is tempted to question why SA and other developing countries have elected to sign a largely voluntary agreement in the case of the Fourth Protocol. The world was not faced with the same paucity of international capital for telecommunications investment in the mid-1990s that it did in the post-dot-com recession. The short answer suggests that the telecoms policy process was underway and the Department of Trade and Industry, wanting other GATS concessions, succeeded in getting the Department of Communications on board. They had little to lose, as the parameters of concession were easily circumscribed by domestic policy intentions. More theoretically, we argue that accession to the GATS can be understood as a transaction cost-reducing exercise and as a form of credible commitment[96] to foreign (and local) investors that ex post discretionary and opportunistic behavior will be constrained ex ante, and that the rules for liberalization will be phased in according to a specified timetable.[97]

The case for participation is clearly more obvious with regard to the ITU, but the process and history of the GATS telecommunications instruments are telling in themselves of the tradeoffs and compromises sought by members in their negotiation. We suggest that SA stands to gain from accession to the GATS Fourth Protocol because the leeway and flexibility the agreement offered in the initial phase required that SA commit to no more than the contours of its domestic sectoral policy on telecommunications. While this perhaps lends evidence to the perception by many SA respondents in the *Louder Voices* study that the government offered mere lip service to the ITU and WTO rather than buy into the spirit of the agreements, the fact remains that participation in these forums offers developing countries a means with which to establish trust in foreign investors and trading partners that the countries' commitments to a broader liberalization program are indeed credible. This is crucial, because evidence indicates that where strong investor uncertainty exists, investors either withdraw or demand increased returns to compensate for higher risk. Thus, the commitment to effect beyond-the-border reforms—like the establishment of an independent regulator; a transparent and fair interconnection policy; licensing requirement; and so on—and the more uniform its rules of operation, the more credible a government's commitments become, lowering potential investor uncertainty.[98] Failure to abide by these commitments will have longer-term reputational implications that no developing country, hungry for foreign investment, can afford. This interpretation might also serve to remove the sting from the allegation by some respondents that the SA government merely engages in posturing in international forums to indicate support for international consensual positions, but does not act in the spirit of these treaties at home.

In practice, however, the untested enforcement procedures and the cost associated with exercising rights in terms of international instruments such as the regulatory Reference Paper, attaching to the GATS Fourth Protocol,[99] has meant that domestic competitors of the monopoly incumbent, who have claimed that their competitive rights in the VANS market in particular have been undermined by policy, regulatory, or operational practices and that proposed policies or licenses would contravene South Africa's basic commitments, are unable to find relief from South Africa's GATS undertakings. They attribute this inability to the broadness and generality of the exemptions to the principles the Reference Paper espouses.[100]

Recommendations

We have outlined some of the considerations regarding the participation of developing countries in the global governance of electronic networks. At best, these agreements and their articulated intention at inclusiveness provide a framework for potential and equitable participation by all stakeholders and members. At worst, the current institutional designs in which these agreements operate preclude any meaningful substantive participation by developing countries in the governance mechanisms affecting ICTs. Resolving the three-pronged constraint model we have highlighted is a complex and an ambitious task. However, it remains crucial that these constraints are addressed and that the ambiguities of participation are resolved in favor of developing countries. As others have so convincingly argued, "Some of the key decisions in the areas of ICT policies that do have an impact on developing countries, including African countries are often taken in distant capitals and in global institutions. [A]lthough not a new phenomenon, the effect of globalization and the increasing complexity of the subject matter and the decision making process of some of these institutions and policy fora is amplifying the significance and the implications of this fact."[101]

Yet as we stated earlier, inclusiveness alone will not produce what Dzidonu and Quaynor identify as "proportional representational universal participation," namely, that a substantial number of the participating groups that are willing and able to participate are likely to be affected by or benefit from the outcome, and comprise a fair proportion of the total number present.[102] We suggest that true participation goes even further, and address the institutional imbalances that inhere in these organizations as a result of historical factors. The above sentiment is amplified when the lack of participation by developing countries is properly located within the context of an ever-widening digital divide. The development of ICTs and the role this might play in broader social and economic transformation places the problems of participation by developing countries in global governance firmly at the center of resolving how best to approach issues of access and affordability, and other issues associated with the digital divide. While we have aimed here to sketch the contours of the problem and suggest some implications if left unresolved, we make the following basic recommendations that require attention to effect participation by developing countries in global electronic governance, if their impact on agenda setting and decision making is ever to be increased.

"Best Practice" Regulation Should Be Applied to the Global Market

With increasing but highly uneven global communication development, effective global regulation seems more urgently required than ever before. As the global market is highly imperfect, concentrated regulation is required to permit fair competition to allow for the efficient allocation of resources, particularly human resources, at the global level. Likewise, regulation is required to regulate global market failure and to ensure access to global electronic networks.

Level the Playing Field

The uneven power dynamics and their inconsistencies need to be addressed. As most developing countries have no investment interests in developed countries, the imbalance of large, dominant players using international forums, where their interests are more securely represented, to tackle domestic liberalization policies in developing countries that are perceived to frustrate their market expansion goals, needs to be remedied.

Institutions of global governance need to be able to counter forum shopping by large multinationals and "legal play" used to challenge legitimate national policies that are perceived to frustrate market expansion by developed countries. That some of these national policies are not meeting these national objectives, especially the primary one of affordable universal access, is a separate matter that needs to be addressed by domestic policy adjustments and not exploited by vested foreign interests.

Foster Accountability and Transparency

This goal needs to be achieved both at the multilateral and domestic-member country level. While exclusion is presumed to occur in global governance arrangements, the case of South Africa highlights the exclusion of national agencies seeking to participate in national forums, which shape domestic policy and determine international positions to be taken. If South Africa and other developing countries seek to participate equally and effectively in global dialogue and if one assumes that the costs of not doing so will be high, then they too must develop and maintain transparent, accountable, and participatory systems of governance and operations. These will not gestate outside a culture of political and economic rights. Africa will have to draw on its resources to meet more innovatively and effectively the needs of citizens, consumers, and entrepreneurs as espoused in the New Partnership for Africa's Development (NEPAD).

Capacity Development

While structural reform to the institutions of global electronic network governance needs to be pursued, developing the domestic capacity to participate effectively in international forums requires responses on a number of fronts. Current capacity building initiatives infrequently assess fully the roots of constraints on performance, tending to focus mainly on how that incapacity is expressed—inept officials and organizations that do not function effectively.[103] At a minimum, it requires that the institutions responsible for governance prioritize capacity development in trade and ICT policy matters and negotiation skills. (Examples include the commitments from the Doha Declaration toward Technical Assistance for Capacity Building).[104] However, it is absolutely crucial that the curricula and training materials used to enhance capacity and the advisors used in offering technical assistance are seen as neutral and operating without an ideological agenda. While this is not always the case, it is difficult to accept that all technical assistance provided on missions to developing countries from organizations such as the WTO, for example, remain without institutional bias.[105] It also requires SA and other developing countries to nurture the ability to "play by the rules of the game" in order to ensure that their concerns are placed and remain firmly on the international agenda, and further, to exploit to the fullest possible extent the system of safeguards and concessions already present in existing structures. Primarily, this requires a satisfactory solution to the question of funding and financing the development of intellectual and capacity resources at both the international and domestic levels.

Extend the Domestic Inner Circle

Linked to the question of capacity development and accountability is the recognition and utilization of local capacity in other arenas outside of government: for example, by harnessing the skills offered by local NGOs, academics, and industry resources, and using these effectively by avoiding "top-down," often uninformed approaches to formulating domestic positions and policy outcomes. Edward Jaycox, vice president of the World Bank in 1993, once noted that "the use of expatriate resident technical assistance was a systemic destructive force which is undermining the development of capacity in Africa." At the same time however, he also accurately pointed out the complicity of African governments in "pushing away" local professional talent.[106] This trend still exists and needs to be addressed. This will also assist in addressing the lack of readily available human capital and experience in many developing country governments that continually lose local skill to the private sector or foreign markets. In many cases, this may require accessing funding mechanisms to ensure the participation of NGOs and academia that would not otherwise be able to assist government in policy development. This mechanism exists in a number of domestic regulatory processes, and should be expanded to the national level. Similarly, global governance organizations might address the democratic deficit by devising such mechanisms to enhance public participation in their own modes of operation.

Negotiate and Strategize en Bloc

With the establishment of a regional trading zone in the Southern African Development Community (SADC) in 2008, the potential for regional approaches on matters of common concern increases, even where no direct trade regional integration benefits or trading opportunities are immediately evident. Regional integration projects were originally initiated to safeguard against the possible failure of the GATT Uruguay Round, but have established themselves as an independent form of economic cooperation.[107]

The EU model offers considerable potential and scope for developing similar "communities of interest" to shift the balance of power within international organizations through regional coordination, resource pooling, caucusing, and lobbying. Similarly, an increase in regional trade through bilaterals, common minimum standards, and regional concessions will bolster en bloc approaches to governance. Even where regional trade opportunities are absent, developing countries should seek to develop communities of interest to generate clear positions and foster cooperation on matters of mutual interest.

The SADC, in which South Africa is a significant player, is regarded as one of the most promising regional blocs on the continent. While some gains have been made in leveraging the collective power of the fourteen countries with the region and even to collaborate with other regional blocs, to date this has not been very effective in the telecommunications sector. Despite model policy and legislation to harmonize telecom reform efforts in the region, reform has been very uneven with little evidence of political will to effect some of the fundamental reforms that would be likely to stimulate investment in the sector. Specifically, although eleven of the fourteen countries have established national regulatory authorities to oversee the liberalization of their markets, few countries have been willing to introduce competition on any significant scale, other than in mobile, which is often dominated by the fixed-line incumbent. Political will has also been absent in ensuring the independence of most regulators, undermining their ability to act effectively.

While regionalism has been a significant tool globally to overcome marginalization of individual nation-states in the global economy, the benefits of regionalism are only apparent, if based on sound, participatory, public-interest policy. If the regional trendsetter or the harmonization policy for the region is flawed, then the effects of regionalism may be detrimental rather than beneficial. The SADC region with regard to telecommunications requires a serious review of its performance indicators to assess this.

However, the regional progress made in Southern Africa and to some degree in Eastern Africa had prompted the development of a regional association in West Africa. Together, within the context of NEPAD and the African Union, even wider continental responses to global marginalization and continental underdevelopment could be mobilized around cross-cutting issues. A case in point is the issue of the crippling cost of international bandwidth as a result of the monopoly of the club consortium on the SAT 3 undersea cable. With policy frameworks at the national level that either continue to protect incumbents or fail to limit their dominance, together with limited regulatory capacity on the continent, the cable operators have been able to restrict access rights to their national landing stations and set prices with impunity. A collective response in terms of ordinary "essential facility" or "bottleneck" regulation with the necessary technical assistance that is currently underway has the potential to make a dramatic impact not only on the regional or continental ICT sector but also on their entire economies.[108]

Likewise, a testament to the power of collective action is the relatively participatory and transparent manner in which the Eastern African Submarine Cable System (EASSy) consortium has been established, with the inclusion of landlocked countries, adoption of an open-access policy to the landing station, and prices anticipated to be a third lower than even the cheapest segments of the SAT 3 cable.

Conclusion

In this chapter, we have outlined some of the main reasons why developing countries do not participate meaningfully in the international governance of global electronic networks. This is a complex problem that can be addressed in multiple ways. We have chosen to suggest a three-pronged constraint model that internalizes the clearest manifestations of the problems of participation. These in turn call for deconstruction and further analysis.

We have also attempted to reflect the granularities inherent in the assertion that developing countries cannot effectively participate in the global governance structures for ICTs, by commenting on South Africa's implementation of its GATS telecommunications commitments, and how we view the flexibility in that treaty as a tool with which developing countries can leverage positive gains for their own domestic policy goals. We have, however, noted that this ability is entirely predicated on addressing as far as possible the constraint model we propose. We then advance a number of recommendations with considerable emphasis on developing countries getting their "domestic policy house in order" and by extension, utilizing regional leverage that en bloc strategizing and coordination can offer countries with less influence in global governance structures.

Finally, we submit that transformation of these institutions of governance to bring them in line with the democratic and good-governance principles they espouse, together with the identification of areas requiring regulation at the international level to create a fairer competitive global market with the associated efficiency gains at the international level, could provide a new framework for international governance that is more effective, participatory, and representative, and genuinely capable of contributing to sustainable global development within the ICT sector.

Developing countries will, however, need to demonstrate more convincingly their commitment to the principles of democratic good governance, including participatory public consultative processes and the development of the skills and resources needed to govern effectively, if they are to have greater credibility in the forums of global governance. It is not only a matter of tactically gaining the moral high ground. The information communications technology sector is so complex and dynamic, and as the backbone to the modern global economy is so strategic, that no country can afford any longer not to deploy all its available and top human resources in all sectors of society to best inform its national and regional policy and to represent its interests effectively in global governance forums.

Notes

1. See Melody 2002.

2. World Trade Organization, *General Agreement on Trade in Services*, contained as Annex 1B to the Agreement Establishing the World Trade Organization, April 15, 1994, Final Act Embodying the Results of the Uruguay Round of Multilateral Trade Negotiations, *Legal Instruments—Results of the Uruguay Round*, 33 I.L.M. 1143, 1168 (GATS).

3. Mukerji 2000, 61.

4. Specific concerns pertain to the triad of documents constituting what critics see as "SA's neoliberal blueprint for Africa," namely, the Millennium Africa Recovery Plan (from mid-2000 to mid-2001), a New African Initiative (from July to October 2001), and its current incarnation, a New Partnership for Africa's Development (NEPAD), launched in Nigeria, October 23, 2001. See Bond 2001.

5. The nine regions, comprising 1,127 square kilometers after the four preexisting provinces that existed during apartheid were redefined by the 1996 Constitution of the Republic of South Africa.

6. The most recent estimates on the size of the South African population are from Stats SA in 2005. Available at http://www.statssa.gov.za.

7. International Monetary Fund, World Economic Outlook Database, April 2006. Available at http://www.imf.org/external/pubs/ft/weo/2006/01/data/index.htm. The comparative figure for GDP per capita excluding PPP is $5,099 (at current prices).

8. Measured at constant 1995 prices. International Monetary Fund, World Economic Outlook Database, see note 7.

9. SA Reserve Bank Time Series Analysis Sheet 1: GDP at market prices 1999–2000.

10. Although separation of races and exploitation of indigenous people existed in South Africa as it did in other colonized countries prior to 1948, it achieved its highest form of expression in the system of apartheid, a set of white supremacist policies developed by the National Party government over more than forty years and covering every aspect of political, social, and economic activity.

11. Foreign Direct Investment in SA in 2001 was R4 billion or 3.2 percent of GDP, World Markets Research Centre, October 2002. Strict exchange control laws still exist for residents, although foreigners can freely transfer funds into SA for investment purposes. SA uses the Harmonized System of Classification and certain import goods require permits. See Mbendi Information for Africa. Available at http://www.mbendi.co.za/land/af/sa/p0005.htm.

12. SA's "Growth Employment and Redistribution" policy (GEAR) was adopted in 1996. Critics, however, identify GEAR as a significant contributor to growth-rate reversal and declining economic outlook. See for example, Weeks, 1999, at 810. GEAR promotes fiscal austerity through an emphasis on deficit reduction via expenditure restraint, coupled with tight monetary policy and rapid liberalization.

13. Cassim et al. 2000, 111.

14. In 2003–2004, the government aimed to generate R6.3 billion from privatization. The 2003 Budget Review indicates that since 1997, the government has earned R27.6 billion through its program of state asset restructuring.

15. Government of South Africa 2001–2002.

16. Unemployment was 29.5 percent as of October 2001, according to SA Census, 2001. This is a conservative estimate. Available at http://www.statssa.co.za.

17. SA has the largest HIV-positive population in the world at conservative estimates of almost five million infected people, most of whom are between the ages of twenty and twenty-nine. See National HIV and Syphilis Sero-Prevalence Survey of Women Attending Public Antenatal Clinics in South Africa, 2001. Department of Health, South Africa, 2002. Civil society studies put the figure closer to 6.5 million. See Dorrington, Bradshaw, and Budlender 2002, 2.

18. The inflation rate for the period of May 2002 to May 2003 was 7.8 percent. Statistics South Africa, Consumer Price Index, P0141.1, May 2003, available at http://www.statssa.gov.za/Products/Releases/CPI/CPICurrent.pdf.

19. While different statistical agencies use different measures, approximately 50 percent of the population lives in poverty, with average household subsistence levels of less than $200 (U.S. dollars) per month.

20. The Human Development Report is published annually to assess global human development and provide a critical analysis of a specific theme each year. Four composite indices for human development are used each year: the Human Development Index (HDI); the Gender-related Development Index (GDI); the Gender Empowerment Measure (GEM); and the Human Poverty Index (HPI). The HDI measures a country's achievements in terms of life expectancy, educational attainment, and adjusted real income. Countries are then ranked numerically, and divided into three classes of development: high, medium, and low. Norway is ranked at 1, boasting the highest standard in the world in terms of these three measures of development. SA falls into the medium human development category. The Human Development Reports are available at http://hdr.undp.org.

21. UN Development Programme (UNDP), "Human Development Report," 2005. The exact figures are 3.5 for the poorest quintile and 62.2 for the richest quintile's share of income.

22. Government has set a growth rate of 5 percent on average between 2004 and 2014 in two phases: between 2005 and 2009, the annual growth rate sought is 4.5 percent or higher. Between 2010 and 2014, an average growth rate of at least 6 percent of GDP is envisioned. See "Accelerated and Shared Growth Initiative for South Africa (ASGISA)" at http://www.info.gov.za/asgisa/asgisa.htm.

23. WTO, 1994b; see note 2.

24. WTO 1996a. This protocol is encompassed in the Schedule of Specific Commitments. The literature varyingly refers to the 1997 telecom pact in a variety of ways: namely, the Fourth Protocol to the GATS; the Basic Telecommunications Agreement (BTA); or the Telecommunications Pact. We prefer the term Fourth Protocol, as the alternative, the BTA—assumes a higher level of substantive consensus than was evident at the time the treaty was concluded.

25. Horwitz 2001 details the process for telecommunications, broadcasting, and the print media.

26. See Horwitz 2001, 6, citing Huntington 1993.

27. Gillwald 2002, commissioned by the United Kingdom Department for International Development to address clause 5 of the DOT Force Plan of Action.

28. Don MacLean et al. 2002, 23.

29. Although it focuses on the development of trade policy, see an interesting study on how best to accommodate nonstate actors in shaping policy and negotiation agendas in Ostry, Hakim, and Taccone 2002.

30. Government of South Africa, "White Paper on Telecommunications Policy," *Government Gazette* No. 16995 GN 291, March 13, 1996, and The Telecommunications Act No. 103 of 1996, as amended by the Telecommunications Amendment Act No. 64 of 2001.

31. Speech delivered by Jay Naidoo, then minister for Posts, Telecommunications and Broadcasting on the occasion of the budget vote in the National Assembly for the Department of Communications, Monday, March 8, 1999, noting studies that reflect a direct, positive correlation between communications infrastructure and per capita growth. The first study exploring a correlation between telephone infrastructure and economic development was published in 1963 by A. Jipp, who interrogated the extent to which telecoms infrastructure can facilitate economic development; See also Röller and Waverman 2001; and Hardy 1980, cited in Cronin et al. 1993, 416.

32. The Independent Communications Authority of South Africa (ICASA) is the agency responsible for regulating both telecommunications and broadcasting. Its mandate, scope, and sphere of application is prescribed by law. See The Independent Communications Authority of SA Act, No. 13 of 2000, as amended.

33. Janisch and Kotlowitz 1998, 61.

34. Telecommunications Amendment Act No. 64 of 2001.

35. See Ministry of Public Enterprises, "Policy Framework: Accelerated Agenda for the Restructuring of State Owned Enterprises," August 10, 2000. Available at http://www.polity.org.za/govdocs/policy/soe/policyframework01.htm (accessed August 26, 2002).

36. Telkom's total assets are worth R55.2 billion. It boasts consolidated operating revenue of R37.6 billion. See Telkom SA Ltd., Group Preliminary Audited Results, March 31, 2003. This makes it the country's third largest parastatal after the electricity and railway utilities.

37. This involved a 30 percent stake valued at 1.26 billion U.S. dollars (USD) (where 1 USD = 4 ZAR). The consortium, Thintana Communications LLC (30 percent) comprised SBC Communications Inc (18 percent) and Telekom Malaysia Berhad (12 percent).

38. A 3 percent stake was offered to Ucingo Investments (Pty) Ltd, a broad-based investment company representing more than twenty empowerment groups across the country. 27.7 percent of the company's share equity was listed in the IPO. The government now owns 39.3 percent of Telkom. Telkom also owns 50 percent of the Vodacom Group (Pty) Ltd (mobile operator); 64.9 percent of Telkom Directory Services; 100 percent of Swiftnet (Pty) Ltd (wireless data provider); 100 percent of Intekom (Pty) Ltd (Internet Service Provider); and 100 percent of Q-Trunk (Pty) Ltd (national trunking services).

39. World Wide Worx 2002. This is broadly accepted as the benchmark from which SA Internet growth has been measured since 1995.

40. International Telecommunication Union (ITU), *International Telecommunication Indicators,* 2003. The Internet user base grew by only 8 percent in 2002, which is the lowest annual growth rate since public Internet access was made available in 1994. The growth rate for 2003 is estimated between 6–7 percent.

41. ITU, "Internet Indicators: Hosts, Users and Number of PCs," 2002. Geneva: ITU.

42. ITU, "World Telecommunication Indicators," 2005. Geneva: ITU.

43. A second INX in Cape Town was shut down in 2003.

44. Section 40, Telecommunications Act 1996 as amended.

45. World Wide Worx 2002.

46. Telkom SA Ltd., Group Preliminary Audited Results, March 31, 2003.

47. There are two basic access platforms for broadband. ADSL is a broadband access standard that uses existing copper lines to offer high-speed digital connections over the local loop. ADSL transmits data asymmetrically, meaning that the bandwidth usage is much higher in one direction than the other. The "downstream" transmission rate to the customer (downloading) is much greater than the "upstream" (sending) rate. Wireless broadband uses mobile spectrum to offer high-speed digital connections. At present, there are three offerings on the market: GRPS (or 2.5G), 3G, and High Speed Downlink Packet Access (HSDPA).

48. Gillwald, Esselaar, and Stork 2006.

49. Approximately 1.6 million people access the Internet through corporate networks and 450,000 people gain access through educational institutions. Dial-up users account for approximately 1.1 million of the 3.1 million total Internet user base in SA. See World Wide Worx, 2002.

50. The Electronic Commerce and Transactions Act of 2002.

51. Ibid., 59–68, chapter X.

52. Interview with M. Silber, member of the domain.name board of directors, June 2006.

53. WTO 1994b; see note 2.

54. WTO 1996a.

55. For a detailed treatment of GATS, see Arup 2000 and Lal Das 1998.

56. WTO 1994a. For a concise history of the Annex on Telecommunications, see Tuthill 1996, 90.

57. Trebilcock and Howse 2000, 296–297.

58. WTO 1996b.

59. WTO 1994a.

60. The Post Office Act, 1958.

61. Drake and Noam 1997.

62. Suavé 1995, 143.

63. For a more detailed treatment of SA's GATS telecommunications commitments, see Cohen 2001, 725–754. An important point warrants attention, however. Telkom's de facto monopoly on basic services and facilities continues largely unaffected, as lengthy administrative delays ensue in the licensing process for a second fixed-line operator to compete with Telkom. However, while Telkom had the legal right to negotiate a sixth year to its exclusivity, this was not sought—possibly indicating that the government was awakening to the adverse impact of the monopoly on the sector, and indeed on reaching Telkom's rollout targets, due to the high costs associated with "policing" monopoly rights.

64. A license was finally granted to Cell-C (Pty) Ltd in February 2001 after a protracted court battle challenging the substance and the procedure involved in the license evaluation and recommendation. A summary of this process is contained in Cohen 2001; see note 63.

65. Section 1377, Omnibus Trade and Competitiveness Act, 1988. The annual review considers the operation and effectiveness of all U.S. trade agreements regarding telecommunications products and services. Its purpose is to determine whether any act, policy, or practice of an economy that has entered into a trade pact with the United States is inconsistent with the terms of the agreement or denies U.S. firms mutually advantageous market opportunities.

66. WTO and GATS, articles XVI and XVII, respectively, and article VIII:1, 1994.

67. Provision 5(a), GATS, Schedule of Specific Commitments (South Africa) GATS/SC/78.

68. See *AT&T Global Networks Services SA (Pty) Ltd and Others v. Telkom SA Ltd and the South African Telecommunications Regulatory Authority*, High Court of SA (TPD), Case No. 27624/99, November 16, 1999.

69. Chayes and Chayes 1995, 4. Although in this case, the notion that compliance follows simply because it is in the country's interests to do so is an equally useful approach. Koehane suggests that compliance generally needs to be evaluated in terms of how international regimes affect calculations of a state's self-interest. Implicit in this, he asserts, is an evaluation of the value of the regime to a state given any feasible alternatives (i.e., in isolation), and an evaluation of the regime in the context of others in world politics (i.e., relative to others). See Keohane 1984, 98–100.

70. This was facilitated by the development of a regional band plan by the Telecommunications Regulators Association of South Africa. See http://www.trasa.org.

71. See, for example, Schaefer 2002, 341.

72. Suavé 1995, 126.

73. Main differences are of course also accountable to varying stages of development and limiting or facilitating factors present in their domestic markets.

74. Examples include Sen 1999, Olsen 1996, and North 1997.

75. Government of South Africa, "Reconstruction and Development Programme: A Policy Framework," available at http://www.polity.org.za/html/govdocs/rdp/rdp.html.

76. See, for example, Weeks 1999, 810.

77. Marais 1998, 150.

78. Works 1998.

79. See Dolowitz and Marsh 2000, 10–11.

80. See S. Wallsten, "Does Sequencing Matter? Regulation and Privatization in Telecommunications," Policy Research Working Paper 2817. World Bank, Washington, DC, 2002; and A. Gillwald, "Good Intentions, Poor Outcomes: Telecom Reform in South Africa," *Telecommunications Policy* 29, no. 4 (2005).

81. J. Stiglitz, "Promoting Competition in Telecommunications," Working Paper No. 2. Centro de Estudios Econónomicos de la regulación, Instituto de Economa, Universidad Argentina de la Empressa, Chile, 1999, 4.

82. "Green room" meetings, named after a room at the WTO office, denote "secret" caucuses by usually dominant Western countries seeking resolution on a deadlocked or resisted matter. These meetings are restricted to invitees, nontransparent, and generally not recorded as official by the WTO media office. The precise extent of their occurrence is difficult to ascertain, although the nature of a side caucus in such situations is common to the process, whether taking place formally within the green room or in corridor negotiations.

83. Helleiner and Oyejide 1999, 112.

84. Ibid.

85. Lodge and Stirton 2002, 667.

86. Dzidonu and Quaynor 2002, 7.

87. See Vogel 1997, 170. Although used in a different context, the idea of regulatory subsidy pertains to the attempt to rig regulation to favor domestic firms.

88. In terms of ITU classification, there are big-M members that comprise the 185 countries that are signatories to the ITU; and small-m members (nonadministration members) that comprise the private sector (including telecommunication operators and networking and software companies).

89. Helleiner and Oyejide 1999, 110.

90. For a critical account of the various approaches to aid followed by the World Bank, and how these have failed in most cases in the developing world, see Easterly 2001. Here Easterly contends that the failure to structure incentives correctly lies at the root of this problem.

91. See Castells 1999, 94–128.

92. See "Report of the Fifth Meeting of Study Group 3," held in Geneva, December 9–13, 2002, COM 3-R11-E, January 2003.

93. *Mexico—Measures Affecting Telecommunications Services* (WT/DS204).

94. See Mukerji 2000, 70.

95. See Government of India, "Communication from India to the WTO," WTO Document WT/GC/W/108, November 13, 1998.

96. North and Weingast 1989, 806.

97. See Williamson 1985, 48.

98. Fine 2000.

99. See notes 54 and 58.

100. The relative merits and demerits of the flexibility inherent in the Reference Paper are not discussed here, save to say that we believe the flexibility serves SA's purposes far better than a rigid template for regulation would have. Moreover, it is a widely held view that less flexibility would have inhibited a critical mass of signatories to the GATS Fourth Protocol. For a defense of the Reference Paper, see Blouin 2000, 139–140.

101. Dzidonu and Quaynor 2002, 3.

102. Ibid., 4.

103. Ibid., 11–14.

104. WTO, Ministerial Conference, Fourth Session, Doha, November 9–14, 2001, *Ministerial Declaration*, WT/MIN(01)/DEC/1, November 20, 2001.

105. In this regard, see Grindle 1997, which focuses on assessing capacity building needs in developing countries and a variety of strategies for capacity development, including the use of nongovernmental agencies within countries and technical assistance from sources outside a country. See particularly Gray 1997.

106. Quoted in Gray 1997, 414.

107. Dieter 1997, 202.

108. See www.fibreforafrica.net.

References

Arup, Christopher. 2000. *The New World Trade Organization Agreements: Globalizing Law through Services and Intellectual Property*. Cambridge: Cambridge University Press.

Baird, Zoë. 2002. "Governing the Internet: Engaging Governments, Business, and Nonprofits." *Foreign Affairs* 81, no. 6 (November/December): 15–20.

Blouin, Chantal. 2000. "The WTO Agreement on Basic Telecommunications: A Reevaluation." *Telecommunications Policy* 24, no. 2: 135–142.

Bond, Patrick. 2001. *Against Global Apartheid: South Africa Meets the World Bank, IMF and International Finance*. Cape Town: University of Cape Town Press.

Cassim, Rashad et al. 2000. "Determinants of Investment in South Africa: A Sectoral Approach." Fianl Report: Investment Study A.1.003.2A. Pretoria: Trade and Industrial Policy Secretariat (TIPS) and Econometric Research South Africa.

Castells, Manuell. 1999. *The Information Age: Economy, Society and Culture, Vol. 3, End of Millennium*. Oxford: Blackwell.

Chayes, Abraham, and Antonia Handler Chayes. 1995. *The New Sovereignty: Compliance with International Regulatory Agreements*. Cambridge, MA: Harvard University Press.

Cohen, Tracy. 2001. "Domestic Policy and South Africa's Commitments under the WTO's Basic Telecommunications Agreement: Explaining the Apparent Inertia." *Journal of International Economic Law* 4: 725–754.

Coopers & Lybrand. 1992. *Telecommunications Sector Strategy Study for the Department of Posts and Telecommunications*. Pretoria: South African Government Printer.

Cronin, Francis J., Elizabeth K. Colleran, Paul L. Herbert, and Steven Lewitzky. 1993. "Telecommunications Infrastructure Investment and Economic Development." *Telecommunications Policy* 17, no. 7: 415.

Dieter, Heribert. 1997. "Regionalization in the Age of Globalization: What Limits Are There to Economic Integration in Southern Africa." In *The Regionalization of the World Economy and Consequences for Southern Africa*. Marburg: Metropolis-Verlag.

Dolowitz, David P., and David Marsh. 2000. "Learning from Abroad: The Role of Policy Transfer in Contemporary Policy-Making." *Governance: An International Journal of Policy and Administration* 13, no. 1: 5.

Dorrington, Rob, Debbie Bradshaw, and Debbie Budlender. 2002. *HIV/AIDS Profile in the Provinces of South Africa: Indicators for 2002*. Cape Town: Center for Actuarial Research, Medical Research Council and the Actuarial Society of South Africa.

Drake, William, and El. Noam. 1997. "Assessing the WTO Agreement on Basic Telecommunications." In *Unfinished Business: Telecommunications after the Uruguay Round*, ed. Gary Clyde Hufbauer and Erika Wada, 27–61. Washington, DC: Institute for International Economics.

Dzidonu, Clement, and Nii Narku Quaynor. 2002. "Broadening and Enhancing the Capacity of Developing Countries to Effectively Participate in the Global ICT Policy Fora and the ICT for Development (ICTFDEV) Process." Accra-North: International Institute for Information Technology (INIIT), Special Working Paper Series, no. 5.

Easterly, William. 2001. *The Elusive Quest for Growth: Economists' Adventures and Misadventures in the Tropics*. Cambridge, MA: The MIT Press.

Fine, Allison. 2000. "Contracting for Credibility in International Telecommunication Investments." Paper presented at the International Society for New Institutional Economics (ISNIE) Conference, September. Available at http://www.isnie.org/ISNIE00/Papers/fine.pdf.

Gillwald, Alison. 2002. "South African Case Study." In *Louder Voices: Strengthening Developing Country Participation in International ICT Decision-Making,* ed. Don MacLean, David Souter, James Deane, and Sarah Lilley. London: Commonwealth Telecommunications Organization and Panos Institute.

Gillwald, Alison, Steve Esselaar, and Christoph Stork. 2006. "South African ICT Sector Performance Review." LINK Centre Public Policy Research Paper No. 8. Available at http://link.wits.ac.za.

Government of South Africa. 2001–2002. *South African Yearbook 2001/2.* Available at http://www.gov.za/yearbook/2001/transport.html#public (accessed February 23, 2002).

Government of South Africa, Ministry of Public Enterprises. 2000. "Policy Framework: Accelerated Agenda for the Restructuring of State Owned Enterprises." Available at http://www.polity.org.za/govdocs/policy/soe/policyframework01.htm.

Gray, Clive S. 1997. "Technical Assistance and Capacity Building for Policy Analysis and Implementation." In *Getting Good Government: Capacity Building in the Public Sectors of Developing Countries,* edited by Merilee S. Grindle. Boston: Harvard Institute for International Development and Harvard University Press.

Grindle, Merilee S. 1997. Getting Good Government: Capacity Building in the Public Sectors of Developing Countries. Cambridge, MA: Harvard University Press.

Hardy, Andrew. 1980. "The Role of the Telephone in Economic Development." *Telecommunications Policy* 5, no. 4: 278.

Helleiner, Gerry, and Ademola T. Oyejide. 1999. "Global Economic Governance, Global Negotiations and the Developing Countries." Background Papers to the 1999 Human Development Report. New York: UN Development Programme.

Horwitz, Robert. 2001. *Communication and Democratic Reform in South Africa.* Cambridge: Cambridge University Press.

Huntington, Samuel. 1993. *The Third Wave: Democratization in the Late Twentieth Century.* Julian J. Rothbaum Distinguished Lecture Series, vol. 4. Norman: University of Oklahoma Press.

International Labor Organization. 2002. *Key Indicators of the Labor Market: Poverty and Income Distribution 2001–2002.* New York: Routledge.

Janisch, Hudson N. and Danny Kotlowitz. 1998. "African Renaissance, Market Romance: Post-Apartheid Privatization and Liberalization in South African Broadcasting and Telecommunications." Paper for the symposium Has Privatization Worked? The International Experience, Center for Tele-Information, Columbia University, June 12.

Jipp, A. 1963. "Wealth of Nations and Telephone Density." *ITU Telecommunication Journal* 30.

Kaplan, David. 1990. *The Crossed Line: South African Telecommunications Industry in Transition*. Johannesburg: Witwatersrand University Press.

Keohane, Robert O. 1984. *After Hegemony: Cooperation and Discord in the World Political Economy*. Princeton: Princeton University Press.

Lal Das, Bhagirath. 1998. *An Introduction to the WTO Agreements*. London: Zed Books.

Lodge, Martin, and Lindsay Stirton. 2001. "Transparency Mechanisms: Building Publicness into Public Service." *Journal of Law and Society* 28, no. 4: 471.

Lodge, Martin, and Lindsay Stirton. 2002. "Embedding Regulatory Autonomy: The Reform of Jamaican Telecommunications Regulation 1988–2001." *Annals of Public and Cooperative Economics* 73: 667.

MacLean, Don, David Souter, James Deane, and Sarah Lilley. 2002. *Louder Voices: Strengthening Developing Country Participation in International ICT Decision-Making*. London: Commonwealth Telecommunications Organization and Panos Institute.

Marais, Hein. 1998. *South Africa Limits to Change: The Political Economy of Transformation*. London and New York: Zed Press.

Melody, William, H. 2002. "The Triumph and Tragedy of Human Capital: Foundation Resource for the Global Knowledge Economy." *Southern African Journal of Information and Communication* no. 3, available at http://link.wits.ac.za/journal/j0301-melody-fin.pdf.

Mukerji, Asoke. 2000. "Developing Countries and the WTO." *Journal of World Trade* 34, no. 6: 61.

North, Douglass C. 1997. "The New Institutional Economics and Third World Development." In *The New Institutional Economics and Third World Development*, ed. John Harriss, Janet Hunter, and Colin M. Lewis, 17–26. London: Routledge.

North, Douglass C., and Barry Weingast. 1989. "Constitutions and Commitment: The Evolution of Institutions Governing Public Choice in Seventeenth Century England." *Journal of Economic History* 49, no. 4: 803.

Olsen, Mancur. 1996. "Big Bills Left on the Sidewalk: Why Some Nations Are Rich and Others Are Poor." *Journal of Economic Perspectives* 10, no. 2: 3.

Ostry, Sylvia, Peter Hakim, and Juan Jose Taccone. 2002. "Trade Policy-Making Process, Level One of the Two-Level Game: Country Studies in the Western Hemisphere." Washington, DC: Inter-American Development Bank, Integration and Regional Programs Department, Occasional Paper No. 13.

Röller, Lars-Henrik, and Leonard Waverman. 2001. "Telecommunications Infrastructure and Economic Development: A Simultaneous Approach." *American Economic Review* 4: 910.

Schaefer, Matthew. 2002. "Sovereignty, Influence, Realpolitik and the WTO." *Hastings International & Comparative Law Review* 25: 303.

Sen, Amartya. 1999. *Development as Freedom*. New York: Knopf.

Suavé, Pierre. 1995. "Assessing the General Agreement on Trade in Services: Half-Full or Half-Empty." *Journal of World Trade* 29, no. 4.

Tomaselli, Ruth, Keyen Tomaselli, and Johan Muller. 1989. *Broadcasting in South Africa*. New York: St. Martin's Press.

Trebilcock, Michael, and Robert Howse. 2000. *The Regulation of International Trade*, 2nd ed. London: Routledge.

Tuthill, Lee. 1996. "'Users' Rights? The Multilateral Rules on Access to Telecommunications." *Telecommunications Policy* 20, no. 2: 89–99.

Vogel, Stephen K. 1997. "International Games with National Rules: How Regulation Shapes Competition in Global Markets." *Journal of Public Policy* 17, no. 2: 169.

Weeks, John. 1999. "Stuck in Low Gear?: Macroeconomic Policy in South Africa 1996–1998." *Cambridge Journal of Economics* 23, no. 6: 795–811.

Williamson, Oliver E. 1985. *The Economic Institutions of Capitalism*. New York: Free Press.

Works, Brandon. 1998. "South Africa Limits to Change: The Political Economy of Transformation." *African Studies Quarterly—The Online Journal for African Studies* 2. Available at www.africa.ufl.edu/asq/v2/v213a9.htm.

World Wide Worx. 2002. "The Goldstuck Report on Internet Access in SA: 2002." Available at http://www.theworx.biz/access02.

WTO. 1994a. *Annex on Telecommunications*. Appendix to GATS, April.

WTO. 1994b. *General Agreement on Trade in Services* contained as Annex 1B to *General Agreement on Tariffs and Trade—Multilateral Trade Negotiations (The Uruguay Round): Final Act Embodying the Results of the Uruguay Round of Trade Negotiations*, December 15, 33 I.L.M. 1167, 1195.

WTO. 1996a. *Fourth Protocol to the Basic Agreement on Trade in Services*, April 30, WTO Doc. S/L/20.

WTO. 1996b. *Reference Paper on Regulatory Principles Used for Consideration as Additional Commitments in Offers on Basic Telecommunications*. WTO Negotiating Group on Basic Telecommunications, April 24.

Young, Oran. 1983. "Regime Dynamics: The Rise and Fall of International Regimes." In *International Regimes*, ed. Stephen Krasner, 277–297. Ithaca, NY: Cornell University Press.

14 Spectators or Players? Participation in ICANN by the "Rest of the World"

Milton Mueller and Jisuk Woo

The Internet Corporation for Assigned Names and Numbers (ICANN) is an important and interesting example of how the Internet and e-commerce are creating new forms of international governance. ICANN is a private corporation, created at the end of 1998. It controls the assignment of domain names and IP addresses, critical resources needed for identifying Internet users and interconnecting them. ICANN could have been even more revolutionary than it turned out to be, as its initial founders originally tried to create a system of individual membership that would conduct global elections to appoint half of its board of directors. This attempt was resisted and eventually defeated, however, by ICANN's management.

ICANN is important as a precedent for three interrelated reasons. First, it was created rather pointedly as an alternative to existing intergovernmental organizations, notably the International Telecommunication Union (ITU). By relying on a private sector corporation that enters into a privileged relationship with the U.S. government, it bypasses and in some ways threatens the hegemony of established intergovernmental institutions. Thus, it reflects a continuing power struggle between actors who want to avoid existing international institutions, and the established international organizations and their constituents who do not want to be bypassed and rendered irrelevant by new developments. Second, and related, ICANN was created as an experiment in private-sector-based international governance and is sometimes put forward as an example of "industry self-regulation." A private, contract-based regime was viewed as a way to create a global jurisdiction for setting and enforcing policy without relying on intergovernmental organizations. Third, ICANN's politics reveal and reflect the contention, now a familiar part of the international environment, between a dominant United States government able to project its economic and political power globally, the Europeans as a countervailing regional power, and a residual category of actors we can refer to as the "rest of the world." It is this third category upon which we wish to focus in this chapter.

The "rest of the world" (ROW) refers to a heterogeneous group of actors: it includes underdeveloped countries in Africa; newly developed economies such as Korea,

Singapore, and Hong Kong; and developing countries such as China, India, and Brazil. They are, in short, somewhat peripheral and not cohesive players, but nevertheless important due to the need of ICANN to gain credibility and legitimacy as an international organization. They are also important because of their potential to become major economic powers some time in the future. Like the concept of "nonaligned nations" in the Cold War, ROW countries are perceived as a unit by virtue of what they are *not* rather than any distinctive, common characteristics such as development levels, ideology, or language. The one thing they have in common is that during the initial, formative stages of the Internet domain name regime's development, which mostly involved U.S. initiatives and European reactions,[1] they had little leverage and joined primarily to avoid being left out. The concept is particularly applicable to developed and developing Asian countries. These nations are strong enough economically and in terms of Internet development to participate in ICANN and be aware of its significance, but are not yet ready to play a role as prime movers. They also are disadvantaged within ICANN's relatively informal processes by cultural and linguistic differences. For that reason, we rely on South Korea as a case study, for Korea epitomizes the dilemmas of ROW status perfectly. It is an active, even aggressive presence on the Internet and has achieved impressive levels of development. But it was not an initial partner to the regime, its government does not seem to like the self-regulatory model, and its civil society participants feel marginalized and overwhelmed by the challenge of keeping pace with ICANN's processes.

This chapter will discuss the participation, policy agendas, and effectiveness of representatives from outside of North America and Europe in ICANN's processes. How does the unique ICANN approach to an international governance problem affect the ROW? Does the reliance on a private sector organization with some channels open to civil society participation make representation of ROW needs and interests easier or harder? How effectively have ROW actors participated in ICANN's regimes? What procedures and substantive policies would improve the benefits received by the ROW in the ICANN process? This chapter will try to establish a general framework for analyzing this problem, but will rely on the case of South Korea for empirical evidence.

The conclusions we reach are clear: ICANN needs to do a lot more work to make its processes and policies less hostile and more encouraging to ROW participants. As steps to that end, we make several concrete suggestions. One is to create a new advisory committee devoted to "Multicultural Awareness and Outreach." This committee would mentor ROW members and encourage their participation. Another is to use more objective and regular procedures for allocating valuable resources such as top-level domains. The goal should be to replace insider lobbying, which favors incumbent actors in the United States and Europe, with procedures such as auctions and random selection, which are less discretionary. ICANN should also explore the possibility of regionalizing control of resources whenever possible. And while we find serious shortfalls in ICANN's processes, we cannot avoid the conclusion that as a mechanism for involving civil society actors, a private sector-based regime such as ICANN's has advantages over traditional intergovernmental organizations. We describe how the agenda and demands of ROW participants from civil society are often in conflict with the agenda of their own governments, and that—as halting as ROW civil society participation has been—it exceeds what occurs in intergovernmental organizations (IGOs). We urge ICANN to broaden and expand the opportunities for civil society participation rather than narrowing them.

ICANN as International Governance Regime

ICANN is the organizational capstone of an international regime for assigning and allocating domain name resources. It was not founded on a treaty or any other formal agreement among states. Rather, it is based on informal bargains among state and nonstate actors.[2] The bargains were mediated by a U.S. government proceeding,[3] and implemented through contracts with the U.S. government that transferred certain functions from direct U.S. governmental management to indirect management through ICANN. The role of the U.S. government as contracting authority for the ICANN "experiment" was supposed to end after two years. Somewhat unexpectedly, the United States has retained its special and increasingly controversial contractor role, and has announced that it has no plans to end it, making it the ultimate and final authority over the fate of the regime.

It is said that "possession is nine-tenths of the law," and that is how the U.S. government obtained its special status in the first place. As the original funder of the researchers who built the Internet, first through the U.S. Defense Department and later through the National Science Foundation (NSF), the U.S. government was in a position to determine, directly or indirectly, who operated the root servers and the top-level domain registries. Most notably, the Defense Advanced Research Projects Agency granted support to Internet pioneer Jon Postel, who assigned protocol parameters, IP address blocks, and country code top-level domains until the mid-1990s. Likewise, the NSF contracted with Network Solutions, Inc., to register names under the .com, .net, .org, and .edu top-level domains in 1993. Until 1998, Postel and Network Solutions had operational control of critical Internet functions, and the U.S. government had indirect authority over both of them, so no regime could have been created without the acquiescence of the United States.

As an international regime, ICANN embodies some contemporary, nontraditional governance principles as well as many traditional principles associated with government regulatory agencies and intergovernmental organizations. We will enumerate these principles and describe their interaction.

Privatization, Contractual Ordering, and Global Jurisdiction

The principles associated with the regime were articulated in the Clinton administration's "A Framework for Global Electronic Commerce," published in July 1997. In forming its policy toward the Internet and global electronic commerce, the Clinton administration and major information technology executives at firms like IBM, MCI WorldCom, and America Online were concerned about the possibility that Internet-based commerce would be undermined by widespread assertions of territorial jurisdiction.[4] National governments might impose upon the naturally global arena of the Internet a patchwork of inconsistent or conflicting national laws and regulations. A private sector governance authority was perceived as a way around this problem. Thus the Clinton administration's policy called for "private sector leadership" and noted "governments should establish a predictable and simple legal environment based on a decentralized, contractual model of law rather than one based on top-down regulation."[5] The governing authority would enter into "private contracts" with industry stakeholders that would be global in scope, rather than subjecting domain name businesses to multiple, potentially conflicting rules based on territorial jurisdictions. Regarding domain names in particular, the United States proposed in mid-1997 that "It may be possible to create a contractually based self-regulatory regime that deals with potential conflicts between domain name usage and trademark laws on a global basis without the need to litigate."[6]

Public Trustee Regulation

Despite its invocation of neoliberal principles such as "private sector leadership," the actual behavior of the ICANN regime incorporates many of the classical regulatory principles we might associate with the ITU or public utility regulation. The domain name space is considered to be a "public resource" subject to regulation in the public interest. Domain name registries that sign contracts with ICANN hold no property rights in their top-level domains. Legally, assignees of IP addresses also lack such formal rights. Holders of these resources are considered trustees rather than owners. A central authority, ICANN, purports to act on behalf of the public as it allocates and assigns the resources. ICANN is organized as a California nonprofit "public benefit" corporation. Many of the principles underlying its governance structure and processes, such as transparency, notice and comment, and decision-making organs based on stakeholder representation, are grounded in the notion that the organization's rule making, policy making, and enforcement roles are basically governmental in nature.

As noted elsewhere,[7] there are striking parallels between ICANN's approach to the award of new top-level domain names and the broadcast licensing regime of the United States. American law considered the airwaves a public resource and the award of a broadcast channel to be a special privilege because of the scarcity of available radio frequencies in a local community. In exchange for the privilege of a limited-duration right to exclusively occupy a broadcast channel, licensees assumed a number of service obligations and conduct regulations, such as carrying a certain amount of children's programs, limiting the number of commercials, and so on. When there were contending applications, the award of licenses was made through elaborate "comparative hearings," an administrative procedure known in the communications industry by the somewhat derisive term "beauty contests." In these beauty contests, prospective licensees vied with each other to stand out and to promise greater public benefits. (In reality, incumbent licensees were almost always renewed. There is, as we shall discuss later, always a gap between the theory and practice of beauty contests because of the enormous discretion they give to the judges.)

Top-level domain name (TLD) awards administered by ICANN have unambiguously taken the form of beauty contests. ICANN issues a call for applications with a few broad, normative criteria. Applicants who respond to these calls strive to convince ICANN's board that they are the most technically sound and virtuous, that their marketing plans and business models are well crafted, that the public really needs their proposed service, and so on. What is missing, however, is the scarcity rationale. ICANN could easily have authorized all of the forty-four applicants in its first round of new TLD authorizations. It chose not to because of the opposition of intellectual property interests, which sees restricting the supply of new domains as a cheap way to reduce its trademark policing costs. The domain name space is flexible enough to accommodate thousands of new top-level domains and faces nothing like the technical and material constraints of radio spectrum use. In ICANN's case, regulation is not a response to a distributional issue caused by scarcity; the regulator and the special interests influencing it *cause* the scarcity.

Geographic Representation

Geographic representation is another key principle in the ICANN process, as it might be in any traditional intergovernmental organization. A demand for the internationalization of Internet governance was one of the political drivers of the proceedings that created ICANN.[8] Although the ICANN process has disappointed many international participants because of the special, dominant role of the U.S. Department of Commerce, the internal processes of ICANN are constrained somewhat by concerns of diversity. ICANN divides the world up into five regions. Most selection processes of its board members and council representatives must distribute their choices over those regions. The board, council, and other policymaking organs are, however, dominated by U.S. and European actors, a fact to which we shall return. And at no point does ICANN engage in one-country, one-vote decision-making processes. Geographic diversity affects the selection of representatives to councils and the board; it does not structure voting directly.

National Sovereignty: An Open Question

The degree to which the principle of national sovereignty plays a role in the ICANN regime is still unclear. At its core, however, the ICANN regime was based on assumptions that were fundamentally hostile to the principle of national sovereignty. In its original charter, the 1998 U.S. Commerce Department White Paper, ICANN was intended to preclude the "participation of national governments acting as sovereigns ... [or] intergovernmental organizations acting as representatives of governments" from representation on its board or policy-making councils.[9] As the self-styled inheritor of the Internet technical community's stateless and self-governing Internet Assigned Numbers Authority (IANA), ICANN views itself, and to some degree was designed to be, a centralized manager and delegator of top-level domains with ultimate authority over what goes into the root zone file, which defines the top of the domain name hierarchy. Even the award of country code domains, in the classical days of the academic/scientific Internet, considered those so delegated to be trustees not of the national government but of the *"local and global Internet communities."*[10]

In the process of bringing the ICANN regime into being, however, a number of compromises with the existing nation-state system had to be made. The European Union, for example, prevailed upon the United States to create a special advisory committee for governments and intergovernmental organizations, the Governmental Advisory Committee (GAC). The GAC has grown in power and influence over time. It now makes policy recommendations and has liaisons placed on ICANN's policy-making organs. The EU and the World Intellectual Property Organization (WIPO) also pushed hard for involving WIPO in the process of formulating domain name dispute resolution policies. In the sweeping reforms of ICANN's representational structure and board selection process made in 2002, ICANN's president recognized its need to gain more support and participation from national governments. At one point he even proposed allowing the GAC to appoint board members.[11]

The issue of national sovereignty arises most significantly around the issue of who decides which organizations operate country code top-level domains (ccTLDs). CcTLDs are top-level domain names based on the ISO-3166-1 list of two-letter country codes (e.g., .us for the United States, .de for Germany, .mx for Mexico, and so on). Each ccTLD name is associated with a specific geographic territory. Most (but by no means all) of the territories are recognized national states.[12] Since there is only one country code on the ISO-3166 list for each geographic territory, delegation of this name to a domain name registry constitutes a kind of exclusive right to register names for anyone who wants to be identified by that country code.[13] In the early days of the Internet's growth, when major governmental actors did not really care or were not paying attention, the technologists who built the Internet delegated the right to operate these country codes to anyone in a country who applied and seemed to be capable of doing the job. Typically this meant that the ccTLDs were delegated to technologists involved in building and operating academic and scientific research networks.

Since those early days, the Internet has become much more important and registration of domain names under country codes has grown into a respectable-sized business in some countries. About one third of the forty-six million domain names are registered in country code TLDs. The German (DENIC) and British (Nominet) ccTLDs are the second and fourth largest domain name registries in the world. Registrations under country codes continue to grow in stable developed and developing economies. The growth of the sector has made control of the country code registry and the definition of policies regarding a ccTLD's name space a matter of much greater interest politically. Many national governments would now like to play a role in the regulation of the domain name business, and involve themselves in many of the same issues ICANN handles, such as dispute resolution, naming conventions, privacy policies, governance representation, and procedural issues.

But it would be incorrect to impose a simple globalist versus nationalist dichotomy on the politics of country codes, as von Arx and Hagen (2002) seem to do. In fact, there is a three-way cleavage in international domain name politics. First, there is ICANN, a private corporation set up to manage the root as an alternative to governmental and territorial control. ICANN considers itself the heir of the Internet technical community, the source of the original delegation of the ccTLD to a local group, and the proper global authority for all changes in delegations. Second, there is the local delegate for the ccTLD. The interests of these local delegates are not always aligned with those of the national government. The ccTLD operator often consists of private sector nonprofits, or even individual Internet pioneers within the country, as opposed to the official national ministry of information or communications. Third, there is the national government and its official policy apparatus for communication and information.

There are tensions in each direction of that triadic relationship. ICANN would like to fully incorporate the ccTLDs into its regime so that it can increase its tax base and give its policies and regulations truly global force. The ccTLD managers view ICANN as a regulating and taxing authority that it is not properly accountable to them. They feel that they are not adequately represented within its decision-making procedures. And at any rate they have little to gain from entering into the contractual relationship: their names are already in the root, their domains are operational, and thus contracts with ICANN can bring loss of control and additional costs but no discernable additional benefit. As for the third leg of the stool, national governments (in many but not all cases) would like to have more direct control over who is delegated the ccTLD representing their country. In some cases, there may be tensions between the "local Internet community" in control of the ccTLD, and the national government ministries or politicians. Or in some cases, ISO-3166 country codes have been assigned to protectorates of a nation and the national government thinks it ought to have more control over how the ccTLD associated with their protectorates are used.

The ccTLDs, led by the largest and most vocal registries in Europe and Australasia, have demanded special forms of representation within ICANN before submitting to its contractual authority.[14] National governments have demanded greater and more direct participation in the ICANN regime and greater say in the delegation or re-delegation of ccTLDs. ccTLDs have been slow to sign contracts with ICANN that fully incorporate them into the regime. Until these tensions are worked out, it is difficult to tell the degree to which the principle of national sovereignty will be operative in the ICANN regime.

Governance Mechanisms

Resource-Based Regulation

Even though it is considered an alternative to government, ICANN is a rather powerful regulator, with sweeping control over the economic structure of those segments of the domain name industry with which it has contractual agreements. ICANN is able to act as a regulator because of its control of a strategic leverage point in a technical system. The U.S. Department of Commerce has given ICANN policy authority over the root zone file. By deciding what names can be entered into the root, ICANN controls and restricts entry into the market for domain name registrations. It uses this control of market entry to impose, by "contract," a number of regulations on suppliers and users of domain names. For example, it caps the price of domain name registrations at the wholesale level. ICANN also regulates the market structure of the domain name industry by preventing generic TLD (gTLD) registries from selling registrations directly to end users and imposing technical standards for access to the registry by competing, retail-level domain name registration companies known as "registrars." It defines and enforces policies that accredit registrars. It defines policies regarding whose rights to names can be enforced in the domain name space; it accredits domain name dispute-resolution service providers. ICANN also plays a relatively minor role in the allocation and assignment of Internet protocol address blocks, but this role might become more important in the future, as the possibility of a transition to a new Internet protocol, IPv6, may open up a huge new address space.

ICANN's decisions also have privacy implications. Domain name registration records contain contact information that can be searched and used for service of process. ICANN sets and enforces policies regarding the accuracy of this information, access to the data, and the supply of this data by registries and registrars. It uses its power to take away a domain name as the method of enforcement.

Note the tension between ICANN's regulatory functions, which affect the rights of consumers of domain names and (indirectly) all Internet users, and the concept of industry self-regulation, which is attuned primarily to the needs of the suppliers of services. Traditional governmental regulators are politically accountable to elected officials, who in turn may through voting be responsive to consumers and users. ICANN now lacks any such mechanism of direct or indirect accountability to consumers and users. Originally, ICANN was supposed to have a membership structure, known as the at-large membership, open to individuals and allowing anyone with an e-mail address to join. In the bargaining surrounding the White Paper process, the Clinton administration's policy makers forced ICANN's management to accede to the creation of a membership that would elect half the members of the board. However, ICANN's management and many of the business and technical interests that influenced it resisted such incursions of democracy because it threatened their dominance of the regime.[15] Political support for the at-large membership came primarily from North American and European advocacy groups with a liberal-democratic ideological bent.[16] Few ROW representatives took a leadership role in advocating for the at-large membership persistently or vocally. However, because the representational structure of the at-large body was based on geographic territories, such a board-selection mechanism over the long term would have greatly empowered populous developing countries such as India and China. After ICANN's "reform" efforts, the at-large body has been demoted to an advisory committee, and lacks the broad interest and participation of civil society organizations and individuals.

Financial Support

ICANN funds itself by imposing, via its contracts with registries and registrars, taxes on the contracting parties. Registrars pay accreditation fees that vary based on how many top-level domains they service, while registries pay both flat annual fees and fees that vary based on how many domains they have registered. ICANN gets most of its revenue from the gTLD registries that run domains such as .com, .info, and .net. It has had trouble getting the "country code" registries to sign contracts with it and provide funds for it.

Enforcement

Enforcement takes place via contract between ICANN and the domain name supply industry. These contracts contain provisions that flow down to end users and regulate their behavior. For example, all registrars must be accredited by ICANN before they can register domain names in the gTLDs. In order to obtain accreditation, the registrars must incorporate by reference into their contract with all of their customers a commitment to subject themselves to the Uniform Domain Name Dispute Resolution Policy. The main end-user sanction of ICANN is to take away a domain name. It cannot levy fines or put people in jail. As noted earlier, ICANN has not established contractual relationships with many ccTLD operators, which means that there is a significant enforcement hole in the regime. It also does not have solid contractual relationships with the operators of ten of the thirteen root servers. These root server operators emerged informally from the Internet technical community in the late 1980s and early 1990s; some of them are still U.S. government agencies.

Power and Policy Making

As a private corporation called into existence by the U.S. Commerce Department, ICANN's structure and policies predominantly reflect the interests of several major private sector actors. These actors include first and foremost the intellectual property lobby (the International Trademark Association, WIPO, Fédération Internationale des Conseils en Propriété Industrielle (FICPI), major international brand holders); major e-commerce firms (e.g., IBM, AOL-Time Warner, and trade associations such as ITAA and WITSA, all of whom also have an interest in the trademark protection aspects of domain name policy); and major multinational telecommunications companies or ISPs (BT, AT&T). It was formed with the acquiescence of certain key members of the Internet Society's governing hierarchy, which hoped that a nongovernmental approach would preserve as much as possible the Internet Engineering Task Force (IETF)-based technical community's historical control over the domain name system (DNS) management. The European Union also participated in and basically approved of the private sector approach. Along with the EU, certain parties in the government of Australia played an important role in carving out a larger role for national governments and traditional international organizations in ICANN's structure.

The issues and stakes of ICANN policy making are pretty straightforward. Fundamentally, ICANN regulates a $3 billion per annum domain name registration industry; it determines what firms are able to enter that market and the terms and conditions under which firms compete. ICANN also defines and enforces rights to names in the domain name space. That is, it decides how trademark laws are applied within the global domain name space and can also define and enforce other exclusive rights to names or words, such as reservations of country names. ICANN's control of the registry industry also gives it a substantial amount of power over the technical standards that might apply to the DNS (although its failure to take control of or sign contracts with the root server operators and address registries undermines this power for now.) For example, in the transition to internationalized domain names, ICANN can affect how or even whether domain name registries are allowed to implement such a service. It could also affect the geographical placement of the root servers, which would have a direct affect on the technical quality of Internet service in many ROW countries. And as noted before, one of the key high-stakes issues is how or whether the ccTLDs are incorporated into the ICANN regime. This will determine whether it is a truly globalist regime and to some extent impinges on traditional notions of national sovereignty. Another similar issue is the degree to which the United States will continue to be the dominant, regulator of last resort in the ICANN regime. That is, will globalism be ushered into existence through the dominance of one superpower or will it be a truly distributed, self-governing regime composed of cooperating members of the global Internet community?

General Barriers to Participation

The most daunting structural barriers facing ROW participation seem to be language, funding, mobilizing expertise, the lack of clearly defined interests, and different historical and cultural experiences. According to resource-mobilization or resource-dependency approaches, in order to survive in a new environment an organization needs to acquire funds, personnel, information, power, specialized expertise, good connections, and authority.[17] ROW countries lack most of these resources, and face a great amount of difficulty due to their subsequent lack of ability to monitor and gather information that is needed to act in their interests. Among these resources, funding and expertise are the biggest obstacles to meaningful participation; lack of interest and of ability to communicate in English compound the problem.

In ROW countries, most of the financial resources to support participation in international agencies are in the public sector. ICANN's focus on private participation thus makes it difficult for people from these countries to gain funding. Some private companies are interested, but they have their own agendas and interests. Thus, the most feasible way for individuals and civil society to gain funding to participate in ICANN is to draw resources from the government. But the irony is that the more financial support the civil society participants receive from the public sector, the less their participation becomes truly private or self-governing. The relationship between the government and the private sector, and its influence on ICANN participation, will be discussed in more detail later.

Language has always been a problem in international participation, but it generates even more difficulty when we try to encourage the participation of civil society, as ICANN did. This is particularly applicable to many citizens of Asian and developing countries, most of whom have never been abroad or trained in English. In the case of Korea, the language problem exists in both the public and private sectors. Most government officials are educated and trained within the country. Those who have been trained in America tend to become academics or business people due to the quite closed bureaucratic system in Korea. Therefore, even government officials who participate in international meetings cannot communicate well in English in many cases, although they do have the financial resources and authority to employ translators or other experts. Those who participate from the private sector, especially from civil society, may or may not have better English ability than government officials, but when they don't, they cannot enjoy the same kind of assistance due to the lack of financial resources.

The language problem is not only applied to Asian or developing countries but also to other ROW countries that have not had much language difficulty in other meetings of the international organizations. Unlike most of the intergovernmental meetings such as those of the ITU, ICANN meetings are not simultaneously translated into multiple languages. In addition, the pace of communication in ICANN meetings is quite fast compared to other international meetings. Non-English speakers find it very difficult to follow the flow of the meetings. The lack of translation, compounded by the speed of communication, places all the countries that do not use English as their official language in a disadvantageous position. The disadvantage is felt not only by Asian countries or developing countries but also by many of the European countries that use French or Spanish as their primary language. Thus, the new governance system of ICANN places not only the developing or Asian countries but also most of the ROW in a disadvantaged position.

Cultural differences related to the style of communication and decision-making processes make it even more difficult for some ROW participants to be active players in ICANN. For example, decision making through presenting arguments, exchanging different views, and criticizing them is not familiar to people in many Asian countries. In Asian societies, most people are not trained or encouraged to talk freely and openly in public, especially regarding sensitive issues that involve conflicting interests. Most Asian people feel more comfortable being instructed by senior people or those with authority than with criticizing or arguing with the existing authority. In the ICANN setting, however, they are pushed to argue and express their positions, which may result in confusion and vague feelings of inadequacy. Also, Asian institutional processes are arguably slower-paced in general than Western ones. ICANN has a great many urgent new issues that require quick discussion and decision making, which places Asian, particularly individual, participants who lack institutional support and incentives in a difficult position.

The widespread use of online communication in ICANN also generates a variety of difficult situations in ROW countries. In ICANN many issues are dealt with through e-mail communication or bulletin boards, with great speed. Some ROW participants do not have the basic resources to have access to this kind of communication. Even with the technical resources, online communication is not so easily learned and acquired in ROW countries. The lack of experience in online discussions and even the time differences between the East and the West contribute to participation problems. The problem is exacerbated when many of the Internet domain name issues are new and cannot be grasped fully within a short period of time. A slow pace of thinking and pondering is not practical when the discussion is heated. A Korean participant said, "I looked at the bulletin board at night, and thought about it, but could not make a decision right away as to what to say myself. So I decided to post my opinion the next morning, but I had a busy day. I went to work, did my job, met people, and came back to my computer that evening, only to find that most of the discussion was made during the day, which was at night in America, and it was almost over already." When the participants were not actively engaged in the online communication, attending the offline meetings often proved to be a lot less meaningful and influential. Thus, both the private sector and the public sector in Korea can make very limited participation in this regard due to their unfamiliarity with the intensive online communication.

ROW countries may also have other problems due to a historical and institutional background that is different from the United States. Notions of self-regulation and the privatization of governance valorize Western ideas of freedom and liberty in cyberspace. However, to ROW societies these values may not be the most important ones, at least not at this time. When countries do not have the historical and philosophical background that makes the private sector's and individuals' participation meaningful, the self-regulatory framework may not be recognized as plausible even if it "sounds" good.

Because the Internet domain name and IP address economy is quite new and dominated (both economically and in terms of expertise) by the United States and Europe, ROW participants face huge levels of uncertainty regarding what they can get out of the process, who they should interact with, and what the real issues are for them. Uncertainty is exacerbated by the unusual form that ICANN has taken, with completely new and relatively informal mechanisms for policy development. The fact that many ROW countries lack clearly defined national or group interests in ICANN-related issues is one of the most significant barriers for them to participate meaningfully in ICANN. In order for a country or a group to consume financial and human resources to participate in a meeting or an organization on a continuing basis, there must be a goal or a purpose that is clearly in line with their interests. The important question then becomes whether the interested parties exist in the society. Many of the issues in the ICANN process are not the most important and urgent issues ROW countries are currently facing. The ROW countries may not yet have developed the technical infrastructures, or the business or cultural interests related to these infrastructures, that need governance.

Even if the ROW participants do not perceive ICANN issues to be directly related to their interests, they do understand that many of the decisions made at ICANN will eventually influence all Internet stakeholders. Hence, by the time the stakeholders in ROW countries finally develop their interests in these areas, many of the important decisions will already have been made for them by the U.S. and European participants. But when the country does not have strong and immediate social or national interests in the issues dealt with at ICANN, it is difficult for its participation to be urgent or genuine, even if it recognizes that the decisions made at ICANN will later influence its society.

South Korea: A Case Study

In this section, we go into more detail about the problems confronting ROW participation in ICANN by analyzing a specific institutional context: South Korea.

Stakes and Stakeholders

In Korea, the existence of material interests in relation to ICANN issues tends to be greater than in many other ROW countries. The development of digital technology is quite advanced and the spread of the Internet is extensive. There are several Korean groups with an interest in participation, such as the country code TLD manager, KRNIC; gTLD registrar businesses; and to a lesser degree noncommercial and governmental actors. Korea has about ten ICANN-accredited registrars. Local registrars for the ccTLD also have some business interests in ICANN policy making, although they do not see themselves directly influenced by it. Korea has also developed a steady, but individual-based participation in the Non-commercial Users Constituency (NCUC), as the civil society in Korea has been interested in social and cultural issues that affect Internet users.

The Korean government has also participated more regularly in GAC compared to other ROW countries. But it tends to oppose private ordering or self-regulation, an attitude that conflicts with the ICANN regime. At the ITU Marrakesh meeting in 2002, the South Korean government suggested that the current activities of ICANN should be delegated to ITU. Within Korea, the government routinely attempts to control or influence the private sector's participation in ICANN. Whereas civil society sees ICANN and the new Internet governance system as a slightly more desirable model of policy making and decision making regarding Internet issues, the Korean government regards it as a nuisance and a threat to its preexisting regulating power and authority. This power struggle between the civil society and the government in Korea contributes to the difficulty and confusion regarding ICANN participation, as discussed earlier.

On the other hand, Korea has very little interest and participation in the Generic Names Supporting Organization (GNSO)'s Registry Constituency, the Intellectual Property Constituency, the Internet Service Providers Constituency, the Business and Commercial Users Constituency, and the Root Server Advisory Committee. Not only does Korea have few resources to participate in these areas, but also it does not have keenly developed interests or experiences. Korea has no registries or root servers, and ISPs and relevant businesses in Korea also do not have developed interests in order to undertake lobbying or active participation, although there is some interest by companies such as Netpia in multilingual domain names. There is some interest in intellectual property issues, but only among individual academics or lawyers, not at the level of law firms or businesses. Korea has not yet developed such specialized services in this area as to participate in ICANN for clearly defined purposes. Even when their interests are relatively clear, the potential participants still have to ponder whether they will receive the appropriate reward by participating. They confront a cost-benefit calculus with a very uncertain result, and may become reluctant to put their limited resources into participating in this international organization.

The Tensions of Self-Governance: KRNIC and the Government

Korea and many other countries have a history of strong government administration and a relatively weak private sector that closely follows the government's instructions. This fundamental difference from the United States in institutional framework and experience makes it more difficult for Korean participants to understand what to do within ICANN. Consider the following example. In the 1990s, the Korean Ministry of Information and Communication (MIC) came back from OECD meetings and told its Internet Service Providers to come up with suggestions for a self-regulatory framework. After considerable pondering, the ISPs and other Internet players came back to the government and said that they would rather have the framework given by the government. This happened for two reasons. First, the Internet actors were worried that the government would later say or do something that contradicted what they would put forward. Second, they did not have a clear idea of what to do. For its part, the government also did not know what to do, because its officials have not had experience with Internet governance.

This incident clearly illustrates the persistence of top-down approaches to policy making and implementation in Korea. That approach dates back to the 1960s, when Korea was undertaking its rapid economic development. Korean governments have little experience in systematically doing research and mobilizing experts and interest groups. The government does not trust the private sector to come up with feasible solutions to the controversial issues or problems in the society. They view the government as being accountable for any problems that arise from policies. In addition, they have experienced criticism and bureaucratic reprisals if any problems occur. As a consequence, in Korea the public sector, not the private sector, acts and is viewed as the responsible body. So self-regulation and private ordering are yet to be meaningful. The attitude of deference to government characterizes many ROW countries, not just Korea. This is one of the reasons why the United States ends up leading within ICANN and in other activities involving Internet governance.

The tension between the government and the private sector, including civil society, in Korea is most keenly revealed in the policy controversies surrounding the Korea Network Information Center (KRNIC). KRNIC is the entity in Korea that centrally coordinates ICANN participation. Arguably, KRNIC is a private entity. It was founded pursuant to the Basic Law of Facilitating Informatization as a nonprofit foundation that deals with matters regarding facilitating the use of information services and the management of Internet addresses. But there is much confusion as to the legal status and the actual nature of KRNIC's activities. Some argue that KRNIC is a private organization because its activities are not those mandated by the government.[18] Others argue that KRNIC is an administrative agency under MIC because it was founded under a law enacted by MIC. And the articles of incorporation of KRNIC were drafted to give MIC authority over many of its activities. This confusion is a continuation of historical struggles over KRNIC's founding. While MIC wanted to retain its regulatory power over matters regarding Internet addresses, due to the initiative of the Ministry of Finance and Economy and the Ministry of Planning and Budget, KRNIC was made a nonprofit foundation instead of a governmental agency. However, the close tie between MIC and KRNIC is yet to be broken. Many KRNIC staff members are former employees of organizations that were founded, funded, and controlled by MIC, such as the National Computerization Agency and the Korea Agency for Digital Opportunity and Promotion. These people lack private sector experience and are keenly sensitive to the stability and conservatism of the public sector. Some KRNIC staff members tend to perceive themselves as half-employees of the government and to acquiesce to the requests of MIC.

The new role for the private sector generates additional tensions. KRNIC supports a number of private committees, such as the Name Committee and the Name and Number Committee, which are the central channel by which the civil society and individual experts and activists participate in the decision-making processes regarding Internet address policy. When the views and recommendations of these committees conflict with those of KRNIC or when the committees make decisions that are not favorable to KRNIC, one of the easiest ways to limit their power is to encourage the intervention of MIC. At the same time, when the committees do not follow the instructions or suggestions made by MIC on policy matters and make other recommendations, or simply fail to follow the instructions immediately, MIC would also encourage KRNIC to ignore the decisions of the committees or to simply implement the policies suggested by MIC. As a result, there is escalating tension among the committees, which are private decision-making bodies; KRNIC, which was designed to support and facilitate this private decision-making process; and MIC, which is the traditional governmental body. One negative result of the growing tension is that many committee participants, who are mostly volunteer experts with their own jobs and many other obligations, observe that their evaluations and decisions are not followed. They become disillusioned about the so-called private rule-making structure, and scale back their participation or simply turn away from this process.

The tension has an unfortunate effect on ICANN participation. Recently, fewer of the Internet experts who were actively involved in the committees since the beginning are turning out for meetings. They cannot be replaced, as there are only so many people in

Korea with the will and expertise to participate. Also, the role of KRNIC in coordinating and supporting private sector participation in ICANN sometimes is not fully realized, even absent any bad will or lack of ability. KRNIC has been supporting those who participate in ICANN meetings since its inception in 1999. Those who receive the most support have formal membership or chairmanship in ICANN, councils, or constituencies. Members of KRNIC committees who appear to participate actively in the policy discussions within Korea are also supported. In the earlier stages of ICANN, and when the meeting was held in a nearby country, a great many people from Korean civil society and educational institutions were supported to attend the meetings. But as the exact role that many of them played was not clear, the support was limited to those who could actively contribute to the discussion and other processes. However, sometimes very active participants did not receive any support from KRNIC. The only apparent reason for their exclusion was that they openly opposed the amendment of a bill suggested by the government.

In addition, attending ICANN meetings does not automatically result in making a meaningful contribution. To participate meaningfully, one must constantly follow the discussions of important issues and understand the policy alternatives. Only a small number of Korean participants have achieved this level of involvement. They did it not because of the strong support of the government or other entities, but because of their own individual dedication and strong will. They also must perform jobs to make a living, however, and are involved with many other matters more urgent than Internet addresses. Thus they find it difficult to keep their involvement strong. In the future, KRNIC must explore how to incorporate this need for constant involvement and support. So far Korean participation in ICANN has remained relatively stable, but it is still scattered and individual-based. It remains to be seen if this tension between the government and civil society could later function in a positive way to facilitate more dialogue and open the way to a new, creative governing mechanism that involves all interested parties.

Korea and the UDRP

The Domain Name Supporting Organization (DNSO) task force on revising ICANN's Uniform Domain Name Dispute Resolution Policy (UDRP) is an example of how ROW participation plays out in the nitty-gritty details of policy making. The DNSO Names Council, which formed the task force, consisted at the time of three seats elected from each constituency. Each of a constituency's three seats must be drawn from a different geographic region. When the UDRP Task Force (TF) was formed, each DNSO constituency got to select one representative; additionally, each dispute resolution service provider (two of which were based in North America and one of which was based in Geneva) nominated two representatives each. These selections were rounded out by a representative of complainants, a representative of respondents, and two independent experts in international law and alternative dispute resolution. The resulting geographic distribution of the TF was sixteen North Americans, three Europeans, one African, one Asia Pacific, and one Latin American/Caribbean. Complaints about the lack of geographic balance were aired. The TF chairs decided to add two more representatives, one from Latin America and one from South Korea.

While the distribution looks awful, it would be a mistake to view geographic representation as the primary goal of the UDRP deliberations. The UDRP involves conflicts between trademark rights and domain names in gTLDs. The vast majority of UDRP cases involve complainants and/or respondents based in North America and domain names registered in .com; the names involved are English or European names involving trademarks registered in the developed world. The issues involved may or may not have a regional or geographical dimension. Most of the time, the problem of domain name dispute resolution pits trademark-holding corporations against individual or small business registrants of domain names, and the nature of these conflicts are often quite similar whether they are in the global North or South.

In a constituency-based or stakeholder system, each interest wants to send the strongest and most informed representative it can to a task force concerned with a high-stakes, contentious issue. On this issue (the UDRP), expertise and strength of advocacy is concentrated in North America and Europe. Thus, even the public interest advocates in the Noncommercial Users Constituency, a part of ICANN that has historically been most sensitive to issues of ROW representation, nominated U.S.-based law professor Michael Froomkin to the UDRP Task Force. Professor Froomkin is an expert in this area, a strong advocate of individual rights, and has served on WIPO panels before. The noncommercial constituency favored him because he could be presumed to know how to vocally battle the dominance of intellectual property interests. To the public interest advocates, task force representation was about trademark interests versus civil society interests, not regional or geographical interests. Nevertheless, this method of developing policy definitely marginalizes the ROW; or perhaps it would be more accurate to say that the method reflects the marginal role that the ROW already plays in domain name trademark disputes.

Being "represented" on a task force is one thing; being influential on it is another. Over the fifteen-month life of the UDRP Task Force, the African representative sent only one message to the TF listserv, the Korean participant sent only two messages, and the Latin American participant sent eight. This pattern is quite typical of ROW participation in ICANN processes. Even when the ROW participants were unusually active, as the Latin American representative was, they were primarily pointing to sources of information or responding to tasks given to them by the TF chair—they were not taking an aggressive role in proposing or shaping policies. Professor Michael Froomkin, in contrast, sent thirty messages to the list and frequently attempted to shape the agenda or promote specific policy ideas.

Another reason ROW participants are not more active is that they are often overloaded with tasks. Expertise in this highly specialized area of Internet policy is very rare in their countries, so people who are knowledgeable and capable typically have numerous responsibilities thrust upon them and must wear multiple hats. Typically, they have heavy domestic responsibilities as well as many international processes to monitor. When sending his one message to the list, for example, the African UDRP TF representative previously mentioned wrote "Apologies for not taking a more active role in this task group . . . been very busy on this end especially with the release of new piece of draft legislation (South Africa—Electronic Communications and Transactions Bill)—It appears our government wishes to regulate a number of issues, especially the ccTLD." By way of contrast, the representatives of intellectual property interests were trademark lawyers fully specialized in trademark issues and in some cases specialized in domain name cases within trademark law and dispute resolution.

If Korean participation in UDRP reform at the ICANN was minimal, within Korea the Uniform Domain Name Dispute Resolution Policy was not even acknowledged as important at first. Even after some scholars tried to alert people to its significance, not many domestic actors understood the implications. Some stakeholders, such as domain name registrants in Korea, did understand and asked the government or KRNIC to do something. Not much could be done, however, because few people even knew about the UDRP, and those who did know were simply too busy to take on additional tasks. Even when Korea attempted to develop its own dispute resolution system there was much confusion. It was not even clear why Korea would need a dispute resolution system apart from the court system and ICANN's UDRP. At first, the new system imitated UDRP completely, leading to many disagreements. Some experts including lawyers and scholars cautioned that UDRP enlarges the trademark holders' interests, especially the big companies in developed countries. They also argued that the UDRP was made for gTLDs with first-come, first-served registration policies, which naturally generates jurisdictional and legal conflicts, and thus is not appropriate for the Korean ccTLD, which has a more restricted registration policy. Many suggested different kinds of dispute resolution policies that are more consistent with the Korean legal system and situations regarding the use of domain names in Korea. Others advocated avoiding the creation of any new system of dispute resolution at all, arguing that national courts and other arbitration centers, using Korean laws and procedures, are enough to deal with domain name disputes. On the other hand, some argued that they did not see why Korea should have a different system than UDRP when UDRP will govern the international domain names and thus will soon govern other domains as well. The cacophony of voices seemed to generate an impasse for a while. Eventually, MIC intervened after it succeeded in amending the relevant law to get the power to create a domain name dispute resolution mechanism.

After a few meetings and discussions led by MIC, the Domain Name Dispute Resolution Center (DDRC) was established in KRNIC in 2002. The dispute resolution policy is quite similar to UDRP in terms of its procedural rules, without similarly specific, substantive policy. The composition of the panelists (judges of domain name disputes) also reflects the influence of the government. Many of those who had been involved with domain name dispute resolution policy from the beginning of the discussion but had different views from that of MIC were excluded from the list of panelists. For example, among the eight Koreans who are working as WIPO UDRP panelists, only one person was included on Korea's national list. In this case, the Korean government excluded the more liberal civil society interests as dispute resolution panelists, pursuing a more pro-business agenda. This may have a negative impact on the domestic and international discussion regarding domain name dispute resolution policy, because most of those who were excluded are the experts in international domain name dispute resolution.

Not many cases have been decided through DDRC so far. There are numerous reasons for this: the higher-than-expected cost (880,000 Korean won for one-member panel and 1,760,000 won for three-member panel); the existence of other dispute resolution bodies such as the national courts, the Korea Commercial Arbitration Center, and the E-Commerce Mediation Committee, which have already been dealing with domain name disputes; and the fact that the short result of the decision is publicly known but the decision itself is not provided for review. It remains to be seen whether the new system is developed as a central body for domain name disputes as it was designed, and whether the existence of this center can facilitate Korea's participation in the UDRP processes and its changes or future developments.

ICANN versus a Traditional Intergovernmental Organization

Now we confront an important question: were the problems we have encountered within ICANN unique to ICANN's peculiar status and structure, or were they a function of ROW countries' relative lack of resources and weaker bargaining power, and hence would occur in *any* international arena? This proved to be a difficult question. The answer was not easy to formulate and difficult for the coauthors to agree on. As we have seen, cultural differences related to experiences with online communication, active discussion, and self-ordering had a detrimental effect on Koreans' ability to participate in ICANN. The peculiar structure of ICANN, as developed and operated mainly by U.S. and European participants, seemed to add problems to the already difficult participation of the ROW countries.

However, the conflict between Korea's Internet-related civil society participants and the Korean government raises an important qualification. Because ICANN is a new institutional framework, it has provided a channel for participation for a different, and broader, set of people in Korea (and probably several other ROW countries) than an institution that relied entirely on traditional governments. Even though some civil society interests in Korea are highly dissatisfied with the ICANN model and feel marginalized within it, they are relatively more empowered within the ICANN framework than they would have been in a traditional intergovernmental organization system. That is, the new private-sector-orientated governance structure shifted power away from governments and toward civil society, even in ROW countries.

What makes this problem even more complex is that, as noted before, in ROW countries most of the financial and human resources available for participation in international organizations are in the public sector. Yet the public sector often lags behind individual users and actors in terms of their understanding and appreciation of the interests at stake in Internet-related issues. If these individual actors outside the government are provided with the necessary support, they can become a great asset to the ICANN process and enhance ROW participation. However, civil society in many ROW countries did not develop independently as in the Western countries, but with very close interconnection with the government. Thus, encouraging this noncommercial private sector in ROW to develop more strength and identity is an important step to be taken before we envision ICANN-like private sector participation. Funding and otherwise supporting relevant NGOs would be a good way to start, but devising a new way to mobilize and organize these potential nongovernmental representatives is also needed.

Clearly, we cannot too easily condemn or ignore the role of government. An ICANN or ICANN-like alternative did not and cannot eliminate or avoid the need for "governance" and government-like representation and procedures.[19] It may be beneficial to fundamentally re-think the usefulness and advantages of the self-governance system, and search for a new, creative system. As a research agenda, it might be useful to do a large survey of different governance systems in other policy domains, countries, and societies, and explore whether we could find a better combination of the participation and initiatives from the public and private sectors.

Recommendations

We put forward four recommendations designed to improve the lot of ROW countries in Internet governance.

Create Open and Regular Procedures for TLD Additions

We must not lose sight of the importance of facilitating open and competitive markets in international governance regimes. It is, after all, the relative weakness of domestic economic players and the consequent lack of expertise about the issues and stakes that most handicaps ROW participants in ICANN processes. ICANN's management of the root to date has created artificial scarcity, making it extraordinarily difficult for new players to enter the market for domain name registration.

In the registrar market, ICANN has implemented a simple accreditation process that allows any business meeting certain qualifications to enter the market and compete. This approach has worked well at making the retail market competitive, driving down prices and improving service. The door is always open to ROW companies to become accredited registrars, and as noted before about ten Korean firms already are.

In contrast to the registrar market, ICANN has failed to create sufficient competition in the far more important registry market.[20] In the gTLD market, 85 percent of the share is controlled by Verisign, even after the divestiture of .org. ICANN has been unable to successfully produce competition at the registry level because it has not defined a routine method for adding top-level domain names and for authorizing new registries to operate them in response to market demand. At present, ICANN has fixed no regular timetable for accepting and deciding upon applications for TLDs. No one knows when ICANN will add new TLDs; no one knows upon what basis it might choose to do so or refuse to do so. ICANN has defined no uniform criteria for evaluating applications or for accrediting registries. This policy vacuum has made the addition of new domains a painfully slow, unpredictable, and entirely discretionary process. The effect has been to substantially raise the costs of entry into the domain name registry market, and to make insider politics rather than economic value the chief determinant of who gets to participate. The delays and costs of this nonpolicy negatively affect potentially competitive ROW countries, such as China, Korea, or India.

A centralized system of resource allocation such as ICANN's puts control over entry into the domain name market in one U.S.-based organization's hands. In that regime, well-connected U.S. and European actors have proven to be consistent winners, and ROW players consistent losers. Incumbent registries with longstanding connections to ICANN, such as Afilias, are most likely to win ICANN beauty contests. Major business associations in developed countries, such as the Information Technology Association of America, Club Informatique des Grandes Entreprises Françaises (CIGREF) of France, and the International Chamber of Commerce are able to restrict new businesses from entering the domain name market in order to make the world safer for their trademarks. They have also been able to delay and possibly regulate to death the prospect of multilingual domain names, an extension of the domain name system that could empower ROW players rooted in indigenous, non-ASCII (American Standard Code for Information Interchange) language groups. Whenever ICANN has made awards of valuable resources, such as new top-level domain names in 2000 or the reassignment of the .org TLD in 2002, it has consistently favored American and European firms.

An open entry regime for new TLDs, such as that proposed by Mueller and McKnight,[21] would greatly improve the status of ROW participants in ICANN. This assertion contradicts the common belief that political, equity-focused processes are better for ROW countries than market mechanisms. However, in the domain name space, control of entry into the market makes the industry subservient to the interests of incumbent registries, trademark interests, and the ASCII–based language groups. A more lightweight, open market procedure that allowed multiple, heterogeneous players to enter and become stakeholders could only improve the status of ROW actors.

Add a "Multicultural Awareness and Outreach Advisory Committee" to ICANN's Structure

In order to alleviate the difficulties generated by historical and cultural differences, an internal educational program that helps ROW participants become more familiar with alien ICANN processes is urgently needed. We propose a Multicultural Awareness and Outreach Advisory Committee within ICANN to help ROW people participate. Advisory committees are well-established features of ICANN's structure; there is one for governments, for example, and one for root server operators. The need for such a new committee is especially urgent given ICANN's failure to employ long-established practices in international meetings, such as language translation and relatively slow-paced discussions. This new advisory committee would be charged with mentoring ROW members and encouraging their participation. It could provide "consciousness-raising" forums where non-English speaking and developing country participants could share experiences in ICANN processes. Lessons about how to participate, such as knowing that making arguments in meetings is important to make their voices heard, could be reinforced.

Regionalize Control of Resources

Another recommendation is to explore the possibility of regionalizing control of DNS resources. Decentralizing some of the authority over top-level domain names by region, while retaining enough coordination to ensure the technical integrity of the network, might reduce some of the participation barriers by placing things in a slightly less heterogeneous cultural, political, and economic environment, and allowing policies to reflect more localized and not just dominant multinational interests. ICANN has already regionalized control of IP addresses through the creation of Regional Internet Registries (RIRs); something similar might be done with domain names.

Balance Representation and Enhance Civil Society Participation

ICANN's abolition of at-large elections in 2002 was not protested much by most ROW countries, and indeed most governments from the ROW viewed a democratic ICANN as a threat to their supremacy over international policy making. Long term, however, those elections could have stimulated greater organization and participation among ROW countries. The catch, of course, is that some of the organization would come from outside traditional governmental channels. The incentive to win board seats via regional elections would have stimulated political organization by Internet-oriented civil society groups that spanned national boundaries. Such organizations would be subject to all the problems and limitations of self-governance we discussed above. But ICANN's potential to become a new, more distributed and democratic form of international governance should not be discarded cavalierly. By creating opportunities for civil society organization and real rewards (i.e., real influence) when they participate, ROW capabilities can only be strengthened. We therefore recommend that ICANN return to some form of direct election of its at-large board members by individuals.

Detach ICANN from the U.S. Government

ICANN operates under agreements with the U.S. government, and the Department of Commerce has policy authority over the content of the Internet DNS root. ICANN will not be a truly international organization and fully self-governing until that control is released. In the short term, this is more of a symbolic issue than a substantive one, because until ICANN's legitimacy is established and its procedures more settled, the U.S. government is unlikely to relinquish its special grip over ICANN's authority; and other governments, no matter what they say in public, do not seem to mind the stabilizing presence of some higher authority. In the long term, however, the special control of the United States over the root servers and ICANN's ability to implement its policies undermines ICANN's legitimacy. The highly political U.S. Congress, for example, could always intervene in some way to exploit the potential leverage that control of the root might give. This issue could flare up at any time, and is most likely to flare up under the worst circumstances (war, terrorism, major political conflict).

Conclusion

While efforts to broaden participation in international institutions are laudable, we should avoid setting unrealistic expectations about how expanding participation might alter major international power relationships. International governance relationships of necessity reflect, as well as reproduce, inequalities in power and wealth among countries. To think that stronger countries will passively surrender advantages to weaker countries, possibly harming their own interests, simply to facilitate more widespread participation in international governance, is not realistic. Structural changes, such as the open entry procedure and regionalization have a better chance of success in the long term. More democratic processes that devolve power to individuals and civil society rather than other states or IGOs also may have the potential to short-circuit traditional political objections to change. But of course implementation of those recommendations must be done over the objections of those who benefit from the current regime.

Nevertheless, there is some hope that the growing participation and influence of civil society actors in international policy arenas can create political pressure to move in the directions we have described. It is noteworthy that during and after the 2002–2005 World Summit on the Information Society process, both civil society actors and business interests resisted calls by nation-states to turn ICANN's functions over to an intergovernmental organization.[22] True, the civil society organizations involved in WSIS, similar to the authors of this report, had ambiguous and unsettled opinions about the problems of representation in ICANN. Over the long term, however, ICANN's evolution may signal important and positive changes in international governance.

Notes

1. We are counting Australia and New Zealand as European countries in this sense.

2. See Mueller 2002, 211–226.

3. NTIA 1998b.

4. "The Internet is emerging as a global marketplace. The legal framework supporting commercial transactions on the Internet should be governed by consistent principles across state, national, and international borders that lead to predictable results regardless of the jurisdiction in which a particular buyer or seller resides." "A Framework for Global Electronic Commerce," The White House, July 1, 1997. Available at http://www.ecommerce.gov/framewrk.htm.

5. Ibid.

6. Ibid.

7. See, for example, Weinberg 2002.

8. See NTIA 1998a,b.

9. NTIA 1998b.

10. See Postel 1994.

11. See Lynn 2002.

12. Many are not, however. For example, there is a .io for the Indian Ocean, and special codes for numerous island protectorates or territories such as .cc for Cocos and Keeling, .ws for Western Samoa, and so on.

13. Of course, there could be multiple TLDs for each country, such as .germany or .france, but ICANN's policy of artificial scarcity and the demands for exclusivity by governments and incumbent registries make that option unrealistic for some time.

14. At the June 2003 meeting in Montreal, an agreement was reached that will allow the country code managers to create their own ICANN supporting organization.

15. A compilation of articles is in Klein 2001.

16. See NAIS 2002.

17. See Lauman and Knoke 1987; Knoke and Wood 1981; Aldrich 1979.

18. See Kim 2001.

19. See Mueller 1999 for a critique of self-regulation rationales as applied to ICANN.

20. Registries are the critical infrastructure of the domain name system. Registrars are just intermediary services built on top of that infrastructure. Ultimately, effective competition in domain services requires open entry and robust competition at both the registry and registrar levels. If there is insufficient competition in the registry market, the whole domain name services industry is not sufficiently competitive. The real technical expertise and development requires knowledge of the registry level.

21. See Mueller and McKnight 2003.

22. See the "Statement of the Civil Society Internet Governance Caucus" at http://prepcom.net/wsis/1058345885001.

References

Aldrich, Howard E. 1979. *Organizations and Environments*. Englewood Cliffs, NJ: Prentice-Hall.

Kim, Ki-Jung. 2001. "A Study on the Participation of Internet Users in Policy-Making and Elections." Paper presented at the Internet Users' Forum, September 1, Seoul, Korea.

Klein, Hans, ed. 2001. *Global Democracy and the ICANN Elections*. A special issue of *Info* 3, no. 2 (August).

Knoke, David, and James R. Wood. 1981. *Organized for Action: Commitment in Voluntary Associations*. New Brunswick, NJ: Rutgers University Press.

Lauman, Edward, and David Knoke. 1987. *The Organizational State: Social Choice in National Policy Domains*. Madison: The University of Wisconsin Press.

Lynn, Stuart. 2002. *President's Report: ICANN—The Case for Reform*. Marina Del Rey, California: Internet Corporation for Assigned Names and Numbers, February 24. Available at www.icann.org/general/lynn-reform-proposal-24feb02.htm.

Mueller, Milton L. 1999. "ICANN and Internet Governance: Sorting through the Debris of Self-Regulation." *Info* 1, no. 6: 477–500.

Mueller, Milton L. 2002. *Ruling the Root: Internet Governance and the Taming of Cyberspace*. Cambridge, MA: The MIT Press.

Mueller, Milton L., and Lee McKnight. 2003. "The Post-COM Internet: A Five-Step Process for Top-Level Domain Additions." Syracuse, NY: The Convergence Center. Available at www.digital-convergence.org.

National Telecommunications and Information Administration. 1998a. *A Proposal to Improve the Technical Management of Internet Names and Addresses (Green Paper)*. January 30. Washington, DC: NTIA.

National Telecommunications and Information Administration. 1998b. *Management of Internet Names and Addresses: Statement of Policy (White Paper)*. June 5. Washington, DC: NTIA. Available at www.ntia.doc.gov/ntiahome/domainname/6_5_98dns.htm.

NGO and Academic ICANN Study (NAIS). 2001. *ICANN, Legitimacy, and the Public Voice: Making Global Participation and Representation Work*. Available at www.naisproject.org/report/final.

Postel, Jon. 1994. "Domain Name System Structure and Delegation." Reston, VA: The Internet Society Request For Comments 1591, March. Available at www.rfc-editor.org/rfc/rfc1591.txt.

von Arx, Kim G. and Gregory R. Hagen. 2002. "Sovereign Domains: A Declaration of Independence of ccTLDs from Foreign Control." Richmond Journal of Law & Technology IX, no. 1 (Fall).

Weinberg, Jonathan. 2002. "ICANN, 'Internet Stability', and New Top Level Domains." Available at www.law.wayne.edu/weinberg/icannetc.pdf.

15 Multistakeholderism, Civil Society, and Global Diplomacy: The Case of the World Summit on the Information Society

Wolfgang Kleinwächter

In the past twenty-five years an academic debate has raged about the future role of the nation-state in the information age. One group of scholars argues that the "Westphalian System" of sovereign nation states, which emerged in the seventeenth century from thirty years of religious wars and was driven in its development to a high degree by the first wave of the industrial revolution and the growth of national economic markets, will weaken amid the globalization that has created global markets even for local products and services. The role of the nation-state will decline and global institutions with transnational corporations will take the lead in policy development and decision making. Governmental hierarchies will be replaced by nongovernmental networks, and the international system will become more and more decentralized, issue-oriented, and bottom-up.

Others argue to the contrary. For them globalization will strengthen the role of the nation-state in the future. While cyberspace has no physical borders and will continue to grow, people will continue to live in real places where they need and enjoy their local roots, communities, and cultures. These scholars introduce the concept of "glocalization" in which both global processes and local activities are interlinked and are seen as two sides of one coin. The role of the state will change; instead of regulating everything, it will be more responsible for creating legal frameworks, facilitating political processes and moderating discussions, and consensus building among nongovernmental stakeholders, in particular between providers and users of goods and services that will undertake growing responsibility for their own problems. According to these scholars the principle of "national sovereignty," as practiced in the nineteenth and twentieth centuries and defined in the United Nations Charter of 1945, is not outdated but needs a broader interpretation.

The four-year-long World Summit on the Information Society (WSIS) process reflected the complex rebalancing underway between states and nonstate actors in the globalized information age. In this chapter, I trace and assess the involvement of global civil society in the evolution of the WSIS. I believe that the WSIS experience indicates that civil society, and also the private sector, will increasingly win a seat at the table and hence could change substantially the practice of multilateral diplomacy on information and communication technology (ICT) issues in the years ahead. If these actors become more mature and better organized and find ways to legitimize their input into global policy development and intergovernmental negotiations, they will have a growing, direct impact on the result of diplomatic conferences. Even more, the nongovernmental stakeholders will be increasingly needed in the implementation phase after the end of intergovernmental negotiations. A growing set of issues on which governments make policy, can be managed only if private industry and civil society become engaged with investment, awareness-raising, education, and expertise.

The WSIS case is certainly a special one because as the process wore on, it became mainly focused on the Internet. The Internet, as a network of networks, is run not by a single entity but by numerous organizations, particularly private corporations and nonprofit technical bodies. By its nature the Internet is international; there is no separate Internet for a single nation. Therefore the WSIS was the ideal testbed to figure out how a new trilateral relationship among governments, private industry, and civil society could be developed and how far this type of multistakeholderism can be implemented. Indeed, there was no alternative to this broader concept for policy development based on the involvement of all stakeholders in their respective roles and responsibilities. However, it will be interesting to see whether the best practices that were developed in the WSIS process will have an effect also on other intergovernmental organizations and diplomatic negotiations in the future.

From Westphalia to Multistakeholderism

Globalists versus Nationalists

The political and economic developments of the past decade indicate that the trends toward a new relationship between governmental and nongovernmental stakeholders and the growing role of civil society in global policy making go far beyond ICTs and the Internet. However, the political processes are rather complex, full of inner contradictions and are moving partly in different directions.

On the one hand, we have seen the growing progress of transnational activities on the global level. More and more decisions with relevance for the day-to-day operations of individuals and institutions are now made in international organizations like the G8, the World Trade Organization (WTO), and the World Bank. In Europe, power has shifted from the individual sovereign member states to the institutions in Brussels. At the same time, new nongovernmental stakeholders, like transnational corporations and nongovernmental organizations, have intervened forcefully in global policy debates.

On the other hand, we have witnessed the revitalization of nationalism in many parts of the world. New small states have separated from their previous "mother states," as with the Balkans and the new republics of the former Soviet Union. Bloody wars have been fought by local separatist groups from Chechnya and Abchasia and the Muslims in the western part of China. Nationalist movements from the Basque country in Spain to Scotland in the United Kingdom and Quebec in Canada have argued for more "nationalism" and independent "sovereignty" based on a special local history, culture, and language.

The Internet that has removed the barriers of time and space interlinks all these processes. Cyberspace, while itself both a result of and a driver of globalization, becomes also the space where both the transnationalists as well as the new nationalists are coordinating and campaigning. While both sides have their points and the discussion continues, the challenge is to go beyond an either-or approach, to take a more holistic view and find new flexible frameworks that allow pluralism and diversity and a broad range of individual solutions for specific groups combining elements of centralization and decentralization at the same time. The same challenge applies in the realm of global ICT governance.

To find complex multilayered solutions for global governance, the concept of "multistakeholderism" is useful. While there is no widely accepted academic definition of a stakeholder or the principle of multistakeholderism, among practitioners, there is a general understanding that stakeholders are groups of actors organized around specific common principles, values, visions, legal status, and organizational structures and that have a certain stake in a process or issue. Such actors are not linked primarily to a special nation-state but are linked in networks operating across borders mainly on the global level. However, the concept of multistakeholderism is also being experimented with on the national and local level in the form of public-private partnerships. So far, governments, private industry, and civil society have been identified as the main stakeholders on the global level. Furthermore intergovernmental organizations and, with regard to the Internet, also the technical and administrative community are seen as special stakeholder groups.

This new terminology began to appear in political United Nations documents in the 1990s. The "Earth Summit" in Rio de Janeiro in 1992, which adopted the "Agenda 21," can be seen as the starting point for this development. While the Rio summit was an intergovernmental meeting of sovereign nation-states, nongovernmental stakeholders like private industry and civil society, and in particular the environmental groups, were invited to make contributions to the process and to the implementation of the conference decisions.

The multistakeholder concept evolved further in the second half of the 1990s when the Internet stimulated the growth of global networks constituted by nongovernmental groups from the private sector and the civil society. The breakthrough came with the WSIS. In the WSIS process, for the first time in global diplomacy, the specific role and responsibility of the different stakeholders was discussed in depth, and the nongovernmental stakeholders contributed actively to the debate. For the United Nations, the appearance of private industry and civil society participants in the conference halls and at the negotiation table was a procedural revolution, which partly divided the governments into groups that supported the participation of new actors in the policy debate and those that rejected private industry and civil society partners for global diplomacy and discredited them as "pressure groups" and "noise makers" without any legitimacy. But in the end, the private sector and civil society succeeded in going beyond their traditional roles as critical observers and lobbyists and influenced to a certain degree the WSIS negotiations process and the final results.

The intensive interaction between stakeholders can make it difficult to figure out exactly which elements of language in the WSIS outcome documents came from the private sector and civil society. Nevertheless, it is certainly no overstatement to claim that their interventions directly affected the whole concept of a global information society, which the 2003 Geneva Declaration of Principles held must be a "people centred, inclusive and development oriented" social formation in which "everyone can create access, utilize and share information and knowledge, enabling individuals, communities and peoples to achieve their full potential in promoting sustainable development and improving the quality of life, premised on the purposes and principles of the Charter of the United Nations and respecting fully upholding the Universal Declaration of Human Rights."

One of the most interesting playing fields and test cases in this context was the battle over Internet governance. The Internet, which is not a unique body but a multilayered network of networks, mainly is managed by players from the nongovernmental sector, such as companies, technicians, user organizations. The Internet itself has grown out of a process that took place mainly in the shadow of governmental and intergovernmental regulation. Its development was largely driven bottom-up by the technical community, the provider and user of Internet applications and services themselves. But the fact that the Internet now is seen as one of the most critical infrastructures of the twenty-first century has mobilized governments to step in. They want to have a say in how the Internet is managed, regulated, and controlled.

Consequently, one group in the WSIS process argued that the Internet should remain in the hands of the global private actors, recognizing the borderless nature of the virtual cyberspace. Vint Cerf, chairman of the board of the Internet Corporation of Assigned Names and Numbers (ICANN) rejected efforts to create new regimes for Internet governance with the argument, if it isn't broken, don't fix it. In contrast, a substantial number of governments preferred the creation of an intergovernmental regime for the management and further development of the Internet. As a result of these different ideas, the issue of Internet governance became one of the most controversial subjects within the WSIS process.

However, it became clear during the WSIS debate that the management, regulation, and control of the Internet cannot be done by one stakeholder group or one organization alone. Individual national solutions based on the traditional concept of state sovereignty will not work in the borderless cyberspace. The management of the Internet needs a global approach where many players on many layers work together in a mechanism of communication, coordination, and collaboration to make the Internet work, to secure its stability and security, and to stimulate innovation and creativity for the development of new applications and services for the benefit of the whole global community.

To understand the new concept of multistakeholderism and its meaning for the diplomacy of the twenty-first century, it makes sense to look deeper into the WSIS case, to examine what experience has been acquired and what lessons must be learned to prepare for the future. Hence, the rest of this chapter will look into the history of the WSIS process and its background, analyze in detail the multistakeholder debate during the 2002–2003 Geneva phase and the 2004–2005 Tunis phase of the WSIS, and make some recommendations for further actions.

The Long Road to WSIS

The WSIS is "unique," said UN Secretary General Kofi Annan in his opening speech to the Geneva summit, on December 13, 2003. "Where most global conferences focus on global threats, this one will consider how best to use a new global asset."[1] And indeed, the WSIS process, which aimed to bring half of humankind online by the year 2015, was less about the threats of the past and more about the opportunities of the future that come with the development and application of new ICTs. It is true that the ways we live, work, teach, do business, research, and entertain are more and more shaped by ICTs. In 2006 there were more than two billion people using a mobile phone and 1.1 billion online. The Internet allows everyone to communicate with everyone everywhere any time. In the borderless cyberspace, which can be seen as the "new territory" of the twenty-first century, everybody is our neighbor, individuals and institutions are just "one click away," and Marshall McLuhan's vision of the "global village" has become a virtual reality.

From the Carlsbad Treaty to the NWICO Debate

Information and communication issues are among the biggest challenges of the twenty-first century, which is also labelled the information age. Hence it is only a natural consequence that this global development calls for a coming together of the international community to discuss the dimensions, the implications, the opportunities, and the risks and threats. Hence, issues of information and communication are not new subjects for diplomats and intergovernmental negotiations. Nearly two hundred years ago, when cross-border communication developed quickly and reached a critical mass, governments began to look into the political and economic implications of media development and information freedoms. The Carlsbad Treaty of 1819 can be seen as the first international information treaty. Unfortunately, governments looked more into the "control" of the information instead of its opportunities. The treaty regulated, inter alia, the transport of printed material across national borders and introduced a system of oversight and censorship for individual freedom of expression. When in the 1830s the electric telegraph was invented, governments jumped also into this sphere to regulate it. In 1850 the Treaty of Dresden between Austria, Prussia, Bavaria, and Saxony put the transborder telegraph traffic under a governmental control regime. Control over telegraphy was seen as part of national sovereignty. This concept was later multilateralized and became the model for the first International Telegraph Convention signed in 1865 in Paris. The day of the signing ceremony of the Paris Convention is seen today by the International Telecommunication Union (ITU) as its "birthday" and it celebrates May 17 every year as the World Telecommunication Day. In the preamble of the ITU Constitution, the member states begin by "fully recognizing the sovereign right of each State to regulate its telecommunication."

Throughout the history of diplomatic conferences on international ICT issues, each new wave of technological development produced a new wave of political negotiations among governments. And all these negotiations were mainly about national sovereignty over the use of communication technology and the flow of information content across borders. This was the case after the invention of wireless technology (Berlin Convention of 1906), of radio broadcasting (Geneva Convention of 1936), and of satellite broadcasting (draft conventions in the UN Outer Space Committee in the 1970s and 1980s).

All these diplomatic conferences were intergovernmental meetings. Nongovernmental groups like the provider of information and communication services, private media, and media consumers often had no voice in these negotiations. And the result of these diplomatic conferences very often were intergovernmental regulatory regimes that expanded governmental control of the flow and content of information and communication rather than creating new opportunities for technological innovation and individual freedoms. It was only after World War II that the United Nations Conference on Freedom of Information (Geneva, 1948) drafted conventions that tried to strengthen the role of the private media and individual users, and the proposed treaties were never adopted. However, the draft of an article on individual "freedom of opinion and expression" and the right "to seek, receive and impart information and ideas through any media and regardless of frontiers" from the Geneva diplomatic conference made its way as article 19 into the Universal Declaration on Human Rights, adopted by the third UN General Assembly in December 1948.

The dominant role of governments in global communication negotiations changed with the evolution of the Internet at the end of the twentieth century. The Internet was seen for nearly three decades as a mainly technical tool for communication among computers and networks, linked via the TCP/IP protocol. When the United Nations Educational, Scientific and Cultural Organization (UNESCO) became the intergovernmental battlefield for a New World Information and Communication Order (NWICO) in the 1970s and 1980s, the Internet was not part of this controversy. Neither the Soviet Union nor the developing countries, fighting for a NWICO, had any idea at this time what the Internet was and could become.

The Reagan administration (1980–1988) in the United States, where the Internet was invented, rejected the NWICO idea, left UNESCO in 1983 to protest against efforts to introduce "global censorship," and was guided in its national policy and international strategies by the idea of "deregulation." Although the research for Internet development was financed until the late 1980s by the U.S. Department of Defense, the U.S. government had no intention to introduce a national regulatory framework for the Internet or to propose an "International Internet Convention." On the contrary, the absence of any kind of specific political regulation for the Internet was later seen as one of the key drivers for the explosive development of the medium.

This coincided in the early 1990s, after the end of the Cold War, with a development in the UN system toward a broader involvement of nongovernmental actors in international conferences. The UN, according to article 71 of its charter, had always kept open a channel to nongovernmental organizations (NGOs) via its Economic and Social Council (ECOSOC). NGOs could get—via a defined procedure—a "consultative status" that allowed them to provide written contributions to UN bodies, and be invited to make presentations in UN conferences. But the level of NGOs' real involvement in policy development or decision making was rather low.

A key event and turning point was the United Nations Conference on Environment and Development in Rio de Janeiro, June 1992. When the Rio summit adopted Agenda 21, it stressed the need for a greater involvement of nonstate actors in achieving the goals set by the conference. Also other UN summits in the 1990s—in particular the Human Rights Summit in Vienna in 1993, the Woman's Summit in Beijing in 1995, and the UN Social Development Summit in Copenhagen in 1995—saw a growing involvement of nongovernmental organizations. However, there was a clear separation in all these summits between intergovernmental negotiations and nongovernmental activities. During the Human Rights Summit in 1993 the NGOs were packed into the basement of the Vienna UN Conference Centre and had no access to the plenary conference hall where governments negotiated. In Beijing, the NGO event took place in another part of the city with no special transportation services among the two conference sites. Nevertheless, one could observe a growing role of NGOs and civil society organizations in the global diplomacy in the 1990s.

In 1996 the ECOSOC adopted a new resolution that specified in detail special procedures for the involvement of NGOs in UN activities but kept them in an observer role. When Kofi Annan became UN secretary general in 1997, he pushed this development further down the road, in particular with regard to the private sector. In his speech to the World Economic Forum in Davos in January 1997 he said that "strengthening partnership between the United Nations and the private sector will be one of the priorities of my term as Secretary General."[2]

The Global Information Infrastructure (GII) Initiative

Kofi Annan's statement reflected the recognition of a changing global economy and the beginning of globalization. It was made also against the background of the emergence of an information economy and the start of the dot-com boom in the middle of the 1990s. After Tim Barners Lee invented the World Wide Web in 1991, the Internet changed from a more technical communication tool into a business platform and a mass medium, in particular in the United States. The Clinton/Gore administration (1993–2000) was pushing this development in 1993 by launching first a National Information Infrastructure (NII) Initiative and later, in 1995, a Global Information Infrastructure (GII) Initiative. The concept behind both the NII and the GII was to encourage the private sector to make more investment and to give industry a greater leadership role in building the "Information Superhighway." At the same time, the Clinton/Gore administration also encouraged civil society organizations, in particular nongovernmental Internet groups, to play a greater role in Internet policy. Principles like openness, inclusiveness, transparency, and bottom-up policy development played a key role in policy discussion around the Internet in the second half of the 1990s, in particular in the United States.

In a G7 ministerial conference in Brussels in 1995, the leading governments of the industrial world concluded that "the information society is not only affecting the way people interact but it is also requiring the traditional organisational structures to be more flexible, more participatory and more decentralized." And they stressed: "To succeed, governments must facilitate private initiatives and investments and ensure an appropriate framework aiming at stimulating private investment and usage for the benefit of all citizens."[3]

The G7 conference was a conference of the "rich North" while the "poor South" was more or less excluded. However, the G7 realized that such a development toward an information society would have a global dimension. They invited South African Vice President M'Beki to Brussels and agreed to have another G7 meeting in 1996 in Midrand in South Africa to discuss the implications of the digital revolution for developing countries.

But the North did not speak with one voice. While the United States and the EU were united in principle with regard to challenges of the future, as reflected in the

"Conclusions of the Chair" document of the Brussels G7 meeting, they had a different approach. The U.S. government concluded that the leadership role would go to the private sector—knowing that the main players in the Internet economy are U.S. companies—while the EU stressed the special role of governments and argued more in favor of public-private partnerships. The EU preferred "co-regulatory regimes," whereas the U.S. favored "industry self-regulation," especially with regard to the Internet.

In September 1997 the EU Commissioner Martin Bangemann proposed during an ITU meeting in Geneva that governments should draft a "Global Communication Charter" to define the framework for the information society. The initiative was not aimed to introduce a new system of governmental control in the information society, but rather to find a new model of coregulation where governments would define the general political and legal framework on the level of principle while the industry would get enough flexibility in day-to-day operations. Bangemann's plan was to adopt such a charter in connection with the fiftieth anniversary of the Human Rights Declaration in December 1998, also making sure that such a Global Communication Charter would be based on the full recognition of human rights.

In contrast, the U.S. government worked toward a reduced role of governments and more private sector leadership, in particular with regard to the Internet. In 1998 the U.S. Department of Commerce created the ICANN as a private not-for-profit corporation under California law. ICANN received a mandate to manage and coordinate the Internet core resources: the root server system, the domain name system, the IP address space, and the Internet protocols. These Internet resources would not be managed by governments but by the provider and user of services themselves. In the original ICANN bylaws, both the private sector and the civil society (in ICANN language, the "at large" membership) got nine voting seats on the board of directors while the role of governments was reduced to serve in an advisory capacity via the Governmental Advisory Committee (GAC). GAC recommendations are not legally binding and the ICANN board, according to the bylaws, is free to reject them and take its own different decisions, based on a bottom-up policy development process among the ICANN nongovernmental constituencies.

Bangemann supported the creation of ICANN but learned that his initiative for a charter was watered down by global industry leaders (and the U.S. government) when they pushed for a "marked driven approach" under "industrial leadership" in the process of drafting more general political and regulatory frameworks for the information society. According to the industry representatives, governments should be invited to intervene only when needed. As a result the Global Communication Charter idea disappeared from the policy agenda and the Global Business Dialogue on eCommerce (GBDe), a global network of CEOs from the ICT and Internet industry, was launched in January 1999. The GBDe became for a couple of years the main source also for regulatory initiatives, not only for e-commerce but also for content regulation, Internet security, privacy in cyberspace, and intellectual property rights.

The debate on the relationship between state and nonstate actors in the information age was broadened when an OECD ministerial conference in Bonn in 1997 invited not only governments and industry leaders but also representatives of users and consumers. The meeting adopted three separate but interlinked declarations: a ministerial declaration, an industry declaration, and a users declaration. Such an approach reflected the end-to-end principle of the architecture of the Internet, where the end user plays a significant and growing role, moving from a passive receiver into a more active player who also can "provide" services. Such a trilateral approach—governments, private industry, and users—signaled that the information society would go beyond the traditional public-private partnership between government and industry and include a new, third partner: the user, the consumer, the Netizen. This new "trilateralism" was reflected later when the G7 launched in 2000 the Digital Opportunity Task Force (DOT Force) which got a mandate to turn the digital divide into digital opportunities and was constituted by a membership representing governments, the private sector, and civil society.

The Emergence of New Stakeholders: Confrontation or Cooperation?

These developments took place against the background of growing conflicts between governments and the private sector on the one hand and civil society on the other hand around globalization issues. The WTO ministerial conference in Seattle in 1999 marked the beginning of a series of street protests of civil society groups against the global policies of the G8, WTO, World Bank, and the World Economic Forum (WEF), which led to violent battles between the police and demonstrators in Göteborg (2000), Genoa (2001), and Geneva (2002). Civil society organizations like ATTAC, Green Peace, and Amnesty International called for more participatory rights in global diplomacy. They created a "World Social Forum" in Porto Allegre as an alternative to the WEF in Davos and challenged the established system of global diplomacy.

In the beginning a lot of governments ignored the protests, but soon it became clear that the street demonstrations were much more than "protests as usual." They reflected a changing reality, where nongovernmental groups no longer feel represented by their national governments. These new multinational groups, representing all kinds of constituents in "rainbow coalitions," organized themselves around issues and were not only protesting against governmental decisions but were also developing and fighting for alternative policies. More and more governments realized that it would be much better to start a dialogue with the protesters than to provoke more confrontation. Also, in the nongovernmental movement two wings emerged from the general protests after Seattle. One group wanted to be included in the policy development process and asked for access rights, speaking rights, and negotiations rights in global negotiations; the other group wanted to continue with street protest and violence.

The UN system was not unaffected by such developments. When UN Secretary General Kofi Annan started his UN reform process, he included building the relationship between UN and civil society in his agenda and appointed the former Brazilian president Fernando Enrique Cardoso to chair a "Panel of Eminent Persons on United Nations—Civil Society Relations" in February 2003. Although the Internet and the so-called digital divide did not play a central role either in the civil society protests in the early 2000s or in the Cardoso report, which was presented to the UN General Assembly in September 2004, the debate made clear that ICT will play a central role in achieving the UN's Millenium Development Goals. Issues related to the information society and the Internet moved higher and higher on the global world policy agenda. All this culminated with the beginning of the WSIS process.

ITU and the Start of WSIS

The WSIS was first proposed at the ITU Plenipotentiary Conference in Minneapolis in 1998. There the ITU recognized, after a controversial discussion about its mandate and scope of competence, the principle of private sector leadership for Internet governance and the special role of ICANN, which was established just one week after the end of the Minneapolis conference. However, the Plenipotentiary Conference also adopted a resolution that invited the ITU to start preparations for WSIS. The justification for this resolution was mainly that governments should agree to develop global policies on how to bridge the digital divide and to turn it into digital opportunities. For the ITU the new challenge of the digital divide was nothing more than a continuation of its efforts to foster global connections between "telecommunication rich" and "telecommunication poor" countries. Thus the ITU felt it had a natural mandate to host the WSIS.

But when the idea of the information society summit was further elaborated, it became clear that such a conference had to go further than the narrow objective of providing new communication services to underdeveloped countries. The information society is about much more than information infrastructure. And numerous implications of new ICTs call for a much broader look into the totality of the challenges of the "information revolution." Such a broad approach went far beyond the mandate, competence, and expertise of the ITU. As a result, while the ITU would serve as the lead agency, the summit was put under the umbrella of the United Nations secretary general.

In 2000 an ECOSOC ministerial conference on the information society launched, inter alia, the establishment of the UN Information and Communication Technologies (UN ICT) Task Force. The UN ICT Task Force—in parts—mirrored the DOT Force of the G7 but was based on global participation. Its membership structure copied the multistakeholder approach of the DOT Force and included representatives from governments, the private sector, and civil society. For the UN this was obviously an innovation that went beyond the traditional practices according to ECOSOC rules. However both the DOT Force and the UN ICT Task Force had no negotiation or decision-making capacity. Both bodies worked on recommendations and proposals for governments that kept their sovereign right to develop policies and to make decisions. In 2002 the DOT Force was integrated into the UN ICT Task Force.

The issue of multistakeholderism came back when the UN General Assembly in 2001 discussed the mandate for the WSIS. On the one hand, there was a consensus that such a summit had to go beyond the traditional practice of the UN intergovernmental summits of the 1990s. On the other hand, there was no consensus on how to integrate nonstate actors into an intergovernmental process. Hence ECOSOC adopted a resolution that not only invited all UN member states and intergovernmental organizations to participate, but also encouraged "non-governmental organisations, civil society and the private sector to contribute to, and actively participate in the intergovernmental preparatory process of the Summit and the Summit itself." For the first time in the long history of UN summits, nongovernmental stakeholders were officially and directly invited not only to "observe" the process but also to "*actively*" participate both in "the preparatory process" and in "the summit."[4]

WSIS became a turning point and a test case for the implementation of the principle of multistakeholderism. It was a challenge for governments to open the door to new stakeholders and to move forward into the new, unknown, and slippery territory of multistakeholderism. But at the same time WSIS was also a challenge for the nonstate actors to understand such an invitation as an opportunity to become players in their own rights in global policy development. They were challenged to rethink their strategies, objectives, and procedures, to accept established rules for global policy making, and to move from outside criticism to constructive intervention. In particular, civil society found itself in a new role. It had to decide whether to continue to be a critical observer and a watchdog with no concrete responsibilities and confrontational inclinations, or to move into the role of a player inside the process and to fight there while also accepting responsibilities.

The WSIS Geneva Phase: 2002–2003

The WSIS was endorsed in UN Resolution 56/183 on December 21, 2001. The resolution defined the broad mandate and the basic structure for WSIS: The first phase would comprise a summit held at Geneva in December 2003 (WSIS I), while the second phase would involve a summit held in Tunis in November 2005 (WSIS II). Both summits would be organized via extensive preparatory meetings and regional gatherings. It would not be a single event but rather a two-phase process.

The Formation of an Organizational Framework

The preparation of both the Geneva and Tunis summits was guided by a High Level Summit Organizational Committee (HLSOC) organized under the patronage of UN Secretary General Kofi Annan, and chaired by ITU Secretary General Yoshio Utsumi. The HLSOC included as members the heads of all other UN specialized agencies. There was no clearly defined role for all these agencies but the general understanding was that each would take a special responsibility for issues under its mandate. As a result, UNESCO became more involved in media, UNDP more in development, WIPO in intellectual property, WHO in e-health, the FAO in e-agriculture, and so on. The HLSOC established a special WSIS Executive Secretariat (ES). The structure of the ES reflected the multistakeholder approach. It had four divisions, one each for governments, intergovernmental organizations, the private sector, and civil society.

The negotiation of procedures and outcome documents was delegated to a Preparatory Committee (PrepCom). The PrepCom was an open-ended body in which all UN member states participated. The PrepCom elected an intergovernmental bureau with nearly fifty members that became the main executive body in the WSIS process.

The road map to the first summit in Geneva included three meetings of the PrepCom and five regional ministerial meetings. The plan was that PrepCom1 would decide on rules and procedures and a first draft of the agenda. The five regional ministerial conferences were to produce content-related input into the agenda to enable PrepCom2 to start the drafting of the envisioned two final documents: a Declaration of Principles and a Plan of Action. PrepCom3 was reserved for the final negotiations and the adoption of the documents for ratification by the heads of states during the summit itself. All preparatory meetings—with the exception of the regional ministerial conferences—were to take place in Geneva.

The WSIS road map worked more or less. However, a special "Intersessional Meeting" (Paris, July 2003) was needed to discuss the unsolved issues from PrepCom2. And PrepCom3 was unable to find consensus within the scheduled time frame. PrepCom3 was three times reconvened in the fall of 2003 (PrepCom3+, PrepCom3++, PrepCom3++bis). Only on the last night before the summit opened could a consensus on the Geneva Declaration of Principles and Plan of Action be reached.

As discussed, the innovative new procedural element in paragraph 5 of resolution 56/183—which invited nongovernmental stakeholders to participate actively in the process—did have substantial consequences and created both controversy and confusion. It created the vision of a new partnership, in which private industry and civil society would work together with governments as partners, at least on the global level. It raised fundamental procedural issues concerning how such a new partnership could be executed and translated into practice, in particular with regard to concrete issues like access rights to working sessions, speaking rights both in plenary and subgroup meetings, and, eventually, negotiation and voting rights.

Resolution 56/183 was not specific about the details. Paragraph 2 stated only that the "open-ended intergovernmental preparatory committee" would decide "on the modalities of the participation of other stakeholders in the Summit." Consequently, the question of how far the involvement of nongovernmental actors could and should go became a controversial issue and overshadowed the first phase of the preparatory process itself.

On the eve of PrepCom1 in July 2002, the expectations of nongovernmental actors, and in particular of civil society, to achieve a new quality of involvement into a global policy development process were rather high. In the months prior, UNESCO, the UN specialized agency with the biggest number of accredited nongovernmental organizations, convened a series of expert conferences that explored opportunities for NGO involvement in information society-related issues in the field of UNESCO's competence. Background for this initiative certainly includes a rivalry between ITU and UNESCO. During the 1980s, when UNESCO was hosting the controversial NWICO debate, the plan for a World Communication Conference, hosted by UNESCO, was one of the projects discussed. After the NWICO debate came to an end, governments feared that if UNESCO were to host a World Communication Conference the result would be another wave of destructive political controversies. Given the unwillingness of the global community to give it the leading role for WSIS, UNESCO was not extremely happy that the ITU moved into such a position. On the other hand UNESCO saw an opportunity to make substantial and innovative contributions to the WSIS by bringing its experiences of collaboration with nongovernmental actors to the process.

Based on the four expert meetings, UNESCO convened a concluding consultative NGO meeting in Paris, April 2002, which was attended by more than 150 NGOs. During the conference a series of "Recommendations for the Participation of Civil Society" in the WSIS process were adopted. The most important one was "Recommendation 3." It said that "civil society actors should, in substantive agenda development, debate and drafting modalities, be treated as peers and equals to nation state and private sector organisations/corporations." This recommendation was later endorsed by a high-level expert meeting of national UNESCO Commissions, which included also numerous governmental representatives, in Mainz, Germany (June 2002), on the eve of PrepCom1.[5]

PrepCom1: A Procedural Conflict

When PrepCom1 started in Geneva (July 2002), the issue of nongovernmental involvement in the WSIS process became a substantial controversy from the very first day. It was much more than a simple technical question of the "rules of procedures." Governments began to realize that the opening up of the negotiation doors for nongovernmental actors to an intergovernmental process represented a fundamental political watershed with far-reaching consequences for the future of diplomacy in the twenty-first century. Introducing new negotiation principles for the information society could go beyond the WSIS itself and—if successful—produce a new model, a blueprint for all coming negotiations in areas of global importance.

The governments were rather divided with regard to the interpretation of the principle of multistakeholderism. In general there was a consensus "in principle," and in their public statements governments welcomed the "new multistakeholder approach" as an innovation. But as soon as it came to concrete consequences, a great majority of governments expressed a lot of reservations. Such reservations had different sources ranging from mistrust and ignorance to special individual interpretations of this new principle. Questions were raised like: Who are these new actors? Where does their legitimacy come from? Do they have any accountability? To whom are they responsible? Representatives of civil society organizations (CSOs) were characterized as "self-nominated noise makers" who do not really represent a social group. Governments feared that the involvement of CSOs could undermine the seriousness and efficiency of the intergovernmental negotiation process with its well-established procedures and methods to find compromise language on difficult, complex, and controversial issues by respecting basic national interest. And some governments feared that such a multistakeholder policy development process would undermine their individual sovereignty and control of the diplomatic process.

The PrepCom and the WSIS executive secretariat (ES) faced a dilemma. The UN Resolution 56/183 obliged them to invite nonstate stakeholders to the summit for "active participation," but there was no model available under which such participation could be executed. To reduce the risk, the ES introduced a screening procedure for civil society and private sector organizations that wanted to participate in PrepCom1. To get an official "ES-accreditation," organizations and institutions had to provide some basic information about their legal status, membership, financial resources, and contact details. Organizations already accredited under ECOSOC could get an automatic accreditation. For PrepCom1 more than 150 civil society representatives were accredited.

Nevertheless PrepCom1 started with turmoil about the procedural arrangements of the conference. While the governmental delegates had their meetings in the conference rooms of the Geneva International Conference Centre (GICC), the civil society groups were packed into the basement of the ITU building. The ITU building is connected to the GICC via a tunnel, but delegates had to cross a "check-point" where access to the GICC for persons without a governmental badge was blocked. Nongovernmental participants could participate in the official opening ceremony of PrepCom1. But after the end of the formal speeches, they were moved out of the room. This exclusion provoked emotional protests by civil society groups, members of which knocked loudly on the closed doors.

Keynote speaker Cees Hamelink, a professor and human rights activist from the University of Amsterdam and a former president of the International Association for Media and Communication Research (IAMCR), one of the accredited NGOs, had expected to address the governments with his prepared speech. But he, like the other 150 civil society participants, had to go to the ITU basement, where the "observers"—official participants without a governmental badge—could use a conference hall for their own separate meeting.

The disappointment on the side of CSOs, which had prepared for a dialogue between governmental and nongovernmental WSIS participants, was high. Civil society realized that the "NGO Paris Recommendation" from April 2002 had had no impact so far, that they were not yet seriously welcomed in the process, and that there were no mechanisms in place for a structured communication between civil society and governments. Their exclusion from the start of the WSIS negotiations provoked a controversial debate among CSOs on the question of whether they should continue to look for cooperation with governments or start an alternative process with a counter summit. This was well noticed by the governments, which had no interest in kicking the new stakeholders out of the process at the risk of other forms of protest in public spaces attracting high media coverage.

Immediately after his election, WSIS President Adama Samassékou, a former minister for education from the Republic of Mali, went to the parallel civil society meeting and stressed the importance of the "new multistakeholder approach" for the success of WSIS. He declared that civil society contributions to the process would be welcomed. And he promised that he as a chair of the preparatory process would investigate in a positive and constructive manner how to improve civil society participation so that WSIS process could become as inclusive as possible.

Regardless of Samassékou's promises, the negotiations about the rules of procedures during PrepCom1 made the limits for innovation visible. In the proposed procedural draft, the key rule 55 had two options. Option A said that accredited NGOs may designate representatives "to sit as observers," and Option B offered NGOs the right "to participate as observers." Option A gave nongovernmental stakeholders the right to access "public meetings of the Preparatory Committee," while Option B proposed active participation "in the deliberations of the Preparatory Committee, and, as appropriate, any other sub-committee within the scope of their activities." In Option A NGO speakers would have only the right to speak "upon the invitation of the presiding officer of the body concerned and subject to the approval of that body." In Option B there was no formal barrier for a right to speak—which could include also a right to negotiate—and only "voting rights" were explicitly denied.

While Pakistan on behalf of the Group of 77 developing countries argued in favor of Option A, the United States and EU preferred Option B. The formal majority was on the side of the Group of 77 and neither the United States nor EU was interested in starting a big battle about this procedural question. As a result, the industrialized countries joined the consensus based on an enhanced version of Option A, with the interpretation that it would be in the hands of the chairperson of a relevant body how far the doors would be opened and how much nongovernmental speakers could say.

PrepCom2: Civil Society Input into Content and Themes

In the series of five regional ministerial meetings, which were organized between PrepCom1 (July 2002) and PrepCom2 (February 2003), a different practice emerged from the bottom. Some governments that would have preferred Option B encouraged civil society to look for new opportunities and to bypass restrictive procedural regulations. In the European meeting (Bucharest, November 2002) the EU governments recommended to have all sessions open to governmental and nongovernmental participants. A large number of experts from civil society and the private sector were invited as speakers. Only the drafting and the formal adoption of the Bucharest Declaration was in the hands of the governments, but even the final round of negotiations was open for observers. In the Asian meeting (Tokyo, December 2002), the WSIS procedures—with the help of the government of Japan—were bypassed by a practice that avoided formal negotiations and voting entirely. The Tokyo Declaration was adopted "by acclamation" of all groups in the room, including civil society and private sector members. Both groups also contributed draft resolutions and diplomatic "language" to the text of the final document during the proceedings of the Tokyo meeting. Similar practices were used and developed in the other three regional meetings in Bamako (May 2002, Africa), Bavaro (January 2003, Latin America), and Beirut (February 2003, West Asia).

During these regional meetings, civil society and governments learned about the strength and weakness of the policy involvement of nongovernmental stakeholders. Both sides realized that civil society can bring substantial and recognized added value to the negotiations process with special expertise, knowledge, and linkage to the people on the ground. But they also understood that to be efficient, civil society participants have to demonstrate their legitimacy and channel all their contributions in an appropriate way to the negotiation table. Both sides realized the need for a workable organizational structure that would give not only individual groups but also the whole civil society as a major stakeholder a higher degree of credibility and the ability to speak with a coordinated voice.

When PrepCom2 convened in February 2003, hundreds of civil society delegates, with the support of the Civil Society Department of the WSIS Executive Secretariat, began to develop an organizational structure. This was not an easy process. The WSIS participants with a civil society badge represented an enormous variety of constituencies with diverse legal status, political weight, and opinions. It was clear from this perspective that there is no "one voice" of civil society. Civil society is based on pluralism and diversity and this includes also varying positions on individual issues. The challenge was to find a procedure to determine where the civil society organizations and individuals—regardless of their diversity—can agree, to find a common ground, and to represent—more or less—a broad mainstream in the middle of the road. What a lot of skeptics did not expect was that in painful and sometimes chaotic discussions, the civil society groups participating in WSIS were able to create workable structures and procedures to develop bottom-up policy proposals and diplomatic language, which produced statements that on the one hand represented the "middle of the road" in civil society, and on the other hand were clear enough to challenge governmental—and partly also private sector—positions in key areas like human rights, development, intellectual property rights, privacy, and Internet governance.

Such a process often was initiated, led, and pushed forward (in particular in the Geneva phase) by individuals who demonstrated their capacity to steer a discussion and champion it through complicated and controversial in-fighting among civil society groups with the aim of reaching a "rough consensus" at the end of the day. This was done both offline in the meetings of the civil society groups on the spot but also online with hundreds of participants in virtual discussion groups on listservs and wikis. As the final result showed at the end of the Geneva phase, the process worked surprisingly well regardless of all the heated debates, and impressed governments and other WSIS stakeholders, which resulted in the enhanced reputation and role civil society was afforded in the second phase of the summit.

During PrepCom1 a WSIS Civil Society Plenary (CS-P) was established as the highest decision-making body. CS-P was designed to be open, transparent, and inclusive. All accredited CSOs could participate. There were no formal membership requirements. CS-P got two cochairs (as a rule one was from the North and one from the South, one male and one female) who rotated from meeting to meeting. The offline meetings were complemented by online forums so that individuals and institutions unable to come to Geneva had an opportunity to participate.

Two main subsidary bodies also emerged: one for content-related issues, called the WSIS Civil Society Content and Themes Group (CS-C&T); and one for procedure-related issues, called the WSIS Civil Society Bureau (CS-B). CS-C&T coordinated the work of issue-based CS working groups and caucuses. CS-B represented some civil society constituencies that were grouped into undesignated "families."

Policy-related aspects were discussed mainly in working groups and caucuses that were created from the bottom-up by engaged individuals and NGOs. Within a short time period such bodies as the Internet Governance Caucus, the Media Caucus, and the Privacy Caucus became virtual entities with broad informal memberships drawn from world-leading NGOs and key individual experts. Recommendations from these groups went to CS-C&T, which forwarded key proposals to CS-P for confirmation. Such a bottom-up policy development process in an open and transparent environment created a new form of legitimacy for proposals made by the civil society. At the end of the WSIS process CS-C&T was composed by about thirty different working groups and caucuses. Membership in the caucuses was open and not formalized. The self-understandings of the groups were rather idealistic. It was expected that members in any given group bring to it competence, relevance, and readiness to do voluntary work. The groups coordinated offline and online via e-mail list-servers, moderated by two cochairs round the clock, seven days a week.

The CS-Bureau emerged in a different way, on the basis of a proposal by the Civil Society Division of the WSIS Executive Secretariat. This proposal was driven by the idea of a "parallel structure" to the Intergovernmental Bureau so that in concrete cases a communication channel between governments and civil society would be available. At the time, when the Civil Society Division made this proposal, the CS working groups and caucuses did not yet exist. It was justified to look for a practical solution to "institutionalize" CS and to enable it to talk with a coordinated voice to the Intergovernmental Bureau and to draft language for official documents. The idea was that such a bureau should not be composed of nonlegitimized individuals or single NGOs but of broader, "major groups." The difficulty was how to define a major group.

It was Alain Clarke and Louise Lassonde from the Civil Society Division who discovered in paragraph 23 of the UN Agenda 21 a list of major groups in society such as "women, children and youth, indigenous people, non-governmental organisations, local authorities, workers and their trade unions, business and industry, the scientific and technological community and farmers." According to Clarke and Lassonde, each of these major groups should have one representative in the bureau. The groups as such should be formed among the present members of the WSIS CS-P. When the Civil Society Division proposed this to the CS-P, there was a mixed reaction. While some groups recognized the need to move forward toward some form of representative body others criticized the proposal as too narrow and top-down.

The compromise that was finally reached in a turbulent and chaotic night session was to accept the proposal in principle but to add more families to the bureau. Numerous amendments reached the chair and within two hours after midnight the list of civic society families grew from ten to twenty-one, including families like media, think tanks, philanthropic institutions and foundations, networks, and five so-called regional families. Furthermore, speakers made clear that they did not want to see the bureau as a governing body in an executive role. Such a mandate would undermine the principles of bottom-up, transparency, and openness. The CS-P decided to have a clear distinction between the roles of the CS-B and the CS-C&T. While CS-B would deal with

Table 15.1
Basic structure of civil society organizations

Civil Society Plenary	
Civil Society Bureau (for procedures)	Civil Society Content and Themes Group (for content)
Civil Society Families	Civil Society Working Groups and Caucuses

Table 15.2
Civil society working groups and caucuses represented in the content and themes group

Regional caucuses	Thematic working groups and caucuses	Cross-thematic working groups and caucuses
Europe and North America	Cities and Local Authorities Caucus	Gender Caucus
Asia Pacific	Community Media Caucus	Youth Caucus
Latin America	Cultural and Linguistic Diversity Caucus	Working Group on Volunteers
Western Asia and Middle East	E-Government/E-Democracy Caucus	Working Group on Values and Ethics
Africa	Education and Academia Caucus	
	Environment and ICTs Working Group	
	Health and ICT Working Group	
	Human Rights Caucus	
	Indigenous Peoples Caucus	
	Internet Governance Caucus	
	Media Caucus	
	Patents, Copyright, and Trademarks Working Group	
	Persons with Disabilities Working Group	
	Privacy and Security Working Group	
	Scientific Information Working Group	
	Trade Union Caucus	

Source: www.wsis-cs.org.

procedural questions like nomination of speakers, arrangement of consultations with other stakeholders, meetings with the intergovernmental bureau, negotiation of speaking slots for sessions, and the like; the caucuses and subsequently CS-C&T would have "sovereignty" over substantial content-related issues. The basic structures and specialized groupings of civil society in WSIS are illustrated in tables 15.1–15.3.

The general plan, which got broad support from all the civil society constituencies involved, was to create an environment that would (a) enable a bottom-up policy development process among the concerned and affected groups, and (b) foster workable and effective communication both within and among the different civil society groups and between CS and other stakeholders in the WSIS process.

The mandate of the CS-B was hence limited to procedural formalities only. For cases where formal questions had content-related implications, the chair of the CS-C&T was brought in by the CS-B. However, both CS C&T and CS-B were obliged to report back to the CS-P, which oversaw the whole process as the highest sovereign body of civil society.

Table 15.3
Civil society "families" represented in the bureau

Thematic and constituency families	Regional families
Education, Academia, and Research	Africa
Science and Technology Community Media	Latin America and the Caribbean
	Asia and Pacific
Creators and Promoters of Culture	Europe, Commonwealth of Independent States, and North America
Cities and Local Authorities	Middle East and Western Asia
Trade Unions	
NGOs	
Youth	
Gender	
Volunteers	
Indigenous People	
Networks and Coalitions	
Multistakeholder Partnerships	
Philanthropic Institutions	
Think Tanks	
People with Disabilities	

Source: www.wsis-cs.org.

Such a formal and structural maturation of the civil society countered governmental efforts to deny direct communication with civil society representatives by referring to the confusing and chaotic diversity of CSOs and the illegitimacy of their spokespersons. Nevertheless, numerous governments continued to refer to the "intergovernmental character" of the WSIS process and pushed toward the consequent adoption of rule 55 in its limited interpretation, which gave the chairman of a session the right to deny access and speaking rights to nongovernmental observers.

At the end of PrepCom2 a greater controversy emerged, when Pakistan called for the removal of all observers from the conference room. The chair of Working Group II, the South African Ambassador Lyndall Shope-Mafole, countered this intervention with the proposal to include all observers as members in their national delegation to enable the nongovernmental representatives to follow the deliberations in the plenary. Pakistan withdrew its proposal.

During PrepCom2 a series of high-level expert round tables had demonstrated that a lot of expertise and capacity for many information society-related issues—from software development via network security to Internet governance—resided in the nongovernmental stakeholders. While governments continued with their traditional diplomacy in the areas with which they were familiar—like human rights and financial mechanisms—they came to recognize some weaknesses in their understanding and knowledge when the discussion moved to more specific issues related to new ICTs, and in particular, to the Internet.

It was difficult to reject the argument that the special expertise of nongovernmental stakeholders was needed to find answers to the challenges of the information revolution. Moreover, the argument that a discussion on the future of the "information society" without representatives of this society would be a contradiction in itself steadily won ground. While some governments insisted that governments adequately represent both the private sector and the civil society of their nations, those civil society groups and private sector members acting transnationally and globally or operating under undemocratic regimes rejected this assertion. They called for an inclusive, open, and transparent process that would allow them to make complementary contributions, to bring another perspective, additional values, and specific interests to the process.

As noted, PrepCom2, in February 2003, did not achieve its target to start the substantial work on the final summit documents. In this situation UNESCO offered to host an additional WSIS intersessional meeting in Paris, in July 2003. This meeting started with a procedural innovation: In every plenary session there was one hour reserved for statements by nongovernmental actors: twenty minutes each for international organizations, the private sector, and civil society. Both the CS-P and the Coordinating Committee of Business Interlocutors (CCBI), the coordinating body of the private sector, nominated their speakers in an internal discussion process and transferred the names to the WSIS Executive Secretariat. No speaker was rejected by the governmental bureau. Furthermore, nongovernmental delegates were allowed to remain in the room, although they had no right to make interventions during the intergovernmental deliberations. Step by step, rule 55 was being executed in a more and more flexible way.

This was the case in particular when the real negotiations started during the second half of the intersessional meeting. Governments agreed to establish five subworking groups to deal with concrete issues like human rights, financing, and Internet governance. It was unclear whether nongovernmental participants would be allowed to go to the meetings of these subworking groups. According to procedural rule 55, observers could sit and speak (by invitation) only in "public meetings." The rule did not say anything about meetings of subworking groups. But when the working group on Internet governance had its first meeting late in the evening of a long conference day, nobody took care of who entered the room.

After a number of governmental participants raised some basic questions, partly of a technical nature with regard to IP addresses, root servers, and domain names, Paul Wilson, president of the Asian-Pacific Information Center (APNIC)—a nongovernmental network that manages the IP numbers for Internet users in the region—took the floor and explained how the allocation of IP numbers and domain names has been developed bottom-up without the involvement of governments. He rejected some statements of governments with regard to political discrimination in the allocation of IP addresses and future shortages of IP addresses. Wilson's intervention was seen as a very constructive contribution. The chair of the subworking group, an ambassador from Kenya, thanked him for the enlightment. Wilson's statement demonstrated what civil society could contribute to the process: expertise on Internet governance, which many governments lacked. This was another stone in building a foundation of credibility with regard to civil society participation in policy development processes.

At the end of the intersessional meeting, a first official joint meeting between the WSIS Intergovernmental Bureau and the WSIS Civil Society Bureau took place. Civil society representatives referred to the positive experiences of the Paris meeting and took them as an indicator that the process had moved from turmoil to trust. While recognizing the intergovernmental character of the negotiations process, civil society speakers called for a next step, an innovative inclusion of observers into the final phase of the drafting of the WSIS documents. They proposed a "stop-and-go-negotiation-mechanism," where observers would not formally have the right to negotiate but could sit in the room and would have a chance to speak to the point at the beginning of the discussion of each individual paragraph.

PrepCom3: The Limits of Multistakeholderism

Before the start of the PrepCom3 in September 2003 in Geneva, WSIS President Adama Samassékou addressed a preconference meeting of civil society. He said that his vision was that civil society "input" would lead to "impact" in the final negotiations and improve the quality of the summit documents. But already the first days of PrepCom3 showed the gaps between visions and realities.

The first meeting of the Internet governance working group was again open to observers. But when civil society activists with their laptops started blogging from inside the conference room—the Geneva International Conference Centre (GICC) is a Wi-Fi zone—while diplomats exchanged handwritten "secret documents," the Chinese delegate intervened and called for an exclusion of nongovernmental observers. According to rule 55 the chair asked the governmental members of the working group to take a position. There was no consensus with regard to the continuing participation of observers. After some chaotic moments, when nobody knew who had the authority to decide whether the private sector and civil society had to leave the room again, the chair proposed having a five-minute open session in the beginning of each meeting for statements by nongovernmental stakeholders with a public briefing to follow each session. But nongovernmental participants would no longer be allowed to observe the negotiations. Speakers from civil society and the private sector used the five minutes to protest this "kick out," characterized it as a reversal from "trust" back to "turmoil," and asked what the governments had to hide in discussions concerning the future of the Internet used by hundreds of millions of individuals around the globe. Civil society participants criticized government officials openly for such a "revolving conference door policy" and blamed them for paying just "lip service" to multistakeholderism.

This criticism was further fuelled by an analysis of the redrafted official final documents. Between the WSIS intersessional meeting and PrepCom3, civil society caucuses and working groups had discussed concrete issues both online and offline. The CS C&T produced a substantial document in "diplomatic language" with eighty-six concrete recommendations to nearly all paragraphs of the draft Declaration of Principles and the draft Plan of Action. The document was endorsed by the CS-P and demonstrated the capacity of the civil society to develop policies bottom-up and to speak with one voice based on a rough consensus among major constituencies. But when the redrafted governmental documents were circulated in the beginning of the second week of PrepCom3, a comparative analysis of the new drafts and the civil society recommendations made clear that the input had no or only little impact. More than 80 percent of the CS recommendations were totally ignored. The rest was reflected in rather vague formulations around issues like human rights, privacy, or Internet governance.

The conflict escalated when some participants in the CS-P started to call again for a "Plan B," that is, to leave the conference, organize protests in the streets of Geneva, and organize a counter summit. Other CS representatives argued for a continuation of a critical dialogue by keeping the option open to produce the CSOs' own declaration, independent of the governmental documents.

Spokespersons for the civil society criticized the governmental drafts as "too technocratic and too bureaucratic." They called for an "information society with a human face" and blamed the governments for avoiding concrete commitments to move forward in bridging the digital divide and securing human rights in the information age.

The civil society document identified the following five primary areas of conflict:

- human rights
- cybersecurity and privacy
- intellectual property
- media
- Internet governance

In the area of human rights, civil society was calling for an advanced version of article 19 of the Universal Declaration of Human Rights from 1948, which would include also information and communication rights for access and participation in cyberspace, including the right "to share information." Some civil society groups, like the Communication Rights in the Information Society (CRIS) campaign, argued in favor of a new "right to communicate," a controversial issue with roots in the UNESCO NWICO debate.

In the area of cybersecurity and privacy, civil society called for comprehensive respect of the individual's rights to privacy and data protection. Participants argued that the fight against cybercrime and cyberterrorism should not undermine the open character of a democratic society that guarantees individual rights.

In the area of intellectual property, civil society stressed the need to recognize nonproprietary free and open source software as equal to other forms of software like Microsoft. Furthermore, civil society argued that free access to knowledge, including the right to freely share information, is a precondition for innovation and creativity, while an ongoing commercialization and limitless extension of intellectual property rights protections could undermine the free flow of information and ideas.

In the area of media, civil society called for effective measures against concentration of media ownership and for the promotion of cultural diversity and pluralism in information content. It supported the existence and development of traditional public media, in particular on the local and community level as well as in rural areas. Multilingualism in new media was stressed as important to allow people to communicate in their own languages.

In the area of Internet governance, civil society argued in favor of a multistakeholder mechanism for the management of the core resources of the Internet—root server, domain names, IP addresses—which would allow individual Internet users and their representatives to participate effectively in bottom-up policy developments and decision making. While critical of ICANN, civil society did not support proposals to put the control of the Internet under an intergovernmental organization like the ITU.

With regard to one of the main intergovernmental controversies—the proposed creation of a Digital Solidarity Fund—civil society had no specific position. While it supported in principle all activities aimed at bridging the digital divide, civil society did not speak clearly in favor of the establishment of a new bureaucratic institution.

In a second joint bureau meeting between the governments and civil society in November 2003, civil society speakers expressed their dissatisfaction with the process and blamed governments for "ignorance." Samassékou and other governmental representatives defended the latest drafts of the WSIS documents and explained that numerous governments did not find their proposals directly expressed in the text. Governments would not "ignore" the positions of civil society but must seek consensus positions among all governments.

During this meeting it became clear that constructive dialogue between governments and civil society had both merits and limits. While it helped on the one hand to improve their mutual understanding, it also showed—when it comes to a negotiated text—that the impact of nongovernmental actors is rather limited.

To avoid further confrontation and to reduce the tensions between the governmental bureau and civil society, Swiss President Pierre Couchepin in his capacity as host of WSIS I invited nongovernmental stakeholders to a private consultation. The meeting in the Swiss Embassy in Geneva in mid-November 2003 was also attended by high-ranking governmental negotiators for Switzerland. Couchepin assured civil society representatives that governments respect key essentials in civil society positions. He stressed that civil society should understand that the consensus-driven intergovernmental process was not directed against civil society groups and their values and principles, but had its own internal dynamics that had to be respected. The dialogue reestablished a constructive climate between the two stakeholders. But it also made clear that it would be impossible to have one final document, negotiated by governments and "endorsed" by civil society, as was intended by the UN Resolution 56/183.

During a joint press conference with the Swiss president, speakers from civil society said that they respected the exclusive intergovernmental character of the final negotiations. They would make no further contributions to the text of the two drafts for the Plan of Action and the Declaration of Principles but would produce their own "Final Declaration," which would reflect the visions and ideas of civil society more directly. Civil society would not withdraw from the WSIS process. Their document would be not an "alternative" declaration but a "complementary" document. "We are on the road, but still in the rain," said one CS speaker during the press conference.

The Geneva Summit and Recognition of the Principle of Multistakeholderism

The formal separation of civil society from the intergovernmental process had no negative effects on the Geneva summit meeting in December 2003. On the contrary, a number of governments encouraged civil society groups to go ahead with their own declaration and to establish positions in critical areas like human rights where a consensus among governments was reachable on a very low level only. The consensus principle limited certainly the flexibility of governments and allowed them in a number of cases only to agree to disagree on key issues. Civil society on the other hand had more space to agree on a text. At the end of the day governments agreed on "what could be done" that day according to the existing political and economic realities, while civil society was able to formulate visions and to say "what should be done" tomorrow.

When the final intergovernmental negotiations started, civil society had no voice and vote anymore. But the concerns that it had raised constantly from PrepCom2 to PrepCom3++ showed, regardless of the public noise and rejection in various drafts of the final documents, a real impact. This was seen in particular in how the principle of multistakeholderism was included in the Geneva Declaration of Principles. In this main summit outcome document, multistakeholderism got formal recognition as a new political principle for the information age. The principle as such and the final formulations were not the subject of negotiations among stakeholders but remained in the hands of the governments. However, it was the process as a whole from PrepCom1 to the Geneva summit that ultimately pushed governments into a position to recognize multistakeholderism as a key principle for the information society. Thus the input by civil society had an impact on the intergovernmental level.

The very first principle of the Geneva Declaration of Principles said clearly: "Governments as well as private sector, civil society and the United Nations and other international organisations have an important role and responsibility in the development of the Information Society and, as appropriate, in the decision making process." And it added: "Building a people centred Information Society is a joint effort which requires cooperation and partnership among all stakeholders."

More specifically, the Plan of Action defined the individual roles of the different stakeholders in paragraph 3 as follows:

> All stakeholders have an important role to play in the information society especially through partnerships:
> a. governments have a leading role in developing and implementing comprehensive, forward looking and sustainable national e-strategies. The private sector and civil society, in dialogue with governments, have an important consultative role to play in devising national e-strategies;
> b. The commitment of the private sector is important in developing and diffusing information and communication technologies (ICTS) for infrastructure, content and application. The private sector is not only a market player but also plays a role in a wider sustainable development context;
> c. The commitment and involvement of civil society is equally important in creating an equitable information society, and in implementing ICT-related initiatives for development.

Also other parts of the two Geneva WSIS documents refer to the specific role of the different stakeholders. Governments are invited "to cooperate with other stakeholders" in nearly all areas covered by the summit.

One very special example are the recommendations with regard to the governance of the Internet. This issue became highly controversial in the final phase of the Geneva summit. While one group of governments—in particular the G77 and China—wanted to bring the governance of the Internet under the umbrella of an intergovernmental organization of the UN system (like the ITU), the United States and other industrial countries supported "private sector leadership" with regard to the management of Internet resources (by ICANN). The final agreement was a compromise by which governments could agree on some framework principles but continue to disagree on substance.

The important element of the framework agreement was that governments recognized that the governance of the Internet has to include all key players—governments, private sector and civil society, providers and users of Internet services, developers and managers of the technical key resources—and has to be executed in new forms in an open and dynamic environment.

Paragraph 48 of the Geneva declaration says that "the international management of the Internet should be multilateral, transparent and democratic with the full involvement of governments, the private sector, civil society and international organisations."[6] In the following paragraph 49, the specific roles of the main stakeholders are further described in more detail. While governments have the "policy authority for internet related public policy issues," the private sector "has had and should continue to have an important role in the development of the Internet, both in the technical and economic fields." Equally the role of civil society in Internet governance is qualified as "important," especially "at community level."

Because no consensus could be reached on the substance of the issue of Internet governance, the WSIS declaration invited UN Secretary General Kofi Annan to establish a Working Group on Internet Governance (WGIG) to study the problem in more depth and to elaborate proposals until the second phase of the summit in 2005. The working group should be established "in an open and inclusive process that ensures a mechanism for the full and active participation of governments, the private sector and civil society both from developing and developed countries, involving relevant intergovernmental and international organisations and forums."

The recognition of an "important role" nongovernmental stakeholders have to play in the governance of one of the key resources of the information society and the official invitation for them to take part as more or less "equal partners" in the WGIG was a landmark political decision that illustrated the powershift in the information age.

In the final summit plenary, for the first time at a UN summit a limited number of representatives of nongovernmental stakeholders were officially invited to speak next to presidents and ministers. And after the formal adoption of the two governmental documents by the heads of states, a representative from the civil society got a chance to officially present the Civil Society Declaration, "Shaping Information Societies for Human Needs," to the WSIS president. The declaration stated, inter alia that: "Building [information] societies implies involving individuals in their capacity as citizens, as well as their organisations and communities, as participants and decision-makers in shaping frameworks, policies and governing mechanisms. This means creating an enabling environment for the engagement and commitment of all generations, both women and men, and ensuring the involvement of diverse social and linguistic groups, cultures and peoples, rural and urban population without exclusions."

The WSIS Tunis Phase, 2004–2005

The second phase of WSIS was organized along the model of the Geneva phase. The plan was to have three PrepComs with similar mandates to phase 1. In addition to regional meetings there were so-called thematic meetings where specific issues were discussed on an experts' level. In all these meetings, the principle of multistakeholderism was taken as a guideline for the procedural arrangements. While governments retained their special territory as the final negotiators of language for the summit documents, all other proceedings and discussions were rather open, transparent, and inclusive. This was evident in particular with regard to the discussion of the most controversial issue in the Tunis phase, Internet governance.

For the two main issues carried over from WSIS I—financing and Internet governance—the UN secretary general established two working groups with the mandate to make proposals for the solution of the conflicts. In both groups civil society representatives were invited as equal members, although in the Task Force on Financial Mechanism (TFFM) civil society was rather underrepresented. But in the WGIG, civil society got nearly one third of the seats in the forty-member group. Four of the authors in this volume—Peng Hwa Ang, William Drake, Wolfgang Kleinwächter, and Don MacLean—were appointed to the WGIG and would play active roles in its work.

In addition to discussing financing and Internet governance, the Tunis summit participants also had to decide on the follow-up and implementation of the WSIS decisions on the road to 2015.

PrepCom1 and the GFC: Multistakeholderism in Practice

PrepCom1 (Hammamet Tunisia, June 2004) was, like PrepCom1 of the Geneva phase, a largely procedural meeting. It decided that PrepCom2 should concentrate on the financial mechanisms and PrepCom3 on Internet governance. It was expected that the report of the TFFM would be available at the beginning of PrepCom2 while the WGIG report was expected before PrepCom3. In parallel, both PrepComs were also to discuss the planned follow-up and implementation. PrepCom1 agreed furthermore not to reopen the Geneva package and to restart debates on human rights, cybersecurity, or intellectual property. The newly elected WSIS president, Janis Karklins, an ambassador from Latvia, proposed to establish a special Group of the Friends of the Chair (GFC) to assist him in preparing the documents for PrepCom2 and PrepCom3.

The GFC was designed as an informal group. But it was unavoidable that the GFC also became the subject of a procedural controversy. A number of governments wanted to have the GFC as a closed intergovernmental body. Private sector and civil society participants argued that an exclusion of nongovernmental stakeholders from the GFC would be a violation of the Geneva documents.

As a compromise, the GFC became a mixed body that had both open and closed sessions. But more important than the formal arrangement was the fact that the chair invited all stakeholder groups to give issue-related input for the drafts of the different parts of the proposed final documents on an equal footing. During the first meeting of the GFC in October 2004 in Geneva, some governmental representatives challenged the participation of private sector and civil society representatives, in particular with regard to the right to deliver "language" for a diplomatic text. They proposed to have a "closed section" where governments would discuss the text among themselves.

Reacting to this proposal, the chair offered two options for written input: Open input, published on the WSIS web site, and closed input, not published. Procedurally this way was correct; substantively it was nonsense. As a result, no government made any "closed proposal" and proposals from nongovernmental stakeholders appeared on the GFC web site in the same form as the governmental proposals.

While this story can be seen as a minor issue, it symbolizes the real political shift that was going on. Governments could and would continue to discuss and negotiate among themselves in closed shops, but this diplomatic mechanism became partly embedded in a broader policy development process that was more open and transparent and included more actors.

At the same time, this small GFC story does not mean there will be no secret diplomatic negotiations anymore. Governments will always find a way to exclude the broader public when it comes to serious political negotiations. But it signals that the whole process is becoming more complex, has different dimensions, and takes place in parallel on different levels that are interlinked in a way not yet specified. As the process leading to the Tunis summit demonstrated, personal relationships between individual representatives of different stakeholder groups can play an important role in reaching acceptable solutions.

The GFC had eight meetings between October 2004 and September 2005. It produced a number of documents that became the basis for the negotiations in PrepCom2 (February 2005) and PrepCom3 (September 2005). In the proposed GFC texts there was no differentiation between governmental or nongovernmental input. Later, when PrepCom2 and PrepCom3 started concrete negotiations on financial mechanisms and Internet governance, documents that circulated among conference participants always had three columns: proposed GFC text, amendments by governments, and amendments by the private sector and civil society. Such a procedure was rather different from that of the Geneva phase, when input from nongovernmental stakeholders often was ignored in official documents and somewhat sidelined in the final negotiations.

In the two final documents adopted by WSIS II—the Tunis Commitment, which was a statement of political intent, and the Tunis Agenda for the Information Society, which included the specific agreements on financing, Internet governance, and follow-up—one can find much more input from nongovernmental stakeholders than in the two Geneva documents, although it is not easy to identify the real source of a special paragraph.

Nine of the forty articles of the Tunis Commitment include a reference to multi-stakeholderism, much more than in the Geneva Declaration of Principles. The document is also more specific with regard to this principle. Article 6 calls upon "governments, private sector, civil society and international organisations to joint together to implement the commitment." In article 8, governments reaffirm "the important roles and responsibilities of all stakeholders as outlined in paragraph 3 of the Geneva Plan of Action," but acknowledge at the same time "the key role and responsibility of governments in the WSIS process." This signals that governments have not yet reached a final common understanding of what "multistakeholderism" means in reality. And what does "key role" mean? Does it mean that the other stakeholders are subordinated in a hierarchy, or does it mean that governments play a "key role" in a network where other players operate more or less as equal partners?

In a similar way, chapter 3 of the Tunis Agenda, which covers implementation and follow-up, reflects this mixed approach. On the one hand, governments recognized that all stakeholders are needed to implement the Geneva Plan of Action and the Tunis Agenda for the Information Society. On the other hand there was no agreement on how far the involvement of nongovernmental stakeholders should go.

When the question was discussed of who should coordinate the relevant activities in the so-called action lines C1 to C11, no nongovernmental actor was nominated as a moderator/facilitator. This role remained in the hands of the intergovernmental organizations like the ITU, the UNDP, UNESCO, UNCTAD, WTO, WIPO, WHO, the Food and Agriculture Organization (FAO), the Universal Postal Union (UPU), and others. There is no formal recognition of the principle of multistakeholderism in any of these intergovernmental organizations. In the majority of them, nongovernmental groups have no rights and are not allowed to observe intergovernmental negotiations, such as in the WTO, which is a "closed shop." Even the ITU, which was the main organizer of the two WSIS phases, effectively excludes civil society from its membership, although it has been open to private sector members since its Plenipotentiary Conference in Kyoto in 1994.

In addition to the action lines the Tunis summit also decided (a) to give the UN Commission on Science and Technology for Development (CSTD) an enhanced mandate to deal also with WSIS issues; and (b) to create a new UN Group on the Information Society within the Chief Executives Board (CEB) of the United Nations where all UN intergovernmental organizations are represented. The CSTD operates under the UN ECOSOC, which is also a purely intergovernmental body, although ECOSOC is the main bridge to the NGOs in the UN system.

When ECOSOC had its annual meeting in July 2006, multistakeholderism again became the subject of a complicated diplomatic discussion. What was achieved in the WSIS process, which was mainly driven by Geneva-based UN organizations, was not yet broadly accepted by UN bodies based in New York, including ECOSOC. However, the ECOSOC resolution 2006/46 acknowledged the multistakeholder principle and allowed—on an exceptional basis for the next two years—the participation of nongovernmental stakeholders in its proceedings. At the same time, the resolution stated that "while using the multistakeholder approach effectively the intergovernmental nature of the Commission on Science and Development should be preserved."

It remains to be seen how the process will develop further. It is interesting to also note here that again the WSIS follow-up on Internet governance is pioneering new forms and methods, as can be seen in the following section.

Financing the Digital Solidarity Agenda

The Geneva phase, which discussed at length financial mechanisms for the so-called digital solidarity agenda—including the proposal of the president of Senegal, Abdoulaye Wade, to establish a digital solidarity fund (DSF)—could not reach an agreement on the issue. The compromise was to ask UN Secretary General Kofi Annan to establish the Task Force for Financial Mechanism (TFFM) to help to find a solution.

The TFFM was established in March 2004 and presented a final report in December 2004. The issue of financing was primarily an intergovernmental conflict between developed and developing countries, and donating and receiving countries. Nongovernmental stakeholders, both from the private sector and civil society, did not intervene deeply into the debate. While civil society in principle supported all efforts to mobilize financial and material resources for the implementation of the digital solidarity agenda, it had no concrete proposal for how to organize "case flow." Ideas like the introduction of a "bit tax" on international phone calls or domain names had been watered down already within the different caucuses and working groups of the civil society itself.

The TFFM included representatives from all stakeholder groups. But compared with the WGIG, there was an imbalance in the composition of the group. Anriette Esterhuysen, president of the Association for Progressive Communication (APC), a leading player in the WSIS civil society family, was the only civil society representative in the TFFM. Civil society protested in the summer of 2004 against the unbalanced composition and its underrepresentation in the TFFM. But the life span of this task force was rather short. The TFFM ended its work in the fall of 2004 and presented its report to PrepCom2 in December 2004.

The TFFM did not produce a comprehensive document with concrete recommendations and proposals. The final TFFM report was more an overview about existing financial mechanisms. It did not include any recommendations for new financial models, nor did it analyze options for establishing a DSF. Governments welcomed the TFFM report as constructive input, but started their negotiations on the chapter on financial mechanism from zero when they reconvened for PrepCom2 in February 2005.

The main controversial subject—the DSF—was settled before PrepCom2. A public-private initiative announced the launch of a DSF in January and established the fund formally in March 2004 in Geneva outside the official WSIS negotiations. A number of governments, local and regional authorities, and private sector members agreed to constitute the DSF as a voluntary fund in the form of a private foundation, incorporated under Swiss law. The interesting point here is that the DSF was not designed as a new governmental mechanism to channel money into third world countries, but as a public-private partnership in which all stakeholders were invited to make contributions, both in cash and in kind. The innovation of the DSF is the so-called Geneva Principle, a voluntary self-obligation by local authorities to transfer to the DSF one percent of the value of an IT contract that an administration signs with a private company.

During PrepCom2 all stakeholders supported the launch of a voluntary DSF. This was later reflected in article 28 of the Tunis Agenda, which welcomed "the Digital Solidarity Fund (DSF) established in Geneva as an innovative financial mechanism of a voluntary nature open to interested stakeholders with the objective of transforming the digital divide into digital opportunities for the developing world by focusing mainly on specific and urgent needs at the local level and seeking new voluntary sources of 'solidarity' finance. The understanding of the role of the fund is that it will complement existing mechanisms for funding the Information Society, which should continue to be fully utilized to found the growth of new ICT infrastructure and services." But it remains to be seen how much in contributions the DSF can mobilize, and what the fund can achieve in the years ahead.

The DSF launch made it much easier for the negotiators in PrepCom2 to agree on the remaining issues. Civil society did not play a central role in that discussion. However it offered contributions in kind, in particular with regard to human capacity building and training and expertise to help implement the digital solidarity agenda.

Governments underlined that public investment in IT infrastructure and services will remain important and market forces alone cannot bridge the digital divide, but they also stressed that the digital solidarity agenda will remain an empty concept without the investment by the private sector. The chapter on financing finally got twenty-five articles. In every fourth article there is a reference to the private sector. In the whole chapter the private sector was invited six times to make financial and material contributions and to take the lead in this arena. Governments promised to reconceptualize the spending of public money via their development policies and promised to reorganize, improve, and innovate existing financing mechanisms, but no group promised to channel additional financial resources into the WSIS process.

The Working Group on Internet Governance (WGIG)

As decided in Geneva, UN Secretary General Kofi Annan established a WGIG in October 2004. WGIG was formed only after a long diplomatic battle about its composition. The traditional procedure for working groups, established by UN world summits, is that an intergovernmental group is formed, composed by member states based on the principle of geographical diversity. WGIG was no such UN group. As previously quoted, article 50 of the Geneva Declaration of Principles obliged the UN secretary general to form the group in an open and inclusive process with representatives from governments, the private sector, and civil society.

In March 2004, Kofi Annan nominated the former Swiss diplomat Markus Kummer as the executive secretary of WGIG. It took another seven months before forty hand-picked individuals were invited by the UN secretary general to participate in WGIG.

WGIG members represented a fair balance of involved constituencies, following exactly the guidelines of article 50. All three stakeholder groups constituted more or less one third of the members. There was a fair balance between South and North. The Indian diplomat Nitin Desai, former UN under-secretary general, became the WGIG chairman.

According to article 13, paragraph B of the Geneva Plan of Action, WGIG got the mandate to "i) develop a working definition of Internet governance; ii) identify the public policy issues that are relevant to Internet governance; iii) develop a common understanding of the respective roles and responsibilities of governments, existing intergovernmental and international organisations and other forums as well as the private sector and civil society from both developing and developed countries; iv) prepare a report on the results of this activity to be presented for consideration and appropriate action for the second phase of WSIS in Tunis in 2005."

Between November 2004 and July 2005, WGIG had four meetings, all linked to a round of "open consultations" with the broader Internet community. The group started to identify the key issues according to its mandate and to draft discussion papers. It prepared an interim report for PrepCom2 (Geneva, February 2005). The final report was drafted after the fourth open consultation in a closed session at the Chateau de Bossey in a neighborhood outside of Geneva. The report and the background documentation were presented to the general public on July 15, 2005.

One of the most interesting questions for the WGIG, from its very early days, was how members of the group composed by different stakeholders from the governmental and the nongovernmental sector, could work together, reach consensus, and present workable recommendations to the governments of the UN member states. WGIG was confronted with a multidimensional conflict situation—much more complex than in a traditional UN working group where positions of governments are more or less known and the challenge is to find compromise language to bridge these differences.

WGIG had to deal not only with conflicts both within and among different stakeholder groups but also with a more culturally based conflict rooted in the different practices and procedures of governmental and nongovernmental groups when it came to decision making, regulation, and standardization for the Internet. While the governmental law-making culture is more or less a top-down process, nongovernmental code-making culture is more bottom-up. These two different approaches could easily coexist as long as they were clearly separated. Intergovernmental organizations like the ITU or the European Telecommunication Standards Institute (ETSI) operated under the principle of national sovereignty, while Internet groups like the Internet Engineering Task Force (IETF) or the World Wide Web Consortium (W3C) place themselves in the borderless cyberspace where national sovereignty is difficult to execute.

However, it is a fact that in realpolitik and in real life, "real places" and "virtual spaces" cannot be separated easily. When it comes to the Internet, political and technical issues, laws, and codes are interdependent and highly interwoven. Without its virtual components, the real world would not be able to produce the extra value that the Internet makes possible. And the virtual world needs the real world to make use of its potential. Every electronic communication among virtual Netizens starts and ends with a real citizen.

The challenge for WGIG members was to recognize the interdependence between the two different worlds of the "borderless cyberspace" and the "bordered real space." The WGIG mandate pushed the group to look into both governmental policy regulation and nongovernmental technical standardization for the Internet.

The problem was to avoid an either-or approach. The question was not whether governmental top-down regulation should be enlarged to the "technical world," or whether it should be substituted by the bottom-up self-regulation of the private sector and civil society. The challenge was to find a way to link the two concepts. It was not "law or code," it was "law and code."

During the ten months of work, in which WGIG had four internal meetings linked to open consultations with all WSIS stakeholders and a final one-week drafting meeting for the final report, the group basically agreed that the governance of the Internet required collaborative efforts and coregulatory models that take into consideration the role of the public administrations as well as the private constituencies. WGIG could agree that there is a need to improve both national and international Internet legislation by sovereign nation-states and the self-regulatory mechanisms of nongovernmental networks. WGIG recommended a productive interaction between these two sets of norms, principles, and procedures.

WGIG avoided creating a hierarchy of norms. It did not decide whether law should top code, or code should top law, or who should take the leadership in Internet governance. Based on a broad definition of Internet governance—which includes dozens of individual issues and various layers and players—they agreed that the overriding basic principle has to be multistakeholderism. WGIG defined Internet governance as "the development and application by Governments, the private sector and civil society, in their respective roles, of shared principles, norms, rules, decision-making procedures, and programmes that shape the evolution and use of the Internet."[7] WGIG acknowledged that there is no one-size-fits-all model: each individual Internet governance issue needs an individual solution. But such a solution has to be embedded in a framework guided by horizontal principles like multistakeholderism, openness, transparency, democracy, and so on. While the formal legal status of an individual solution for a special Internet issue plays a role, at least for the settlement of disputes, it is not decisive. More important is how the solution functions: it has to be adequate, efficient, accountable, predictable, fair, balanced, inclusive, and workable. Somebody has to be "in charge."

WGIG recognized that neither governmental top-down regulation nor private sector or civil society bottom-up self-regulation alone can manage the totality of Internet issues. The weakness of one partner in one area can and has to be compensated for by the strength of the other stakeholder, and vice versa. If a nonbinding legal norm works, it is okay. If a binding norm is needed, this is also okay. The choice of which form of arrangement is needed is determined by the substance of the issue and not by any general theory of law and politics.

In its report, WGIG tried to define the specific roles and responsibilities of the stakeholders in a more detailed way. However, it did not recommend any principles and procedures for interaction among the stakeholder groups.

After ten months of hard work, there was a high degree of consensus among the WGIG members. Nevertheless, fundamental conflicts both within the governmental stakeholder group and between some governments and nongovernmental stakeholders from the private sector and civil society could not be bridged. This became apparent when the group started to discuss models for Internet oversight.

WGIG was not a negotiation body. This status allowed WGIG to offer a choice of options when the group could not reach a consensus. For the oversight issue, WGIG offered four models that reflected the basic positions of the different "political camps." The spectrum of the models reached from "status quo minus" to "status quo plus plus," with some middle-of-the-road variations in between. A group of nongovernmental representatives preferred the status quo minus model, where the role of governments, including the U.S. government, in managing the Internet core resources would be close to zero and activated only in cases where a clear public policy component is involved and/or the security and stability of the Internet is at stake. The other end of the spectrum was represented by a number of governmental WGIG members from the third world who proposed the establishment of a UN-like Intergovernmental Internet Organization, where an intergovernmental council would include a decision-making capacity. In such a model nongovernmental stakeholders would have an advisory capacity for the proposed decision-making Intergovernmental Internet Council.

Within WGIG all stakeholder groups participated on an equal footing. There was no "governmental leadership" in the drafting of the final report. On the contrary, the members from civil society and the private sector were the main drivers in the discussions. They influenced the internal agenda of WGIG, they dominated the online discussion groups, and they produced the majority of the issue papers for the background report. The working definition of Internet governance and one of the WGIG's key proposals, the creation of a new multistakeholder discussion space called the Internet Governance Forum (IGF), came from the WGIG members representing civil society.

Toward the Internet Governance Deal of Tunis

When WSIS was reconvened for PrepCom3 (Geneva, September 2005), the negotiators had the final WGIG report at their disposal. The WGIG report

- proposed a definition for Internet governance;
- identified main public policy issues related to Internet governance;
- defined the specific roles and responsibilities of the involved stakeholders;
- recommended the establishment of an Internet Governance Forum (IGF) as an open space for discussion without decision-making capacity;
- provided four models with regard to Internet oversight for discussion; and
- made thirty-four individual recommendations for specific Internet governance-related issues.

The open question was whether governments would take the WGIG report as a starting point for their negotiations or reopen the debate from a purely governmental perspective, as they did with the report of the TFFM. Another key question was of a procedural nature: would the negotiations be organized in an open plenary, accessible to all stakeholders, or in a closed working group?

PrepCom3 decided to negotiate Internet governance in Sub-Committee II. The Committee, which mainly met in plenary session, was chaired by Masood Khan, an ambassador from Pakistan. Ambassador Khan proposed to take the WGIG report as the basis for the negotiations that was accepted by all governments. He expressed his recognition of the multistakeholder principle as the basis for solutions with regard to Internet governance. And he invited representatives of the private sector and civil society to individual consultations and encouraged them to make written contributions with "diplomatic text."

His approach was to avoid unneeded ideological discussions, to respect sovereignty of member states as well as the legitimate interests of stakeholders, and to look for acceptable compromises on key issues. One way to improve the confidence among stakeholders was to continue the negotiations in the open plenary of the Sub-Committee II. This avoided a conflict that overshadowed the final Internet governance negotiations during the Geneva phase of WSIS, where a closed subworking group hammered out the final resolution.

The only restriction nongovernmental stakeholders had to accept was that they reduce their interventions to one hour in the daily morning session. For civil society and the private sector this was an acceptable arrangement. It was seen as a step forward, to build more trust among the stakeholders. And it opened the opportunity to comment on the progress of negotiations in real time and to make concrete language proposals.

The negotiations moved rather fast in areas where all sides agreed to the controversial points. The majority of the WGIG recommendations were taken as a sufficient base for an intergovernmental agreement. Governments recognized the WGIG definition for Internet governance. The WGIG proposal to launch an IGF got a positive evaluation and was accepted. But with regard to the Internet oversight function, none of the four proposed models from the WGIG report got a majority.

After days of controversial discussions, the conflicts were narrowed to two alternatives. The model of a new Internet UN, or intergovernmental Internet organization (status quo plus plus), did not get enough support and was not pushed forward by governments like China or Brazil. But the model "status quo minus," which argued in favor of a reduced role of governments in the management of the Internet core resources (and included the call for a withdrawal of the U.S. government from ICANN and Internet root oversight) and was proposed by nongovernmental stakeholders in the WGIG report, also was not further investigated.

The two remaining models on the negotiation table came from the European Union and the United States. The EU proposal called for a "new cooperation model," something like a public-private partnership arrangement, where governments should take the leadership on the "level of principle" while the private sector should take the lead in the day-to-day operations. The United States argued for a continuation of the existing model with private sector leadership, with ICANN as the main international organization for the management of the technical resources of the Internet and a limited role for governments within the ICANN's Governmental Advisory Committee (GAC). The U.S. government also reiterated its position that it intends to preserve the security and stability of the Internet DNS by maintaining its historic role in authorizing changes or modifications to the authoritative root zone file.

For civil society, neither the EU proposal nor the U.S. position was a good answer to the challenge. In the EU proposal, civil society criticized the unspecified role of governments, which included the risk that governments, in particular in nondemocratic countries, could interfere in daily Internet communication and restrict human rights, such as the right to freedom of expression or data protection. But civil society also was not satisfied with the U.S. position. To continue with unilateral oversight was seen as undemocratic and unacceptable in the long run.

When PrepCom3 ended after two weeks of controversies around Internet governance, no consensus was reached in the Sub-Committee II. Consequently, a PrepCom3+ was convened for November 2005 on the eve of the summit in Tunis.

In the meantime, the conflict between the U.S. government and the EU governments escalated. High-level politicians intervened, including members of the U.S. Congress and Secretary of State Condoleeza Rice. When EU President Barroso visited the White House in October 2005, U.S. President Bush raised the issue as a question of central importance for the U.S. government.

The Chinese government, which was one of the main opponents of the status quo in the Geneva phase, did not interfere in the EU-U.S. battle. It also did not call for "governmental leadership," as it had in the Geneva phase. The Chinese government gave highest priority to the recognition of national sovereignty over its virtual domain name space, that is the management of the country code top-level domain (ccTLD).

This approach was not in conflict with the U.S. position. In its statement on DNS principles from June 30, 2005, the U.S. government recognized that governments have legitimate interest in the management of their ccTLDs.

This statement did not include a unilateral legally binding declaration that the U.S. government would never interfere in the management of a ccTLD, and would not use its oversight function to stop the authorization of the publication of a ccTLD zone file in the Hidden Server that is at the apex of the domain name system; yet it was obviously satisfactory for the Chinese government in that it gave the PRC (and other countries) a free hand to develop and execute a specific national policy within its own domain name space. In the case of China, this includes the option to develop—within its own borders—an internationalized domain name (iDN)-based Internet root server system.

The sovereignty question was not a significant controversy in the final negotiations during PrepCom3+. In article 63 of the Tunis Agenda, governments could agree that "countries should not be involved in decisions regarding another country's ccTLD. Their legitimate interests, as expressed and defined by each country, in diverse ways, regarding decisions affecting their ccTLDs, need to be respected, upheld and addressed via flexible and improved frameworks and mechanisms."

The key conflict for PrepCom3+ was whether the United States and the EU could bridge their different positions, and whether third world countries would become part of a global policy development process with regard to Internet governance.

In the final stage of the negotiations, nongovernmental stakeholders had no speaking rights in the small group meetings drafting particular sections of text, but could listen to the governmental statements. The presence of nongovernmental stakeholders in the negotiation room had an indirect effect. The openness and transparency of the negotiations pushed governments that would have preferred a more restrictive Internet arrangement to be silent or to water down radical proposals. Obviously, such an open environment was helpful to finding solutions acceptable to all stakeholders. At the end of the day, the Tunis Internet governance deal more or less satisfied all stakeholders.

With regard to Internet oversight, the compromise, to launch a "process of enhanced cooperation," enabled both the EU and the United States to agree. For the EU, such a process was expected to lead in the long run to a "new cooperation model," as proposed in PrepCom3; for the United States, such a process would simply improve the cooperation among existing organizations. The compromise also got support from developing countries and China, which got an equal seat at the table. Article 68 says: "We recognize that all governments should have an equal role and responsibility for international Internet governance and for ensuring the stability and security and continuity of the Internet."

While this process of "enhanced cooperation" is seen primarily as an intergovernmental process, it must also include the private sector and civil society. According to article 71, the UN Secretary General, tasked with starting such a process, is obliged "to involve all stakeholders in their respective roles." The process of enhanced cooperation is embedded in a framework of principles and procedures and includes, as said in article 70, "the development of globally applicable principles on public policy issues associated with the coordination and management of critical internet resources." As a consequence, more time for discussion and clarification of what "enhanced cooperation" will mean in practice is needed both within and among the stakeholders groups.

The other main decision of Tunis was the creation of the IGF. The IGF, which held its first meeting in October–November of 2006, is designed to be a "forum for multistakeholder policy dialogue." Its mandate is "to facilitate discourse between bodies dealing with different cross-cutting international public policies regarding the Internet" and "to identify emerging issues, bring them to the attention of the relevant bodies and the general public and, where appropriate, make recommendations."

The IGF can be seen as a very interesting innovation in international politics. It is not an intergovernmental organization, as some governments wanted it to be. But it is also not a private corporation like ICANN. It has no decision-making capacity but offers a space for communication, coordination, and cooperation among interested, involved, and concerned constituencies. It has low entry barriers and is subject-oriented. It does not fix anything, but pulls everybody into a discussion process. As already noted, the IGF was proposed by civil society representatives in WGIG. It is a visionary concept that will pull all stakeholders into a holistic process to explore the unknown territory of cyberspace.

Other arrangements with regard to Internet governance deal with general principles for a broad range of issues, from generic top-level domains to IP numbers, from cybercrime to multilingualism.

A comparison between the WGIG recommendations and the Tunis agreement proves that the governments followed in principle what the WGIG proposed. This is also true with regard to the principle of multistakeholderism. In twenty-five of the fifty-three articles of the Internet governance chapter of the Tunis Agenda for the Information Society, the principle of multistakeholderism is reconfirmed, reiterated, or specified.

In article 33 governments recognize that the WGIG report has enhanced the "understanding of the respective roles and responsibilities of governments, intergovernmental and international organisations and other forums as well as the private sector and civil society both from developing and developed countries." In the same article it is reaffirmed "that the management of the Internet encompasses both technical and public policy issues and should involve all stakeholders." Similarly, article 80 states that "the development of multistakeholder processes at the national, regional and international levels" are encouraged.

In other words, the principle of multistakeholderism has been accepted as a horizontal principle that constitutes the basis for all elements related to Internet governance. This does not necessarily mean that each stakeholder has the same role and responsibility in all individual Internet governance issues; different forms of governance may apply to different issues. In the struggle against cybercrime, for example, governments have to play the leading role. In the management of technical Internet core resources, it is the private sector that will have the leading role. But leadership does not mean exclusivity. In both cases, all stakeholders, according to the Tunis agenda, should be included. With regard to the struggle against cybercrime, article 40 calls upon governments "to develop necessary legislation for the investigation and prosecution of cybercrime . . . in cooperation with other stakeholders." And while in article 65 private sector leadership in the management of technical core resources is recognized, in article 69 governments reserve their right to carry out their roles and responsibilities in international public policy issues pertaining to the Internet.

These guidelines and compromises reflect a big step forward. They constitute de facto recognition of a coregulator model for Internet co-governance, where individual issues related to Internet governance are regulated and managed differently in detail but follow the same general principle. The multistakeholder partnership of government, the public sector, and civil society can take different forms according to the specific needs of an individual issue, but the regulation and the management needs the involvement of all three partners.

Governance becomes more issue driven and less interest driven. This new "division of labor" among the stakeholders can lead to a governance system of shared responsibilities where the different functions of different stakeholders are complementary but each stakeholder participates on an equal footing.

The Tunis summit avoided pushing Internet governance into a box. On the contrary, it opened the door for a process that allows further development and innovation, both technically and politically. The first mile of this long road is paved, now the stakeholders have to move.

Looking Forward

Putting WSIS in the context of world politics, it can be concluded that the summit process indeed pioneered a new governance model. During the four years of the numerous WSIS conferences from the first PrepCom1 to the second summit, the principle of multistakeholderism steadily evolved. It is now accepted as a basic principle for the information society and Internet management.

The WSIS process has demonstrated that in the information age

- international politics involves other actors in addition to governments;
- each of the three main stakeholder groups—governments, private industry, and civil society—has a specific role and responsibility;
- no stakeholder can replace another stakeholder; and
- it is necessary for all three stakeholders to work together.

The rather simple ICT global governance system of the nineteenth and twentieth centuries with a limited number of players in global diplomacy is no longer feasible. There is no alternative any more to a broader involvement of the concerned and affected stakeholders if it comes to political and economic negotiations on global issues. Although each stakeholder is rather different, they need each other. To achieve political goals, inclusion is essential. Exclusion is counterproductive and will fail. If the private sector and civil society were to exclude governments, they would risk losing the stability a governmental system can provide. If governments and civil society were to exclude the private sector, they would risk losing the material capacity that is needed to stimulate development. And if governments and the private sector exclude civil society, they are confronted with demonstrations and street protests and lose legitimacy for their actions. In other words, a new system of bilateral relationships among the three major stakeholders has to be developed, and will shape the way forward into the year 2015.

Between 2005 and 2015, when the objectives described in the Geneva Plan of Action are supposed to be achieved, a complex follow-up and implementation process will offer more opportunities to experience new forms of multistakeholderism. This is true both for the IGF and the digital solidarity fund, but it is also relevant for the seventeen action lines, where under the leadership of relevant intergovernmental organizations of the UN system the various stakeholders will work together around concrete issues—from e-health to e-agriculture, from e-government to e-commerce, from security to privacy, from intellectual property to human rights, from infrastructure development to media.

Recommendations

What should be done? The question goes to all stakeholders, but each group has to do different things according to its specific roles and responsibilities. Here are some key recommendations for all stakeholder groups:

Governments have to develop new procedures and principles for the collaboration with nongovernmental stakeholders. They have to learn and to understand the motives, visions, and plans of nongovernmental stakeholders. And they have to enable the private sector and civil society—according to the national circumstances—to take on more and concrete responsibilities for issues that go beyond the capacity of governments or the intergovernmental system.

The private sector has to learn that its involvement in global policy development is needed, but must be driven by more motives and criteria than the normal business aimed at satisfying shareholders. Transnational corporations have to understand the private sector's responsibility as an important actor that has to serve the global community as a whole, to make effective contributions to bridging the digital divide and managing the Internet to the benefit of the local and global Internet community.

Civil society has to continue to provide expertise, knowledge, and concrete project activities on the ground. But civil society groups must also improve their organizational structures to get more legitimacy and credibility, to establish a sound material and financial basis that would enable them to participate in a sustainable manner in the follow-up and implementation phase of the WSIS process. Civil society needs engaged individuals who in historical moments can play a crucial role to effectively position civil society politically; this requires also a more stable organizational framework that is not yet in place. Furthermore, civil society has to intensify its conceptual work to position itself in global policy making both as a critical watchdog and counterforce to the establishment, and a constructive, equal partner in new, diverse coalitions that will emerge from such innovative processes.

One key recommendation for the technical and academic communities is to intensify study and research on new governance models and new technologies that will enable in particular individual Internet users and civil society groups to become active Netizens in cyberspace, able to participate constructively and innovatively in the global policy development of the information age.

Conclusion

It remains to be seen what this road toward 2015 will produce. But in analyzing the lessons learned from the WSIS debate so far, one can make numerous conclusions with regard to the development of a new governance concept and regimes:

1. The principle of multistakeholderism, even if it is still defined in vague language, is broadly accepted as a basic principle for the management of the problems related to the future development of the information society and in particular to the Internet. There is a clear consensus that to meet the challenges of the information age requires the full involvement of governments, the private sector, and civil society in their specific roles and responsibilities.
2. The specific role and responsibility of the civil society is to provide expertise and knowledge to the process, as well as linkage to real people involved in projects on the ground.

3. The involvement of nongovernmental stakeholders in diplomatic negotiations produces extra value that enriches the outcome of conferences. Although the proceedings before and during the intergovernmental negotiations are getting more complex and producing new conflicts, at the end of the day the final result is of a higher quality.
4. The involvement of nongovernmental stakeholders in UN summits does not change the intergovernmental nature of the final negotiations process. Politically and legally binding commitments can be made only by governments.
5. One of the main weaknesses of civil society is its vaguely recognized and demonstrated legitimacy. In the WSIS debate process it was argued that there are more and broader sources for legitimacy in international negotiations than democratic elections that legitimize governments to act on behalf of their people—such as expertise, knowledge, market acceptance, consumer confidence, and so on. However, stable and transparent procedures are required for nongovernmental civil society organizations to improve recognition of their legitimacy to represent groups of people in a democratic way.
6. Another weakness in civil society is lack of organizational stability. Civil society groups are very often dependent on engaged individuals and lack the financial resources needed to secure a sustainable involvement in governance processes.
7. While the emergence of an organizational structure within the WSIS process can be seen as a big achievement and a step forward to become more mature, stable, and representative, the overall CS structure is still rather fragile and not yet sustainable. The risk is high that it will collapse now that WSIS is over. It remains to be seen how an advanced and more stable civil society structure will emerge from the process of WSIS follow-up and implementation, in which CS is more included and can participate with higher status and recognition as a stakeholder.

There are also a number of risk factors that cannot be overlooked in the future:

1. While civil society is probably more active on the ground than private industry, it does not have the substantial influence of big corporations. A further broadening of the multistakeholder principle within the UN system could lead to an unbalanced triad in which the private sector and business, which are primarily responsible to their shareholders and not to the international community as a whole, get more and more influence and decision-making capacity in global policy making.
2. The system of accountability with regard to nongovernmental stakeholders is in its infant stage. A huge number of actors in the WSIS process still have to clarify to whom they are accountable at the end of the day, and how their commitments can be seen as serious, obligatory, and implementable in the practical process.
3. There is a risk of fragmentation of the political process and the global governance system as a whole if specific issue-oriented partnerships or—in the context of the IGF—"dynamic coalitions" appear that take the lead in a specific arena and ignore other groups and stakeholders.

4. There is no stable financial system for a sustainable development of multistakeholderism. Neither private industry nor civil society organizations have legally binding commitments to allocate part of their budgets to a process like WSIS. Thus a loss of material resources could lead to a quick collapse for partnerships with nobody in charge.
5. Governments could use the principle of multistakeholderism to escape from their responsibilities and to ignore issues that normally fall under their public policy duties.
6. There is a risk that a new multilayer governance system would emerge not in the form of a network but a new hierarchy with elite groupings at the top both from governments and from private industry and civil society, which would dominate and/or ignore other networks lacking comparable resources, reputations, and global recognition.

Despite these risks, there is no way back to a diplomatic world of the twentieth century, when governments were the only players in international policy making. The challenge of the twenty-first century is to broaden the understanding of the new principle of multistakeholderism, to conceptualize its meaning for diplomatic negotiations and the management of global issues and resources as well as to define in more detail the specific roles and responsibilities. It seems in particular important to develop procedures for the interaction among the individual stakeholders.

A lot can be done through research, but it is more efficient to test new forms of cooperation among stakeholders in practice. In the Joint Project Agreement that the U.S. government signed with ICANN in September 2006, it is proposed that ICANN become a model for a multistakeholder organization. Indeed, when ICANN was established in 1998, all stakeholders were involved. ICANN introduced in its bylaws a number of new principles like openness, transparency, inclusion, and bottom-up policy development, although rather often ICANN did not meet these standards in practice. However, it makes sense to strengthen efforts to move forward and to experiment with new forms of policy development in the ICANN framework.

To deepen and broaden our understanding of the new principle of multistakeholderism, the specific roles and responsibilities of stakeholders and in particular the potential opportunities and power of civil society organizations (the at-large membership in ICANN language) requires more studies, discussions, and experiences on the ground. The door is opening to the future, but the territory beyond is still uncharted. Innovation and creativity are needed to develop new global governance models that lead humanity into the future information age.

When history reaches a turning point, it is always useful to look back. A lot of authors have compared the information revolution of today with the industrial revolution of the eighteenth and nineteenth centuries. The interesting question we have before us is whether the information revolution like the industrial revolution will have political consequences for the system of governance: The industrial revolution and the economy that emerged based on its inventions paved the way for the bourgeois revolutions. In that era, new stakeholders such as private investors challenged the absolutist power of their rulers and called for more rights to participate in policy development and political decision making. As a result, a new governance model was developed that partly substituted new republics for the old kingdoms. Parliaments emerged, democratic elections were introduced, and a system of representative democracy developed. Such a governance system, based on a parliamentarian democracy, was obviously much more complex than a traditional kingdom, but there was no alternative to keep the old system alive. In some countries a "constitutional monarchy" emerged as something like a coregulatory system, where both the old and new stakeholders agreed on a division of power. The model that emerged from this kind of power struggle has worked more or less efficiently for the last two hundred years.

It remains to be seen whether the challenges emerging with the information revolution are also challenges for the newly established governance model. The German philosophers Hegel and Marx instructed us long ago that things are moving from simple to complex structures. Obviously, a governance model based on multistakeholderism is much more complex than a model of individual, separate governmental institutions. But as the WSIS process has shown, the power struggle between the old governmental stakeholder and the new nongovernmental stakeholders has already started. The new stakeholders representing different constituencies are asking for more participatory rights in policy development and decision making. And this will probably go beyond the issue of Internet governance. Will the further development and implementation of the principle of multistakeholderism ultimately lead to a new global coregulatory governance model where power is shared and decentralized in the form of a multilayer multiplayer governance mechanism of communication, coordination, and collaboration? Will the Internet governance model become a blueprint for the governance of other global issues and challenges of the twenty-first century?

Notes

1. Address by the UN Secretary General to the World Summit on the Information Society, Geneva, December 10, 2003. Available at www.itu.int/wsis/geneva/coverage/statements/opening/annan.doc.

2. Kofi Annan, Address to the World Economic Forum, UN Press Release SG/SM6153, January 31, 1997.

3. Chair's Conclusions of the G7 Summit, "Information Society Conference," Brussels, February 26, 1995, doc/95/2. Available at http://europa.eu.int/ISPO/docs/services/docs/1997/doc_95_2_en.doc.

4. UN Resolution 56/183, World Summit on the Information Society, December 21, 2001. Available at www.itu.int/wsis/docs/background/resolutions/56_183_unga_2002.pdf.

5. See "Preparing the World Summit on the Information Society: Final Report of the Consultations with Non-Governmental Organisations," UNESCO, Paris, April 28, 2002. The Mainz Declaration states in paragraph 5 of its preamble: "Recognizing the special role civil society plays in the Information Society and supporting the 'Recommendations on the Participation of Civil Society' which have been elaborated in UNESCO's consultation process with professional NGOs (4/02) as part of its preparation for the WSIS." See "Information Cultures and Information Interests: European Perspectives for the Information Society," UNESCO Regional Pre-Conference for the World Summit on the Information Society (WSIS), Mainz, Germany, June 29, 2002. Available at portal.unesco.org/ci/en/ev.php-URL_ID=2539&URL_DO=DO_TOPIC&URL_SECTION=201.html.

6. See World Summit on the Information Society, 2003, p. 6. All four of the Geneva and Tunis summit outcome documents are available at www.itu.int/wsis.

7. See Working Group on Internet Governance (WGIG), 2005, p. 4.

References

Ang, Peng Hwa. 2005. *Ordering Chaos: Regulating the Internet.* Singapore: Thomson.

Cairncross, Francis. 1997. *The Death of Distance: How the Communication Revolution Is Changing our Lives*. Cambridge, MA: Harvard Business School Press.

Castells, Manuel. 2000. *The Rise of the Network Society*. Oxford: Blackwell Publishers.

Castells, Manuel. 2001. *The Internet Galaxy: Reflections on the Internet, Business and Society*. Oxford: Oxford University Press.

Drake, William J., ed. 2005a. *Reforming Internet Governance: Perspectives from the Working Group on Internet Governance*. New York: UN ICT Task Force.

Drake, William J. 2005b. "Collective Learning in the World Summit on the Information Society." In *The World Summit on the Information Society: Moving from the Past into the Future*, ed. Daniel Stauffacher and Wolfgang Kleinwächter, 135–146. New York: UN ICT Task Force.

Drucker, Susan J., and Gary Gumpert. 1999. *Real Law@Virtual Space: Regulation in Cyberspace*. Cresskill, NJ: Hampton Press.

Giddens, Anthony. 1998. *The Third Way: The Renewal of Social Democracy*. Cambridge, UK: Polity Press.

Goldsmith, Jack, and Tim Wu. 2006. *Who Controls the Internet: Illusions of a Borderless World*. Oxford: Oxford University Press.

Kahin, Braid, and Ernest Wilson. 1997. *National Information Infrastructure Initiatives: Vision and Policy Design*. Cambridge, MA: The MIT Press.

Kleinwächter, Wolfgang. 2001. *Governance in the Information Age*. Aarhus: Aarhus University Press.

Kleinwächter, Wolfgang. 2004. *Macht und Geld in Cyberspace: Wie der Weltgipfel zur Informationsgesellschaft die Weichen für die Zukunft stellt*. Hannover: Heise Verlag.

Stauffacher, Daniel, and Wolfgang Kleinwächter, eds. 2005. *The World Summit on the Information Society: Moving from the Past into the Future*. New York: UN ICT Task Force.

Lessig, Lawrence. 2002. *The Future of Ideas: The Fate of the Commons in a Connected World*. New York: Vintage Books.

Lessig, Lawrence. 2004. *Free Culture: How Big Media Uses Technology and the Law to Lock Down Culture and Control Creativity*. New York: Penguin Books.

Loader, Brian D. 1997. *The Governance of Cyberspace*. New York: Routledge.

MacLean, Don, ed. 2003. *Internet Governance: A Grand Collaboration*. New York: UN ICT Task Force.

Miller, Steven E. 1996. *Civilizing Cyberspace: Policy, Power and the Information Superhighway*. New York: ACM Press.

Mueller, Milton. 2002. *Ruling the Root: Internet Governance and the Taming of Cyberspace*. Cambridge, MA: The MIT Press.

Raboy, Marc. 2002. *Global Media Policy in the New Millennium*. Luton: University of Luton Press.

Stieglitz, Joseph E. 2006. *Making Globalization Work*. New York: W. W. Norton & Company.

Tapscott, Don. 1995. *The Digital Economy: Promise and Peril in the Age of Networked Intelligence*. New York: McGraw Hill.

Working Group on Internet Governance. 2005. "Report of the Working Group on Internet Governance." Geneva: WGIG, July. Available at www.wgig.org.

World Summit on the Information Society. 2003. "Declaration of Principles—Building the Information Society: A Global Challenge in the New Millennium." Geneva: WSIS, December. Available at www.itu.int/wsis.

16 Conclusion: Governance of Global Electronic Networks: The Contrasting Views of Dominant and Nondominant Actors

Ernest J. Wilson III

Introduction

Intellectual Barriers to Entry

For serious students of the governance of global electronic networks these are the best of times and the worst of times: the best of times because a whole string of major theoretical and practical questions have arisen in the past few years about the relationship of the information revolution to matters of major scholarly concern—about information and communications technologies (ICTs) and the structure and exercise of world power; about ICTs and economic efficiency; about ICTs and democracy; and about ICTs and distributional equity. And the worst of times because theorists and good analysts must build their work by pointing out lasting causal relationships that do not change, or change only slowly; yet early in this new century, many fundamentals of the international system are changing simultaneously before our very eyes. In studying ICT there are far fewer anchors to which we can secure our theoretical frameworks. Ceteris paribus—all things being equal—is cited with less frequency today than decades ago. Everything everywhere seems to be moving, and fast—human migration, capital flows, weapons of mass destruction, broadcast images, terrorism, and technological innovations.

Under these circumstances scholars confront a tough double imperative. They must reach down and dig through the gritty details and minutiae of rapid global transformations, and try to master the empirical realities, while simultaneously searching for the elusive answers to the larger "so what" questions. They must discern patterns before they are fully fixed and seek out trends in inherently unpredictable human behaviors (Aronson 2002).

One common challenge is to decide where to cut into this continuum between big-picture and minute details. Yet theoretical sophistication has not kept pace with empirical detail. The explanatory models scholars now employ to master the minutiae as well as the big patterns of new information and communication technologies remain rather modest. There is a lot of churning within the disciplines and professions, with tensions across and within them over the best ways to poke and probe the ways of the digital world. In the absence of a good match among big questions, good integrative analytic tools, and accessible, intelligible information and data, we are all in a bind.

With such field properties, ICT remains a tough domain, where intellectually inquisitive but less informed researchers confront what economists would call high "barriers to entry." These include the conceptual issues just cited, but also data sets that are either highly aggregated; guided by blunt, uninteresting questions; or case study materials that are scattered and often incommensurate with one another. These conditions combine to keep researchers from exploring ICT-related topics in greater depth or sophistication, or from sustaining their interests once they get involved. In this regard the study of ICTs may not be too different from trying to master the politics, rules, and performance of other high-technology issues such as biotechnology.

Later in this chapter I will suggest a meso-level framework I term the Quad that attempts to provide an analytical pathway into the domestic bases for global governance, and that encourages theorizing and conceptual clarity as well. This Quad model helps fill in some of the conceptual and theoretical gaps in the literature on ICTs nationally and globally, and can prompt better empirical work. It seeks to capture the dynamic that many of the authors note, such as Henry Farrell, who in chapter 10 writes that ICT issues like privacy are entering a distinctive period in which new issues arise not from the actions of states or private actors working in isolation, but from the growing relations *between* states and private actors, including private actors carrying out tasks on behalf of public authorities. Increasingly, public, private, and nonprofit stakeholders seek such cross-sector alliances to gain new knowledge and to broaden their political support.

Four Guiding Questions

The major purpose of this chapter is to address these and related epistemological "so what" questions in the area of the governance of global ICT, and to report on the answers offered by the authors in this volume. We believe these questions receive far too little attention as sustained and serious objects of inquiry. Of course, in this volume not every author addresses all concerns equally, and I will try to highlight some of the differences across the chapters. The four initial questions we posed are as follows:

1. Is there a Washington consensus separate from the preferences of nondominant actors?
2. Are the current ICT governance mechanisms working well or are they broken?
3. What is the impact of the current global governance of electronic networks (GGEN) arrangements on nondominant actors?
4. What can scholars and researchers do to help practitioners in the field of ICTs?

It would be naïve in the extreme to assume that these are the only important questions to be posed, or that we can perfectly answer each in a single chapter. However, asking and beginning to answer them reduces some of the entry barriers in this emerging field of inquiry, and can improve both scholarship and practice. In the world of politics, the main global stakeholders forcefully pushed their favorite issues in the two sessions of the World Summit on the Information Society (WSIS) held in Geneva in 2003 and Tunis in 2005. I will discuss them in greater detail.

Some Definitions

Before answering these four questions we need common definitions of our core concepts, especially the meaning of governance. Governance is a term with many meanings, perhaps most usefully defined by the Commission on Global Governance as: "the sum of the many ways individuals and institutions, public or private, manage their common affairs. It is a continuing process through which conflicting or diverse interests may be accommodated and co-operative action may be taken. It includes formal institutions and regimes empowered to enforce compliance, as well as informal arrangements that people and institutions either have agreed to or perceive to be in their interest" (O'Siochru and Gerard 2002, 2).

Robert Keohane and Joseph P. Nye Jr. offer a complementary minimalist definition of governance as "the processes and institutions, both formal and informal, that guide and restrain the collective activities of a group.... Governance need not necessarily be conducted exclusively by governments.... Private firms, associations of firms, nongovernmental organizations... all engage in it, often in association with governmental bodies, to create governance; sometimes without governmental authority" (2002, 12).

Finally, in his introduction to this volume, in chapter 1, William Drake defines global governance as the development and application of shared principles, norms, rules, decision-making procedures, and programs intended to shape actors' expectations and practices and to enhance their collective management capacities in world affairs. In some areas, governance is quite explicit and institutionalized with clear enforcement mechanisms. These arrangements may take the form of a regime, authoritatively defined by Stephen Krasner (1991) as a "set of implicit or explicit principles, norms, rules and decision-making procedures around which actors' expectations converge in a given area of international relations." Krasner begins his 1991 article with the sentence "There is no single international regime for global communications. [The different technologies] are governed by a variety of principles, norms, rules and decision-making procedures—or in some cases, no regime at all. Variation in outcomes can be explained by the interests and relative power capabilities of the actors in each case" (ibid., 336). Ten years later, Randi Bessete and Virginia Haufler concurred with Krasner about the absence of an overarching governance regime in this domain of action (2001). Krasner's second point is especially worth noting here—the ICT domain (as distinct from a regime) is best explained by an interest- and power-based model, not a neoliberal institutionalist one, a theme to which I return.

There are industry or technology-specific rules and expectations, of course. For example, Milton Mueller, John Mathiason, and Lee McKnight defined Internet governance as "collective action, by governments and/or the private sector operators of TCP/IP networks, to establish rules and procedures to enforce public policies and resolve disputes that involve multiple jurisdictions" (2004, 4).

Much of the original theorizing on international regimes focused on developed countries and most of the work done on the information revolution at the global level also concentrates on the more privileged and powerful countries where ICT penetration rates are very high. There is far less work on nonpowerful and nonprivileged stakeholders, thereby limiting interesting questions of political economy and further marginalizing two-thirds of mankind.

In this project, we were especially interested in the position of nondominant actors in processes of governance, that is, those outside the circle of powerful actors in what other theorists might call "the center of the center" (Galtung 1971). Internationally, nondominant actors encompass virtually all the developing world and the "transitional" political economies of central and eastern Europe. Domestically, nondominant actors include nongovernmental organizations (NGOs) that typically lack the power to influence important decisions about the allocation of scarce ICT resources in their own countries (much less globally).

Question Number One

Is there a Washington consensus separate from the preferences of nondominant actors? Seeking to find a collective consensus in a large set of actors on any complicated global issue is always difficult. In our initial project discussions there was divergence of opinion about the existence of a Washington consensus on GGEN. Some claimed that the differences of opinion and policy orientation across the World Bank, U.S. Agency for International Development (USAID), the Federal Communications Commission, and powerful corporations are so great as to make moot any coherent notion of a common "Northern" agenda. But others argued that the definition of consensus was largely a function of purposes. If the purpose is to compare and contrast institutions inside the United States, then a consensus is elusive. When comparing the entire U.S. (and selected other G8 nations) on the one hand with other, quite distinct nondominant actors on the other (the "global South," for example), then a common position does emerge.

If one can establish the existence of a Washington consensus held by America's most powerful bilateral, private, and multilateral actors, it is far more difficult to define a single consensus for the "rest of the world" (ROW). By ROW we mean nondominant international actors, especially the developing and transitional countries, which are more likely to be rule takers than rule makers. For us, ROW is roughly equivalent to "transitional and developing countries." It also includes most NGOs in the core states. We conceive of a ROW grouping distinct from and sometimes at odds with the Washington consensus. The ROW lack power, playing only small parts in establishing the rules of the game of the international ICT market or regimes.

Differences at Two Levels

The differences between the "conventional wisdom" of Washington, DC, and other global actors is best understood at two levels—agenda and action. First is the matter of setting the global ICT agenda by determining which items should receive top priority and which should be addressed later. Second, at the level of collective action, stakeholders determine their own preferences for precisely what should be done about any given priority issue.

Most informed and responsible actors, whether North or South, private or public, probably agree on the top dozen or so ICT-related issues that are critically important to global governance, and must be seriously addressed *at some point, in some way, by some set of global actors*. These same issues surfaced in a post-WSIS forum held in Spain in 2006. Based on discussions with the authors, participation in international meetings, and the available literature, in alphabetical order I submit the key global issues are capacity building, the digital divide, financing and investment, governance reform (including expanded participation), infrastructure, innovation, intellectual property rights, market openness, privacy (and pornography), regulatory issues, security, sustainability, and trade in ICT goods and services.

However, the authors in this volume and other experts also agree that different stakeholders hold quite different priorities. That there are such differences between the Washington consensus and the ROW is not a surprise. The fundamental structural differences between Washington, DC, and Kigali, Rwanda, for example, are huge. The saturation of infrastructure and almost universal availability of ICT applications in the United States means that most basic information needs have been largely met in the United States and other G8 countries, while remaining hugely problematic in poor countries. There are fundamental differences in the core interests of the Washington consensus and the typical preoccupations of the rest of the world, at virtually every level of analysis, from the micro level of individual concerns to the most global and macro levels. These differences shape how stakeholders allocate political attention and other scarce resources to the issues.

What Are the Washington Consensus Priorities?

The antecedent to this contemporary search for the outlines of a Washington consensus arose in the 1980s during an earlier controversy surrounding the preferred policy packages to achieve growth and financial stability in developing countries. In his thoughtful and important work, economist and Washington think tank insider John Williamson described the common policy positions around which the multilateral banks and others coalesced, which came to be known as structural adjustment programs (SAPs) (Williamson 1990). They were based on a prescribed policy sequence of stabilization and then privatization, liberalization, exchange rate reform, foreign investment, fiscal conservatism, and a few other elements.

A similar intellectual consensus—in some respects an extension of the earlier liberalism into a new policy domain, mainly telecommunications policy—has again emerged among the leading Washington-based multilateral and bilateral institutions. The dominant consensus has several core elements: the priority order in which the policy issues should be addressed, the preferred substantive positions, and preferences about the ideal forums within which governance of global electronic network (GGEN) negotiations should occur.

One expression of the American ICT consensus is the series of publications by the consummate Washington insider institution, the Aspen Institute. Aspen is a think tank devoted to big ideas, with a well-respected special program on ICT that brings together thought leaders from industry, academia, and government. Their publications list is a perfect indicator of Northern policy priorities. One recent publication, *People-NetworksPower: Communications Technology and the New International Politics* (Bollier 2004), is a sophisticated treatment of how the United States can most effectively promote national power in the information age. Invitees to the Aspen meeting that produced the report included not only ICT cognoscenti and "geeks," but former Secretary of State Madeline Albright, Queen Noor of Jordan, President Bill Clinton, and other leaders outside the usual community of ICT experts. This high-level group addressed ICT as a tool of statecraft and an instrument of power to be wielded by government officials in a rapidly globalizing world that hampers the use of more traditional instruments of influence.

Other topics in this publications series include *The Rise of Netpolitik: How the Internet is Changing International Politics and Diplomacy* (Bollier 2002); *Uncharted Territory: New Frontiers of Digital Innovation* (Bollier 2002); and *Ecologies of Innovation: The Role of Information and Communications Technologies* (Bollier 2000). While these works indicate a broad appreciation of the power implications of ICT innovations, the titles also express the particular combination of Washington policy preoccupations and the strategic priorities of corporations that are globally dominant. I submit that these issues are not the typical policy preoccupations when nondominant stakeholders meet to discuss their preferred ICT preferences.

In the early and middle years of the 1990s, the Clinton-Gore administration began to argue forcefully that the information revolution could only reach its full potential if four major shifts occurred, shifts that included elements of governance. To a large extent these four shifts extended the earlier liberal Washington consensus into the ICT domain, holding that government power needed to give way to private; monopoly conditions must move toward competition; domestic markets should be opened to foreign participation; and distributed management of ICT markets was preferable to centralized controls. Furthermore, the consensus held that government policies should be technology-neutral. As with the original Washington consensus, the most radical proposals of telecommunications and other ICT reforms put forward by the administration were actively opposed by many others, especially in the developing world. While the political leadership and the technical experts in Britain and Japan were somewhat supportive, the French and other Europeans were far more reluctant to embrace a new telecoms consensus. Initially then, the Washington consensus was very much only an "inside the Beltway" consensus pushed by the White House and a few other institutions. Eventually, more and more of the G8 and then other developed nations and international organizations came to accede to if not accept the core tenets of the consensus, which came to include these priorities: security, intellectual property rights, privacy and pornography online, and innovation and market openness.

What Are the ROW Priorities?

The agenda of the ROW differs from the agenda of the Washington consensus in several important ways. The governments of the least privileged nations give pride of place to ICT financing, infrastructure development, and capacity building. These are sometimes bundled under the rubric of "digital divide." Matters of intellectual property rights, market access, security, or liberalization are far less likely to be on the list of LDC and transitional country priorities. Other nondominant actors (NGOs or nonprofits) that operate domestically in the global North are typically concerned with democratic participation, human rights, and access. The authors in this book provide a number of examples of the ROW agenda. Boutheina Guermazi and Christopher May, for example, devote much of their chapters to delineating North-South differences in substantive topics like ICT and trade or intellectual property rights (IPR), while Don MacLean and David Souter analyze how the North and South differ in their political strategies toward governance. And while MacLean, Souter, Cees Hamelink, and Peng Hwa Ang treat topics traditionally embraced as important by the global South, Peter Cowhey et al., Henry Farrell, Ian Hosein, and Rob Frieden point to critical nontraditional topics of global governance like third-generation mobile services, privacy, crime, and satellite slots where Northern and Southern interests diverge. For example, Frieden shows that while equity norms suggest equal formal access to international airwaves, efficiency norms (and political economy realpolitik) push in opposite directions, with dominant and nondominant actors expressing different initial preferences. In general, the chapters show that governments and policy intellectuals in transitional and developing nations are more likely to rely on state-led than on market-led solutions to ICT governance problems. This tracks the approach of such states in other substantive areas and is hardly surprising given the weakness of their markets and their statist traditions.

North-South differences come through quite clearly in the North American literature on governance. Joseph Nye Jr. has held senior positions inside and out of government, including assistant secretary of defense, and dean of the prestigious Kennedy School of Government at Harvard University. His writing and thinking have influenced a generation of scholars and practitioners, and in recent years he has devoted himself to issues of globalization, and especially the growing power of the new communication technologies. His edited volume *Governance in a Globalizing World* is a touchstone of contemporary American thinking on the problems this essay addresses, and he includes two chapters devoted to the governance of global electronic networks, written by Kennedy School colleagues. Their focus in these two fine essays is entirely on core Washington consensus issues. Like many others in ICT-saturated North America, the authors are entirely preoccupied with issues of individual choice and freedom, reflecting the cultural orientation of North America and parts of Europe.

Beyond Ranking—The Substantive Differences

Beyond the critical issue of priorities is policy substance—once issues are prioritized from most to least important, how should they be resolved? What precisely should government (and other stakeholders) do about these issues? What are the "best" policies? Even if North and South could agree roughly on the same hierarchy of particular issues, they would not easily agree on how the issue should be resolved. The intellectual property issue is a case in point, as Christopher May underscores. He traces the evolution of intellectual property governance from mainly a domestic matter, then becoming international, and ultimately global, with developing nations eventually incorporated into restrictive global regimes that do not always serve their interests.

Big global companies like Disney or trade associations like the Motion Picture Association of America vehemently insist on the primacy of the private property rights aspects of IPR. Indeed, they seek aggressively to extend the time period and the scope of content ownership. By contrast, developing countries emphasize the community access elements of IPR—they seek shorter time periods of privileged protection, and insist that collective welfare considerations should trump commercial ones, as with pharmaceuticals to combat diseases like AIDS. They believe the North's liberal IPR demands on LDCs today are excessive and unfair constraints on their capacities to develop socially and economically, constraints not imposed on currently developed countries during their own earlier periods of industrialization. Sharp disagreements pop to the surface too in IPR debates over trademark, patents, and privacy, even among rich countries. As Ian Hosein and Henry Farrell demonstrate in their chapters on transborder flows and privacy, there are important agenda and substantive differences even among the OECD nations, especially between Europe and the United States.

It is worth noting that for most of the 1990s and into the 2000s the larger, more sophisticated developing countries did not take radical or aggressive policy positions opposing the ICT powerhouses of the developed world. They fought and lost key battles earlier in the 1970s and 1980s under the rubric "New International Information Order," but between 1990 and 2005 on global ICT issues, Brazil, China, and India did not act collectively and were generally playing "below their weight." Few LDCs took the hyperengaged and supercritical stance that the nonaligned nations took earlier toward the New International Information Order. It was not until 2003 and beyond that Brazil and others seized the issue and tied it to other core concerns like access to developed country markets. Why the big LDCs should have followed this path and not another is a topic that should be pursued by scholars.

In summary, the authors in this volume consistently identified the core components of the Washington consensus, which privileges market solutions over government, and market actors over government ones. Where regulation was called for, it was more likely in the service of market efficiency than social equity, and private self-regulation was far more preferred in the North than in the South. As we see in the following discussion of the WSIS, these tenets are not widely held in the global South.

World Summit on the Information Society (WSIS) as a Site of Contestation over Global ICT Governance

If any single phenomenon epitomizes the clash of competing North and South visions for the future of global ICT governance, it was certainly the WSIS. Held in two parts—Geneva in December 2003 and Tunis in November 2005—most of the issues analyzed in this essay were joined in that two-step process: conflicts between the Washington consensus and the ROW consensus; between Northern and Southern priorities; and among the views of governments, private sector actors, and civil society. (See *Information Technologies and International Development*, which devoted an entire double issue to the WSIS phenomenon, Spring–Summer 2004, http://mitpress.mit.edu/itid.)

Several of this book's authors tackle these tough political and substantive issues that continue to roil international organizations, including MacLean and Souter. For them, the political and the substantive are closely tied since without adequate opportunities for voice and representation the agenda and the substantive outcomes will be unrepresentative. Our authors concentrated heavily on the theme of partnership, but recognized serious structural and institutional impediments to effective cross-sector partnerships. Still, several authors and project participants like Wolfgang Kleinwächter conclude that WSIS was the kind of laboratory marked by experiments with "multistakeholderism" where a new triangular relationship among governments, private sector, and civil society is emerging.

At the end of the Geneva meeting emerged a list of top-priority issues that had remained contentious from the beginning and were never resolved. One could not imagine a more perfect representation of the concerns of South, North, and civil society—each got one favorite issue, and they all agreed Internet governance was critical, though for very different reasons. Internet governance; financing mechanisms to reduce the digital divide; human rights and democracy; and intellectual property rights—these four issues were carried forward to the second WSIS session in Tunis.

The WSIS was conceived in the heyday of the digital divide debates and before the IT and telecom collapse when ICT was the hottest topic. It was to be a global conference of heads of states and governments to discuss how ICTs could be brought into the service of development and organized under the formal authority of the International Telecommunication Union (ITU), a UN agency. Over time, the conference became a curious bifurcated beast with two main venues at one site. In one venue dozens of heads of state gave speech after speech declaiming their policy preferences and their commitment (often rhetorical) to "ICT for development" (ICT4D); in the other venue hundreds of NGOs, nonprofits, and a handful of private corporations held colloquia and exhibitions extolling their visions and showcasing their accomplishments and their wares.

The four troublesome, unresolved issues—Internet governance, financing, democratization and human rights, and IPR—represented in one neat package the major points of disagreement between North and South, between the "Northern consensus" with the Washington consensus at its core, and the ROW consensus. Bundled together were the fundamental Northern concern of protection of intellectual property rights, as well as the Southern preoccupation with funding activities to reduce the digital divide, an initiative led by the West African country of Senegal. A third issue, human rights and free press, was promoted especially by another nondominant group of stakeholders, the increasingly vocal and assertive nonprofits, including groups like the Association of Progressive Communication (APC) and Computer Professionals for Social Responsibility (CPSR). As the Internet moves more and more to the center stage of commerce, finance, trade, and government services, the various stakeholder groups, each with its own reasons, have grown increasingly concerned that the current institutional arrangements are inadequately representing their interests. (ICANN was especially problematic, but there were sharp concerns about the ITU, especially from the private sector and G8 governments.)

The Internet governance proposals stretched from the minimalist conservative positions of the ICANN greybeards and market liberals who strongly supported the status quo, to ITU backers who preferred a radical restructuring under their own aegis.

The discordant preferences expressed at the twin conferences also reflected different preferences about participation and governance within the summit process. Most notable was the self-consciousness and aggressive mobilization among nondominant actors like nonprofit, civil society-based organizations, mainly from the developed North, seeking to expand their circle of influence in discussions of global governance at the Geneva and Tunis summits and by extension to other venues. On the other hand some important LDC governments like China tried to exclude NGOs from meaningful participation in the formal sessions, and to restrict the summit to interactions among government delegations. These same points of contention arose in the 2006 forum meeting in Spain.

The final declaration in Tunis emphasized two of these elements and downplayed the others. After high-level and hard-driving lobbying by the U.S. government, ICANN's authority over Internet governance was confirmed. At the same time, a new global forum on ICT issues was agreed to. The digital divide and other redistributive topics were swept from a central place in Tunis. This author believes the biggest consequence of the WSIS was less the substantive policy outcomes and much more the shift in a critical element of governance—which stakeholders get to come to the table legitimately and help (re)shape the rules of the game. To a considerable extent the vigor of the NGO nondominant actors at WSIS was the result of earlier seeds of activism planted around (and in reaction to) the 2000 and 2001 G8 summits in Japan and Italy, respectively. In 2000, the Japanese hosts selected "digital divide" as a central issue of the meeting, and the Clinton-Gore administration, in its last year in office, backed the idea of bringing in more NGOs into the summit preparatory process, along with a group of prominent private companies. This process was cochaired by an executive from AOL and the president of the nonprofit Markle Foundation, Zoe Baird, who struck an alliance with some White House staff members to incorporate NGOs more prominently into the summit process than they had been in the past (Wilson 2004, chapter 7).

Thus, matters of Washington and ROW consensus and dissensus will not go away, and analyzing shifting positions and underlying interests will remain essential to understanding governance of global electronic networks.

Question Number Two

Are the current ICT governance mechanisms working well or are they broken? There are a plethora of ICT-relevant organizations with overlapping, competing, and complementary authorities. The editors of this volume employed a singular schema to categorize this unruly and complicated set of authorities—*multilateral organizations* typically defined by treaties among many states; *minilateral organizations* that encompass particular geographic or functional subgroupings (Africa, all rich countries, etc.); and *self-governing institutions*, especially in the private sector. In this essay I add governance through *market mechanisms*.

There was a general consensus among the practitioners and scholars of our group, though not equally shared by all, that

- each type of global governance institution holds certain advantages and disadvantages for ROW that are inherent in their principles of organization;
- none of the institutions was performing as well as needed for ROW stakeholders, but some were doing better than others;
- there are some particular, special concerns about governance authorities that are especially relevant for ROW countries, and they should be directly addressed; and
- problems and solutions are relatively easy to identify analytically, but difficult to implement practically.

Beyond these few general statements, one has to turn to the particular, concrete governance arrangements to determine the extent to which they are working well or badly for nondominant actors.

Multilateral

Multilateral organizations include such ICT-specific bodies as the ITU and the World Intellectual Property Organization (WIPO), as well as general-purpose international bodies that affect ICT issues (such as the World Bank). Multilateral bodies have a number of important strengths, some of which have become much more evident now that some major bodies like the ITU are in relative decline compared to newer bodies like ICANN whose power has been growing; these shifts have particular impacts on nondominant stakeholders.

Most of the authors in this book concentrated on the more visible and mainstream organizations of the global ICT domain including the ITU, one of the very oldest organizations, and more recent entrants like the newly relevant World Trade Organization. Other authors analyzed multilateral organizations to the degree that they intersected with their main topic, such as WIPO. The essays revealed several strengths for multilateral bodies: they are usually very inclusive and participatory; with their adherence to transparency and widespread membership for nearly all states, they can promote international buy-in and reinforce the norm of universality of the rule of law and universalistic goals. These bodies generally have high legitimacy among ROW states. They may also reduce some transaction costs for nondominant stakeholders. Because the same norms or rules will tend to be accepted more widely than those achieved through other, more limited channels, ICT agreements struck there may increase the likelihood of wider conformity when they are implemented.

In practical terms for ROW stakeholders, especially LDCs, agreements achieved through this channel probably reduce organizational costs as well, since one meeting may cover a variety of topics; by contrast as we see below, the private or self-governance process has more meetings and attendees are expected to pay their own way and provide their own expertise. For nondominant, less wealthy stakeholders, the multilaterals help them monitor outcomes.

Of course, the multilaterals also have drawbacks from the ROW perspective of non-dominant states and of nonprofit organizations in developed countries. The most significant is that while nominally quite participatory and democratic, in point of fact there are huge power disparities within multilaterals that reflect the real-world power differences between the North and the South; some institutional arrangements not only reflect these differences, but exacerbate them as well. Second, while the multilaterals can write "universal" rules, no solution is perfectly universal. One size never fits all.

Third, state-centric forums like the ITU often exclude private sector actors or non-profits from effective participation. More and more these stakeholders believe that their participation is not commensurate with their stakes and their weight in global markets in determining ICT outcomes. Finally, these bodies are inherently slow-moving and bureaucratic.

Minilateral

Minilateral bodies are those like the European Union or the Economic Commission for Africa with members drawn from particular well-defined subpopulations, and are designed to address the needs of that particular group of stakeholders. They are not meant to be universal. Sometimes there are ICT functions attached to preexisting regional organizations. Sometimes the organizations are entirely ICT-specific.

Minilateral organizations are more likely to share certain fundamental values and interests, reflecting shared material, social, and political conditions. Any given member of the European Union is more likely to share ICT interests with other members of the EU than with any single country in the Economic Commission for Africa or the Economic Community of West African States, or vice versa. In complicated areas of ICT governance, it is easier to reach agreement in smaller bodies with like-minded members; in general agreements can be reached faster. They have heightened possibilities for consensus. Minilaterals also have the advantage of being closer to local conditions and local concerns.

Some global issues can be put up for discussion and negotiated at the regional level and agreements reached prior to taking them forward to truly global forums. This can provide a positive clustering of regional issues, and by speaking with one voice, ROW representatives may carry more weight.

The weaknesses of minilateral bodies are the flip side of their strengths. They tend to be parochial in the pursuit of their interests. They risk creating or reinforcing international fragmentation, which is particularly problematic in the era of globalization and the worldwide spread of new communication and information technologies. "Regional" Internet protocols raise severe and sometimes impossible problems with a global technology. Political regionalism and technological universalism seem to pull in different directions. Regional myopia, or selfishness, may serve to block the more rapid spread of ICTs to precisely those regions that most need them. In the run-up to global negotiations, minilateral deals may actually delay achieving full international agreement.

There are also aspects of realpolitik at the regional level (or functional level), as there are internationally. All regions of the world have at least one regional hegemon with disproportionate power to set and enforce minilateral priorities (think China and India in Asia; Brazil in Latin America). They may use their regional clout to influence outcomes unduly.

In a very useful presentation at the conference that launched the GGEN project, prepared for this project, Tim Kelly of the ITU suggested areas where multilateral action are almost always better than bilateral, regional, or small-group actions. Most notably, he cited trade negotiations, and stewardship of international resources like satellite slots, frequencies, and numbering systems. Conversely, he noted, there are areas where small-group actions work better, for example network security arrangements, "first round" standards-making, and cross-border licensing agreements.

Private Sector Self-Governance

This last of our three initial categories has become highly problematic for the ROW. Whereas multilateral and minilateral bodies have had substantial authority in the ICT domain for decades, and are familiar to the ROW, self-governing institutions are much more recent and unfamiliar, especially for LDCs and NGOs. According to the authors in this volume, they pose some advantages but also real problems.

Milton Mueller and Jisuk Woo (chapter 14) describe the body that has become the largest international experiment—and the largest lightning rod—in this category: the International Corporation for Assigned Names and Numbers (ICANN). In an earlier work, Mueller described in great detail the origins and evolution of this remarkably hybrid form of nonstate governance, and underscored the shifting political dynamics among governments, firms, NGOs, and technocrats of all types that prevented ICANN from having a purely private or purely public form of governance (Mueller 2002).

The advantages of self-governing institutions are similar to those of the other non-universal type—minilaterals. They are close to the problem, and encompass many of the actors most affected by technological change. Their smaller size and greater flexibility help reduce transaction costs, and promote micro-efficiency, enabling them often to reach agreement and act on their agreements quickly.

These arrangements also suffer shortcomings. Bodies like ICANN or the International Chamber of Commerce (ICC) suffer from their exclusivity and absence of transparency to outsiders, and hence lack legitimacy for some stakeholders. Nor is there much representativeness (and hence legitimacy) afforded to these bodies since participation by nondominant actors is limited. Indeed, because they may lack legitimacy even among their own members, it can be difficult for such bodies, especially when they are advisory, to enforce agreements.

For outsiders, there is also the risk that these arrangements smack too much of the "fox guarding the chicken coop." Indeed, some of these self-governance efforts are established precisely to thwart other initiatives launched by governments, NGOs, or other interests that private stakeholders seek actively to avoid. They calculate which forums are most likely to provide favorable and unfavorable rulings for which proposals, and act accordingly. The self-interest of the private actors will not always be coincident with the public interest at large.

It is important to distinguish between the self-government of firms and corporations in the private sector, and the self-governance of the nonprofit, nongovernmental sectors. ICT-oriented bodies like CPSR have become much more active in recent years and are becoming more assertive in a variety of forums, with a sharp spike in visibility around the WSIS. However, they lack the comparable long experience of cross-organizational cooperation that marks private sector cooperation through bodies like the ICC. These two types of stakeholders bring very different interests, resources, and experience to the process of ICT governance.

The Fourth Sphere of Governance: Market Powers

For the most part the editors of this volume asked the authors to concentrate on the opportunities and constraints that flow from the actions of public institutions like national governments and international bodies like the WTO or ITU. Therefore, the chapters devote less attention to the constraints and enabling powers of market institutions.

Yet among the most powerful trends in the new global ICT environment has been the massive shift from public authority and power toward private. In both new media and old we see growing legitimate exercise of power by private firms and national and international trade associations like the Global Business Dialogue on Electronic Commerce [GBD(e)], as well as growing authority within the traditional intergovernment international organizations. Most prominently is the straightforward if politically charged transformations of state assets and state control into private assets and management through privatization (around $15 billion in Brazil's telecom sector alone).

These shifts significantly alter the logic of global ICT governance. Today governments have ceded their first place to firms in service provision and the introduction of new products. Today the private sector exceeds the public in capital investment and in number of clients; the public sector now plays second fiddle to the sheer market power of the private. Decisive power over property rights, culture, innovation, job creation, and other valuable resources has been enhanced not only because of the shifts of responsibility within the telecommunications and broadcast sectors, but also because they have occurred across the board in other sectors of the economy as well, as correctly pointed out by scholars like Virginia Haufler in her book *A Public Role for the Private Sector: Industry Self-Regulation in a Global Economy* (2001). Thus, the private global investment decisions of an Intel in Costa Rica or Texas Instruments in Bangalore or IBM in Brazil have literally changed the ICT trajectories for those nations and others.

As a result of this global rebalancing of public-private relations, the GGEN today is much more the product of private as well as public stakeholders jockeying to advance their interests, with private stakeholders gaining more and more influence. Of course, which particular actor or coalition dominates any particular negotiation is largely a matter of the immediate political, institutional, and policy context in which the negotiation is embedded (Wilson and Wong 2006). These contexts are quite fluid and unsettled because of the rapid changes in technological and commercial convergence, concentration, and privatization around the globe (Ó Siochrú and Gerard 2002). We do not deny that in the last instance it is states that possess the authority to set the rules and structure new regimes, as Krasner reminds us in his essay (1991). But the growing role of private power should not be minimized.

Lawrence Lessig has written at great length and convincingly on the implications of the trend not only toward privatization, which he supports, but also toward tremendous corporate concentration as well, both domestically and internationally, a theme that the authors repeatedly address in this volume (Lessig 2000). He argues persuasively that the gargantuan multimedia companies are selfishly shrinking the space for public civic engagement and social coherence, and that steps need to be taken to protect the commonweal in the age of digitalization. Lessig's arguments focus mainly on private-nonprofit interactions in developed economies, but they are equally applicable to nondominant nations, and to NGOs and communities within poor countries.

In the words of one author, the international governance system is "screwed up and screwy." None of the authors in this book believed it is working particularly well as a whole, and all identified areas of weakness and suggested where it could be strengthened substantially. However, none of the authors claimed that the system was completely ineffectual or that the status of the ROW stakeholders was hopeless. Instead, the view tended to prevail that ICT global governance is complex and difficult for all governments and stakeholders and especially difficult for developing countries; and that it is essential to make the institutions more inclusive at every level.

We have seen that the big, important treaty-based arrangements like the ITU have the advantages of inclusiveness, continuity, legitimacy, and wide coverage of issues. On the other hand these bodies are big and clumsy with slow decision-making procedures. They tend to be stodgy and pay too much attention to the past and not enough to the future. They also have suffered from a lack of expertise in critical areas of innovation, and lack the trust of some of the most powerful players like the U.S. government and much of the global private sector. Other types of governance arrangements are more nimble and forward looking, but often lack the legitimacy of the multilaterals. The challenge is to make these four distinct governance approaches complementary to one another, and simultaneously to advance the interests of the entire global community, the nondominant actors as well as the dominant.

Question Number Three

What is the impact of the current global governance of electronic networks (GGEN) arrangements on nondominant actors? This was the most problematic question to answer of the entire exercise. All the authors found it quite difficult to point consistently and convincingly to the most likely national impacts of the international governance arrangements they analyzed. Two chapters focused explicitly on external-internal linkages in a single country—South Africa and South Korea.

The problem is partly rooted in the absence of the necessary empirical work carefully reviewing on-the-ground outcomes in Africa, Asia, or Europe; partly in the absence of appropriate conceptual frameworks linking diffusion and impacts (Attewell 1998). Another major problem in the literature—both academic and policy—is that analysts consistently conflate changes in ICT and their impacts *within* the confines of the sector with impacts *beyond* the sector, upon society more broadly. It is useful therefore to differentiate between the sectoral and the societal impacts.

Let me rephrase this question about impacts more explicitly: *What is the impact of global governance arrangements on the institutions and processes that most directly affect the activities and conditions most people care about in their daily lives—their jobs, their health, their security, their education, their general well-being?*

Impact Findings

Several critical analytic, methodological, and theoretic issues emerged out of the authors' treatments that bear on the extent of international institutions and governance arrangements' impacts on ROW stakeholders.

1. Problem of multi-causality Even if we can identify particular changes in the domestic production and allocation of resources that seem to derive from causes beyond their borders, it is difficult to show which of many external conditions "caused" a particular

domestic outcome. For example, if a poor country's government liberalized its telecoms sector soon after adopting the 1996 WTO global telecoms accords we cannot be confident that it was only, or even mainly, the formal signing that provoked the domestic reforms. A partial list of other potentially relevant factors must include demonstration effect of reforms in other countries; pressure from multilateral lending bodies like the World Bank requiring structural adjustment programs, including privatization; direct bilateral pressures from powerful Northern governments with domestically influential telecoms corporations; commercial and political pressures from individual local firms on government decision makers; indirect influence to reform exercised through business associations like the International Chamber of Commerce; pressure from local entrepreneurs seeking greater opportunities in the local markets; and recognition by domestic actors that liberalization, competition, and privatization have improved services to their consumers as seen in reduced waiting periods for telephones, lower prices, better service, and wider choice.

2. Lag times The domestic impacts of changes in international governance do not always flow swiftly to the local level, even when implemented quickly. Investments local or foreign do not spring forth overnight when policies suddenly change, nor do sales and purchases.

3. Slow implementation These lags reflect tardiness in policy and project implementation; they may also result from deliberate attempts by a government to drag its feet, resisting putting in place policies it does not like but was forced to accept. All authors point out that domestic, local interests have been able to resist the most severe pressures of the new regime to sign agreements the North prefers, and to delay official timetables the North prefers. The principal conclusion of chapter 13 in this volume by Tracy Cohen and Alison Gillwald is that while the political leaders in South Africa's transition from apartheid were already moving toward a more economically liberal system, prior to the WTO agreements, they were unwilling to go much beyond the parameters of what their own domestic experts and constituents seemed to demand of them. Authors do believe that stalling or reneging will become more difficult through time, but for now ROW officials can and do delay implementation.

4. All agreements are different No two countries have perfectly identical arrangements. Guermazi (chapter 5), Cohen and Gillwald (chapter 13), and others point out that it is virtually impossible to review the general language of a global deal and have a clue about its specific national impacts, since each agreement is individually negotiated. There is no simple cookie-cutter pattern; in effect each country signs a different accord, so the impacts will differ as well.

5. Limited ICT diffusion Impacts of IC governance rules beyond the sector will be sharply limited in poor countries because ICTs themselves are not widely distributed in most developing countries—less than 1 percent of Africans have access to the Internet. Less than 5 percent of Latin Americans do.

6. ICTs' relative unimportance Other global rules and treaty requirements weigh much more heavily on ROW decision makers than the Internet or telecoms. National leaders and their constituencies are much more deeply concerned with trade in agricultural commodities and Northern government subsidies to their farmers; or debt renegotiation; or structural adjustment requirements by the World Bank or IMF. Relative to these other very burdensome challenges, ICT outcomes may seem rather unimportant and will rarely attract the same high-level leadership and institutional attention and support.
7. Data shortcomings Finding accurate, meaningful, reliable indicators of domestic-international links is not easy.

If it would be extremely naïve to claim a one-to-one relationship between external cause and internal effect, it would be equally naïve to argue the opposite—that these arrangements have had no impact, whether through a single channel or in the aggregate.

The impact of the external on the internal is an issue where scholars need to "import" frames available in the mainstream social science literature to explicate the ICT case. The long tradition of critical analysis by dependency theorists, Marxists, and even neorealists would assume that the external linkages are inherently asymmetrical and likely to be exploitative, at least in the short term, and that governance arrangements are imposed by powerful external actors on unknowing or powerless ones. Neoliberals might assume the opposite—new regimes and governance arrangements will enhance international exchanges which will by definition enhance the well-being (and probably global status) of nondominant actors. Alternatively, other scholars might be less confident of paradigmatic predictions and assume greater indeterminacy requiring very careful analysis of each and every case of governance-mediated ICT impacts and the structure and dynamics of negotiations that surround them.

Most of the authors included in this book take a fourth perspective—assuming the power advantages of the dominant states, but insisting that *nondominant stakeholders can improve their future standing through smarter international negotiations based on better domestic governance arrangements*.

Question Number Four

What can scholars and researchers do to help professionals in the field of ICTs? A major purpose of this project was to explore the multiple intersections between scholarship and practice in a rapidly evolving domain of growing interest and importance. To achieve this aspect of our project the organizers consulted with and included in our early discussions in Washington and Budapest an equal number of practitioners from the worlds of government, NGOs, and the private sector as well as scholars from universities and think tanks. All had demonstrated experience working across their institutional and sectoral borders, which contributed immensely to the project's success. We hoped this rich mixture of practitioners and scholars would yield benefits not otherwise achievable. The engaged scholars came from universities and research centers in India, Korea, South Africa, the United Kingdom, the United States, and the thoughtful practitioners hailed from private companies like AOL, Fujitsu, and Hong Kong–based PCCW; NGOs like the Brazilian populist ICT group called RITS, and the Association for Progressive Communication (APC) based in South Africa. The group also included government representatives (Russia). In terms of their countries of origin, the project drew people from Canada, China, Hungary, India, Japan, Korea, Russia, Singapore, South Africa, Tunisia, the United Kingdom, and the United States. These diverse perspectives sparked unique observations and insights we could not obtain otherwise.

Exactly What Are Scholars Asked to Contribute by Practitioners?

As I argued at the start of this essay, periods of rapid, widespread, and deeply felt changes complicate the role of the scholar. Scholars can be valuable, however, because they have the potential to help others sort through the chaff to uncover what is truly of great and lasting importance as distinct from the ephemeral and superficial; they can also help identify those issues that are ripe for immediate attention and resolution, while recommending others that can be addressed in the fullness of time. Yet these potential contributions are highly problematic precisely because the usual signposts have been shaken up, knocked down, and sometimes point in the wrong direction. Data are scarce and unreliable, concepts unclear, and links between causes and consequences seem newly problematic. For scholars, such conditions call for extreme caution, with occasional dramatic leaps of faith.

If working conditions inside the scholarly community are confusing, relations between scholars and practitioners are also highly problematic. The two groups sometimes seem to occupy parallel universes that intersect only rarely, each with different purposes, priorities, reward structures, and with a different sense of urgency and time. Yet the field of ICT is an area where those who must act quickly frequently seek useful knowledge, guidance, and information from others, and where researchers seek new and interesting topics of investigation.

Several distinctive themes emerged from the discussions among practitioners and scholars. One had to do with the channels, mechanisms, and processes through which interactions occur, and how to improve them. The other had more to do with categories of knowledge and their relevance for action—theories, concepts, definitions, and so forth.

While in this essay I pay more attention to the latter than the former, let me address briefly the question of channels. Most scholars contribute to the advancement of science and action through a variety of means that create and diffuse new knowledge.

These channels include their classroom teaching, scholarly writing in academic journals, participation in professional meetings and associations, training, and providing advice to groups or individuals beyond the campus and through their public engagements like speaking. A consistent theme from the practitioners is their need for greater access to new scholarly knowledge through a much wider array of channels, in different formats and with language accessible to the informed layperson. These could include everything from web sites to regular face-to-face briefings to short concise research summaries available by e-mail.

Frames, Concepts, Cause and Effect, Details, Dynamics, Downstream Linkages, and Freedom of Choice

Practitioners demand help on substantive matters as well as processes and products. Let me try to translate the requests of the practitioners into terms of art more familiar to scholars. These I would characterize as frames, concepts, cause and effect, details, dynamics, downstream linkages, and freedom of choice.

Frames

Again and again practitioners in Budapest and Washington explicitly asked scholars for help in "putting the issues in context." Contextualization meant, in part, situating the particular issue at hand into its most relevant societal setting. Of what broader whole is ICT a component part? To what does it relate? How best should questions about the global governance of ICTs be broadly framed? This means pointing to clear as well as subtle linkages between the particular ICT issue at hand (broadband, digital divide) and broader issues of society, culture, or economy. Take the question of how to frame Internet governance. At the most fundamental level, is Internet governance best understood as a matter of neutral, nonpartisan experts cooperatively setting global technical standards in everyone's interest? Or are stakeholders really engaged in pursuing narrow agendas, best understood as a matter of high politics and power struggles among competing interests? If the Internet is political, is it a global struggle among states, or among companies, or between both with the involvement of civil society organizations and interstate institutions? Realist political science theorists such as Krasner frame Internet governance quite differently than Mueller and Woo, for example.

One recent example of creating a new frame to discuss ICTs is the way in which some scholars have taken what are typically seen as separate and distinct technical issues and reframed them into a broad new category of public policy. David Bollier and others have seized upon rather dry communications and information issues like spectrum allocation, patents, and copyright and bundled them together under the overarching frame of "information commons" (Bollier 2004). Under this rubric heretofore technical issues are redefined as single manifestations of a larger issue—scarce, valuable, publicly held resources—and revealed to share a collective importance to citizens, analogous to the way common pasture land was important to herders and other citizens of an earlier age. (See Kranich 2004 for a statement of this process.) The title of a book by Lawrence Lessig reveals the thrust of the reframing: *The Future of Ideas—The Fate of the Commons in a Connected World.*

Concepts

During this tumultuous period of deep and far-reaching changes in the ICT sector, new terms are introduced in cascading numbers, and old terms quickly lose their meaning and are reinvented, as Farrell demonstrates with privacy or May with IPR. This occurred with concepts like universal service and digital divide. Under simultaneous pressures from rapid technological change and growing standards of living in many countries, the meaning of universal service began to change, and the concept of universal access grew in popularity. Universal access came to mean that Internet connectivity was available to citizens, although not necessarily in their homes, as with telephones, but within a "reasonable" proximity to their homes. While admitting the huge challenge of defining reasonable proximity, there arose the challenge of distinguishing between "formal" access and "effective" access. Should access be conceptualized mainly as access to basic communications infrastructures? Or did it also include access to the training, cognitive skills, financing, or relevant content that would transform formal into effective access? The concept of digital divide was also defined differently by different actors. For some, it was interpreted as a growing gap between information haves and have-nots, a definition promulgated by some international bodies like the UN. For others, especially international business groups, the most appropriate conceptualization was "digital opportunity." Digital have-nots were defined positively as a potential business opportunity. The ways in which these and other institutions acted on the digital divide substantially reflected the priorities and perspectives of the different stakeholders, as codified in competing conceptualizations. Reaching a commonly accepted conception proved impossible.

Cause and Effect

Practitioners also want to know about what works and what doesn't, and under what particular circumstances. In other words, they want to know about cause and effect in the ICT sector. Statements that point out relations among cause and effect are theories. A wise man once pointed out that nothing is as practical as a good theory: if this condition occurs, then this thing will happen. Will ICT "cause" development? (No.) Does development cause ICT diffusion? (Sort of.) Is the Internet reducing hierarchy inside formal organizations? (Yes, under some circumstances.) The biggest challenge in our project and in general is that the practitioners strongly prefer what scholars are hard-pressed to give them—a cause and effect rule, a "best practice" that is universally true and universally applicable with the same outcome all the time. It is the scholar's job, however, to resist such over-easy generalizations and to point out these theories only work when the conditions are specified. Then the question becomes *Under what circumstance* is this or that a best practice? Under what conditions of supply and demand is this a best practice? Under what institutional conditions? A "lesson learned" or best practice in the presence of an effective telecoms regulator may not be a best practice in the absence of one. As several multilaterals have discovered recently (infoDEV, housed at the World Bank), as well as bilateral agencies, capturing best practices is both difficult and expensive. It cannot be done post hoc but must be built into the front end of projects.

Details

Scholars are also frequently asked to contribute empirical details about global governance. What are the exact responsibilities of INTELSAT? What position did the EU take on the governance of privacy relative to Japan? Freestanding case studies of aspects of global governance of electronic networks should not be underestimated during periods of great change. Accumulation of concrete facts about the world, facts which then can be agreed upon by the relevant actors, is important. The chapter by Robert Frieden in the volume on the sometimes arcane details of international satellite services is a case in point of a scholar addressing the minute details, but also linking them to broader issues.

Dynamics

There is a great temptation in studies of ICT to capture details analytically by holding everything else constant, providing a kind of static snapshot. A tremendous contribution to the field by scholars has been and will remain identifying dynamic regularities in GGEN, and in the construction of plausible stories out of them. Constructing good ICT policy is hard absent a sure sense of the dynamic trends of the relevant technology, demand and supply, and political timing.

Several years ago a group of ICT practitioners and analysts in a dozen African countries met and insisted that what they really needed in order to do their jobs well was to have descriptions of the political and institutional dynamics that surround technology diffusion in sometimes hostile territories. Concretely, they needed stories, specifically "war stories." They wanted nuanced narratives about how stakeholders maneuvered and negotiated, won and lost, in other settings. They knew that ICT successes and failures hinged as much on *timing and sequencing* as technology. Successes require early backing by political figures, the timely mobilization of resources and staffing, and on-time implementation. In response, this author and a team of colleagues developed a dynamic negotiation framework focused on a dozen "critical negotiating issues" that appear sequentially in the dynamic process of ICT diffusion (Wilson and Wong 2006).

We came to understand that these practitioners were asking for two distinct things. One was a simple list of best practices and lessons learned. But they were also making a very human request—give us a story with a beginning, a middle, and an end that we can recognize; a story that corresponds roughly to our own realities; a generic story line onto which we can then hang our own local experiences. Good analytic stories also provide milestones of what to expect next. Having a story line of how things unfolded in other settings provides one with expectations of how things might unfold at home. If Internet diffusion has four stages in most countries, then maybe it will still have four in the next country. When these things happen, and one arrives at the threshold of phase 2, then one knows what to look for and perhaps even some things to do. This topic is very important, but not one we devoted a lot of space to in this volume.

Downstream Linkages

How do these technologies link to downstream applications like health or tax collection where most practitioners work and most citizens and customers seek services? Scholars can indeed provide great insights into the specific linkages between this particular ICT domain of practice and other substantive areas, like health, education, or ports administration. This is especially important for more senior ICT policy makers, since the higher up the chain of command, the more important it is for executives to anticipate and recognize cross-issue linkages, as with ICT and trade or tax collection. This is where their responsibilities intersect with those in other sectors, markets, and institutions.

Freedom of Choice

Finally, at the end of the day, all practitioners whether policy makers, bureaucrats, entrepreneurs, or grassroots NGO activists want to know how much freedom of action they really have to pursue their interests. In scholarly terms this is a "structure-agency" problem. That is, what percentage of our possible action is already determined unalterably by the situational givens like income per capita or educational levels? Is our freedom of action illusory in the face of poverty, ignorance, and globalization? What can actually be done *inside* these constraints, taking them as given in the short to medium term?

For example, our NGO colleagues stated forcefully that by far the biggest contribution scholars could make to their practical work would be to help them distinguish the political from the technical. That is, they are often told by "experts" that in solving some immediate practical problem certain institutional and authority arrangements are absolutely required by the imperatives of the technology: If you want to use *this* application, then you *must* buy this equipment, pay that much, and restructure your organization along the prescribed lines. NGO practitioners wanted to know how they could do a better job of recognizing what the technology actually "requires" to distinguish between the required and the optional. They want to know where politics and power enter the equation.

The matter of personal autonomy and the practitioner's scope for action leads back to our first issue of framing. This book project began in earnest when we noticed that at one international meeting after another the options for action of nondominant actors seemed to be grossly underspecified. LDC options were presented that made nondominant stakeholders either thoroughly choiceless or, equally unrealistic to us, perfect masters of their fate. The "options" were stated so narrowly as to rob most actors (especially nondominant actors) of scope for movement that could be judged achievable. Either the recommendations said that ICT diffusion was entirely the consequence of technological and economic imperatives, so that LDC managers need not worry much, or insisted ICTs brought the capacity for LDCs to leap-frog into the future, with choices nearly infinite for leaders who would simply seize the time. Both positions were wrongheaded, and this volume is in part an effort to reframe the issues to introduce more realistic options for nondominant stakeholders, and hence enhance their freedom of action.

Reframing the Question: Introducing the Quad

Let me offer one pathway into these intersections of scholarship and practice through a conceptual framework I term the Quad, which seeks to capture the heart of the social, political, and institutional dynamics of the governance of global electronic networks. The considerable literature cited in these chapters, and the chapters themselves, indicate widespread recognition that the road to enhancing the capacities of ROW for effective participation in global governance must start at home enhancing domestic capacities. Experts on the South African ICT scene like Tracy Cohen and Alison Gillwald (chapter 13) and others like Derrick Cogburn do a skillful job of tracing the processes of domestic contestation and concord over IT policies, pointing to interests excluded as well as included. They show that as the South African antiapartheid freedom struggle came to an end, the African National Congress (ANC) consolidated power as the dominant political party.

ICT policy decision making became increasingly insular in South Africa, excluding experts from civil society and small enterprises (e.g., in the ISP business). To remedy it Cohen and Gillwald call for LDCs to "increase the size of the domestic inner circle." Mueller and Woo criticize the problematic relations among public, private, and nonprofit stakeholders in Korea. Hosein too insists on wider domestic participation in policy making. However, simply calling for greater domestic capacity to achieve better global governance is inadequate theoretically and practically. A "call to arms" is not an action agenda. A more theoretically robust and strategic construct would need to

1. identify the most important stakeholders whose capacities must be enhanced;
2. specify the dominant patterns of relations among them that thwart more effective capacity to participate in global governance;
3. build a theoretical construct to capture their structural and dynamic aspects;
4. recommend specific interventions to enhance capacities.

Based on my own research in Brazil, China, Ghana, Hungary, India, Malaysia, South Africa, and the United States, I developed a model of cross-sector capacities that identifies the four key actors and the dynamic, sustained interactions among them, linking the robustness of the relationship among the four nodes with the technical and economic performance of the community or city in which they are embedded (Wilson 2003, 2004). Thus, the more robust the relations among leaders in the public, private, research, and nonprofit sectors, the higher the performance in knowledge-intensive activities. The Quad model finds that at the domestic level small groups of individual innovators in one sector or stakeholder community are sometimes motivated to reach out to like-minded "information champions" in other sectors to overcome certain institutional limitations, seeking two key objectives: to obtain more and better knowledge about the new technologies and their applications, and to gain institutional and political support in a hostile political and regulatory environment. Rules, laws, and eventually norms must be retooled to be more supportive of the new distributed technologies. In many countries, from a core "conspiracy" of about a dozen information activists across the four sectors a genuine, broad-based coalition of domestic reformers eventually emerges, promoting rapid, grassroots ICT diffusion that empowers disempowered stakeholders. Eventually, these conspirators begin to build support networks internationally as they do domestically. The local private sector entrepreneurs find international counterparts through bodies like the International Chamber of Commerce. NGO activities act through sympathetic bodies like the Association for Progressive Communication (APC). Researchers and university-based innovators gain political authority and material support through bilateral bodies like the International Development Research Centre (IDRC), while government officials create global communities of practice through the multilateral and minilateral channels I described earlier. Thus, the national Quad nodes have their direct counterparts at the global level that, when effectively mobilized, can provide material and moral support to help local champions articulate and amplify their positions in global forums as well as local ones. Domestic stakeholders understand that their efforts to influence some international rules of the game may redound back to enhance their status domestically, and vice versa.

However, this arrangement seems to work well for rich countries but badly for poor ones. This book's authors reveal that in poor countries the domestic governance structure for science and technology, as for economic governance more broadly, is exclusionary and balkanized. A small handful of government agencies dominate the local scene through conservative and exclusive "iron triangles" that generally oppose innovation. As such, when it comes time to design new rules and regulations to govern the Internet, for example, in South Africa, Korea, and other developing countries government officials are reluctant to consult and coordinate with local NGO, university, or private sector experts. Similarly, when governments assemble their teams to attend global governance meetings they rarely include domestic stakeholders from other sectors, even when those stakeholders possess superior knowledge and experience, with their own professional networks that are invaluable to participate effectively in global rule making. Instead of arriving in Geneva or New York with a fully loaded negotiating team, LDCs tend to exclude businessmen, NGOs, and researchers, thereby slashing their analytic and action capacities by three-fourths. Instead of capacity building, they commit capacity busting. Meanwhile, the G8 nations often arrive with all four nodes of the Quad well represented in their delegation. This is partly a question of money, but mostly a question of political will.

Ian Hosein does note that domestic actors in different sectors are being pushed into more interaction with one another than in the past, as each seeks to adjust to the new global challenges. He points out that transjurisdictional data flows now raise much broader questions that fall beyond the traditional interest and competence of the usual actors in rather narrow techno-legal debates. Civil liberty groups and law enforcement agencies are two examples. Though this dynamic is mainly in OECD nations, he argues that as a consequence, mass publics are becoming involved in this arcane field, which requires that the relevant international organizations accept more open discourse involving nonstate actors.

Cross-sectoral cooperation is a sine qua non for success in the modern world. The global power of the United States rests increasingly on the huge successes of public-private partnerships. The same might be said of the remarkable ICT successes of Sao Paolo or Bangalore and even Beijing.

A final question concerns the nature of the relationships between the governance of global ICT and the governance of other sectors and the international system as a whole. Shaped by technology, markets, and the preferences of powerful international actors, the ICT sector is marked by its own specialized norms, expectations, rules, regulations, and institutions. These are not perfectly identical with those found in other sectors—think for example of the rules of the game of the international petroleum sector. Cross-sectoral differences in regimes remain insufficiently explored in the literature. What are the continuities and discontinuities between international ICT norms, practices, and institutions, and those of other sectors like finance or biotechnology? By definition, the governance of global electronic networks is a subset of all governance mechanisms that structure behaviors more broadly across sectors and space. The degree of continuity between the governance arrangements in the ICT sectors and in other sectors of the international system is both a theoretical and empirical issue that deserves more attention.

Concretely, scholars could review the particular networks and ties that link ICT to other international activities. What precisely are the channels and pathways between the ICT sector and others, and how will these linkages affect the evolution of the global information revolution? Again, links between finance and biotechnology come to mind, but other global activities are also directly relevant.

Conclusions

Let me restate the original questions with a short summary of the authors' findings.

1. Is there a Washington consensus separate from the preferences of nondominant actors? Yes. All the usual caveats apply. Observers must recognize that nations start with very different policy priorities and substantive positions that reflect in part their unique position in the global system, and especially their own domestic priorities. The Washington consensus emphasized market solutions and efficiency, while the ROW interests stressed equity, access, and financing.
2. Are the current ICT governance mechanisms working well or are they broken? Some work better than others. Our authors generally agreed that institutions like ICANN and the ITU badly need to be fixed, and that the current system for consultation, policy design, and implementation doesn't work well for nondominant actors, LDCs, and NGOs. At the same time, there was equally strong agreement that there is enough space in the architecture of the international ICT governance system to allow the poorer, less powerful players greater freedom of action, and greater scope for getting things they want. The WSIS process created a multistakeholder forum that ventilates LDC opinions but possesses little real clout.
3. What is the impact of the current GGEN arrangements on nondominant actors? We can't say precisely because there are so many confounding methodological uncertainties of data availability, multiple causality, interactive effects, time lags, and so forth. We can say however, that the North's preferred balances among public-private ownership, monopoly, and competition; market access by domestic and foreign interests; and centralized and distributed government intervention have been conveyed unambiguously and forcefully through a variety of channels and affected local leaders' incentives to restructure their entire national ICT systems.
4. What can scholars and researchers do to help practitioners in the field of ICTs? Scholars can engage in a variety of complementary activities that range from reframing the most basic issues of importance to stakeholders, to providing accurate details and empirical information, to articulating and trying to confirm or disconfirm statements of cause and effect, in order to provide practitioners with practical guidelines about the way the world works.

Common Policy Recommendations

An essential element of this Social Science Research Council project was to identify "useful knowledge" (Calhoun 2004). This meant asking the participants to make concrete recommendations for action in their areas of expertise, and specific recommendations are found in each chapter. Happily, beyond the domain- or topic-specific recommendations emerged some more general, policy-relevant insights that bear on the design of more transparent, democratic, and efficient governance arrangements. Some common threads run through the recommendations.

1. Today's rules governing access to essential ICT resources reflect current balances of power within the world system as a whole—they reflect the policy preferences of the powerful. There are real constraints imposed by international institutions and their rules on nondominant actors. However, LDCs do possess substantial leeway to gain additional benefits under existing rules, but most have failed to take adequate advantage of them. "Excess capacity" exists to exploit more fully the current flexibilities of, for example, TRIPS (May, chapter 9), yet without fighting to change the basic rules of the game.

2. At the same time, additional rule making and amendments to current arrangements should be aggressively pursued that recognize (and compensate for) the distinctive needs and circumstances of transitional and LDC economies—low levels of effective demand, lack of adequate infrastructure, the need to achieve greater market access to developed country markets, high levels of poverty, and inexperience in participating in global ICT forums. The nondominant suffer from unequal abilities to benefit equally from markets. International organizations do need to change their structures and ways of doing business. Organizations like ICANN, for example, could "add a multicultural awareness and outreach advisory committee" to their extant structures (chapter 14).

3. The current "universal" arrangements (whatever they may be) should be fairly applied so *all* countries, and not just the rich, may take advantage of them. That is, the playing field should be leveled, sometimes by arranging the rules so that weaker players can benefit. Global distributional issues receive insufficient attention.

4. Even pro forma, formal participation levels by ROW and stakeholders in international organizations are still far too low in terms of representativeness (and hence long-term legitimacy), and the current channels for gauging LDC preferences are flawed. Improvements in technical, representational, and consultative capacities also should be applied at the regional as well as the global level.

5. Specific institutional and policy solutions that work in developed country markets are unlikely to work in LDC markets. As Peter Cowhey, Jonathan Aronson, and John Richards argue in chapter 4, developed countries and their firms need to be more humble and encourage the LDCs to experiment with clever, innovative microsolutions to innovation.

6. Most important, there was a strong, unequivocal conclusion among the authors that a major, if not *the* major roadblock to further progress without which the other problems cannot be adequately resolved is *the weak capacity of ROW stakeholders to organize themselves domestically to draw on their local potential strengths*. Citing the *Louder Voices* report, MacLean (chapter 2) insists that the key to enhanced developing country participation is to build technical and policy capacity at the national and regional levels. Absent this, changes to global governance structures to create special spaces for developing countries may amount to little.

David Souter insists that LDCs need a program to advance understanding of the ICT governance issues not just among the handful of ICT cognoscenti, but also among ROW media and research institutes "with the aim of improving the quality of public debate about ICT policy issues and establishing self-supporting networks of expertise within individual countries" (451).

There has been provocative speculation and some analysis about the impacts the Internet and other ICT innovations might have on reducing *domestic* democratic deficits. There has been far less interest in possible ICT impacts on the global democratic deficit. This could be a fruitful line of investigation.

Whether or not ICTs directly alter authoritative global governance institutions through the likes of ICANN or other bodies, the larger and ultimately more important question is whether these new technologies will differentially alter the underlying power capabilities in the global system. Will Singapore find its net influence increase relative to other nation states like Thailand or France because it has so much more effectively incorporated ICT into its production systems? To what extent will China or India's economic rise and their ability to sustain and project global influence be shaped by ICT diffusion? How will they channel that influence to shape global governance rules? Beyond states, more and more scholars believe that the power of NGOs to affect global outcomes is substantially enhanced by new ICT tools that have the effect of expanding their power relative to states. During and after WSIS we have seen NGOs beginning to flex their muscles and take aim at domestic and global ICT governance arrangements.

The capacities of ROW state and nonstate actors to restructure global governance relations will hinge substantially on whether or not they are able to restructure their own domestic cross-sector relations along lines that give them more effective leverage in the WTO, ICANN, OECD, and other global institutions. Thus, students of the future evolution of information and communications technologies and of their uses and impacts must become more sophisticated about understanding the links between domestic power and global governance, including the ways they are modified by the new technologies.

The world system today requires more and better forms of global governance to become more efficient, and people in the poor and underdeveloped regions deserve better forms of global governance as a matter of justice. Better governance in the emerging knowledge society of the future requires dismantling the many institutional barriers between the producers and users of knowledge, and facilitating the free flow of knowledge across national and other borders. The authors in this project, including the author of this chapter, have tried to provide some helpful signposts toward a more open, participatory, and equitable global community.

Notes

The author wishes to thank Tracy Cohen, William Drake, Alison Gillwald, Robert Latham, Don MacLean, and Rafal Rohozinski for their helpful comments on this chapter. The author alone is responsible for any errors or omissions.

{Portions of this chapter appeared in *Information Technologies and International Development*, vol. 2, no. 4, 2005.}

References

Aronson, Jonathan. 2002. "Global Networks and Their Impact." In *Information Technologies and Global Politics: The Changing Scope of Power and Governance*, ed. James Rosenau and J. P. Singh, 39–62. Albany: State University of New York Press.

Attewell, Paul. 1998. "Research on Information Technology Impacts." In *Fostering Research on the Economic and Social Impacts of Information Technology*, 133–137. Washington, DC: National Research Council.

Baird, Zoe. November/December 2002. "Governing the Internet: Engaging Government, Business, and Nonprofits." *Foreign Affairs* 81, no. 6 (November/December): 15–20.

Bessete, Randi, and Virginia Haufler. 2001. "Against All Odds: Why There Is No International Information Regime." *International Studies Perspectives* 2, no. 4: 69–92.

Bollier, David. 2004. *PeopleNetworksPower: Communications Technology and the New International Politics*. Washington, DC: Aspen Institute.

Calhoun, Craig. 2004. "Word from the President: 'Toward a More Public Social Science.'" *Items and Issues* 5, no. 1–2 (Spring–Summer): 12–15.

Commission on Global Governance. 1985. *Our Global Neighborhood: The Report of the Commission on Global Governance*. Oxford: Oxford University Press. Quoted in Sean O'Siochru and Bruce Gerard with Amy Mahan, *Global Media Governance*. Lanham, MD: Rowman & Littlefield, 2002.

Duff, Alistair S. 2000. *Information Society Studies*. London: Routledge.

Faulhaber, Gerald R., and William J. Baumol. 1988. "Economists as Innovators: Practical Products of Theoretical Research." *Journal of Economic Literature* XXVI (June) 577–600.

Frederich, Howard H. 1993. *Global Communications and International Relations*. Belmont, CA: Wordsworth.

Galtung, Johan. 1971. "A Structural Theory of Imperialism." *Journal of Peace Research* 8, no. 2: 81–117.

Gore, Charles. 2000. "The Rise and Fall of the Washington Consensus as a Paradigm for Developing Countries." *World Development* 28, no. 5: 789–804.

Halabi, Yakub. 2004. "The Expansion of Global Governance into the Third World: Altruism, Realism, or Constructivism?" *International Studies Review* 6, no. 1.

Haufler, Virginia. 2001. *A Public Role for the Private Sector. Industry Self-Regulation in a Global Economy*. Washington, DC: Carnegie Endowment for International Peace.

Keohane, Robert O., and Joseph S. Nye, Jr. 2002. "Power and Interdependence in the Information Age." In *Governance.com: Democracy in the Information Age*, ed. Elaine C. Karmack and Joseph S. Nye Jr. Washington, DC: Brookings Institution Press.

Kranich, Nancy. 2004. "The Information Commons: A Public Policy Report." Available at www.fepproject.org/policyreports/infocommons.contentsexsum.html.

Krasner, Stephen D. 1991. "Global Communications and National Power. Life on the Pareto Frontier." *World Politics* 43 (April): 336–366.

Lessig, Lawrence. 2000. Code and Other Laws of Cyberspace, New York: Basic Books.

MacLean, Don. 2004. "Herding Schrodinger's Cats: Some Conceptual Tools for Thinking about Internet Governance." Background paper for the ITU Workshop on Internet Governance, Geneva, February 26–27.

Markle Foundation. 2000. *Toward a Framework for Internet Accountability*. New York: Markle.

May, Christopher. 2002. *The Information Society. A Skeptical View*. Cambridge, UK: Polity Press.

Michling, Jerry. 2002. "Information Age Governance: Just the Start of Something Big?" In *Governance.com: Democracy in the Information Age*, ed. Elaine C. Kamarck and Joseph S. Nye. Washington, DC: Brookings Institution Press.

Mueller, Milton. 2002. *Ruling the Roost: Internet Governance and the Taming of Cyberspace*. Cambridge, MA: The MIT Press.

Mueller, Milton, John Mathiason, and Lee W. McKnight. 2004. "Making Sense of 'Internet Governance': Defining Principles and Norms in a Policy Context." Typescript, Internet Governance Project, Syracuse University.

National Research Council. 1998. *Fostering Research on the Economic and Social Impacts of Information Technology*. Washington, DC: National Academy of Sciences.

Negroponte, Nicholas. 1995. *Being Digital*. New York: Knopf.

NGO and Academic ICANN Study (NAIS). Available at www.naisproject.org.

Nye Jr., Joseph S., and John D. Donahue, eds. 2000. *Governance in a Globalizing World*. Washington, DC: Brookings Institution Press.

Ó Siochrú, Sean, and Bruce Girard with Amy Mahan. 2002. *Global Media Governance*. Lanham, MD: Rowman & Littlefield.

Peizer, Jonathan. 2003. "Cross-Sector Information and Communications Technology Funding for Development: What Works, What Does Not, and Why." *Information Technologies and International Development* 1, no. 2 (Winter): 81–89.

Putnam, Robert D. 2003. "The Public Role of Political Science, APSA Presidential Address." *Perspectives* 1, no. 2 (June): 249–255.

Rosenau, James N., and J. P. Singh, eds. 2002. *Information Technologies and Global Politics: The Changing Scope of Power and Governance*. Albany: State University of New York Press.

United Nations. 2000. *Report of the High-Level Panel on Information and Communication Technology*. New York.

Williamson, John. 1990. "What Washington Means by Policy Reform." In *Latin American Adjustment: How Much Has Happened?*, ed. John Williamson, 5–20. Washington, DC: Institute for International Economics.

Wilson III, Ernest J. 2003. "Scholarship and Practice in the Transition to a Knowledge Society." *Items and Issues*, vol. 4, no. 2–3 (Spring–Summer): 25–29.

Wilson III, Ernest J. 2004. *The Information Revolution and Developing Countries*. Cambridge, MA: The MIT Press.

Wilson III, Ernest J., and Kelvin Wong, eds. 2006. *Negotiating the Net in Africa: The Politics of Internet Diffusion*. Boulder, CO: Lynne Reinner.

Zacher, Mark W., and Brent A. Sutton. 1996. *Governing Global Networks: International Regimes for Transportation and Communications*. Cambridge: Cambridge University Press.

Contributors

Peng Hwa Ang is dean and associate professor at the Wee Kim Wee School of Communication and Information and director of the Singapore Internet Research Centre at Nanyang Technological University, Singapore. He is also a central committee member of the Consumers' Association of Singapore, and is the legal adviser to the Advertising Standards Authority of Singapore. He has consulted for government and private bodies in Singapore as well as for international agencies such as the UN Development Programme, and was a member of the UN Working Group on Internet Governance. Among his publications is *Ordering Chaos: Regulating the Internet* (2005).

Jonathan D. Aronson is professor of communications at the USC Annenberg School for Communication and also professor of international relations at the University of Southern California. Previously he was chair of the School of International Relations, and executive director of the Annenberg Center for Communication at USC, where he helped launch the *International Journal of Communications*. He has written extensively on trade in services, telecommunications and information services, and international intellectual property rights. Among his publications are *When Countries Talk: International Trade in Telecommunications Services* (1988) and *Managing the World Economy: The Consequences of Corporate Alliances* (1993), both coauthored with Peter F. Cowhey.

Byung-il Choi is dean and professor of international trade and negotiations at the Graduate School of International Studies, Ewha Womans University, in Seoul. He is also the director of the Institute for International Trade and Cooperation. Previously he was a trade negotiator representing the Korean government, including in the Korea–U.S. telecommunication talks, and was Korea's chief negotiator for the WTO basic telecommunications negotiations. He also worked on Korean negotiations with the European Union and the Asia-Pacific Economic Cooperation, and was the convener of the latter's telecommunications working group. Among his publications are *International Negotiations: Theory and Practice*, coauthored with Brendan Howe (2007), and *The Success and Failure of Trade Negotiations of Korea* (2004), which earned him the Korean Academy of Arts and Sciences Award.

Tracy Cohen serves on the Council of the Independent Communications Authority of South Africa. In addition to this four-year appointment, which ended in June 2008, she is an Honorary Research Fellow at the LINK Centre, Witwatersrand University; serves on the editorial board of the *South African Journal of Information and Communication*; and is an advisory panel member for Privacy International. She held posts at the LINK Centre, the Institute for Tele-Information at Columbia University, the Centre for Innovation Law and Policy at the University of Toronto, and the Centre for Analysis of Risk and Regulation at the London School of Economics and Political Science. She also has worked as a consultant to such organizations as the International Telecommunication Union and the Commonwealth Telecommunication Organization.

Peter F. Cowhey is associate vice chancellor and dean of the Graduate School of International Relations and Pacific Studies at the University of California, San Diego. He is also a research fellow at the California Institute on Telecommunications and Information Technology. Previously he was, inter alia, director of the Institute on Global Conflict and Cooperation and professor of political science at UCSD; and senior counselor for international economic policy to the chairman, as well as chief of the International Bureau at the U.S. Federal Communications Commission, where his responsibilities included international standard setting and spectrum negotiations. Among his publications are *When Countries Talk: International Trade in Telecommunications Services* (1988) and *Managing the World Economy: The Consequences of Corporate Alliances* (1993), both coauthored with Jonathan D. Aronson.

William J. Drake is director of the Project on the Information Revolution and Global Governance in the Center for International Governance at the Graduate Institute of International and Development Studies in Geneva, Switzerland. Previously, he was, inter alia, president of Computer Professionals for Social Responsibility, a global civil society organization; senior associate and director of the Project on the Information Revolution and World Politics at the Carnegie Endowment for International Peace; associate director of the Communication, Culture and Technology Program at Georgetown University; and assistant professor of communication at the University of California, San Diego. He also has served as a member of the High-Level Panel of Advisors to the UN Global Alliance for ICT and Development, and of the UN Working Group on Internet Governance. Among his publications are the edited volumes *Reforming Internet Governance: Perspectives from the UN Working Group on Internet Governance* (2005) and *The New Information Infrastructure: Strategies for US Policy* (1995).

Henry Farrell is assistant professor in the Department of Political Science and in the Elliot School of International Affairs at George Washington University. Previously, he served as assistant professor at the University of Toronto and as a senior research fellow at the Max Planck Institute on Common Goods in Bonn, Germany. Among his publications are "Constructing the International Foundations of E-Commerce: The EU-US

Safe Harbor Agreement," in *International Organization* (2003); "Trust, Distrust, and Power," in *Distrust*, ed. Russell Hardin (2004); and "Trust and Political Economy: Comparing the Effects of Institutions on Inter-Firm Cooperation," in *Comparative Political Studies* (2005).

Rob Frieden is Pioneers chair and professor of telecommunications in the College of Communications at Penn State University. Previously he provided a broad range of business development, strategic planning, and policy analysis for the IRIDIUM mobile satellite venture; held senior policy-making positions in international telecommunications at the U.S. Federal Communications Commission and the National Telecommunications and Information Administration; practiced law in Washington, DC; and served as assistant general counsel at PTAT System, Inc., where he handled corporate, transactional, and regulatory issues for the nation's first private undersea fiber optic cable company. Among his publications are *Managing Internet-Driven Change in International Telecommunications* (2001) and *The International Telecommunications Handbook* (1996).

Alison Gillwald is associate professor in the Graduate School of Public and Development Management and the research director of the LINK Centre at the University of the Witwatersrand in South Africa. She launched the LINK Centre in 1999 to accelerate ICT policy and regulatory training in Southern Africa. She is responsible for Research ICT Africa!, an Africa-wide research network, and serves on the board of the South African Broadcasting Corporation. Previously she was appointed to the founding council of the South African Telecommunications Regulatory Authority; established the Policy Department at the Independent Broadcasting Authority; and chaired the national Digital Broadcasting Advisory Body to the Minister of Communications. She is the founding editor of the *Southern African Journal of Information and Communication*.

Boutheina Guermazi is a regulatory specialist in the Public Sector Policy and Operations Division of the Global Information and Communications Technologies Department in the World Bank Group. She is involved in telecommunications sector reform and the elaboration of procompetitive legal and regulatory frameworks in a number of African and Middle Eastern countries. Previously, she was assistant professor at the faculté des Sciences Juridiques Politiques et Sociales of Tunis; an advisor in the Sector Reform Unit and an analyst in the Strategic Planning Unit at the International Telecommunication Union; a consultant for the UN Development Programme; and a research scholar at the Center of Studies for Regulated Industries in Canada. She has held a Fulbright Scholarship and has published a number of articles and book chapters on trade law, telecommunications policy, and regulatory reform.

Cees J. Hamelink is professor emeritus of international communication at the University of Amsterdam, and professor of globalization, human rights, and public health at the Vrije Universiteit in Amsterdam. He is also the editor-in-chief of the *International Journal for Communication Studies: Gazette*; president of the Dutch Federation for Human Rights; and a board member of the international news agency, Inter Press Service. In addition, he has worked as a journalist and a consultant to several international organizations and national governments, was the founder of the People's Communication Charter, and president of the International Association for Media and Communication Research. Hamelink has guest-lectured in over forty countries and has published over 250 articles, papers, and chapters in academic publications. Among his sixteen books are *Human Rights for Communicators* (2004) and *The Ethics of Cyberspace* (2000).

Ian Hosein is a visiting fellow in the Information Systems and Innovation Group of the Department of Management at the London School of Economics and Political Science. He is also a senior fellow at Privacy International, where he directs the Terrorism and the Open Society program; a visiting scholar at the American Civil Liberties Union project on Technology and Liberty; and coordinator of the Policy Laundering Project. With Simon Davies, he cofounded the Policy Engagement Network, which conducts projects on Internet governance, identity policy, and constitutional change. He has written widely on technology issues and consults to governmental and intergovernmental institutions.

Wolfgang Kleinwächter is professor of Internet policy and regulation in the Department for Media and Information Sciences at the University of Aarhus in Denmark. He is also special adviser to the chair of the Internet Governance Forum, a member of the High-Level Panel of Advisers to the UN Global Alliance for ICT and Development, and a member of ICANN's nominating committee. Previously he was director of the Institute for International Studies at the University of Leipzig; chair of the Management Board of the Interregional Information Society Initiative of the European Commission; a visiting scholar at the University of Tampere in Finland and at American University in Washington, DC; and a member of the UN Working Group on Internet Governance. Among his publications are *Global Governance in the Information Age* (2001) and *Macht und Geld im Cyberspace* (2005).

Don MacLean is an independent consultant on ICT-related policy, strategy, and governance issues based in Ottawa, Canada. At the national level, his work has included projects on access to broadband networks and services, online delivery of government services, measures to counter spam, innovation strategies for the e-economy, and reform of Canadian telecommunications policy and regulation. At the international level, he has worked on projects for, inter alia, the Global Knowledge Partnership, the G8 Digital Opportunity Task Force, and the World Summit on the Information Society. He has been chief of strategic planning and external affairs at the International Telecommunication Union; worked in the Canadian government's Department of Communications; and was a member of the UN Working Group on Internet Governance. Among his publications is the edited volume *Internet Governance: A Grand Collaboration* (2004).

Christopher May is professor of political economy and head of the Department of Politics and International Relations at Lancaster University in the United Kingdom. In addition, he is series coeditor of the International Political Economy Yearbook, and works with the UK's National Consumer Council to expand their coverage of intellectual property issues. Before becoming an academic he worked in the music industry, as a bookseller, and for the political pressure group Charter 88. Among his publications are *The World Intellectual Property Organisation: Resurgence and the Development Agenda* (2006) and *Intellectual Property Rights: A Critical History*, coauthored with Susan Sell (2005).

Milton Mueller is professor, director of the Program in Telecommunications and Network Management, and codirector of the Convergence Center in the School of Information Studies at Syracuse University. He is also a partner in the Internet Governance Project, a research network, and was the cofounder of the Noncommercial Users Constituency of the Internet Corporation for Assigned Names and Numbers. Among his publications are *Ruling the Root: Internet Governance and the Taming of Cyberspace* (2002) and *Universal Service: Competition, Interconnection and Monopoly in the Making of the American Telephone System* (1997).

John E. Richards is a visiting scholar at the School of International Relations and Pacific Studies at the University of California, San Diego. He is also the director of the Windows Live Platform at Microsoft Corporation. Previously, he was a consultant at the Palo Alto office of McKinsey & Co., where he served a variety of high tech and telecommunications clients. He has published extensively on international telecommunications, ICT, and aviation services markets.

David Souter is a visiting professor of communications management at the University of Strathclyde in Scotland and a visiting research fellow in the Media and Communications Department at the London School of Economics. In addition, he is the managing director of ict Development Associates ltd, a consultancy working at the interface between development policy and information and communication issues. He has consulted widely to governments, international organizations, NGOs, and other organizations. Previously, he was, inter alia, chief executive officer of the Commonwealth Telecommunication Organisation. Among his publications are *Whose Summit? Whose Information Society? Developing Countries and Civil Society at the World Summit on the Information Society* (2007).

Ernest J. Wilson III holds the Walter H. Annenberg chair in communication and is dean at the Annenberg School at the University of Southern California. In addition, he is an adjunct fellow at the Pacific Council on International Policy; founding coeditor-in-chief of The MIT Press journal *Information Technologies and International Development*; and the ranking senior member of the board of directors of the Corporation for Public Broadcasting. Previously, he was director of the Center for International Development and Conflict Management and professor in the Department of Government and Politics and in the Department of African-American Studies at the University of Maryland, College Park; a faculty member at the University of Michigan and the University of Pennsylvania; director of International Programs and Resources on the National Security Council at the White House; director of the Policy and Planning Unit, Office of the Director, U.S. Information Agency; and deputy director of the Global Information Infrastructure Commission. Among his publications are *The Information Revolution and Developing Countries* (2004) and *Negotiating the Net in Africa: The Politics of Internet Diffusion*, coedited with Kelvin R. Wong (2006).

Jisuk Woo is associate professor in the Graduate School of Public Administration at Seoul National University. She is also a legal advisor to and member of several government bodies in Korea, including the Korean Broadcasting Commission, the Ministry of Information and Communication, and the Ministry of Foreign Affairs and Trade; and is a domain name panelist at the World Intellectual Property Organization. She has published widely on such issues as the fair-use standard in the digital environment, the authorship construct in computer software copyright law, and the concept of harm in computer-generated images of child pornography. Among her publications is *Copyright Law and Computer Programs: The Role of Communication in Legal Structure* (2001).

Index